Hepatitis C Virus I

Tatsuo Miyamura • Stanley M. Lemon •
Christopher M. Walker • Takaji Wakita
Editors

Hepatitis C Virus I

Cellular and Molecular Virology

Springer

Editors
Tatsuo Miyamura
National Institute of Infectious Diseases
Tokyo, Japan

Stanley M. Lemon
Departments of Medicine and Microbiology & Immunology
The University of North Carolina
Chapel Hill, North Carolina
USA

Christopher M. Walker
Center for Vaccines and Immunity
The Research Institute at Nationwide Children's Hospital
Columbus, Ohio
USA

Takaji Wakita
National Institute of Infectious Diseases
Tokyo, Japan

ISBN 978-4-431-56096-8 ISBN 978-4-431-56098-2 (eBook)
DOI 10.1007/978-4-431-56098-2

Library of Congress Control Number: 2016951214

Printed on acid-free paper

This Springer imprint is published by Springer Nature
The registered company is Springer Japan KK

Contents

Contributors

Thomas F. Baumert, M.D., Inserm U1110, Institute de Recherche sur les Maladies Virales et Hépatiques, Strasbourg, France

Université de Strasbourg, Strasbourg, France

Pôle Hépato-digestif, Institut Hospitalo-Universitaire, Strasbourg University Hospitals, Strasbourg, France

Inés Romero Brey, Department of Infectious Diseases, Molecular Virology, University of Heidelberg, Heidelberg, Germany

Maria Teresa Catanese, Department of Infectious Diseases, King's College London, London, UK

Steven K. H. Foung, Department of Pathology, Stanford University School of Medicine, Stanford, CA, USA

Stanford Blood Center, Palo Alto, CA, USA

Thomas R. Fuerst, Institute for Bioscience and Biotechnology Research, University of Maryland, Rockville, MD, USA

Department of Cell Biology and Molecular Genetics, University of Maryland, College Park, MD, USA

Jenna M. Gaska, Department of Molecular Biology, Princeton University, Princeton, NJ, USA

Takayuki Hishiki, Laboratory of Primate Model, Experimental Research Center for Infectious Diseases, Institute for Virus Research, Kyoto University, Kyoto, Japan

Michael Houghton, Ph.D., Department of Medical Microbiology and Immunology, Li Ka Shing Applied Virology Institute, University of Alberta, Edmonton, Canada

Zhen-Yong Keck, Department of Pathology, Stanford University School of Medicine, Stanford, CA, USA

Stanley M. Lemon, Departments of Medicine and Microbiology & Immunology, Lineberger Comprehensive Cancer Center, University of North Carolina, Chapel Hill, NC, USA

Kui Li, Ph.D., Department of Microbiology, Immunology and Biochemistry, University of Tennessee Health Science Center, Memphis, TN, USA

Volker Lohmann, Department of Infectious Diseases, Molecular Virology, University of Heidelberg, Heidelberg, Germany

Joseph Marcotrigiano, Department of Chemistry and Chemical Biology, Rutgers, The State University of New Jersey, New Brunswick, NJ, USA

Roy A. Mariuzza, W. M. Keck Laboratory for Structural Biology, Institute for Bioscience and Biotechnology Research, University of Maryland, Rockville, MD, USA

Department of Cell Biology and Molecular Genetics, University of Maryland, College Park, MD, USA

Yoshiharu Matsuura, Department of Molecular Virology, Research Institute for Microbial Diseases, Osaka University, Osaka, Japan

Masashi Mizokami, Genome Medical Science Project, National Center for Global Health and Medicine, Ichikawa, Japan

Kohji Moriishi, Department of Microbiology, Division of Medicine, Graduate School of Medicine and Engineering, University of Yamanashi, Chuo-shi, Japan

Hironori Nishitsuji, Research Center for Hepatitis and Immunology, National Center for Global Health and Medicine, Ichikawa, Japan

Alexander Ploss, Department of Molecular Biology, Princeton University, Princeton, NJ, USA

Glenn Randall, Ph.D., Department of Microbiology, The University of Chicago, Chicago, IL, USA

Catherine Schuster, Inserm U1110, Institute de Recherche sur les Maladies Virales et Hépatiques, Strasbourg, France

Université de Strasbourg, Strasbourg, France

Yuko Shimizu, Research Center for Hepatitis and Immunology, National Center for Global Health and Medicine, Ichikawa, Japan

Kunitada Shimotohno, Research Center for Hepatitis and Immunology, National Center for Global Health and Medicine, Ichikawa, Japan

Tadasu Shin-I, Genome Medical Science Project, National Center for Global Health and Medicine, Ichikawa, Japan

Ana Shulla, Department of Microbiology, The University of Chicago, Chicago, IL, USA

Masaya Sugiyama, Genome Medical Science Project, National Center for Global Health and Medicine, Ichikawa, Japan

Ryosuke Suzuki, Ph.D., Department of Virology II, National Institute of Infectious Diseases, Tokyo, Japan

Tetsuro Suzuki, Ph.D., Department of Infectious Diseases, Hamamatsu University School of Medicine, Shizuoka, Japan

Rajiv G. Tawar, Inserm U1110, Institute de Recherche sur les Maladies Virales et Hépatiques, Strasbourg, France

Université de Strasbourg, Strasbourg, France

Saneyuki Ujino, Research Center for Hepatitis and Immunology, National Center for Global Health and Medicine, Ichikawa, Japan

Markus von Schaewen, Department of Molecular Biology, Princeton University, Princeton, NJ, USA

Takaji Wakita, M.D., Ph.D., National Institute of Infectious Diseases, Shinjuku, Japan

Daisuke Yamane, Departments of Medicine and Microbiology & Immunology, Lineberger Comprehensive Cancer Center, University of North Carolina, Chapel Hill, NC, USA

Part I
Hepatitis C Virus

Towards the Control of Hepatitis C

Michael Houghton

Abstract The discovery and characterisation of the hepatitis C virus (HCV) genome using a bacteriophage expression screening approach in 1989, quickly led to the development of blood tests to protect the blood supply and to diagnose and facilitate management of HCV patients. The viral-encoded serine protease and replicase then became major drug targets that in combination with viral NS5a-targeting drugs, that were facilitated by the use of in vitro genome replicon systems, has now led to most HCV patients being curable after just short treatment regimens. Natural immunity has been demonstrated in multiply-exposed individuals along with the identification of cellular immune correlates of protection. A growing role for neutralising antibodies in protection has also been indicated following the ability to grow HCV and viral pseudoparticles in cell culture. This knowledge has led to the pre-clinical and clinical testing of various promising vaccine candidates. Approval of HCV vaccines along with the development of much cheaper antiviral drugs will eventually lead to the effective global control of this virus which currently infects an estimated 150 million carriers around the world.

Keywords Non-A, non-B hepatitis • HCV virus discovery • HCV replication • Blood screening • Therapeutic approaches • Daclatasvir • HCV vaccine

1 From Non-A, Non-B to C

Following the discovery of the hepatitis B virus (HBV) in 1968 (Bayer et al. 1968) and the hepatitis A virus in 1973 (Feinstone et al. 1973), it became clear in 1974 (Prince et al. 1974) and 1975 (Feinstone et al. 1975) that most transfusion cases of hepatitis were due to neither virus from which the term Non-A, Non-B hepatitis (NANBH) was born. Studies from Harvey Alter at the NIH (Alter 1980) and separately from an independent consortium (TTTV) of collaborators (Hollinger et al. 1980) indicated that the risk of transfusion-associated NANBH could be as

M. Houghton, Ph.D. (✉)
Department of Medical Microbiology and Immunology, Li Ka Shing Applied Virology Institute, University of Alberta, Edmonton, Canada
e-mail: michael.houghton@ualberta.ca

T. Miyamura et al. (eds.), *Hepatitis C Virus I*, DOI 10.1007/978-4-431-56098-2_1

high as 10 % and that the resulting liver disease persisted frequently. Over the course of many years, this initial mild disease could progress into severe clinical liver diseases such as liver cirrhosis (Dienstag and Alter 1986) and hepatocellular carcinoma (Kiyosawa et al. 1984). Unfortunately, methods used to identify HAV and HBV were unsuccessful at isolating and identifying the etiological agent of NANBH (Shih et al. 1986) due essentially to the much lower titer of NANBH as compared with the known hepatitis viruses (Prince 1983). This problem was solved eventually by the application of sensitive molecular cloning and screening methods. Working with my colleagues Qui-Lim Choo and George Kuo at the Chiron Corporation in the USA, prior to the routine use of PCR amplification technology, we cloned all nucleic acids from an ultracentifuged pellet of infectious chimpanzee plasma into the bacteriophage expression vector lambda gt11 resulting in large proteomic libraries. This plasma, of relatively high titer for NANBH, was obtained from my collaborator Daniel Bradley at the Centers for Disease Control (CDC). While antibodies to NANBH had not then been identified, we assumed their existence and screened the proteomic libraries with NANBH patient sera and identified antibody-tagged clones using a radioactive second antibody. Fortunately, after many failed attempts, this inherently risky approach finally succeeded in identifying one small clone (5-1-1) that we were able to show was not derived from the chimpanzee genome, hybridised to a large RNA molecule present only in NANBH-infected chimpanzee and human samples and which encoded a protein reactive with antibodies found only in NANBH-infected blood. Furthermore, the sequence of the RNA-derived cDNA clones indicated that that it was novel and very distantly related to flaviviruses. This novel method for identifying an infectious agent resulted in the identification of the hepatitis C virus (HCV; Choo et al. 1989; Kuo et al. 1989). This work was a result of a team effort from Qui-Lim Choo, a most thorough molecular biologist working in my own laboratory and George Kuo, who had his own protein chemistry and immunochemistry laboratory adjacent to mine. George provided me with an analysis of the likely limiting concentrations of NANBH antigen in vivo which was largely responsible for persuading me to attempt the precarious recombinant DNA expression screening approach using patient sera as a presumptive, albeit unproven source of NANBH antibodies. Daniel Bradley at the CDC was my long-term external collaborator who provided numerous characterised chimpanzee-derived NANBH samples throughout the course of this HCV virus discovery project in my laboratory from 1982 to 1989.

2 The Hepatitis C Virus (HCV)

A very distant relative of the flaviviruses and pestiviruses, HCV contains a positive-stranded RNA genome of around 10,000 nucleotides which encodes a large polyprotein of over 3000 amino acids which is cleaved co-and post-translationally into virion structural proteins (nucleocapsid (C) and envelope glycoproteins gpE1 and gpE2) and a plethora of non-structural proteins involved in viral replication and assembly (Lindenbach and Rice 2013; Paul et al. 2014). Notable features of HCV replication include it's ability to induce a membranous web within the e.r (Moradpour et al. 2003) in which virus is replicated (Romero-Brey et al. 2015), assembled on lipid droplets (Barba et al. 1997; Miyanari et al. 2007) and secreted using the low-density lipoprotein secretory pathway resulting in the production of apolipoprotein-associated viral particles of very light density (Huang et al. 2007). HCV induces the recruitment of nuclear pore proteins into the membranous web possibly to facilitate a protective subcellular environment from the cell's innate immune response and to create a custom-made virus factory (Neufeldt et al. 2013). Meanwhile, the innate immune response is down-regulated via cleavage of MAVS and TRIF by the viral serine protease (Foy et al. 2005; Li et al. 2005).

Being a RNA virus replicated via it's own RNA-dependent RNA polymerase that lacks proof-reading activity, mutations are common in every replication cycle leading to a highly fluid and heterogenous viral genome currently comprising at least seven basic genotypes, whose distribution varies around the world with numerous subtypes (Smith et al. 2014). As such, each HCV strain comprises a highly adaptable quasi-species of RNA genomes, a feature that quickly emerged from the laboratory of Tatsuo Miyamura who using our original strain, went on to identify the world's most common HCV genotype, the 1b subtype (Kubo et al. 1989).

Despite intense efforts for many years, HCV could not be grown efficiently in tissue culture until 2005 when Takaji Wakita and collaborators identified a Japanese 2a strain that could complete the entire replication cycle in vitro with significant yields of progeny virus (Wakita et al. 2005). Prior to this, infectious HCV pseudoparticles (HCVpp) could be produced in cell cultures co-expressing defective HIV and lentiviral genomes along with HCV envelope glycoproteins (Flint et al. 2004; Sandrin et al. 2005). Earlier, replicons of HCV were produced in Ralf Bartenschlager's (Lohmann et al. 1999) and Charles Rice's (Blight et al. 2000) laboratories in human hepatoma cell-lines. These have proven to be of great value in HCV drug discovery programs and in basic research. An immunodeficient SCID mouse model for HCV infection involving transplantation of human hepatocytes has also proven valuable in research, drug development and virus neutralisation studies (Mercer et al. 2001) and an immunocompetent mouse model shows much promise in future vaccine studies (Dorner et al. 2013). Visualisation of the virion proved difficult and became possible only recently, many years after it's molecular isolation (Wakita et al. 2005; Fig. 1).

3 Prevention and Therapy

Once we identified the viral genome, it was possible to quickly develop blood tests to capture and detect circulating HCV-specific antibody in infected individuals (Kuo et al. 1989). In addition, with the advent of PCR- and TMA-nucleic acid

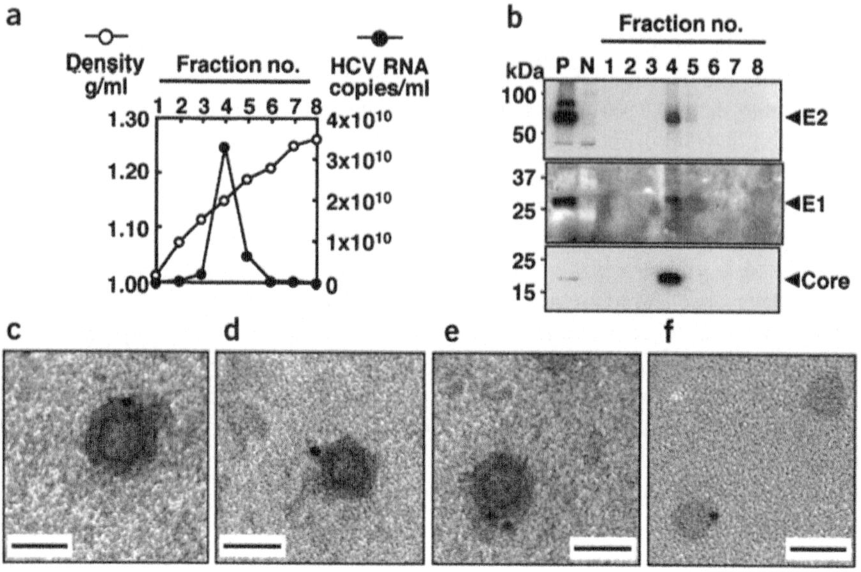

Fig. 1 Density gradient and electron microscope analysis of recombinant HCV particles. (**a**, **b**) Co-sedimentation of viral RNA and structural proteins. (**a**) Concentrated culture medium collected from JFH1/E2HA RNA–transfected cells was fractionated using a 10–60 % sucrose density gradient. HCV RNA titer in each fraction was determined. (**b**) Density gradient fractions were further concentrated and analyzed by western blotting for core, E1 or E2-hemagglutinin. P, cell lysate prepared from JFH1/E2HA RNA–transfected Huh7 cells; N, cell lysate from untransfected Huh7 cells. *Arrowheads* indicate positions of HCV proteins. (**c**–**f**) Electron micrograph of spherical structures shown by immunogold labeling. Grids were incubated with a concentrated JFH1 virus stock and then with the E2 monoclonal antibody CBH5 (Hadlock et al. 2000). Bound antibodies were detected with Protein A coupled to gold particles 10 nm in diameter. (**c**–**e**) Three representative examples showing the same structure. (**f**) Control grid coated with concentrated cell-free supernatant derived from mock-transfected cells. In rare cases, we observed gold particles attached to unstructured protein aggregates. *Scale bar*, 50 nm (Reprinted from Wakita et al. (2005))

amplification technologies, these tests were quickly approved around the world and have effectively eliminated transfusion-associated HCV transmission where blood screening is performed.

Successful therapeutic approaches were initiated prior to the identification of HCV with the seminal discovery of alpha interferon able to cure a small minority of NANBH patients (Hoofnagle et al. 1986). Stabilising the interferon by conjugation with polyethylene glycol led to better potency as did the addition of ribavirin, a guanosine analogue that works to prevent relapse using an unknown mechanism (Manns et al. 2001; Feld 2012). This combination was the mainstay of therapy for many years although it was limited by high toxicity and by only partial potency (McHutchison and Fried 2003).

Surprisingly, it took more than 20 years to develop approvable drugs targeting the HCV protease and polymerase, despite these targets being evident from the initial HCV genomic sequence (Choo et al. 1991). This long delay was caused primarily by many pre-clinical and clinical failures due to drug toxicity rather than from lack of antiviral potency. In addition, unlike the HIV protease, the serine protease encoded by HCV has a rather shallow active site rendering it inherently difficult to design specific drugs with high affinity of binding and of high specificity. In addition, it has proven difficult to develop nucleoside inhibitors specific for the active site of the HCV polymerase. However, this was finally achieved recently and the drug Sofosbuvir now represents one major corner-stone in effective therapy (Alqahtani et al. 2015).

Using the HCV replicon systems, outstanding drug development work identified the symmetrical drug Daclatasvir as the most potent antiviral drug ever developed for any virus (Gao et al. 2010; Belema et al. 2014). Targeting non-structural protein 5a (NS5a), this drug could not have been discovered without use of the replicon system and it has a unique mode-of-binding (Belema et al. 2014; Barakat et al. 2015; Fig. 2).

Various combinations of Sofosbuvir, non-nucleoside polymerase inhibitors, Daclatasvir or Daclatasvir-like drugs along with second and third generation protease inhibitors have now transformed the therapy of HCV such that it has become the only curable chronic viral infection, with nearly all patients now curable after short regimen treatments and with little resistance reported in combination therapy (Nyalakonda and Utay 2015). Following the imminent emergence of drugs with more activity against genotype 3, the situation will become even better. Unfortunately, this excellent situation which is a testament to the skills and dedication of thousands of researchers, drug developers and clinicians is now limited not by lack of drug potency but by the high price of therapy with cures costing between US $50,000 and 90,000 per patient. Additional drugs are urgently required in order to lower the cost (via market competition) to patients and healthcare systems around the globe. Unfortunately, this process may take many more years and will need the input of governments and philanthropists in order to make low cost, non-infringing HCV drug cocktails accessible to all HCV patients around the world. In the meantime, it will be important to identify all HCV carriers and to take some simple steps to minimise disease progression such as avoiding co-factors like alcohol intake, while drinking coffee regularly which has been shown in a number of studies to slow down the rate of liver fibrosis, liver cirrhosis and hepatocellular carcinoma (Ohfuji et al. 2006; Inoue et al. 2009; Modi et al. 2010; Lai et al. 2013; Khalaf et al. 2015). If very cheap drugs became available to the estimated 150 million carriers of HCV around the world, the epidemic could be quickly curtailed and the virus eventually eradicated. However, this ideal may not be possible for many decades and until then, a vaccine is required to help control HCV.

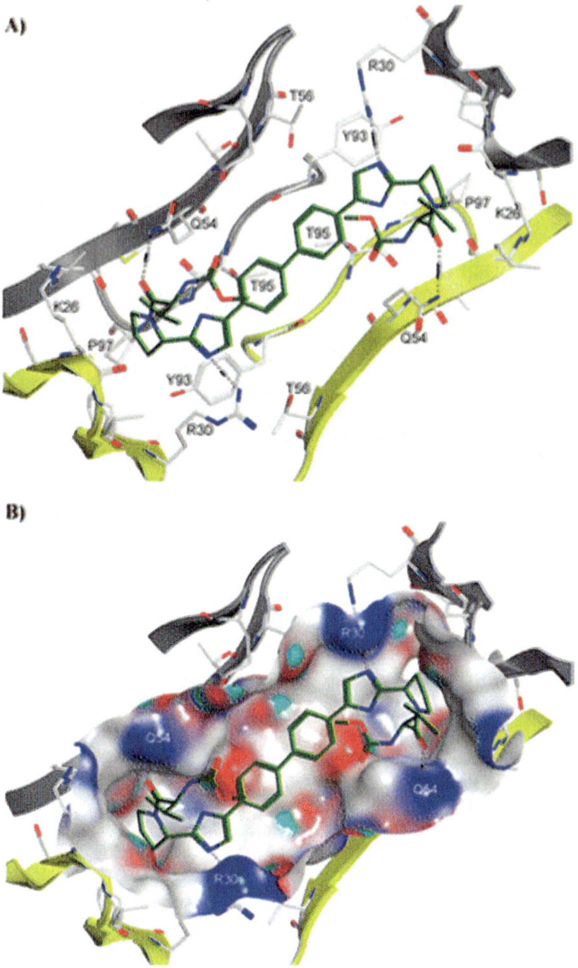

Fig. 2 Binding mode for Daclatasvir. (**a**) Carbon atoms are colored in *green* for Daclatasvir and in white for the protein residues. Hydrogen bonds are shown as *dotted lines*, and the protein boundaries of the binding site are shown in cartoon representation, where the first monomer is shown in *yellow* while the second monomer is shown in *gray*. (**b**) The binding site is shown in surface representation (Reprinted from Barakat et al. (2015))

4 HCV Vaccine Status

A vaccine candidate is currently undergoing phase 2 clinical efficacy trials in the USA and is based on the i/m delivery of non-structural genes via a pair of replication defective viral vectors (Swadling et al. 2014). Based on definitive animal studies from Chris Walker's laboratory showing the required presence of HCV-specific CD4+ and CD8+ T cell responses for protection (Grakoui et al. 2003; Shoukry et al. 2003), this vaccine could be the first demonstration of an efficacious vaccine relying solely on elicited cellular immunity. A different vaccine candidate developed over many years by myself and colleagues (Ralston et al. 1993; Choo et al. 1994) is a heterodimer comprising the two HCV envelope glycoproteins gpE1/gpE2 which has been shown to elicit broadly cross-neutralising antibodies in mice, guinea pigs, chimpanzees, goats and humans (Stamataki et al. 2007; Ray et al. 2010; Meunier et al. 2011; Law et al. 2013; Wong et al. 2014) along with strong CD4+ T helper responses in humans (Frey et al. 2010). It is the only HCV vaccine candidate for which a large and statistically-significant reduction in the carrier rate of vaccinated chimpanzees has been demonstrated (Choo et al. 1994; Houghton and Abrignani 2005; Houghton 2011). There is growing evidence for a protective role of cross-neutralising HCV antibodies and these may be needed for optimal vaccine efficacy (Ishii et al. 1998; Lavillette et al. 2005; Pestka et al. 2007; Meuleman et al. 2011; Osburn et al. 2014; de Jong et al. 2014). An improved version of this vaccine candidate is scheduled for clinical testing in the near future. Other recombinant HCV vaccines are under active pre-clinical investigation (Garrone et al. 2011; Drummer 2014). If successful, a HCV vaccine would be invaluable in HCV seronegative intravenous drug users who account for at least 60–70 % of new infections in the developed world (Cox and Thomas 2013). Other high-risk groups such as healthcare workers, paramedics and police officers would also benefit from vaccination.

5 The Future

There is an urgent need for cheaper drugs to cure HCV patients otherwise the majority of the world's 150 million carriers will continue to be at risk for end-stage liver disease and hepatocellular carcinoma. Because of the high cost of HCV drugs, many countries prioritise the treatment of patients who exhibit significant liver fibrosis leaving the majority of patients still at risk of developing liver cirrhosis and liver cancer. The solution to this unacceptable situation will require the approval of additional drugs from the corporate sector in order to drive prices down but the combined intervention of governments and philanthropists will also likely be required to develop novel drug combinations that are of low cost and which do not infringe corporate patents. The emergence of both cheap new drugs and vaccines are likely to effectively control global HCV infection over the coming decades.

Acknowledgements I would like to thank Amy Weiner, Kang-Sheng Wang and the late Lacy Overby for their many valuable inputs and contributions and I dedicate this review to the memory of Lacy, a mentor, friend, colleague and a great enthusiast of medical science.

References

Alqahtani SA, Afdhal N, Zeuzem S, Gordon SC, Mangia A, Kwo P, Fried M, Yang JC, Ding X, Pang PS, McHutchison JG, Pound D, Reddy KR, Marcellin P, Kowdley KV, Sulkowski M (2015) Safety and tolerability of ledipasvir/sofosbuvir with and without ribavirin in patients with chronic hepatitis C virus genotype 1 infection: analysis of phase III ION trials. Hepatology 62(1):25–30. doi:10.1002/hep.27890. Epub 2015 Jun 1

Alter HJ (1980) The dominant role of non-A, non-B in the pathogenesis of post-transfusion hepatitis: a clinical assessment. Clin Gastroenterol 9:155–170

Barakat KH, Anwar-Mohamed A, Tuszynski JA, Robins MJ, Tyrrell DL, Houghton M (2015) A refined model of the HCV NS5A protein bound to daclatasvir explains drug-resistant mutations and activity against divergent genotypes. J Chem Inf Model 55(2):362–373. doi:10.1021/ci400631n. Epub 2014 Apr 14

Barba G, Harper F, Harada T, Kohara M, Goulinet S, Matsuura Y, Eder G, Schaff Z, Chapman MJ, Miyamura T, Bréchot C (1997) Hepatitis C virus core protein shows a cytoplasmic localization and associates to cellular lipid storage droplets. PNAS USA 94(4):1200–1205

Bayer ME, Blumberg BS, Werner B (1968) Particles associated with Australia antigen in the sera of patients with leukaemia, Down's syndrome and hepatitis. Nature 218:1057–1059

Belema M, Nguyen VN, Bachand C, Deon DH, Goodrich JT, James CA, Lavoie R, Lopez OD, Martel A, Romine JL, Ruediger EH, Snyder LB, St Laurent DR, Yang F, Zhu J, Wong HS, Langley DR, Adams SP, Cantor GH, Chimalakonda A, Fura A, Johnson BM, Knipe JO, Parker DD, Santone KS, Fridell RA, Lemm JA, O'Boyle DR 2nd, Colonno RJ, Gao M, Meanwell NA, Hamann LG (2014) Hepatitis C virus NS5A replication complex inhibitors: the discovery of daclatasvir. J Med Chem 57(5):2013–2032. doi:10.1021/jm401836p. Epub 2014 Feb 12

Blight KJ, Kolykhalov AA, Rice CM (2000) Efficient initiation of HCV RNA replication in cell culture. Science 290(5498):1972–1974

Choo QL, Kuo G, Weiner AJ, Overby LR, Bradley DW, Houghton M (1989) Isolation of a cDNA clone derived from a blood–borne non-A, non-B viral hepatitis genome. Science 244:359–362

Choo QL, Richman KH, Han JH, Berger K, Lee C, Dong C, Gallegos C, Coit D, Medina-Selby R, Barr PJ et al (1991) Genetic organization and diversity of the hepatitis C virus. PNAS USA 88 (6):2451–2455

Choo QL, Kuo G, Ralston R, Weiner A, Chien D, Van Nest G et al (1994) Vaccination of chimpanzees against infection by the hepatitis C virus. PNAS USA 91:1294–1298

Cox AL, Thomas DL (2013) Hepatitis C virus vaccines among people who inject drugs. Clin Infect Dis 57(Suppl 2):S46–S50. doi:10.1093/cid/cit329

de Jong YP, Dorner M, Mommersteeg MC, Xiao JW, Balazs AB, Robbins JB, Winer BY, Gerges S, Vega K, Labitt RN, Donovan BM, Giang E, Krishnan A, Chiriboga L, Charlton MR, Burton DR, Baltimore D, Law M, Rice CM, Ploss A (2014) Broadly neutralizing antibodies abrogate established hepatitis C virus infection. Sci Transl Med 6(254):254ra129. doi:10.1126/scitranslmed.3009512

Dienstag JL, Alter HJ (1986) Non-A, non-B hepatitis: evolving epidemiologic and clinical perspective. Semin Liver Dis 6:67–68

Dorner M, Horwitz JA, Donovan BM, Labitt RN, Budell WC, Friling T, Vogt A, Catanese MT, Satoh T, Kawai T, Akira S, Law M, Rice CM, Ploss A (2013) Completion of the entire hepatitis C virus life cycle in genetically humanized mice. Nature 501(7466):237–241. doi:10.1038/nature12427. Epub 2013 Jul 31

Drummer HE (2014) Challenges to the development of vaccines to hepatitis C virus that elicit neutralizing antibodies. Front Microbiol 5:329. doi:10.3389/fmicb.2014.00329. eCollection 2014

Feinstone SM, Kapikian AZ, Purcell RH (1973) Hepatitis A: detection by immune electron microscopy of a viruslike antigen associated with acute illness. Science 182:1026–1028

Feinstone SM, Kapikian AZ, Purcell RH, Alter HJ, Holland PV (1975) Transfusion-associated hepatitis not due to viral hepatitis type A or B. N Engl J Med 292:767–770

Feld JJ (2012) Is there a role for ribavirin in the era of hepatitis C virus direct-acting antivirals? Gastroenterology 142(6):1356–9. doi:10.1053/j.gastro.2011.12.064

Flint M, Logvinoff C, Rice CM, McKeating JA (2004) Characterization of infectious retroviral pseudotype particles bearing hepatitis C virus glycoproteins. J Virol 78(13):6875–6882

Foy E, Li K, Sumpter R Jr, Loo YM, Johnson CL, Wang C, Fish PM, Yoneyama M, Fujita T, Lemon SM, Gale M Jr (2005) Control of antiviral defenses through hepatitis C virus disruption of retinoic acid-inducible gene-I signaling. PNAS USA 102(8):2986–2991. Epub 2005 Feb 14

Frey SE, Houghton M, Coates S, Abrignani S, Chien D, Rosa D, Pileri P, Ray R, Di Bisceglie AM, Rinella P, Hill H, Wolff MC, Schultze V, Han JH, Scharschmidt B, Belshe RB (2010) Safety and immunogenicity of HCV E1E2 vaccine adjuvanted with MF59 administered to healthy adults. Vaccine 28(38):6367–6373. doi:10.1016/j.vaccine.2010.06.084. Epub 2010 Jul 7

Gao M, Nettles RE, Belema M, Snyder LB, Nguyen VN, Fridell RA, Serrano-Wu MH, Langley DR, Sun JH, O'Boyle DR 2nd, Lemm JA, Wang C, Knipe JO, Chien C, Colonno RJ, Grasela DM, Meanwell NA, Hamann LG (2010) Chemical genetics strategy identifies an HCV NS5A inhibitor with a potent clinical effect. Nature 465(7294):96–100. doi:10.1038/nature08960. Epub 2010 Apr 21

Garrone P, Fluckiger AC, Mangeot PE, Gauthier E, Dupeyrot-Lacas P, Mancip J, Cangialosi A, Du Chéné I, LeGrand R, Mangeot I, Lavillette D, Bellier B, Cosset FL, Tangy F, Klatzmann D, Dalba C (2011) A prime-boost strategy using virus-like particles pseudotyped for HCV proteins triggers broadly neutralizing antibodies in macaques. Sci Transl Med 3(94):94ra71. doi:10.1126/scitranslmed.3002330

Grakoui A, Shoukry NH, Woollard DJ, Han JH, Hanson HL, Ghrayeb J, Murthy KK, Rice CM, Walker CM (2003) HCV persistence and immune evasion in the absence of memory T cell help. Science 302(5645):659–662

Hadlock KG, Lanford RE, Perkins S, Rowe J, Yang Q, Levy S, Pileri P, Abrignani S, Foung SK (2000) Human monoclonal antibodies that inhibit binding of hepatitis C virus E2 protein to CD81 and recognize conserved conformational epitopes. J Virol 74(22):10407–10416

Hollinger FB, Mosley JW, Szmuness W, Aach RD, Peters RL, Stevens C (1980) Transfusion-transmitted viruses study: experimental evidence for two non-A, non-B, hepatitis agents. J Infect Dis 142:400–407

Hoofnagle JH, Mullen KD, Jones DB, Rustgi V, Di Bisceglie A, Peters M, Waggoner JG, Park Y, Jones EA (1986) Treatment of chronic non-A, non-B hepatitis with recombinant human alpha interferon. A preliminary report. N Engl J Med 315(25):1575–1578

Houghton M (2011) Prospects for prophylactic and therapeutic vaccines against the hepatitis C viruses. Immunol Rev 239(1):99–108. doi:10.1111/j.1600-065X.2010.00977.x

Houghton M, Abrignani S (2005) Prospects for a vaccine against the hepatitis C virus. Nature 436 (7053):961–966

Huang H, Sun F, Owen DM, Li W, Chen Y, Gale M Jr et al (2007) Hepatitis C virus production by human hepatocytes dependent on assembly and secretion of very low-density lipoproteins. PNAS USA 104:5848–5853

Inoue M, Kurahashi N, Iwasaki M, Shimazu T, Tanaka Y, Mizokami M, Tsugane S, Japan Public Health Center-Based Prospective Study Group (2009) Effect of coffee and green tea consumption on the risk of liver cancer: cohort analysis by hepatitis virus infection status. Cancer Epidemiol Biomarkers Prev 18(6):1746–1753. doi:10.1158/1055-9965.EPI-08-0923

Ishii K, Rosa D, Watanabe Y, Katayama T, Harada H, Wyatt C, Kiyosawa K, Aizaki H, Matsuura Y, Houghton M, Abrignani S, Miyamura T (1998) High titers of antibodies inhibiting the binding of envelope to human cells correlate with natural resolution of chronic hepatitis C. Hepatology 28(4):1117–1120

Khalaf N, White D, Kanwal F, Ramsey D, Mittal S, Tavakoli-Tabasi S, Kuzniarek J, El-Serag HB (2015) Coffee and caffeine are associated with decreased risk of advanced hepatic fibrosis among patients with hepatitis C. Clin Gastroenterol Hepatol 13(8):1521–1531.e3. doi:10.1016/j.cgh.2015.01.030. Epub 2015 Mar 14

Kiyosawa K, Akahane Y, Nagata A, Furuta S (1984) Hepatocellular carcinoma after non-A, non-B posttransfusion hepatitis. Am J Gastroenterol 79(10):777–781

Kubo Y, Takeuchi K, Boonmar S, Katayama T, Choo QL, Kuo G, Weiner AJ, Bradley DW, Houghton M, Saito I et al (1989) A cDNA fragment of hepatitis C virus isolated from an implicated donor of post-transfusion non-A, non-B hepatitis in Japan. Nucleic Acids Res 17 (24):10367–10372

Kuo G, Choo QL, Alter HJ, Gitnick GL, Redeker AG, Purcell RH et al (1989) An assay for circulating antibodies to a major etiologic virus of human non-A, non-B hepatitis. Science 244:362–364

Lai GY, Weinstein SJ, Albanes D, Taylor PR, McGlynn KA, Virtamo J, Sinha R, Freedman ND (2013) The association of coffee intake with liver cancer incidence and chronic liver disease mortality in male smokers. Br J Cancer 109(5):1344–1351. doi:10.1038/bjc.2013.405. Epub 2013 Jul 23

Lavillette D, Morice Y, Germanidis G, Donot P, Soulier A, Pagkalos E, Sakellariou G, Intrator L, Bartosch B, Pawlotsky JM, Cosset FL (2005) Human serum facilitates hepatitis C virus infection, and neutralizing responses inversely correlate with viral replication kinetics at the acute phase of hepatitis C virus infection. J Virol 79(10):6023–6034

Law JL, Chen C, Wong J, Hockman D, Santer DM, Frey SE, Belshe RB, Wakita T, Bukh J, Jones CT, Rice CM, Abrignani S, Tyrrell DL, Houghton M (2013) A hepatitis C virus (HCV) vaccine comprising envelope glycoproteins gpE1/gpE2 derived from a single isolate elicits broad cross-genotype neutralizing antibodies in humans. PLoS One 8(3):e59776. doi:10.1371/journal.pone.0059776. Epub 2013 Mar 19

Li K, Foy E, Ferreon JC, Nakamura M, Ferreon AC, Ikeda M, Ray SC, Gale M Jr, Lemon SM (2005) Immune evasion by hepatitis C virus NS3/4A protease-mediated cleavage of the toll-like receptor 3 adaptor protein TRIF. PNAS USA 102(8):2992–2997. Epub 2005 Feb 14

Lindenbach BD, Rice CM (2013) The ins and outs of hepatitis C virus entry and assembly. Nat Rev Microbiol 11(10):688–700. doi:10.1038/nrmicro3098. Epub 2013 Sep 10

Lohmann V, Körner F, Koch J, Herian U, Theilmann L, Bartenschlager R (1999) Replication of subgenomic hepatitis C virus RNAs in a hepatoma cell line. Science 285(5424):110–113

Manns MP, McHutchison JG, Gordon SC, Rustgi VK, Shiffman M, Reindollar R, Goodman ZD, Koury K, Ling M, Albrecht JK (2001) Peginterferon alfa-2b plus ribavirin compared with interferon alfa-2b plusribavirin for initial treatment of chronic hepatitis C: a randomised trial. Lancet 358(9286):958–965

McHutchison JG, Fried MW (2003) Current therapy for hepatitis C: pegylated interferon and ribavirin. Clin Liver Dis 7(1):149–161

Mercer DF, Schiller DE, Elliott JF, Douglas DN, Hao C, Rinfret A, Addison WR, Fischer KP, Churchill TA, Lakey JR, Tyrrell DL, Kneteman NM (2001) Hepatitis C virus replication in mice with chimeric human livers. Nat Med 7(8):927–933

Meuleman P, Bukh J, Verhoye L, Farhoudi A, Vanwolleghem T, Wang RY, Desombere I, Alter H, Purcell RH, Leroux-Roels G (2011) In vivo evaluation of the cross-genotype neutralizing activity of polyclonal antibodies against hepatitis C virus. Hepatology 53(3):755–762. doi:10.1002/hep.24171. Epub 2011 Feb 11

Meunier JC, Gottwein JM, Houghton M, Russell RS, Emerson SU, Bukh J, Purcell RH (2011) Vaccine-induced cross-genotype reactive neutralizing antibodies against hepatitis C virus. J Infect Dis 204(8):1186–1190. doi:10.1093/infdis/jir511

Miyanari Y, Atsuzawa K, Usuda N, Watashi K, Hishiki T, Zayas M, Bartenschlager R, Wakita T, Hijikata M, Shimotohno K (2007) The lipid droplet is an important organelle for hepatitis C virus production. Nat Cell Biol 9(9):1089–1097. Epub 2007 Aug 26. Erratum in: Nat Cell Biol. 2007 Oct;9(10):1216

Modi AA, Feld JJ, Park Y, Kleiner DE, Everhart JE, Liang TJ, Hoofnagle JH (2010) Increased caffeine consumption is associated with reduced hepatic fibrosis. Hepatology 51(1):201–209. doi:10.1002/hep.23279

Moradpour D, Gosert R, Egger D, Penin F, Blum HE, Bienz K (2003) Membrane association of hepatitis C virus nonstructural proteins and identification of the membrane alteration that harbors the viral replication complex. Antiviral Res 60(2):103–109

Neufeldt CJ, Joyce MA, Levin A, Steenbergen RH, Pang D, Shields J, Tyrrell DL, Wozniak RW (2013) Hepatitis C virus-induced cytoplasmic organelles use the nuclear transport machinery

to establish an environment conducive to virus replication. PLoS Pathog 9(10):e1003744. doi:10.1371/journal.ppat.1003744. Epub 2013 Oct 31
Nyalakonda H, Utay NS (2015) A new era of therapy for hepatitis C virus infection. Curr Opin Infect Dis 28(5):471–478
Ohfuji S, Fukushima W, Tanaka T, Habu D, Tamori A, Sakaguchi H, Takeda T, Kawada N, Seki S, Nishiguchi S, Shiomi S, Hirota Y (2006) Coffee consumption and reduced risk of hepatocellular carcinoma among patients with chronic type C liver disease: a case-control study. Hepatol Res 36(3):201–208. Epub 2006 Aug 17
Osburn WO, Snider AE, Wells BL, Latanich R, Bailey JR, Thomas DL, Cox AL, Ray SC (2014) Clearance of hepatitis C infection is associated with the early appearance of broad neutralizing antibody responses. Hepatology 59(6):2140–2151. doi:10.1002/hep.27013. Epub 2014 Apr 29
Paul D, Madan V, Bartenschlager R (2014) Hepatitis C virus RNA replication and assembly: living on the fat of the land. Cell Host Microbe 16(5):569–579. doi:10.1016/j.chom.2014.10.008. Epub 2014 Nov 12
Pestka JM, Zeisel MB, Bläser E, Schürmann P, Bartosch B, Cosset FL, Patel AH, Meisel H, Baumert J, Viazov S, Rispeter K, Blum HE, Roggendorf M, Baumert TF (2007) Rapid induction of virus-neutralizing antibodies and viral clearance in a single-source outbreak of hepatitis C. PNAS USA 104(14):6025–6030. Epub 2007 Mar 28
Prince AM (1983) Non-A, non-B hepatitis viruses. Annu Rev Microbiol 37:217–232
Prince AM, Brotman B, Grady GF, Kuhns WJ, Hazzi C, Levine RW, Millian SJ (1974) Long-incubation post-transfusion hepatitis without serological evidence of exposure to hepatitis-B virus. Lancet 2(7875):241–246
Ralston R, Thudium K, Berger K, Kuo C, Gervase B, Hall J, Selby M, Kuo G, Houghton M, Choo QL (1993) Characterization of hepatitis C virus envelope glycoprotein complexes expressed by recombinant vaccinia viruses. J Virol 67(11):6753–6761
Ray R, Meyer K, Banerjee A, Basu A, Coates S, Abrignani S, Houghton M, Frey SE, Belshe RB (2010) Characterization of antibodies induced by vaccination with hepatitis C virus envelope glycoproteins. J Infect Dis 202(6):862–866. doi:10.1086/655902
Romero-Brey I, Berger C, Kallis S, Kolovou A, Paul D, Lohmann V, Bartenschlager R (2015) NS5A domain 1 and polyprotein cleavage kinetics are critical for induction of double-membrane vesicles associated with hepatitis C virus replication. Mol Biol 6(4). pii: e00759-15. doi:10.1128/mBio.00759-15
Sandrin V, Boulanger P, Penin F, Granier C, Cosset FL, Bartosch B (2005) Assembly of functional hepatitis C virus glycoproteins on infectious pseudoparticles occurs intracellularly and requires concomitant incorporation of E1 and E2 glycoproteins. J Gen Virol 86(Pt 12):3189–3199
Shih JW, Mur JI, Alter HJ (1986) Non-A, non-B hepatitis: advances and unfulfilled expectations of the first decade. Prog Liver Dis 8:433–452
Shoukry NH, Grakoui A, Houghton M, Chien DY, Ghrayeb J, Reimann KA, Walker CM (2003) Memory CD8+ T cells are required for protection from persistent hepatitis C virus infection. J Exp Med 197(12):1645–1655
Smith DB, Bukh J, Kuiken C, Muerhoff AS, Rice CM, Stapleton JT, Simmonds P (2014) Expanded classification of hepatitis C virus into 7 genotypes and 67 subtypes: updated criteria and genotype assignment web resource. Hepatology 59(1):318–327. doi:10.1002/hep.26744
Stamataki Z, Coates S, Evans MJ, Wininger M, Crawford K, Dong C, Fong YL, Chien D, Abrignani S, Balfe P, Rice CM, McKeating JA, Houghton M (2007) Hepatitis C virus envelope glycoprotein immunization of rodents elicits cross-reactive neutralizing antibodies. Vaccine 25 (45):7773–7784. Epub 2007 Sep 14
Swadling L, Capone S, Antrobus RD, Brown A, Richardson R, Newell EW, Halliday J, Kelly C, Bowen D, Fergusson J, Kurioka A, Ammendola V, Del Sorbo M, Grazioli F, Esposito ML, Siani L, Traboni C, Hill A, Colloca S, Davis M, Nicosia A, Cortese R, Folgori A, Klenerman P, Barnes E (2014) A human vaccine strategy based on chimpanzee adenoviral and MVA vectors that primes, boosts, and sustains functional HCV-specific T cell memory. Sci Transl Med 6 (261):261ra153. doi:10.1126/scitranslmed.3009185

Wakita T, Pietschmann T, Kato T, Date T, Miyamoto M, Zhao Z, Murthy K, Habermann A, Kräusslich HG, Mizokami M, Bartenschlager R, Liang TJ (2005) Production of infectious hepatitis C virus in tissue culture from a cloned viral genome. Nat Med 11(7):791–796. Epub 2005 Jun 12. Erratum in: Nat Med. 2005 Aug;11(8):905

Wong JA, Bhat R, Hockman D, Logan M, Chen C, Levin A, Frey SE, Belshe RB, Tyrrell DL, Law JL, Houghton M (2014) Recombinant hepatitis C virus envelope glycoprotein vaccine elicits antibodies targeting multiple epitopes on the envelope glycoproteins associated with broad cross-neutralization. J Virol 88(24):14278–14288. doi:10.1128/JVI.01911-14. Epub 2014 Oct 1

Hepatitis C Virus Genotypes and Their Evolution

Tadasu Shin-I, Masaya Sugiyama, and Masashi Mizokami

Abstract The hepatitis C virus (HCV) genome is highly heterogeneous. Its genetic variability (genotypes and subtypes) is related to its biological and clinical properties. In 2005, a consensus was reached regarding a unified nomenclature system for HCV genotypes and subtypes. Since then, many complete genome sequences have been reported, resulting in the identification of new genotypes and subtypes. To determine the current status of HCV genotypes, complete genome sequences and their annotations were retrieved from public databases. These viral sequences were arranged according to genotype/subtype and geographical distribution and analyzed phylogenetically to determine the relationships between classification and geography. In addition, the relationships between the HCV genome and the genomes of various related viruses were analyzed phylogenetically to determine the HCV origin. These analyses showed that the viruses evolved along with their hosts and that, worldwide, HCV should be classified into seven major genotypes with their serial subtypes.

Keywords Hepatitis C virus • Genotype • Subtype • Geographic distribution • Phylogenetic analysis • Evolution • Complete genome • Database

1 Introduction

The hepatitis C virus (HCV) is a single-stranded RNA virus that requires an RNA-dependent RNA polymerase to reproduce. The error rate of this polymerase is very high, and, thus, the viral genomes are highly heterogeneous. These viruses have been classified into a hierarchy of genotypes and subtypes, which have different biological and clinical properties. For example, patients infected with HCV subtype 2a show a higher response rate to standard interferon therapy than those infected with subtype 1b, which is a major subtype in Japan. In addition, subtype 3a is more common among patients who are thought to have been infected

T. Shin-I, Ph.D. • M. Sugiyama, Ph.D. • M. Mizokami, M.D., Ph.D. (✉)
Genome Medical Science Project, National Center for Global Health and Medicine, 1-7-1, Kohnodai, Ichikawa 272-8516, Japan
e-mail: mizokami0810@gmail.com

T. Miyamura et al. (eds.), *Hepatitis C Virus I*, DOI 10.1007/978-4-431-56098-2_2

with HCV through intravenous drug use than patients who were infected through other means.

Molecular sequencing of the core, E1, and NS5 regions of the HCV genome and analyses based on pair-wise comparisons show that variations in HCV sequences do not follow a normal distribution and can be classified according to three levels (Ohba et al. 1995). *Genotypes* differ from each other by 30–33 % at the nucleotide level, while *subtypes* within a genotype differ by 20–25 %, and isolates or quasispecies within a subtype have been estimated to differ by less than 10 %. Despite such sequence diversity, all genotypes have an identical gene structure within the large open reading frame, and each gene in the genotypes has an almost identical size. In addition, genetic inter-relationships between variants are highly consistent throughout the genome.

The discovery of an increasing number of HCV variants has resulted in a coordinated international effort to standardize the nomenclature of these variants and define the requirements for designating a new variant as a new genotype or subtype. In 2005, a group of experts in HCV genetic variability, including representatives from the Hepatitis Virus Database (HVDB, Japan) (Shin-I et al. 2008), euHCVdb (France) (Combet et al. 2004), and the Los Alamos hepatitis C sequence database (United States) (Kuiken et al. 2005), proposed a consensus nomenclature system for HCV genotypes (designated by Arabic numbers) and subtypes (designated by lower-case letters) (Simmonds et al. 2005). This group also proposed a revised criterion for the assignment of any newly discovered HCV genotypes. Six genotypes of HCV were proposed to represent the six genetic groups that were defined by the phylogenetic analysis. Any HCV variant should be classified as one of these genotypes, with subtype assignment either confirmed or provisional depending on the availability of complete or partial nucleotide sequences or unassigned when fewer than three examples of a new subtype have been described (Simmonds et al. 2005). This group is now part of the International Committee on Taxonomy of Viruses, which is responsible for assigning genotypes and subtypes (Smith et al. 2014).

Since 2005, a large number of complete HCV genome sequences have been reported, and several provisionally assigned subtypes have been confirmed. In addition, the complete genome sequence of a candidate seventh genotype has been submitted to the DDBJ/EMBL/GenBank (Benson et al. 2014; Brooksbank et al. 2014; Kosuge et al. 2014). Of particular interest is the identification of recombinant forms (RFs) between genotypes. For example, a viable and rapidly spreading recombinant containing structural genes derived from subtype 2k and nonstructural genes derived from subtype 1b has been isolated from intravenous drug users in St. Petersburg, Russia (Kalinina et al. 2004). Intersubtype and intragenotype recombinations have also been reported in Peru (Colina et al. 2004); however, although these are thought to be recombinants of subtypes 1a and 1b, their complete genome sequences are not yet available.

The genetic diversity of HCV has also been assessed in studies examining viral evolution and HCV transmission among human populations. Here, we describe the genetic diversity of HCV, the classification of its viral genotypes and subtypes, and their geographic distribution. In addition, we investigated the origin of HCV and its transmission route.

2 Methodology to Study HCV Genotypes and Evolution

The basic strategy used to study HCV genotypes and subtypes is to classify their nucleotide sequences with phylogenetic methods. However, the amount of data has grown quickly in recent years, and the number of HCV entries in the DDBJ/EMBL/GenBank reached 150,000 in 2014. Thus, it is reasonable to use HCV-specific databases. For example, HVDB (http://s2as02.genes.nig.ac.jp), which is the only site that currently provides routinely updated HCV-specific datasets, provides all of the published HCV sequences that are arranged according to the genome and phylogenetic relationships of each locus among the variants. In the database, the sequences are combined and aligned in multiples with ClustalW (Thompson et al. 1994). The genetic distances between the genomes are estimated from the alignment with Gojobori's six-parameter method (Gojobori et al. 1982), and a phylogenetic tree is constructed from the distance matrix with the neighbor-joining method (Saitou and Nei 1987). Bootstrap resampling (Felsenstein 1985) is also performed to evaluate the confidence of the tree branches. The results can be depicted visually along with the annotation of other information, such as viral hosts and the geographic origin of the viruses. The HVDB also provides various data analyses, so that the analysis described above can be done on users' own datasets.

It is ideal to classify HCV genotypes based on their complete genome sequences because the existence of RFs might confuse the relationships among viral isolates in studies that are based on partial sequences. Thus, we will discuss their genotypes based on a complete genome comparison. Complete genome-based analyses are also applicable in studies of evolution among closely related species. For example, the relationships among various species in the Flaviviridae family can be determined this way (Figs. 1 and 2).

In contrast, comparisons of amino-acid sequences of highly conserved proteins or domains are suitable for determining distant evolutionary relationships between HCV and other viruses. For this, sequences are aligned in multiples with ClustalW. Their genetic distances are then estimated with Kimura's two-parameter method (Kimura 1983), and a phylogenetic tree is constructed with the neighbor-joining method. For example, the relationships among various species that contain Flaviviridae and plant viruses can be determined with the RNA helicase domain (Fig. 3).

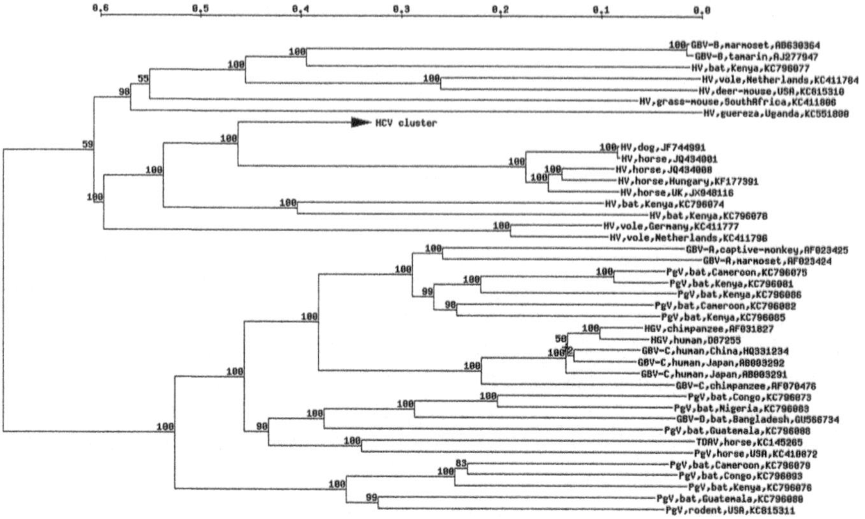

Fig. 1 Phylogenetic tree of complete genome of various species in the two genera in the Flaviviridae family, Hepacivirus and Pegivirus respectively. The tree contains HV (non primate heapaciviruses), GBV-A/B/C/D (GB viruses A/B/C/D), HGV (hepatitis G viruses), PgV (pegivirus), TDAV (Theilers disease associated virus), and HCV (hepatitis C viruses). Species, hosts, geographic origin (if available), and accession numbers are described in OTUs (leaves of the tree). Number described at each branching point shows bootstrap re-sampling rate, which quantifies the confidence of the branch. The cluster corresponding to HCV species is omitted in the tree and shown in Fig. 2 in detail

3 HCV Genotypes

Table 1 lists the HCV subtypes that have been reported and confirmed by the International Committee on Taxonomy of Viruses and their geographical distribution.

Subtype 1a is distributed throughout the US and Northern Europe, whereas subtype 1b is widely distributed throughout the world and is a major subtype in Japan. We collected more complete genome sequences for subtypes 1a and 1b than for any of the other genotypes, and this reflected their presence in the majority of infected patients. Although subtypes other than 1a and 1b have been described, their geographic distribution is uncertain, and the total number of reported complete genome sequences is rather small.

Genotype 2 is present essentially in the same regions as genotype 1, but fewer individuals are infected with genotype 2 than with genotype 1. Subtypes other than 2a and 2b have been described and are present mainly in Europe.

Subtype 3a is widely distributed in South Asia and Oceania, whereas subtype 3b is mainly found in East Asia. Subtypes other than 3a and 3b are found mainly in Canada, but few samples of each of these have been described. These subtypes may have evolved following immigration from Asia.

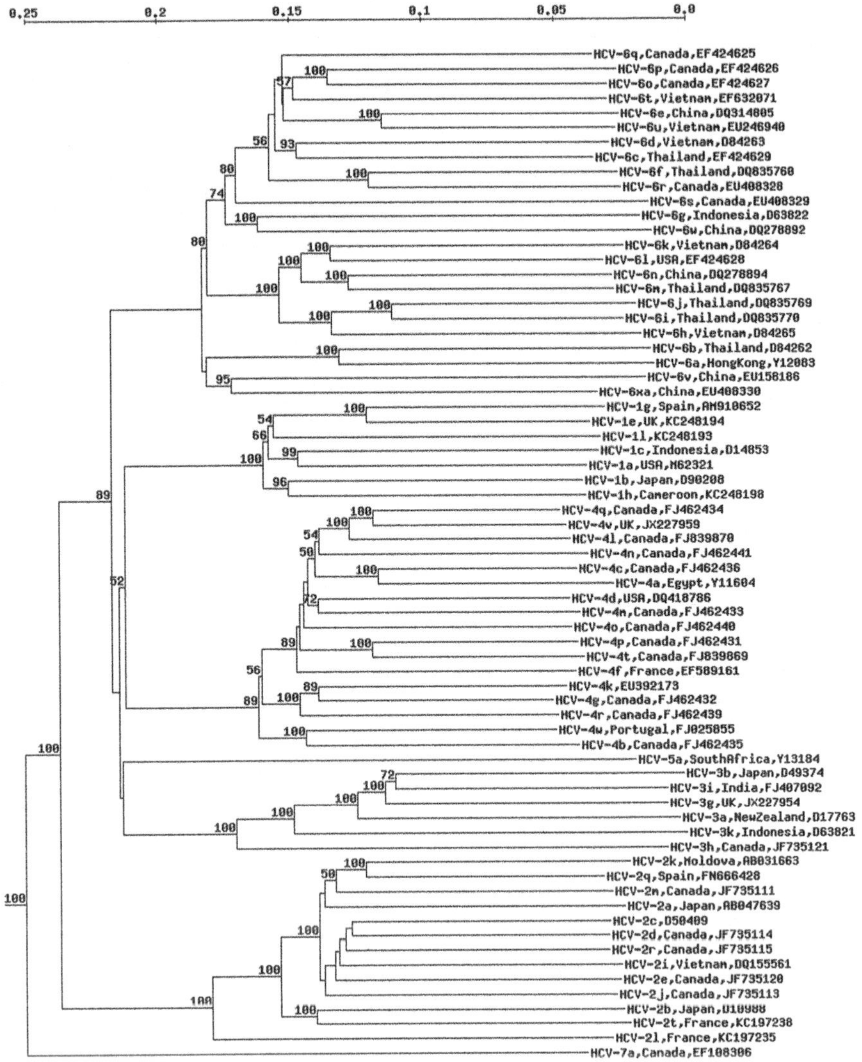

Fig. 2 Phylogenetic tree of complete genome of various HCV subtypes. It contains a representative sequence of each subtype, so that the subtype, geographic origin, and accession number is described in each OTU. This is a part of Flaviviridae tree depicted in Fig. 1, located at the node named "HCV cluster" with an *arrow*

Genotype 4 is mainly distributed in the Middle East, Northern to Central Africa, and Europe. This genotype contains many subtypes, with few examples of each. Some of these, which are found in Canada, may have arisen following immigration from Africa.

Subtype 5a is present mainly in South Africa, but very limited data are available. Thus, its phylogenetic relationship to other genotypes is not clear (Fig. 2).

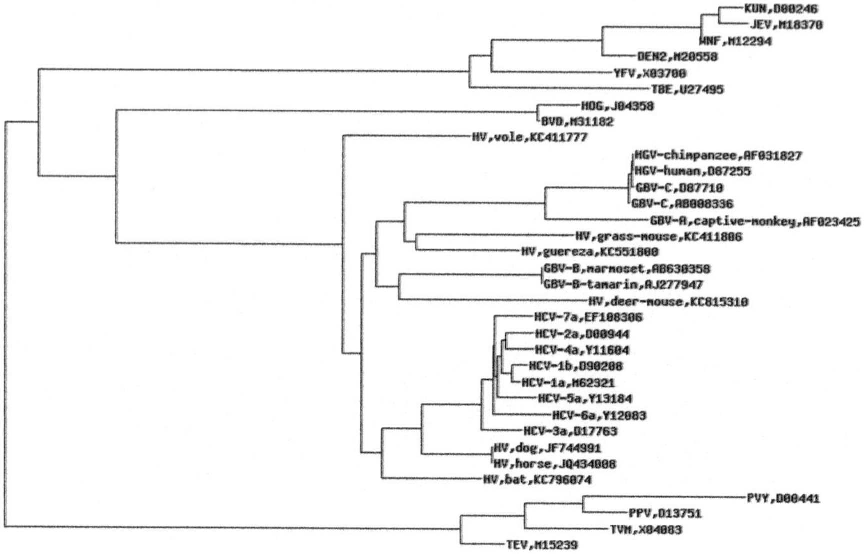

Fig. 3 Phylogenetic tree of amino acids sequences of RNA helicase region of HCV and other various viruses. The tree contains the following species: HCV, HV, GBV-A/B/C, HGV, two pestiviruses (bovine viral diarrhea virus (*BVD*) and hog cholera virus (*HOG*)), six flaviviruses (tick-borne encephalitis virus (*TBE*), yellow fever virus (*YFV*), dengu virus type 2 (*DEN2*), Japanese encephalitis virus (*JEV*), West Nile fever virus (*WNF*), and Kunjin virus (*KUN*)), four plant viruses (tobacco vein mottling virus (*TVM*), tobacco etch virus (*TEV*), potato virus Y (*PVY*), and plum pox virus (*PPV*))

Genotype 6 is mainly distributed throughout East and South-East Asia. This genotype is divided into many subtypes. Several subtypes have been found in the US and Canada, which again is thought to be a result of immigration from Asia.

A new genotype, which was identified in a Canadian immigrant from the Congo, has been classified as subtype 7a. Phylogenetic evidence indicates that this is a new genotype (Fig. 2), although no detailed information has yet been published.

Several intergenotypic RFs have been recorded in the HVDB. Interestingly, all of these RFs, which are listed in Table 2, show different crossover points of recombination. Only one RF, 2k/1b, has been well studied, and it is thought to have originated in Europe-Russia and then to have expanded to Central Asia (Kurbanov et al. 2008). The importance of other RFs is still unknown because a very small number (usually one for each RF) of sequences have been reported.

Table 1 Number of complete genome entries and geographical distribution of reported HCV subtypes

Sub-type	Accession	Isolate	No. ent.	Countries	Reference
Genotype-1					
1a	M62321	HPCPLYPRE	499	USA, Switzerland, Germany, Japan, China, Brazil	Choo et al. (1991)
1b	D90208	HPCJCG	382	USA, Japan, Switzerland, Ireland, China, Germany, Brazil, Russia, Turkey, France	Kato et al. (1990)
1c	D14853	HPCCGS	4	India, Indonesia, China	Okamoto et al. (1994)
1e	KC248194	148636	1	UK	Li et al. (2013)
1g	AM910652	1804	1	Spain	Bracho et al. (2008)
1h	KC248198	EBW443	2	Cameroon	Li et al. (2013)
1l	KC248193	136142	2	UK, Cameroon	Li et al. (2013)
others			12	USA, UK, Equatorial Guinea	
Genotype-2					
2a	AB047639	JFH-1	32	Japan, China, USA, Denmark	Kato et al. (2001)
2b	D10988	HPCJ8G	88	Japan, Denmark, France, China	Okamoto et al. (1992)
2c	D50409	BEBE1	8	UK, France, Denmark	Nakao et al. (1996)
2d	JF735114	QC259	1	Canada	Li et al. (2012)
2e	JF735120	QC64	1	Canada	Li et al. (2012)
2i	DQ155561	D54	3	France	Noppornpanth et al. (2006)
2j	JF735113	QC232	4	France, Venezuela, Canada	Li et al. (2012)
2k	AB031663	VAT96	2	UK, France	Samokhvalov et al. (2000)
2l	KC197235	MRS89	2	France	Jordier et al. (2013)
2m	JF735111	QC178	3	Canada	Li et al. (2012)
2q	FN666428	963	1	Spain	Martro et al. (2011)
2r	JF735115	QC283	1	Canada	Li et al. (2012)
Others			13	Canada, France, China, Moldova	
Genotype-3					
3a	D17763	HPCEGS	30	India, China, UK, Switzerland, Pakistan, NewZealand, Italy, Germany	Sakamoto et al. (1994)
3b	D49374	HPCFG	3	China, Japan	Chayama et al. (1994)
3g	JX227954	BID-G1243	2	UK, Canada	Newman et al. (2013)

(continued)

Table 1 (continued)

Sub-type	Accession	Isolate	No. ent.	Countries	Reference
3h	JF735121	QC29	1	Canada	Lu et al. (2013)
3i	FJ407092	IND-HCV	3	UK, India, Canada	(unpublished)
3k	D63821	HPCJK049E1	2	Indonesia, Canada	Tokita et al. (1996)
Others			2	Pakistan, Canada	
Genotype-4					
4a	Y11604	ED43	11	USA, Egypt, Japan, Canada	Chamberlain et al. (1997a)
4b	FJ462435	QC264	4	Portugal, Canada	Li et al. (2009a)
4c	FJ462436	QC381	1	Canada	Li et al. (2009a)
4d	DQ418786	03-18	5	USA, China, Canada	Timm et al. (2007)
4f	EF589161	IFBT88	5	France, Cameroon	Hmaied et al. (2007)
4g	FJ462432	QC193	3	UK, Canada	Li et al. (2009a)
4k	EU392173	PS3	3	Canada	Kuntzen et al. (2008)
4l	FJ839870	QC274	3	UK, Canada	Li et al. (2009a)
4m	FJ462433	QC249	3	UK, Canada	Li et al. (2009a)
4n	FJ462441	QC97	2	UK, Canada	Li et al. (2009a)
4o	FJ462440	QC93	4	UK, Canada	Li et al. (2009a)
4p	FJ462431	QC139	1	Canada	Li et al. (2009a)
4q	FJ462434	QC262	3	Canada	Li et al. (2009a)
4r	FJ462439	QC384	3	UK, Canada	Li et al. (2009a)
4t	FJ839869	QC155	1	Canada	Li et al. (2009a)
4v	JX227959	BID-G1248	2	UK	Newman et al. (2013)
Others			10	Canada, UK	
Genotype-5					
5a	Y13184	EUH1480	4	SouthAfrica, India, China	Chamberlain et al. (1997b)

(continued)

Table 1 (continued)

Sub-type	Accession	Isolate	No. ent.	Countries	Reference
Genotype-6					
6a	Y12083	EUHK2	21	HongKong, China, Vietnam, Thailand	Adams et al. (1997)
6c	EF424629	Th846	1	Thailand	Lu et al. (2007b)
6e	DQ314805	GX004	4	Vietnam, USA, China	Li et al. (2006)
6f	DQ835760	C-0044	3	Thailand	Lu et al. (2007a)
6g	D63822	HPCJK046E2	2	Indonesia, HongKong	Tokita et al. (1996)
6i	DQ835770	Th602	3	Thailand	Lu et al. (2007a)
6j	DQ835769	Th553	3	Thailand, Canada	Lu et al. (2007a)
6k	D84264	VN405	9	China, Vietnam, Canada	Tokita et al. (1998)
6l	EF424628	537796	6	Vietnam, USA, Laos	Lu et al. (2007b)
6m	DQ835767	B4/92	5	Thailand, China	Lu et al. (2007a)
6n	DQ278894	KM42	6	Thailand, China, Malaysia	Lu et al. (2005)
6o	EF424627	QC227	3	Vietnam, USA, Canada	
6p	EF424626	QC216	1	Canada	Lu et al. (2007b)
6q	EF424625	QC99	1	Canada	Lu et al. (2007b)
6r	EU408328	QC245	1	Canada	Li et al. (2009b)
6s	EU408329	QC66	1	Canada	Li et al. (2009b)
6t	EF632071	VT21	3	Vietnam	Lu et al. (2008)
6u	EU246940	D83	3	China	Noppornpanth et al. (2008)
6v	EU158186	NK46	4	China	Lu et al. (2008)
Others			10	Vietnam, Taiwan, China	
Genotype-7					
7a	EF108306	QC69	1	Canada	(unpublished)
Recombinant					
2k/1b	AY587845		1	Russia	Kalinina et al. (2004)

(continued)

Table 1 (continued)

Sub-type	Accession	Isolate	No. ent.	Countries	Reference
2i/6p	DQ155560		1	Vietnam	Noppornpanth et al. (2006)
2b/1b	DQ364460		1	Philippines	Kageyama et al. (2006)
2/5	AM408911		1	France	Legrand-Abravanel et al. (2007)
2b/6w	EU643835		1	(unknown)	Lee et al. (2010)
2b/1a	JF779679		1	USA	Bhattacharya et al. (2011)

Representative entry is selected for each subtype and its accession number, isolate, publication are described

Table 2 Recombinant forms (RF) and their breakpoints

RF	Accession	Breakpoint	Locus
2k/1b	AY587845	3186	NS2
2i/6p	DQ155560	3405–3464	NS2–NS3
2b/1b	DQ364460	3456	NS3
2/5	AM408911	3366–3389	NS2
2b/6w	EU643835	3429	NS3
2b/1a	JF779679	3429–3440	NS3

Each RF is described by contributory subtypes separated by "/" (slash) in the order in which they appear in the complete genome sequence. Locus in which the breakpoint locates is also described for each RF. Breakpoints of RFs 2i/6p, 2/5, and 2b/1a are not determined strictly

4 HCV Evolution

In order to determine the origin of HCV, as well as the evolutionary relationships between HCV subtypes, we performed a phylogenetic analysis of the complete genome sequences of all of the various species, including representative sequences of all HCV subtypes that are currently available, nonprimate hepaciviruses, GB viruses, hepatitis G virus (HGV), and pegiviruses from various hosts (Figs. 1 and 2).

Figure 2 shows the phylogenetic relationships between various HCV subtypes. In this tree, each genotype forms one cluster. The bootstrap value is rather high, which indicates that the genotype classification is highly consistent with the phylogenetic evidence. However, the relationship between the geographic distribution of subtypes and the genotype-subtype hierarchy is not as clear. The tree indicates that genotype 2 is the origin of HCV. The newly discovered genotype

7 could be the origin according to the topology and bootstrap value of the tree, but there are too few published sequences to conclude it.

Figure 1 shows the phylogenetic relationship between HCV and other related viruses. In this tree, HCV and nonprimate hepaciviruses from various hosts form one cluster, and various pegiviruses constitute a second cluster. Analysis of the cluster that is composed of HCV and nonprimate hepaciviruses showed that the viruses from horses and dogs were closely related to HCV, whereas viruses from rodents were more distantly related. These findings suggest that hepaciviruses were passed from rodents to equine and canine mammals and then passed from these species to primates. However, the relationship between HCV and horse or dog hepacivirus still appears distant, which suggests that there might be an unknown host(s) between them.

In order to determine the distant evolutionary relationships between HCV and other viruses, we performed a phylogenetic analysis of the amino-acid sequences of the RNA helicase regions of the HCV subtypes, nonprimate hepaciviruses, GB viruses, HGV, pestiviruses, flaviviruses, and plant viruses. Figure 3 shows the phylogenetic tree of the species listed in the caption. HCV, GBV, HGV, and HV, which are all derived from mammals, including humans, are clustered first. Second, they cluster with the pestiviruses that infect mammals. Third, they form a cluster with the flaviviruses that are transmitted among vertebrates by blood-sucking arthropods. Finally, the cluster is connected to that of plant viruses. Interestingly, the host branching order of these viruses is plants, insects, mammals, and humans, which suggests that these viruses may have evolved from an ancestor virus in this evolutionary direction (Ohba et al. 1996).

5 Future

The number of HCV sequences has increased rapidly over the past 20 years, with the trend continuing to accelerate (Fig. 4). In addition, the appearance and popularity of next-generation sequencers, which allow extremely high-throughput DNA sequencing, will result in the easy and rapid acquisition of massive amounts of sequencing data. For example, quasispecies and their distribution within a patient can be easily analyzed, thus increasing the number of reported sequences. Under such circumstances, it may be difficult to obtain information about HCVs of virological or clinical interest from general databases, such as the DDBJ/EMBL/GenBank. Thus, the use of an HCV-specialized database with a well-arranged architecture (e.g., HVDB) will become more important.

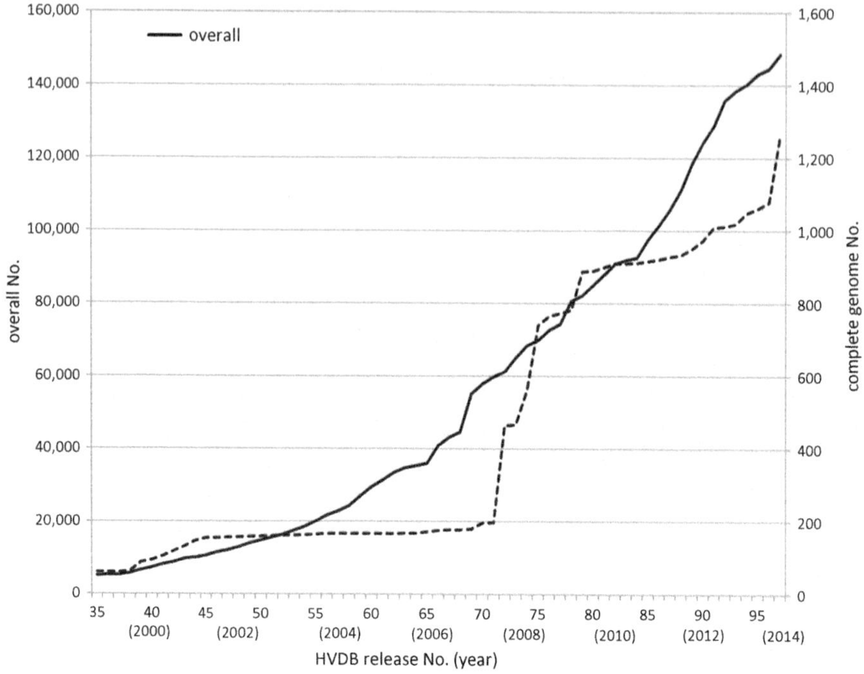

Fig. 4 Progress of HCV data amount in HVDB. Both number of overall entries and complete genome entries are shown in the graph

References

Adams NJ, Chamberlain RW, Taylor LA et al (1997) Complete coding sequence of hepatitis C virus genotype 6a. Biochem Biophys Res Commun 234:393–396

Benson DA, Clark K, Karsch-Mizrachi I et al (2014) GenBank. Nucleic Acids Res 42:D32–D37

Bhattacharya D, Accola MA, Ansari IH et al (2011) Naturally occurring genotype 2b/1a hepatitis C virus in the United States. Virol J 8:458

Bracho MA, Saludes V, Martro E et al (2008) Complete genome of a European hepatitis C virus subtype 1g isolate: phylogenetic and genetic analyses. Virol J 5:72

Brooksbank C, Todd Bergman M, Apweiler R et al (2014) The European bioinformatics institute's data resources 2014. Nucleic Acids Res 42:D18–D25

Chamberlain RW, Adams N, Saeed AA et al (1997a) Complete nucleotide sequence of a type 4 hepatitis C virus variant, the predominant genotype in the Middle East. J Gen Virol 78 (Pt 6):1341–1347

Chamberlain RW, Adams NJ, Taylor LA et al (1997b) The complete coding sequence of hepatitis C virus genotype 5a, the predominant genotype in South Africa. Biochem Biophys Res Commun 236:44–49

Chayama K, Tsubota A, Koida I et al (1994) Nucleotide sequence of hepatitis C virus (type 3b) isolated from a Japanese patient with chronic hepatitis C. J Gen Virol 75(Pt 12):3623–3628

Choo QL, Richman KH, Han JH et al (1991) Genetic organization and diversity of the hepatitis C virus. Proc Natl Acad Sci U S A 88:2451–2455

Colina R, Casane D, Vasquez S et al (2004) Evidence of intratypic recombination in natural populations of hepatitis C virus. J Gen Virol 85:31–37

Combet C, Penin F, Geourjon C, Deleage G (2004) HCVDB: hepatitis C virus sequences database. Appl Bioinformatics 3:237–240

Felsenstein J (1985) Confidence limits on phylogenies: an approach using the bootstrap. Evolution 39:783–791

Gojobori T, Ishii K, Nei M (1982) Estimation of average number of nucleotide substitutions when the rate of substitution varies with nucleotide. J Mol Evol 18:414–423

Hmaied F, Legrand-Abravanel F, Nicot F et al (2007) Full-length genome sequences of hepatitis C virus subtype 4f. J Gen Virol 88:2985–2990

Jordier F, Deligny ML, Barre R et al (2013) Evidence for two phylogenetic clusters within hepatitis C virus (HCV) genotype 2 inferred from analysis of complete coding sequences of 15 HCV strains. J Med Virol 85:1754–1764

Kageyama S, Agdamag DM, Alesna ET et al (2006) A natural inter-genotypic (2b/1b) recombinant of hepatitis C virus in the Philippines. J Med Virol 78:1423–1428

Kalinina O, Norder H, Magnius LO (2004) Full-length open reading frame of a recombinant hepatitis C virus strain from St Petersburg: proposed mechanism for its formation. J Gen Virol 85:1853–1857

Kato N, Hijikata M, Ootsuyama Y et al (1990) Molecular cloning of the human hepatitis C virus genome from Japanese patients with non-A, non-B hepatitis. Proc Natl Acad Sci U S A 87:9524–9528

Kato T, Furusaka A, Miyamoto M et al (2001) Sequence analysis of hepatitis C virus isolated from a fulminant hepatitis patient. J Med Virol 64:334–339

Kimura M (1983) The neutral theory of molecular evolution. Cambridge University Press, Cambridge, pp 65–97

Kosuge T, Mashima J, Kodama Y et al (2014) DDBJ progress report: a new submission system for leading to a correct annotation. Nucleic Acids Res 42:D44–D49

Kuiken C, Yusim K, Boykin L, Richardson R (2005) The Los Alamos hepatitis C sequence database. Bioinformatics 21:379–384

Kuntzen T, Berical A, Ndjomou J et al (2008) A set of reference sequences for the hepatitis C genotypes 4d, 4f, and 4k covering the full open reading frame. J Med Virol 80:1370–1378

Kurbanov F, Tanaka Y, Avazova D et al (2008) Detection of hepatitis C virus natural recombinant RF1_2k/1b strain among intravenous drug users in Uzbekistan. Hepatol Res 38:457–464

Lee YM, Lin HJ, Chen YJ et al (2010) Molecular epidemiology of HCV genotypes among injection drug users in Taiwan: full-length sequences of Two New subtype 6w strains and a recombinant form_2b6w. J Med Virol 82:57–68

Legrand-Abravanel F, Claudinon J, Nicot F et al (2007) New natural intergenotypic (2/5) recombinant of hepatitis C virus. J Virol 81:4357–4362

Li C, Fu Y, Lu L et al (2006) Complete genomic sequences for hepatitis C virus subtypes 6e and 6g isolated from Chinese patients with injection drug use and HIV-1 co-infection. J Med Virol 78:1061–1069

Li C, Lu L, Wu X et al (2009a) Complete genomic sequences for hepatitis C virus subtypes 4b, 4c, 4d, 4g, 4k, 4l, 4m, 4n, 4o, 4p, 4q, 4r and 4t. J Gen Virol 90:1820–1826

Li C, Lu L, Zhang X, Murphy D (2009b) Entire genome sequences of two new HCV subtypes, 6r and 6s, and characterization of unique HVR1 variation patterns within genotype 6. J Viral Hepat 16:406–417

Li C, Cao H, Lu L, Murphy D (2012) Full-length sequences of 11 hepatitis C virus genotype 2 isolates representing five subtypes and six unclassified lineages with unique geographical distributions and genetic variation patterns. J Gen Virol 93(Pt 6):1173–1184

Li C, Njouom R, Pepin J et al (2013) Characterization of full-length hepatitis C virus sequences for subtypes 1e, 1h and 1l, and a novel variant revealed Cameroon as an area in origin for genotype 1. J Gen Virol 94(Pt 8):1780–1790

Lu L, Nakano T, He Y et al (2005) Hepatitis C virus genotype distribution in China: predominance of closely related subtype 1b isolates and existence of new genotype 6 variants. J Med Virol 75:538–549

Lu L, Li C, Fu Y et al (2007a) Complete genomes for hepatitis C virus subtypes 6f, 6i, 6j and 6m: viral genetic diversity among Thai blood donors and infected spouses. J Gen Virol 88:1505–1518

Lu L, Li C, Fu Y et al (2007b) Complete genomes of hepatitis C virus (HCV) subtypes 6c, 6l, 6o, 6p and 6q: completion of a full panel of genomes for HCV genotype 6. J Gen Virol 88:1519–1525

Lu L, Murphy D, Li C et al (2008) Complete genomes of three subtype 6t isolates and analysis of many novel hepatitis C virus variants within genotype 6. J Gen Virol 89:444–452

Lu L, Li C, Yuan J et al (2013) Full-length genome sequences of five hepatitis C virus isolates representing subtypes 3g, 3h, 3i and 3k, and a unique genotype 3 variant. J Gen Virol 94 (Pt 3):543–548

Martro E, Valero A, Jordana-Lluch E et al (2011) Hepatitis C virus sequences from different patients confirm the existence and transmissibility of subtype 2q, a rare subtype circulating in the metropolitan area of Barcelona, Spain. J Med Virol 83:820–826

Nakao H, Okamoto H, Tokita H et al (1996) Full-length genomic sequence of a hepatitis C virus genotype 2c isolate (BEBE1) and the 2c-specific PCR primers. Arch Virol 141:701–704

Newman RM, Kuntzen T, Weiner B et al (2013) Whole genome pyrosequencing of rare hepatitis C virus genotypes enhances subtype classification and identification of naturally occurring drug resistance variants. J Infect Dis 208:17–31

Noppornpanth S, Lien TX, Poovorawan Y et al (2006) Identification of a naturally occurring recombinant genotype 2/6 hepatitis C virus. J Virol 80:7569–7577

Noppornpanth S, Poovorawan Y, Lien TX et al (2008) Complete genome analysis of hepatitis C virus subtypes 6t and 6u. J Gen Virol 89(Pt 5):1276–1281

Ohba K, Mizokami M, Ohno T et al (1995) Classification of hepatitis C virus into major types and subtypes based on molecular evolutionary analysis. Virus Res 36:201–214

Ohba K, Mizokami M, Lau JY et al (1996) Evolutionary relationship of hepatitis C, pesti-, flavi-, plantviruses, and newly discovered GB hepatitis agents. FEBS Lett 378:232–234

Okamoto H, Kurai K, Okada S et al (1992) Full-length sequence of a hepatitis C virus genome having poor homology to reported isolates: comparative study of four distinct genotypes. Virology 188:331–341

Okamoto H, Kojima M, Sakamoto M et al (1994) The entire nucleotide sequence and classification of a hepatitis C virus isolate of a novel genotype from an Indonesian patient with chronic liver disease. J Gen Virol 75(Pt 3):629–635

Saitou N, Nei M (1987) The neighbor-joining method: a new method for reconstructing phylogenetic trees. Mol Biol Evol 4:406–425

Sakamoto M, Akahane Y, Tsuda F et al (1994) Entire nucleotide sequence and characterization of a hepatitis C virus of genotype V/3a. J Gen Virol 75(Pt 7):1761–1768

Samokhvalov EI, Hijikata M, Gylka RI et al (2000) Full-genome nucleotide sequence of a hepatitis C virus variant (isolate name VAT96) representing a new subtype within the genotype 2 (arbitrarily 2k). Virus Genes 20:183–187

Shin-I T, Tanaka Y, Tateno Y, Mizokami M (2008) Development and public release of a comprehensive hepatitis virus database. Hepatol Res 38:234–243

Simmonds P, Bukh J, Combet C et al (2005) Consensus proposals for a unified system of nomenclature of hepatitis C virus genotypes. Hepatology 42:962–973

Smith DB, Bukh J, Kuiken C et al (2014) Expanded classification of hepatitis C virus into 7 genotypes and 67 subtypes: updated criteria and genotype assignment web resource. Hepatology 59:318–327

Thompson JD, Higgins DG, Gibson TJ (1994) CLUSTAL W: improving the sensitivity of progressive multiple sequence alignment through sequence weighting, position-specific gap penalties and weight matrix choice. Nucleic Acids Res 22:4673–4680

Timm J, Neukamm M, Kuntzen T et al (2007) Characterization of full-length hepatitis C virus genotype 4 sequences. J Viral Hepat 14:330–337

Tokita H, Okamoto H, Iizuka H et al (1996) Hepatitis C virus variants from Jakarta, Indonesia classifiable into novel genotypes in the second (2e and 2f), tenth (10a) and eleventh (11a) genetic groups. J Gen Virol 77(Pt 2):293–301

Tokita H, Okamoto H, Iizuka H et al (1998) The entire nucleotide sequences of three hepatitis C virus isolates in genetic groups 7–9 and comparison with those in the other eight genetic groups. J Gen Virol 79(Pt 8):1847–1857

The Structure of HCV

Joseph Marcotrigiano and Maria Teresa Catanese

Abstract According to nature's functional aesthetics principles, hepatitis C virus (HCV) came to acquire a certain structure to serve a specific purpose. Structure explains mechanisms, reveals strategies. In viruses, optimization of the shape is dictated by the number of encoded structural proteins and constant pressure from the host immune system. The result is a durable particle, built to perfection. While perfection is often associated with symmetry, HCV shows how successful an irregular virus can be. This virus compromised its morphology, evolving to mask itself; a beautiful example of exploitation of the host to maintain a chronic infection. This chapter summarizes our current knowledge of the HCV virion and its peculiar properties that explain the challenges that we faced in isolating it and that we keep facing in trying to develop effective vaccine strategies.

Keywords Enveloped virus • Hepatitis C virus • Virus ultrastructure • Electron microscopy • Virion assembly • Virus-host interactions • Lipo-viro-particle • Lipoprotein

Abbreviations

Å	Angstrom
aa	Amino acid
AP2M1	Adaptor-related protein complex 2 μ1 subunit
Apo	Apolipoprotein
BVDV	Bovine viral diarrhea virus
C	Capsid protein
CE	Cholesteryl esters

J. Marcotrigiano (✉)
Department of Chemistry and Chemical Biology, Rutgers, The State University of New Jersey, New Brunswick, NJ, USA
e-mail: jmarco@cabm.rutgers.edu

M.T. Catanese (✉)
Department of Infectious Diseases, King's College London, London, UK
e-mail: maria.catanese@kcl.ac.uk

T. Miyamura et al. (eds.), *Hepatitis C Virus I*, DOI 10.1007/978-4-431-56098-2_3

Chol	Cholesterol (unesterified)
CM	Chylomicrons
cryo-EM	Cryo-electron microscopy
cryo-ET	Cryo-electron tomography
CSFV	Classical swine fever virus
DENV	Dengue virus
DGAT1	Diacylglycerol O-acetyltransferase 1
E	Envelope glycoprotein
EM	Electron microscopy
ER	Endoplasmic reticulum
HA	Hemagglutinin
HCV	Hepatitis C virus
HCVcc	Cell culture-derived HCV
HDL	High-density lipoprotein
HSPGs	Heparan sulfate proteoglycans
Huh	Human hepatoma
HVR	Hypervariable region
IDL	Intermediate-density lipoprotein
iEM	Immuno-EM
IFN	Interferon
Ig	Immunoglobulin
JFH	Japanese fulminant hepatitis
Kb	Kilobase
LD	Lipid droplet
LDL	Low-density lipoprotein
LDL-R	LDL receptor
LPL	Lipoprotein lipase
LVP	Lipo-viro-particle
MTP	Microsomal triglyceride transfer protein
NaBr	Sodium bromide
NANBH	Non-A non-B hepatitis
NDV	Newcastle disease virus
NMR	Nuclear magnetic resonance
NS	Non-structural
PC	Phosphatidylcholine
PE	Phosphatidylethanolamine
PHH	Primary human hepatocytes
PL	Phospholipids
PS	Phosphatidylserine
RIG-I	Retinoic acid-inducible gene I
RNAi	RNA interference
SCID	Severe combined immunodeficiency
SM	Sphingomyelin
SR-BI	Scavenger receptor class B type I

$TCID_{50}$	Tissue culture infectious $dose_{50}$
TG	Triglycerides
uPA	Urokinase plasminogen activator
VLDL	Very low-density lipoprotein
VLP	Virus-like particles
WNV	West Nile virus
YFV	Yellow fever virus

1 Peculiar Biophysical Properties of HCV Particles

For a long time, HCV remained the mysterious etiological agent for the so-called non-A, non-B hepatitis (NANBH) (Feinstone et al. 1975). Unlike most human viruses that were discovered by electron microscopic analysis of clinical samples, this positive-sense, single-stranded RNA virus was only identified in 1989 by molecular cloning using a cDNA library obtained from infectious NANBH plasma (Choo et al. 1989). Based on sequence similarities with flaviviruses and pestiviruses, HCV was subsequently classified into the family *Flaviviridae*, genus *Hepacivirus* (Miller and Purcell 1990). Other members of the *Flaviviridae* family include dengue (DENV), yellow fever (YFV) and West Nile virus (WNV) of the *Flavivirus* genus, while bovine viral diarrhea (BVDV) and classical swine fever virus (CSFV) belong in the *Pestivirus* genus. In addition to similarities in genome organization, all *Flaviviridae* are enveloped viruses whose virion structural proteins consist of a single nucleocapsid (C) protein and two or three (for *Pestiviruses*) envelope (E) glycoproteins (Table 1).

Virion morphogenesis is thought to begin with the encapsidation of a single copy of the viral genome by multiple C molecules. It remains unclear what triggers this process, since a packaging signal in the genomic viral RNA has not been identified for any of the *Flaviviridae*. With only one nucleocapsid protein, capsids of *Flaviviridae* members were hypothesized to have fairly simple structures, possibly icosahedral, fixed by geometric constraints (Strauss 2007). However, so far, cryoelectron microscopy (cryo-EM) image reconstructions of flaviviruses have revealed that the basic C protein condenses around the viral RNA genome without forming an ordered structure (Zhang et al. 2007; 2003a, b Mukhopadhyay et al. 2003; Kuhn et al. 2002). Since fully assembled capsids cannot be found in infected cells, encapsidation has been suggested to occur concomitantly with budding, a process that enables the acquisition of the lipid envelope. The cellular membrane that *Flaviviridae* utilize for viral budding derives from the endoplasmic reticulum (ER), where the viral glycoproteins reside, but it remains unclear what force drives this process. In fact, there is no evidence for nucleocapsid-glycoprotein interactions. Possibly some free energy for the budding process is provided by lateral interactions between the glycoproteins (Ferlenghi et al. 2001).

Importantly, there is no evidence of direct contact between the cytoplasmic tails of the glycoproteins and the nucleocapsid protein, implying that the ratio of

Table 1 Structural features of viruses in the family *Flaviviridae*

Genus/ species	Genome size	Capsid morphology	Glycoproteins	Size (nm)	Symmetry	Density (g/ml)
Hepacivirus (HCV)	9.6 Kb	Undetermined	E1 and E2	40–100	Unclear	1.03–1.25
Flavivirus (DENV, YFV)	11.0 Kb	Disordered	E and prM	40–65	Icosahedral	1.20–1.23
Pestivirus (BVDV)	12.5 Kb	Undetermined	E(rns), E1, E2	40–60	Unclear	1.11–1.15

structural proteins in the virion may not necessarily be one-to-one and may accommodate variations. The envelopment of the capsid by host-derived membranes introduces further degrees of flexibility in building a 3D structure because lipids enable irregularities in construction (Strauss 2007).

Nevertheless, viral envelope glycoproteins can arrange themselves in repetitive arrays, giving rise to highly symmetrical virions (Strauss and Strauss 2001). This is the case for the flaviviruses DENV and WNV, whose EM structures reveal that 90 homodimers of glycoprotein E assemble in an antiparallel fashion to form an icosahedral protein scaffold around the lipid bilayer (Kuhn et al. 2002; Mukhopadhyay et al. 2003). These virions in their mature form display a very homogeneous morphology, with a diameter of approximately 50 nm and a smooth surface. Initial studies in insect cells showed that HCV structural proteins assembled into enveloped virus-like particles (VLPs) with a predominant size of 50 nm, suggesting that HCV might adopt similar structures to flaviviruses (Yu et al. 2007; Baumert et al. 1998). Nevertheless, with the advent of cell culture systems to propagate infectious HCV particles in vitro (Wakita et al. 2005; Lindenbach et al. 2005; Zhong et al. 2005), increasing evidence indicated that, despite their relatedness, HCV virions assembled differently from flaviviruses.

Cell culture-derived HCV (HCVcc) was reported to be very heterogeneous in size, with a diameter between 30 and >100 nm. Furthermore, secreted virions displayed an unusually broad buoyant density profile compared to other enveloped viruses (Lindenbach et al. 2005; Wakita et al. 2005; Zhong et al. 2005). This was in agreement with what observed in plasma of infected individuals, where viral RNA could be isolated across a wide range of densities (1.03–1.25 g/ml; Table 1) (Kanto et al. 1994). In contrast, HCV particles retrieved from the intracellular compartment have a significantly higher and narrower buoyant density (1.15–1.20 g/ml) (Gastaminza et al. 2006). This suggests that the composition of virions is altered during egress and particles become less dense, possibly as a result of an increased lipid to protein ratio.

Notably, the infectivity of HCV particles inversely correlates with their density (Bradley et al. 1991; Carrick et al. 1992) and the HCV RNA in the lighter material (<1.10 g/ml) was the only fraction capable of transmitting infection to chimpanzees (Bradley et al. 1991). Such buoyant density was considered surprisingly low for what was thought to be a small-enveloped RNA virus (Miyamoto et al. 1992;

Carrick et al. 1992; Thomssen et al. 1992). While the heavier fractions were precipitated by anti-human IgGs, suggesting the formation of immune complexes (Thomssen et al. 1993), HCV RNA in the lighter material co-eluted with very low density lipoproteins (VLDLs) (Fig. 1) (Prince et al. 1996). That observation might have simply reflected a similarity in size between VLDLs (ranging from 30 to 80 nm) and circulating HCV particles. Indeed, in humans the latter were estimated to be 30–60 nm by filtration (He et al. 1987) and ultracentrifugation analysis of chimpanzee-derived HCV suggested a diameter <80 nm (Bradley et al. 1985). Nevertheless, mere co-sedimentation would have not explained how HCV RNA could be precipitated by antibodies to β-lipoproteins (i.e. VLDL and LDL) (Thomssen et al. 1993). These initial studies favored the hypothesis of association of HCV virions with lipoproteins. These virions were referred to as lipo-viro-particles (LVPs) because they appeared to be rich in triglycerides and positive for apolipoprotein (apo)B, HCV RNA, and core protein (Andre et al. 2002). Since then,

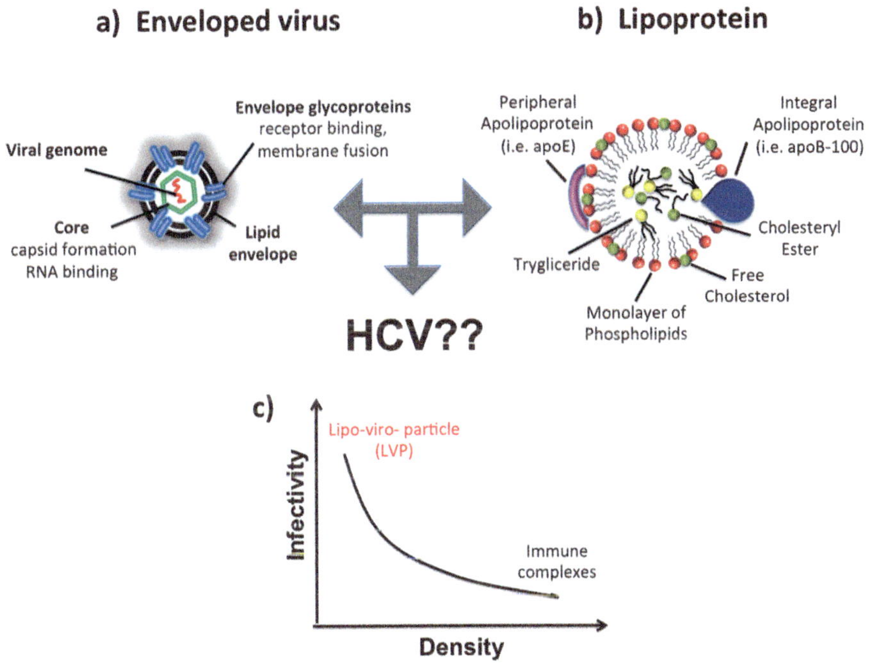

Fig. 1 The hybrid nature of HCV particles. Schematic representation of the structure of an enveloped virus (**a**) and a lipoprotein (**b**). The structure of HCV was proposed to be something in between, since several lines of evidence suggest that infectious particles circulate in the bloodstream as a hybrid lipo-viro-particle (*LVP*). (**c**) HCV particles display a broad buoyant density profile: the density of virions inversely correlates with their infectivity. The association of HCV to host lipoproteins may explain the atypically low buoyant density (<1.10 g/mL) of the most infectious material, that is referred to as LVP. In contrast, the most dense fractions are poorly infectious and can be precipitated by anti-human IgG, suggesting that they might represent immune complexes

a growing body of literature has emphasized how the morphogenesis of HCV is tightly linked to the VLDL biosynthetic pathway (Gastaminza et al. 2010; Merz et al. 2011; Benga et al. 2010; Jiang and Luo 2009; Huang et al. 2007; Chang et al. 2007a).

Another peculiarity about HCV is that the buoyant density of the infectious particles vary depending on the host cell in which the virus is assembled. Interestingly, this impacts on the specific infectivity of the virions, which is intended as the number of infectious particles per genome equivalent. Sedimentation studies with cell culture-produced virus (HCVcc) showed that the most infectious material has a density of approximately 1.14 g/ml and a specific infectivity of 1 in a 1,000. Conversely, when inoculated into chimpanzees or mice with a chimeric human liver, the ex vivo-derived infectious HCV display a density <1.10 g/ml and a hundred fold higher specific infectivity (Lindenbach et al. 2006). This difference has been attributed to the inability of Huh-7.5 cells to produce fully lipidated VLDLs, which might in turn affect the extent of modifications that HCV virions undergo during egress.

After budding from the ER, HCV particles are transported through the cellular secretory pathway to be released in the extracellular milieu. As they mature, in addition to decreasing in density, HCV virions acquire complex-type glycans on their envelope. Intracellular envelope glycoproteins display mainly high-mannose-type glycans, consistently with their accumulation in the ER. However, complex-type glycans were found on E2 purified from secreted HCVcc particles (Vieyres et al. 2010). These sugars are hallmarks of protein transit through the Golgi apparatus since they result from the maturation of high-mannose-type glycans by Golgi glycosidases and glycosyltransferases (Helenius and Aebi 2001). Furthermore, while E1 and E2 on the ER membranes assemble as non-covalent heterodimers, virion-associated glycoproteins form oligomeric complexes that are stabilized by disulfide bonds (Vieyres et al. 2010). This maturation may contribute to the acid-resistance of extracellular HCV virions and have implications for the mechanisms of fusion. Indeed, cell surface-bound HCV needs to be incubated for long periods at 37 °C for low-pH-mediated entry to occur. This suggests that post-binding events are required to prime the HCV envelope proteins for fusion (Tscherne et al. 2006). This contrasts with what was observed with *Flaviviruses* in which the glycoproteins undergo dramatic rearrangements when exposed to acidic pH.

As discussed later in this chapter, HCV seems to incorporate much more than a cell-derived membrane to form progeny virions (Lussignol et al. 2016; Merz et al. 2011; Parent et al. 2009). Its unusual composition might explain the unique biophysical properties, structure and assembly/entry routes of the infectious particles.

2 Challenges in Isolating HCV Virions

The availability of high numbers of purified and well-preserved particles represents an essential first step to structural analysis of any virus. HCV has been a very challenging pathogen to study in these respects, given the difficulties in isolating it from infected patients or chimpanzees, its low titres in cell culture and the lack, until very recently, of a susceptible small animal model. The direct observation of virus particles by EM from infected sera (Kanto et al. 1994; Prince et al. 1996; Schmidt et al. 2000; Trestard et al. 1998; Kaito et al. 1994) or liver tissue (De Vos et al. 2002; Le Sage and Mouland 2013) proved very difficult, likely because of their association with antibodies (Kanto et al. 1994; Hijikata et al. 1993) or lipoproteins (Prince et al. 1996) and the very low levels of replication both at the single cell level and in the organ as a whole. Indeed, only 7–20 % of hepatocytes are found to be positive for viral antigens in patients with chronic hepatitis C (Hiramatsu et al. 1992; Liang et al. 2009). Plausible explanations for this are that innate immune responses contribute to keep viral replication at bay. Moreover, it might be sufficient and advantageous for HCV to only replicate in a small number of cells at any one time to maintain infection over the course of many years while minimizing damage to the host.

Due to the lack of appropriate cell culture systems to grow patient-derived isolates, it remains unclear to what extent the copies of HCV genomic RNA in the serum, defined as viremia, truly reflect the number of infectious virions in circulation or the extent of infection in the liver. This gap also hampered structure-function correlation studies, as it was not possible to determine whether infectious virions produced in vivo shared a specific morphology. The HCV field struggled for a long time to develop a system that recapitulated the complete virus life cycle in cell culture. Although the first HCV sequences were isolated in 1989 (Choo et al. 1989) a full length, infectious genome was only assembled in 1997 (Kolykhalov et al. 1997). RNA transcribed from those molecular clones was able to transmit infection to chimpanzees after direct intrahepatic inoculation but did not yield infectious particles in cell culture. In 1999, subgenomic replicons provided a powerful means to study HCV RNA replication (Lohmann et al. 1999), but it was only in 2005 that infectious virions could be produced in cell culture (HCVcc) (Wakita et al. 2005; Lindenbach et al. 2005; Zhong et al. 2005). This was made possible through the discovery of a genotype 2a HCV isolate from a Japanese patient with acute fulminant hepatitis (JFH)-1. When full length JFH-1 genome RNA was transfected into human hepatoma (Huh)-7 cells, viral particles were released, albeit with low titers, which were capable of infecting naïve cells. Improvements in viral titers followed by using Huh-7.5 cells and derived sub-lines (Lindenbach et al. 2005; Zhong et al. 2005) that are highly permissive for HCV replication (Blight et al. 2002) due to a defect in the retinoic acid-inducible gene I (RIG-I) innate immune response pathway (Sumpter et al. 2005). Furthermore, the development of a chimeric genotype 2a full-length genome, expressing the core through NS2 region of the HCJ6 HCV isolate cloned into the JFH-1 genome (J6/JFH) produced higher titers post transfection than the full length JFH-1 genome (Lindenbach et al. 2005). HCVcc exhibited characteristics predicted for an HCV virion: the infectivity of these particles was blocked by antibodies against E2 and by purified soluble CD81; virion density was similar to that found in sera of infected individuals. Importantly, cell culture grown HCVcc could successfully infect chimpanzees and uPA-SCID mice transplanted with human hepatocytes (Lindenbach et al. 2006). In both cases rising viral loads and sustained infection ensued, confirming that they were bona fide infectious HCV particles. Moreover, virus recovered from HCVcc-inoculated animals (ex vivo-HCVcc) was infectious in cell culture.

Despite substantial progress in growing viral particles in cell culture, infectious HCVcc titers are still in the range of 10^5 tissue culture infectious $dose_{50}$ ($TCID_{50}$) per ml −100 to a 1,000-fold lower than other enveloped viruses for which the ultrastructure has been determined. Even more challenging is the task of producing infectious virus particles in primary human hepatocytes (PHH). Primary cultures represent a more biologically relevant system to study HCV infection compared to hepatoma cells (Podevin et al. 2010). Firstly, they polarize as they normally would in the liver, as shown by the typical polygonal cell morphology, the formation of tightly sealed islands and bile canaliculi. Secondly, from a functional standpoint, their gene and protein expression profile more closely resemble physiological conditions (Pan et al. 2009), enabling more meaningful studies on the interplay between virus propagation and cellular biological processes, such as lipoprotein assembly or innate immune responses to infections. However, PHH are difficult to obtain, being usually isolated from patients undergoing liver tumor resections (Lecluyse and Alexandre 2010) or aborted fetal livers (Lazaro et al. 2003). Additional drawbacks are that they are very difficult to infect both with HCVcc and patient-derived infectious sera and typically yield much lower titers than hepatoma cells. Recent studies have shown that blunting the innate immune response promotes productive HCV infection in PHH. This was achieved by either overexpressing paramyxovirus V proteins to antagonize interferon (IFN) induction and antiviral signaling (Andrus et al. 2011) or by treating cultures with inhibitors of the JAK/STAT pathway (Catanese et al. 2013). Those approaches, in addition to demonstrating the importance of cellular IFN responses in suppressing viral spread in primary cultures, led to a significant improvement in virus production with titers approaching 10^5 $TCID_{50}$/ml.

Nevertheless, attempts by several groups to enrich for viral particles that were structurally preserved using ultracentrifugation protocols optimised for other enveloped viruses proved unsuccessful for HCV. Specifically, poor stability of virions had been observed. While the majority of HCV RNA could be recovered in sucrose and NaBr gradient solutions, the density of the particles increased. Since no parallel structural studies were conducted, it was not possible to determine whether that was solely due to dissociation of host lipoproteins from HCV virions or if there was also damage to the viral envelope (Nielsen et al. 2006). In contrast, iodixanol gradients that are isotonic and isosmotic with blood, best preserved the biophysical properties of virus particles, their interaction with lipoproteins and infectivity (Nielsen et al. 2006; Merz et al. 2011; Catanese et al. 2013). HCV stability also seems temperature-dependent as inactivation to almost background levels was detected at 37 °C in time course experiments (Ciesek et al. 2010).

To complicate further morphological studies of HCV, both HCVcc and HCV from PHH seemed to bind poorly to conventional EM grids and often displayed positive rather than negative staining, indicating that particles had been damaged (Catanese et al. 2013; Gastaminza et al. 2010; Merz et al. 2011). Notably, several groups observed that even potent HCV neutralizing antibodies against the envelope glycoproteins only captured a very small percentage of the HCV RNA secreted in the extracellular milieu. Accordingly, when these antibodies were conjugated to protein A-coated EM grids, only a few extracellular HCV particles could be visualized (Catanese et al. 2013). Overall, these data suggested that envelope glycoproteins on HCV virions were poorly accessible, irrespective of the neutralizing ability of the antibody. In fact, neutralization assays are performed in cell culture, where a dynamic remodeling of the particles is likely to take place prior to and during binding to the plasma membrane, thus making epitopes more accessible on the surface of the virion compared to in silico capture assays. In keeping with the hypothesis that E1–E2 epitopes were not readily exposed on the surface of virions, capture efficiency of HCV RNA and particles was significantly more efficient with antibodies targeting host apolipoproteins or when tags were inserted in envelope glycoproteins. In all cases, tags were fused at the N-terminus of E2, upstream of the hypervariable region 1 (HVR-1) and were rather hydrophilic (6x-His, FLAG). Those features might have contributed to a more outward orientation of these regions compared to the rest of the protein, away from virion-associated lipids. Indeed, this tagging approach enabled purification of enough virions for lipidomic (Merz et al. 2011) and ultrastructural (Catanese et al. 2013) analyses.

Using tag-specific affinity purification methods, exosome-like structures were also isolated. These particles ranged between 50 and 100 nm in diameter and displayed a very smooth surface and a distinguishable bilayer. Since these structures could not be found in wild-type HCV preparations, they likely contain tagged E2 on their surface, rather than being contaminants. However, exosomes could not be found in fractions highly enriched in infectious particles (Catanese et al. 2013). Similarly, Gastaminza et al. observed large, smooth vesicles in HCV preparations purified by sucrose gradient centrifugation but showed that these structures did not significantly contribute to total HCV infectivity (Gastaminza et al. 2010). Interestingly, subviral particles devoid of viral nucleocapsid but positive for HCV glycoproteins were also isolated from the blood of patients, raising the hypothesis that they might act as decoy for neutralizing antibodies and play an important role in HCV chronicity (Scholtes et al. 2012).

In spite of all these difficulties, much progress has been made in determining the properties of infectious HCV virions. The following paragraphs present an overview of our current understanding of the structure of infectious HCV particles, starting from the viral and host-derived components. In particular, we will discuss recent advances concerning the structure of the E2 viral glycoprotein and low-resolution 3D reconstruction of infectious virions. These findings will be integrated with the state of the art on HCV assembly, how that intersects with lipoprotein biogenesis and its implications for HCV structure, entry and ultimately pathogenesis.

3 Viral Structural Components

The first three HCV proteins generated by processing of the viral polyprotein include core, the nucleocapsid protein, and the envelope glycoproteins, E1 and E2 (Fig. 2a). These are also called structural proteins as they are thought to serve as components for the assembly of progeny virions, although the exact composition of the HCV particle remains to be determined.

The core protein is the only HCV protein associated with nucleocapsid formation; as such it is expected to self-associate into an as yet undefined array to encapsulate the viral RNA genome. However, very little is known about the

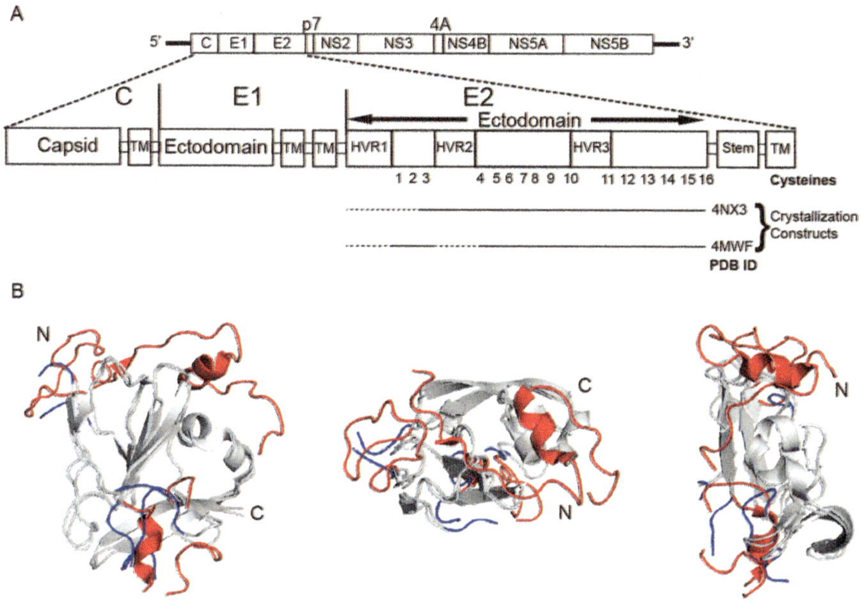

Fig. 2 The HCV structural proteins. (**a**) Schematic overview of the HCV polyprotein, highlighting the structural proteins: C, E1, and E2. The transmembrane regions of each protien are noted as TM. The location of the conserved cysteine residues and hypervariable (*HVR*) regions of E2 are provided. The two different crystallization constructs used to solve the structure of the E2 ectodomain (PDB ID 4MWF and 4NX3) are denoted. (**b**) Superposition of the two E2 ectodomain structures. Regions of structural similarity (r.m.s.d. >0.8 Å) are colored in *grey* while dissimilar regions are depicted in *red* and *blue* for 4MWF and 4NX3, respectively. Three different orientations are provided. The *central* and *right panels* are 90° rotations about a horizontal and vertical axis, respectively, relative to the panel on the *left*

mechanism of nucleocapsid formation. The core protein consists of the first 191 amino acids in the HCV polyprotein. The C-terminal transmembrane segment serves the signal sequence for E1 and is removed by the action of host signal peptide peptidase. The mature protein is approximately 21 kDa, consisting of the first 177 amino acids which are needed for infectious virion production (Kopp et al. 2010). The NMR structure of a synthetic core-E1 peptide is mostly in an α–helical conformation (PDB ID 2LIF; (Oehler et al. 2012)). The first 117 amino acids of core contain numerous basic residues and has been implicated both in RNA binding and oligomerization. Overall the core protein appears to be highly dynamic with regions of α-helical content (Boulant et al. 2005).

E1 and E2 assemble as heterodimers and are thought to be the functional unit on the virion that engages host factors at the cell surface and facilitates membrane fusion during entry (Deleersnyder et al. 1997; Dubuisson et al. 1994; Op De Beeck et al. 2004). E2 is involved in cell targeting through specific interactions with host receptors like CD81 and the scavenger receptor class B, type I (SR-BI). In addition to its role in cell targeting, E2 has three HVR regions that are critical for immune evasion. The role of E1 in HCV entry remains enigmatic. Interestingly, E1, but not E2, is able to bind directly to apoE and apoB (Mazumdar et al. 2011), suggesting that E1 may contribute to the association of virions with host-derived lipoproteins. Based on data from the corresponding envelope proteins of flaviviruses, E1 and E2 may also be important for virus assembly. E2 consists of an amino-terminal ectodomain, a predicted membrane proximal region predicted to be α-helical in nature (Drummer and Poumbourios 2004), and a carboxyl terminal membrane-associating segment. During translation, E1 and E2 are targeted to the ER lumen by signal sequences located at the C-terminus of core and E1, respectively. The transmembrane region is thought to be involved in ER retention (Cocquerel et al. 1998, 1999; Flint and McKeating 1999) and the formation of E1/E2 heterodimers (Cocquerel et al. 2000; Op De Beeck et al. 2000). Since HCV is hypothesized to bud into the ER, retention of the glycoproteins at the ER membrane ensures their placement on virions.

The E1 and E2 ectodomains (eE1 and eE2) have been mapped to the first 160 and 334 amino acids (aa), respectively (Michalak et al. 1997). The ectodomains have been defined as the minimal deletion that results in secretion of a properly folded protein. The ectodomains of E1 and E2 are heavily glycosylated and contain several intramolecular disulfide bonds. E1 and E2 contain as many as 5 and 11 predicted N-linked glycosylation sites (Asn-X-Ser/Thr) with 8 and 18 absolutely conserved cysteine residues, respectively.

High-resolution, atomic structures of HCV glycoproteins have long been sought. Recently, two groups successfully obtained the first high-resolution structure by X-ray crystallography of the core eE2 in complex with different Fab fragments (PBD ID 4MWF and 4NX3) (Kong et al. 2013; Khan et al. 2014) (Fig. 2). In order to obtain crystals, both groups used a similar strategy by forming an E2/Fab complex and making deletions within eE2. Although the methodologies were similar, the deletions and antibodies were different. In 4MWF, a neutralizing human antibody (AR3C) that binds close to the amino-terminus of the ectodomain and prevents CD81/E2 interaction was used. Furthermore, the crystallization construct has both an amino-terminal (HVR-1) and internal deletion (HVR-2). In 4NX3, a non-neutralizing mouse monoclonal antibody (2A12) was used and a larger amino-terminal region was deleted, consisting of HVR-1 through HVR-2. The construct was defined based on the results of solution-based studies of limited proteolysis and hydrogen-deuterium exchange, which demonstrated that the first 80 aa of the ectodomain are disordered. This region includes conserved sequences implicated in binding to the cellular receptors (SR-BI and CD81) as well as several epitopes for neutralizing antibodies (Wahid and Dubuisson 2013; Keck et al. 2012; Kong et al. 2012a, b; Deng et al. 2013), which suggest that these binding sites are flexible in the absence of E1, lipids or cell factors. AR3C stabilizes the amino-terminal portion of the E2 ectodomain, enabling modeling of the CD81 binding site. However, the authors speculated that AR3C binding might distort the N-terminus and CD81 binding site (Kong et al. 2013). In contrast, 2A12 recognizes a linear epitope at the C-terminus of the ectodomain, does not interfere with E2 binding to CD81 and SR-BI, and does not neutralize infection.

The E2 core domain has a globular fold, consisting of mostly β strands and random coil. The protein contains two, four-stranded antiparallel β sheets (termed sheets A and B), that form an IgG-like and novel folds. Structural comparison of the HCV E2 core domain with all known folds in the Protein Data Bank using the Dali server (Holm and Rosenstrom 2010) identified proteins with IgG-like folds similar to the amino-terminal, β sheet region of E2, none of which are class II fusion glycoproteins. Interestingly, the protein with the highest similarity to the eE2 IgG-like fold is domain II (DII) of the Newcastle disease virus (NDV) fusion protein (F) in the post-fusion conformation (PDB ID 3MAW) with a Z-score of 3.4 and RMSD of 2.6 Å for 51 aligned aa (Swanson et al. 2010) (Fig. 3). NDV is an enveloped, negative-strand RNA virus belonging to the *Paramyxoviridae* family, which includes measles virus, mumps virus, and respiratory syncytial virus (Lamb and Parks 2007). The F protein is classified as a member of the class I viral fusion proteins that include influenza hemagglutinin (HA) and HIV Env proteins. Class I fusion proteins have a pre-fusion, metastable conformation that undergoes a large rearrangement upon membrane fusion.

The flavivirus envelope glycoprotein (E) belongs to the class II fusion proteins (White et al. 2008; Vaney and Rey 2011). According to its classification within the *Flaviviridae* family, HCV was predicted to also harbor a class II fusion protein. All class II fusion envelope glycoproteins possess a common elongated structure, containing predominantly β sheets and existing as homo- or hetero-dimers with the membrane-fusion, hydrophobic peptide buried at the dimer interface at neutral pH (Fig. 4). Upon receptor binding and/or exposure to low pH, these proteins undergo self-rearrangement into stable trimers exposing the fusion peptide that drives viral and host membrane fusion. The pestivirus BVDV E2 glycoprotein protein has a similar elongated topology forming a tail-to-tail homo-dimer stabilized by a disulfide bond. BVDV E2 contains no apparent fusion peptide, suggesting that it is unlikely to be a class II fusion protein (El Omari et al. 2013;

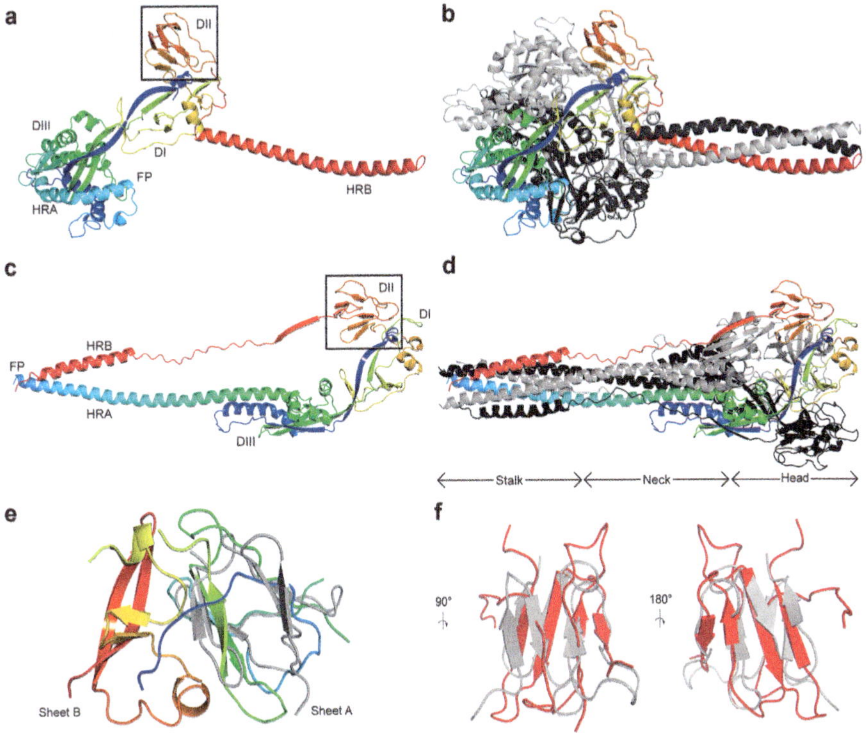

Fig. 3 Structural comparison between HCV E2 and NDV F protein. The pre-fusion (**a**, **b**) and post-fusion (**c**, **d**) forms of NDV F protein are shown in monomeric (**a**, **c**) and trimeric (**b**, **d**) forms. The *rectangle* highlights the location of domain II within the F protein. The head, neck and stalk region are labeled on the post-fusion form. (**e**, **f**) Superposition of HCV E2 with domain II of the NDV F protein colored in rainbow and grey, respectively. The view in f is rotated 90° around a vertical axis from (**e**)

Li et al. 2013). HCV E2 has a globular shape, does not contain a discernable fusion peptide, and does not undergo significant conformational or oligomeric rearrangements upon incubation of low pH, ruling out a direct involvement of E2 in the fusion process. This is consistent with the extensive hydrophobic core and numerous disulfide bonds holding the two regions together. These results implicate either E1 alone or an E1/E2 hetero-dimer in the fusion process.

The eE2 structures are highly similar with an RMSD of less that 0.8 Å for similar carbon-α positions, however there are some interesting divergences that are worth discussing. Despite the high sequence variability in E2, the cysteine residues are absolutely conserved across genotypes, suggesting an important structural and/or functional role. Of the 18 conserved cysteine residues in E2, 16 are within the ectodomain; 1 in the putative stem region; and 1 in the linker connecting the stem and ectodomain. Both structures have three disulfide bonds formed between C5–C9, C7–C8 and C14–C16 that appear to be critical for the fold of the core domain as

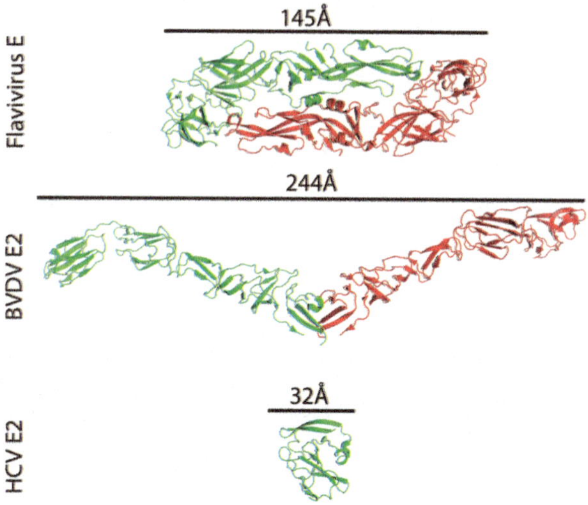

Fig. 4 Structural comparison of Flavivirus E, BVDV E2 and HCV E2 ecotodomains. At neutral pH, flavivirus E and BVDV E2 form elongated, dimeric structures, whereas HCV E2 is monomeric and globular. The longest dimension of each protein is provided in angstrom (Å)

they link secondary structure elements (Table 2). In 4MWF, C1 is bonded to C6, while C6 is free in 4NX3 since the first two cysteine residues were deleted in that construct. All of the disulfide bond discrepancies are in loop regions, most notably around HVR-3. The reasons for these differences are currently unclear. The E2 protein sequences in the two structural studies were from different HCV genotypes; produced in HEK293T cells, which are known to yield functional envelope glycoproteins in form of HCVpp; and have various deletions. Both production methods yielded functional eE2 as the full-length ectodomains can inhibit HCVpp or HCVcc infection (Kong et al. 2013; Khan et al. 2014). It is worth noting that eE2 produced in a Drosophila expression system resulted in a different disulfide bonding pattern as determined by mass spectrometry (Krey et al. 2010), although disulfide bond shuffling can result during this procedure. Moreover, it is possible that E2 has different disulfide bonding patterns on the surface of the viron. E2 has been detected as non-covalently associated with E1 in the ER, whereas E1/E2 appear to be covalently bound on secreted virus particles (Drummer et al. 2003; Flint et al. 2004; Vieyres et al. 2010). In fact, mutagenesis studies have suggested that C2, C3, C11, and C12 are critical for E1/E2 hetero-dimerization (McCaffrey et al. 2012). Furthermore, limited disulfide bond reduction is compatible with HCV entry as well as antibody and CD81 binding (Fenouillet et al. 2008).

Although originally classified as a non-structural (NS) protein, p7 is required for the production of infectious virions (Jones et al. 2007; Sakai et al. 2003; Steinmann et al. 2007; Wozniak et al. 2010). Accordingly, specific inhibitors have been shown to potently suppress particle release (Foster et al. 2014). This small integral membrane protein oligomerizes to form cation-specific ion channels (Griffin et al. 2003; Clarke et al. 2006). Convincing evidence indicates that p7 neutralizes acidic compartments within the secretory pathway, acting as a viroporin to protect virus particles from premature exposure to low pH during virus maturation and

Table 2 The redox state and disulphide bonds of the conserved cysteine residues in eE2 crystals. Cysteine residues (C) are numbered starting from the amino terminus

4MWF (Kong et al. 2013)	4NX3 (Khan et al. 2014)
C1–C6	C1 and C2 not included in construct
C2–C15	**C3 no density**
C3 no density	C4–C15
C4 no density	**C5–C9**
C5–C9	C6 free
C7–C8	**C7–C8**
C10–C11	C10–C13
C12–C13	C11 and C12 no density
C14–C16	**C14–C16**

Similarities between the two structures are shown in bold

egress (Wozniak et al. 2010). However, the jury is still out as to whether this protein is packaged in HCV particles.

4 Host Factors Incorporated in HCV Virions

Prompted by the initial observations made with infectious sera, once the HCVcc system became available, several groups began to investigate the link between lipoprotein and HCV morphogenesis in order to shed light on the LVP model and understand which lipoprotein components were packaged in infectious virions. Lipoproteins are complexes of proteins (apolipoproteins) and lipids that transport lipids in the bloodstream. The apolipoproteins serve as structural components, helping to solubilize neutral lipids in circulation, as well as cofactors for enzymes and ligands for cell-surface receptors. Lipoproteins are classified based on their buoyant density which is determined by the relative content of proteins and lipids and is inversely correlated with their diameter (Table 3). Despite big differences in their composition, different classes of lipoproteins share a similar structural organization. Apolipoproteins, amphipathic lipids [mainly phospholipids (PL) and unesterified cholesterol (Chol)] form the outer soluble shell, a monolayer that is approximately 20 Å-thick. Inside this shell, there is a core of neutral lipids that is made for the most part of triglycerides (TG) and cholesteryl esters (CE) (Fig. 1). Lipoproteins are synthesized either in the intestine [chylomicrons (CM)] or in the liver [very low-density lipoproteins (VLDL) and high-density lipoproteins (HDL)] but undergo dramatic remodeling once in the circulation, giving rise to additional classes of lipoproteins (Table 3). Specifically, CM are converted into chylomicron remnants, and VLDLs are first transformed into intermediate-density lipoproteins (IDL), then low-density lipoproteins (LDL) (Jonas 2008).

For the purpose of this chapter, we will focus on the two classes of lipoproteins assembled in the liver: VLDL and HDL. These particles have very different

Table 3 Properties of lipoproteins and HCV particles

Particle	Density (g/ml)	Diameter (nm)	Origin	Apo	Prot (%tot)	TG (%tot)	Chol (%tot)	PL (%tot)
CM	<0.95	80–1,000	Intestine	B48, A1, A2, C, E	1	90	5	4
VLDL	0.95–1.006	30–80	Liver	B100, C, E, A1?	10	65	13	13
IDL	1.006–1.019	25–30	Blood	B100, C, E	18	34	22	22
LDL	1.019–1.063	20–22	Blood	B100	20	10	45	23
HDL	1.063–1.21	5–15	Liver	A1, A2, C, E	50	10	15	25
HCV_{plasma}	1.03–1.25	30–80	Liver	B100, C, E	n/a	n/a	n/a	n/a
HCVcc	1.08–1.16	40–100	Hepatoma cells	B100, C, E, A1	n/a	n/a	20 (Chol) 44 (CE)	33 (mostly PC, SM)
HCV_{PHH}	n/a	40–100	Primary hepatocytes	B100, E, A1	n/a	n/a	n/a	n/a

compositions that reflect opposite roles in lipid transport. VLDLs distribute lipids from the liver to peripheral tissues, whereas HDLs mediate the so-called "reverse cholesterol transport" by taking up lipids in excess from the periphery and delivering them back to the liver. Accordingly, VLDLs are made of 90 % lipids (by weight), while HDLs only contain 50 % lipids, the lowest amount across the whole range of lipoproteins (Table 3). These two classes are also characterized by different apolipoproteins: apoB-100 is the main VLDL structural protein and mediates the interaction with the LDL receptor (LDL-R); ApoA-I fulfills a similar function for HDL, by interacting with SR-BI (Williams et al. 2000). VLDL and HDL have both been linked to HCV infection: most of the literature on HDL and HCV revolves around virus entry and these aspects will be discusses later in this chapter. In contrast, numerous studies have shown a connection between VLDL and HCV biogenesis, supporting the hypothesis of virions being assembled as LVP.

A range of approaches was used to determine the nature and role of host-derived lipids and apolipoproteins involved in HCV particle formation. Some groups evaluated the impact of lipoprotein lipase (LPL) on HCV density and infectivity. LPL is an extracellular enzyme that hydrolyzes triglycerides in lipoproteins converting them to particles of higher density. It can circulate in the bloodstream but is mostly attached to the cell surface via heparan sulfate proteoglycans (HSPGs). In addition to its catalytic activity, the cell-bound form also exerts a bridging function by interacting with ApoC-II on lipoproteins, ultimately mediating their uptake (Mead et al. 2002). Similarly, LPL was shown to shift HCV particles to higher densities, suggesting a catalytic action of LPL on HCV (Thomssen and Bonk 2002), and to mediate cell binding of LVPs (Andreo et al. 2007). Nevertheless, this did not result in a higher infection rate, but rather diminished HCV infectivity, probably by blocking virions at the cell surface and preventing their internalization (Maillard et al. 2011; Shimizu et al. 2010). These observations were corroborated by Young et al, who reported that LPL lipolytic activity in clinical samples inversely correlates with HCV viremia (Sun et al. 2013). Overall, these data further support the association of HCV with lipoproteins and that it is important for the infectivity of virions.

Accordingly, the lipid composition of HCV purified from infected patients and cell culture was similar to that of lipoproteins, rather than reflecting the membrane composition of the cells that the virus was assembled in (Merz et al. 2011; Scholtes et al. 2012). Interestingly, both HCV particles from cell culture and plasma were virtually devoid of phosphatidylserine (PS) and had very little amounts of phosphatidylethanolamine (PE). These PL make up the inner leaflet of cellular membranes and are normally present in enveloped virions such as HIV and VSV that acquire their envelope from these membranes. In contrast, for HCV, as for lipoproteins, phosphatidylcholine (PC) and sphingomyelin (SM) that form the outer layer of membranes constitute >95 % of the total PL, making it difficult to reconcile with a classical virus envelope. Moreover, CE that are normally absent from virions as they are non-bilayer lipids were the most abundant class of lipids in HCV particles, accounting for >40 % of the total lipid composition (Table 3) (Merz et al. 2011).

As for host proteins incorporated into HCV virions, much evidence has been gathered for apoE. This apolipoprotein is found virtually on all classes of lipoproteins, with the exception of LDL, as it can be exchanged from one particle to another. It plays a fundamental role in lipoprotein catabolism by mediating their binding to LDL-R (aa 136–150). The same region (aa 142–147) is also responsible for the interaction of lipoproteins with HSPGs (Hatters et al. 2006; Libeu et al. 2001). ApoE is also able to bind to SR-BI (N-terminus, aa 1–165), promoting cholesterol efflux from lipoproteins (Chroni et al. 2005). Using RNA interference (RNAi) as well as apoE-specific antibodies in immuno-EM (iEM), immuno-capture of HCV RNA and neutralization assays, several groups showed that apoE localizes on the outer shell of HCV particles and is required both for their production and infectivity (Catanese et al. 2013; Gastaminza et al. 2010; Merz et al. 2011; Chang et al. 2007a).

In addition to apoE, other apolipoproteins, such as apoA-I, apoB and apoC were proposed to associate with serum- and/or cell culture-derived HCV (Meunier et al. 2008; Chang et al. 2007b; Dreux et al. 2007; Thomssen et al. 1992; Sun et al. 2013). Using both iEM and western blot analysis of density gradient fractions enriched in infectious particles, a recent study showed incorporation of apoA-I and apoB-100 into both HCVcc and primary-derived HCV (Catanese et al. 2013). This was in agreement with previous reports showing that downregulation of apoA-I induces a significant decrease in HCV particle production (Mancone et al. 2011) and that apoB-specific antibodies could capture viral genomes (Merz et al. 2011; Thomssen et al. 1992). Finally, two members of the apoC family that are found on CM, VLDL and HDL have been implicated in early steps of HCV infection, suggesting that they may also be packaged within LVPs. Specifically, apoC-III was shown to suppress LPL lipolytic activity, thus reversing the LPL-mediated inhibition of HCV infection (Sun et al. 2013), whereas apoC-I was found to increase the fusion rate between viral and target membranes (Dreux et al. 2007). These may be very intriguing solutions for HCV to deal respectively with enzymatic inactivation in the bloodstream and with membrane fusion, especially since a functional fusion peptide in its envelope glycoproteins has not been identified.

It should be noted that the list of host-derived components incorporated in HCV virions extends beyond apolipoproteins and neutral lipids. Recent proteomics analyses of HCV particles revealed a much more complex scenario, with more than 40 candidate virus-associated human proteins (Lussignol et al. 2016; Parent et al. 2009). So far, only two of these proteins, the heat shock cognate protein 70 (HSC70; Parent et al. 2009) and the nucleoporin 98 (Nup98; Lussignol et al. 2016) have been followed up on, both playing a role in virus assembly/release. Research in this field holds promise to further unveil additional virus-host interactions that are crucial for HCV morphogenesis and propagation.

5 Possible Mechanisms of Virion Morphogenesis

In the most simplistic view, the biogenesis of nascent virions requires: (1) a viral genome, (2) core, the capsid-forming protein and (3) the viral envelope glycoproteins, E1 and E2. For this process to take place, the various structural components have to come together in a coordinated fashion. This is particularly crucial for core since after synthesis in the ER, it is trafficked to lipid storage organelles called droplets (LDs) (Miyanari et al. 2007). Over the last few years, with the advent of a cell culture system to propagate HCV in vitro, tremendous progress has been made in our understanding of the assembly of infectious particles and the host factors participating in it. For example, it is known that the enzyme diacylglycerol O-acetyltransferase 1 (DGAT1) is required for transferring core onto LDs (Herker et al. 2010). This is likely achieved concomitantly with the biogenesis of these organelles. In fact, LDs are formed by inserting neutral lipids (mostly TG and CE) in the ER bilayer; DGAT1 contributes to this process by synthesizing TG that accumulates between the two leaflets (Cases et al. 1998). To initiate the process of virion assembly, core must be retrieved from the surface of LDs and AP2M1 (Adaptor-related Protein complex 2, μ1 subunit) was shown to be involved in trafficking core from LDs to the site of viral budding at the ER (Neveu et al. 2012). During the budding of nucleocapsids into the ER, HCV acquires its lipid membrane as well as E1/E2. In hepatic cells, the same ER membranes are heavily utilized to synthesize lipoproteins that, as discussed earlier, share many features with HCV particles. Moreover, HCV biogenesis relies on key components of VLDL assembly. Specifically, the microsomal triglyceride transfer protein (MTP) was shown to play a crucial role in HCV particle production (Gastaminza et al. 2008; Huang et al. 2007; Jiang and Luo 2009; Nahmias et al. 2008). This was an important observation that further confirmed a tight link between HCV and lipoprotein assembly.

In order to better understand where and how these pathways might intersect, we will briefly summarize what is known about the morphogenesis of VLDLs. The first step involves MTP-mediated transfer of lipids to apoB-100 in the ER lumen, as it gets translated. ApoB-100 is a 4536 aa-long, water-insoluble protein and represents the primary structural component of nascent VLDLs (Yang et al. 1986; Knott et al. 1986). Because of these properties, apoB-100 is a non-exchangeable apolipoprotein, meaning that it remains bound to the same VLDL particle as that which gets converted into IDL and LDL in the bloodstream and is eventually cleared via LDL-R mediated liver uptake. This is in contrast with exchangeable apolipoproteins (e.g., apoA-I, apoC, apoE) that are much smaller and more soluble in water in their delipidated state and can thus be transferred between lipoprotein particles. The primordial apoB particle is a quite dense, spherical emulsion droplet, with a diameter of approximately 25 nm. In the second step of VLDL assembly, this precursor particle gets further lipidated with lipids from luminal LDs to form mature VLDL particles that range between 30 and 80 nm in diameter (Shelness and Sellers 2001). It is not clear whether the conversion into mature VLDL occurs in the ER through a fusion step with lumenal ER LD or if secondary lipidation takes place instead in a post-ER compartment, most likely the Golgi (Yao et al. 2013). The second hypothesis is currently favored as there is no compelling evidence for LD fusion under physiological conditions. As for other apoliproteins that are associated with VLDLs, such as ApoE and apoC, these are thought to be either added directly onto primordial apoB particles or transferred from lumenal ER LDs. If apoB and MTP are absolutely required for VLDL synthesis, apoE was shown to modulate VLDL secretion, though the precise mechanism has yet to be elucidated (Mensenkamp et al. 1999). What is clear though is that VLDLs exit from the ER in COPII-coated vesicles and traffic through the Golgi apparatus where they acquire additional lipids before being secreted, resulting in a lower buoyant density than VLDL precursors (Hebbachi et al. 1999; Gusarova et al. 2003, 2007).

Similarly, nascent HCV particles transit through the Golgi. This is supported by several observations: firstly, Brefeldin A, an inhibitor of ER-Golgi transport, causes intracellular accumulation of HCVcc (Gastaminza et al. 2008); secondly, E1 and E2 on virions display complex glycans that can only be generated by Golgi-resident enzymes (Vieyres et al. 2010); finally, HCVcc particles were shown to acquire their characteristic low buoyant density in a post-ER compartment (Gastaminza et al. 2006, 2008). The latter evidence also supports the hypothesis that HCV particles mature similarly to VLDLs, undergoing post-assembly lipidation.

This takes us back to the concept of the LVP and the debate about where and how the lipids and apolipoproteins found in HCV virions are acquired. The molecular basis for the interaction between HCV and lipoproteins is currently unknown but the association appears to occur during assembly, because HCV assembly and release rely on components of the VLDL biosynthetic pathway, including MTP and apoE. Furthermore, E1 was shown to interact directly with apoE and apoB-100, raising the possibility that this viral envelope glycoprotein may be responsible for recruiting apolipoproteins onto nascent virions (Mazumdar et al. 2011). While apoB-containing LVPs were identified in patients, controversial results have been instead obtained with the HCVcc system. Those incongruences, like differences in buoyant density and specific infectivity, may be explained by the inability of hepatoma cells to completely recapitulate lipoprotein assembly (Gastaminza et al. 2008; Huang et al. 2007; Jiang and Luo 2009; Nahmias et al. 2008). Nevertheless, recently apoB was visualized on purified HCV particles produced both in hepatoma cells and primary human hepatocytes, suggesting that indeed HCV exploits the VLDL assembly pathway and might therefore structurally resemble a lipoprotein more than a classical enveloped virus (Catanese et al. 2013).

A related question concerning the structure of HCV virions is whether infectious LVPs are true hybrid particles, sharing the same envelope or if the virus and serum lipoproteins interact, however tightly, as separate entities. Based on the cell biology data discussed earlier and the EM images collected so far, the most plausible scenario seems that the LVP is a single particle. Indeed, since apoB-100 is a very large, non-exchangeable apolipoprotein, the only possibility for it to be found associated with HCV virions is that it gets incorporated in the intracellular compartment, quite early on during the assembly process. The one particle model would also fit with initial observations made with patient-derived HCV, which was unable to be reassociated with purified lipoproteins after treatment with deoxycholate (Prince et al. 1996). This, however, does not exclude the possibility that HCV virions may still acquire lipoprotein components, be it lipids or exchangeable apolipoproteins, post-secretion in the extracellular milieu. In keeping with this, the buoyant density of HCV in vivo is affected by dietary TG income (Felmlee et al. 2010) and the addition of purified lipoproteins to the cell culture media impacts HCV infectivity, as discussed below.

Cryo-EM studies of purified HCV particles have shown that virions are spherical and highly heterogeneous in size (40–100 nm in diameter), with no obvious symmetry and heavily covered by electron-dense globular material. When highly infectious fractions were analyzed by cryo-electron tomography (cryo-ET), again particles varied in size, although a significant enrichment of particles with 81–85 nm diameter was observed compared to non-purified HCVcc (Catanese et al. 2013). Surprisingly, tomogram reconstruction showed no evidence of a continuous bilayer, as one would expect for an enveloped virus. Although this could be due to the low resolution of the images, these data might also indicate that HCV is embedded in a non-unique amount of lipoprotein, and therefore delimited by a lipid monolayer (Fig. 5). Spike projections were visible, however they are unlikely to be viral glycoproteins. In fact, HCV virions purified from both hepatoma cells and primary human hepatocytes were poorly stained by anti-envelope glycoproteins, whereas they displayed a thick coat of host-derived apolipoproteins. In particular, apoE is incorporated in the majority of virions and seems to decorate the surface of HCV, with multiple apoE-immunoreactive gold particles per virion (Catanese et al. 2013; Gastaminza et al. 2010; Merz et al. 2011). In contrast, most HCVcc- and PHH-derived virions displayed only one apoB-specific immunogold particle (Catanese et al. 2013). This would be consistent with the stoichiometry observed in VLDLs that carry a single apoB molecule per lipoprotein particle. As for apoA-I, a more variable staining was noted (Fig. 5). ApoA-I and apoB-100 are the main structural proteins of HDL and VLDL, respectively. ApoB-100 cannot be exchanged, suggesting that viral particles must acquire it in the liver before being released, whereas apoA-I can exchange. Therefore, it is conceivable that it may be transferred onto LVPs post-secretion. However, it should be noted that proteomics studies have identified ApoA-I as a component of VLDL and LDL particles and silencing of apoA-I was shown to affect HCV particle production, suggesting that apoA-I could also be added onto LVPs as they assemble in the hepatocyte (Mancone et al. 2007, 2011; Karlsson et al. 2005). Finally, apoC-I specific antibodies can efficiently capture and neutralize intracellular virions, indicating that a good proportion of this host protein is packaged into HCV particles prior to being released (Meunier et al. 2008).

Presumably, apolipoproteins incorporated in HCV particles would assume a similar topology to what is observed on lipoproteins. The exchangeable apolipoproteins are composed of amphipathic α-helices with large loops or turns. In the absence of lipids, a portion of the apolipoprotein will form a four or five helix bundle with a simple up and down morphology and the hydrophobic surface of

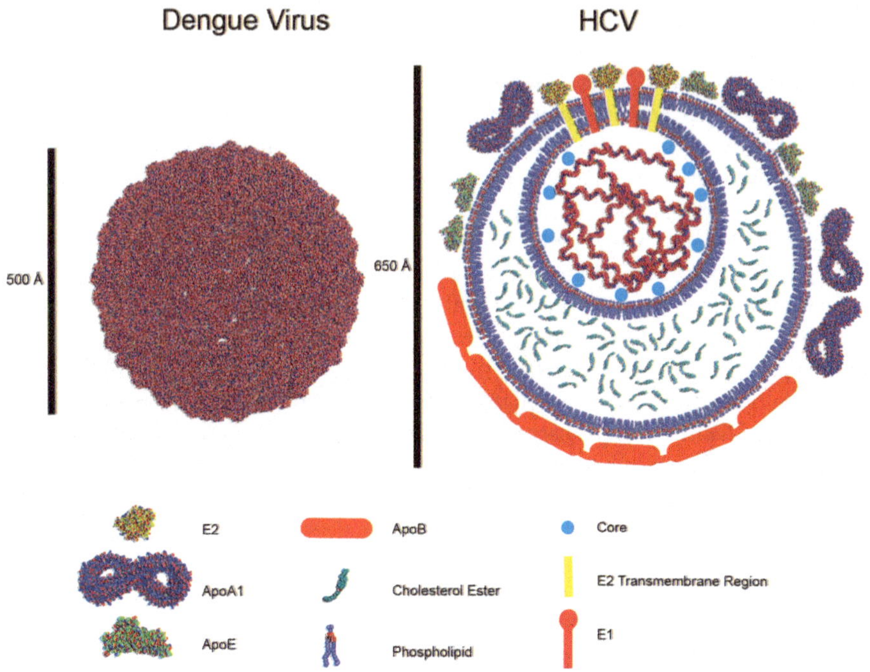

Fig. 5 Model of the structure of infectious HCV virions. Schematic representation of a dengue virus particle and an infectious HCV virion. The *black bars* represent the diameter of the particles in angstrom (Å). A mature dengue virus particle displays a smooth surface and a diameter of 500 Å. 90 homo-dimers of the envelope glycoprotein E assemble in an antiparallel fashion to confer the virion an icosahedral symmetry (*left*). Altough quite variable in size, the average diameter for HCV particles was found to be approximately 650 Å. Here, HCV is depicted as a LVP, delimited for the most part by a lipid monolayer (*right*). A bilayer-containing region is shown where the nucleocapsid is in close proximity of the lipoprotein-delimiting monolayer and E1/E2 are inserted. This particle is spherical, with no obvious symmetry and heavily covered by host-derived apolipoproteins. LVPs seem to incorporate multiple copies of apoE and (to a lesser extent) apoA-1 molecules but only one apoB-100 molecule; nevertheless, the precise stoichiometry with structural viral proteins has yet to be determined. Even though this model shows apoA-1, apoB-100 and apoE on the same viral particle, there is currently no evidence that each LVP harbors all of them. Underneath the phospholipid monolayer, there is the nucleocapsid wrapping the viral genome and a core of neutral lipids, consisting for the most part of cholesteryl ester. All the components of HCV virions were drawn to scale, to give an idea of the relative sizes and occupancy on the particle. When possible, available structures were depicted as space filling models. See text for further details

amphipathic helices packing into the interior of the protein (Fig. 5). The exchangeable apolipoproteins have similar gene structures, suggesting that these proteins have evolved by gene duplication from an ancestral gene. Given their evolutionary relatedness, the helix bundle fold is thought to be found in all the exchangeable apolipoproteins. In the presence of lipids, these proteins form an extended helix circumscribing the discoidal monolayer, adopting the so-called "belt" configuration, with the helical axes oriented perpendicular to the radial axis of particle.

Currently, there is very limited structural information for apoB-100 owing to its large size and limited solubility. However, this apolipoprotein has been proposed to arrange itself as a flexible string wrapped around the surface of VLDLs with an approximate contour length of 70 nm (Phillips and Schumaker 1989), which is approximately a third of the circumference of a 65 nm particle.

The fact that virions are heavily coated with apolipoproteins, in conjunction with the surprisingly high amount of neutral lipids found in infectious particles, makes it unlikely for HCV to be completely surrounded by a lipid bilayer, like a canonical enveloped virus. This poses the question: how are the HCV glycoproteins arranged on virus particles? One possibility is that patches of bilayer might form at the points where the nucleocapsid is in close proximity of the lipoprotein-delimiting monolayer, enabling E1/E2 to acquire a transmembrane topology. EM images of highly infectious HCV fractions would be in agreement with this hypothesis (Catanese et al. 2013). Such a scenario might at least partly explain the low level of incorporation of E1/E2 on HCV particles. Alternatively (or in addition), E1 and E2 might position their transmembrane domains parallel to the lipid monolayer, as suggested by Bartenschlager et al. (2011).

Thus, HCV does not share structural similarities with other members of the *Flaviviridae*, such as DENV or WNV, which instead in their mature stage display a smooth surface, icosahedral symmetry and no association with circulating lipoproteins.

6 Implications of HCV Structure for Virus Entry and Pathogenesis

The association of HCV with host-derived lipoproteins has a number of significant advantages for the virus and important implications for its pathogenesis. As discussed earlier, HCV exploits the lipoprotein biosynthetic machinery to assemble and release nascent virions. An additional intriguing aspect of this interaction is that it might contribute to the hepatotropism of this virus, since the liver plays a key role in lipoprotein clearance. By circulating in the bloodstream disguised as lipoviral particles, HCV virions may have found a unique and efficient mode of viral entry into liver cells. Indeed, HCV attachment to hepatocytes and entry involves lipoproteins and their receptors (Zeisel et al. 2013).

To infect a new host, the virus needs to reach the liver by crossing the fenestrated endothelium and to interact with attachment factors like HSPGs on the basolateral side of hepatocytes. This step is mediated by virion-associated apoE (Jiang et al. 2012; Liu et al. 2012). Interestingly, this host-derived protein is more exposed on the viral envelope than the viral glycoproteins themselves (Catanese et al. 2013), which would be consistent with apoE playing a role in the initial attachment of HCV to target cells via HSPGs.

Subsequently, to gain access into the hepatocyte, HCV utilizes a surprisingly large number of receptors. This list currently comprises eight different molecules, two of which are lipoprotein receptors: SR-BI (Scarselli et al. 2002) and LDL-R (Agnello et al. 1999) that internalize HDL and LDL/VLDL, respectively. The interaction of HCV with those molecules, at least initially, is likely to occur through apolipoproteins exposed on the surface of virions (Owen et al. 2009; Dao Thi et al. 2012), even though direct binding of SR-BI to E2 was shown (Scarselli et al. 2002; Dao Thi et al. 2012). In fact, apoE and apoB-100 have the ability to bind to LDL-R (Zaiou et al. 2000; Knott et al. 1986) and apoA-I to SR-BI (de Beer et al. 2001). As for lipoproteins, the remodeling of LVPs via exchange of lipids and apolipoproteins may be a continuous, very dynamic process accounting for the variable composition and biophysical properties of HCV virions. For instance, SR-BI has lipid-exchange function that might play a role in unveiling E2 determinants on the LVPs for a direct virus-receptor interaction. Further modifications of LVP could be mediated by hepatic and lipoprotein lipases (Andreo et al. 2007; Maillard et al. 2011; Shimizu et al. 2010). The addition of HDL to the cell culture media was shown to enhance virus infectivity (Bartosch et al. 2005; Dreux et al. 2006; Voisset et al. 2005; Catanese et al. 2007). This effect could be due to a number of reasons: exogenous HDLs might stimulate the lipid-transfer function of SR-BI thus promoting parallel LVP uptake; alternatively, HDLs could be a source of exchangeable apoA-I molecules that, once transferred onto LVPs, would mediate direct virion binding to SR-BI. Finally, apoC-I increases HCV infectivity by promoting the fusion between the viral envelope and cell membranes (Dreux et al. 2007). Overall, host-derived apolipoproteins on viral particles seem to play key roles in HCV attachment and entry, explaining the higher infectivity of the lipoprotein-associated HCV.

In keeping with this, ultrastructural studies revealed that purified HCV virions are densely coated with apolipoproteins. Although the stoichiometry of those host proteins relative to the viral glycoproteins has not been established yet, both iEM and immuno-capture data suggest that apolipoproteins are more readily accessible on the surface of viral particles compared to E1 and E2. An important implication in this regard is that host-derived proteins likely mask viral epitopes, thus providing a mechanism for viral escape from the humoral immune response. This would contribute to explaining the weak immunity that follows HCV infection (Farci et al. 1992).

7 Concluding Remarks

Since 2005, with the development of cell culture systems to propagate HCV in vitro and produce bona fide infectious virions, much progress has been made in terms of dissecting mechanisms of virus assembly, defining the required viral and cellular factors, characterizing the close interaction of HCV with the lipoprotein biosynthetic machinery. Furthermore, improvements in our ability to grow and purify particles, not only from hepatoma cells but also primary human hepatocytes, led to significant insights into the biophysical, biochemical and ultrastructural properties of HCV virions. We have learned a great deal about the unusual features of this virus, that doesn't look anything like a canonical enveloped virus. Indeed, HCV represents a unique example of intimate co-evolution with its host, resulting in a hybrid lipo-viro-particle that is heavily coated with different host-derived components and displays a very unique lipid composition, reminiscent of endogenous lipproproteins. This peculiar structure makes for a very deceiving pathogen and has significant implications for receptor usage and entry pathways, likely providing HCV with multiple options for infecting hepatocytes. This same morphology also aids immune escape from circulating neutralizing antibodies and contributes to successful persistence of this pathogen within its host. These are important aspects that we must keep account of in our quest for developing effective vaccine strategies.

Acknowledgements The authors would like to thank Dr. Matthew Miller (Rutgers University) for his assistance with the preparation of the figures.

References

Agnello V, Abel G, Elfahal M, Knight GB, Zhang QX (1999) Hepatitis C virus and other flaviviridae viruses enter cells via low density lipoprotein receptor. Proc Natl Acad Sci U S A 96(22):12766–12771

Andre P, Komurian-Pradel F, Deforges S, Perret M, Berland JL, Sodoyer M, Pol S, Brechot C, Paranhos-Baccala G, Lotteau V (2002) Characterization of low- and very-low-density hepatitis C virus RNA-containing particles. J Virol 76(14):6919–6928. doi:10.1128/jvi.76.14.6919-6928.2002

Andreo U, Maillard P, Kalinina O, Walic M, Meurs E, Martinot M, Marcellin P, Budkowska A (2007) Lipoprotein lipase mediates hepatitis C virus (HCV) cell entry and inhibits HCV infection. Cell Microbiol 9(10):2445–2456. doi:10.1111/j.1462-5822.2007.00972.x

Andrus L, Marukian S, Jones CT, Catanese MT, Sheahan TP, Schoggins JW, Barry WT, Dustin LB, Trehan K, Ploss A, Bhatia SN, Rice CM (2011) Expression of paramyxovirus V proteins promotes replication and spread of hepatitis C virus in cultures of primary human fetal liver cells. Hepatology 54(6):1901–1912. doi:10.1002/hep.24557

Bartenschlager R, Penin F, Lohmann V, Andre P (2011) Assembly of infectious hepatitis C virus particles. Trends Microbiol 19(2):95–103. doi:10.1016/j.tim.2010.11.005

Bartosch B, Verney G, Dreux M, Donot P, Morice Y, Penin F, Pawlotsky JM, Lavillette D, Cosset FL (2005) An interplay between hypervariable region 1 of the hepatitis C virus E2 glycoprotein, the scavenger receptor BI, and high-density lipoprotein promotes both enhancement of infection and protection against neutralizing antibodies. J Virol 79(13):8217–8229. doi:79/13/8217 [pii]. 10.1128/JVI.79.13.8217-8229.2005

Baumert TF, Ito S, Wong DT, Liang TJ (1998) Hepatitis C virus structural proteins assemble into viruslike particles in insect cells. J Virol 72(5):3827–3836

Benga WJ, Krieger SE, Dimitrova M, Zeisel MB, Parnot M, Lupberger J, Hildt E, Luo G, McLauchlan J, Baumert TF, Schuster C (2010) Apolipoprotein E interacts with hepatitis C virus nonstructural protein 5A and determines assembly of infectious particles. Hepatology 51 (1):43–53. doi:10.1002/hep.23278

Blight KJ, McKeating JA, Rice CM (2002) Highly permissive cell lines for hepatitis C virus genomic and subgenomic RNA replication. J Virol 76:13001–13014

Boulant S, Vanbelle C, Ebel C, Penin F, Lavergne JP (2005) Hepatitis C virus core protein is a dimeric alpha-helical protein exhibiting membrane protein features. J Virol 79 (17):11353–11365. doi:10.1128/JVI.79.17.11353-11365.2005

Bradley DW, McCaustland KA, Cook EH, Schable CA, Ebert JW, Maynard JE (1985) Posttransfusion non-A, non-B hepatitis in chimpanzees. Physicochemical evidence that the tubule-forming agent is a small, enveloped virus. Gastroenterology 88(3):773–779

Bradley D, McCaustland K, Krawczynski K, Spelbring J, Humphrey C, Cook EH (1991) Hepatitis C virus: buoyant density of the factor VIII-derived isolate in sucrose. J Med Virol 34 (3):206–208

Carrick RJ, Schlauder GG, Peterson DA, Mushahwar IK (1992) Examination of the buoyant density of hepatitis C virus by the polymerase chain reaction. J Virol Methods 39(3):279–289

Cases S, Smith SJ, Zheng YW, Myers HM, Lear SR, Sande E, Novak S, Collins C, Welch CB, Lusis AJ, Erickson SK, Farese RV Jr (1998) Identification of a gene encoding an acyl CoA: diacylglycerol acyltransferase, a key enzyme in triacylglycerol synthesis. Proc Natl Acad Sci U S A 95(22):13018–13023

Catanese MT, Graziani R, von Hahn T, Moreau M, Huby T, Paonessa G, Santini C, Luzzago A, Rice CM, Cortese R, Vitelli A, Nicosia A (2007) High-avidity monoclonal antibodies against the human scavenger class B type I receptor efficiently block hepatitis C virus infection in the presence of high-density lipoprotein. J Virol 81(15):8063–8071. doi:10.1128/JVI.00193-07

Catanese MT, Uryu K, Kopp M, Edwards TJ, Andrus L, Rice WJ, Silvestry M, Kuhn RJ, Rice CM (2013) Ultrastructural analysis of hepatitis C virus particles. Proc Natl Acad Sci U S A. doi:10.1073/pnas.1307527110

Chang KS, Jiang J, Cai Z, Luo G (2007a) Human apolipoprotein e is required for infectivity and production of hepatitis C virus in cell culture. J Virol 81(24):13783–13793. doi:10.1128/JVI.01091-07

Chang KS, Jiang J, Cai Z, Luo G (2007b) Human apolipoprotein e is required for infectivity and production of hepatitis C virus in cell culture. Journal of virology 81 (24):13783-13793. doi:JVI.01091-07 [pii]. 10.1128/JVI.01091-07

Choo QL, Kuo G, Weiner AJ, Overby LR, Bradley DW, Houghton M (1989) Isolation of a cDNA clone derived from a blood-borne non-A, non-B viral hepatitis genome. Science 244 (4902):359–362

Chroni A, Nieland TJ, Kypreos KE, Krieger M, Zannis VI (2005) SR-BI mediates cholesterol efflux via its interactions with lipid-bound ApoE. Structural mutations in SR-BI diminish cholesterol efflux. Biochemistry 44(39):13132–13143. doi:10.1021/bi051029o

Ciesek S, Friesland M, Steinmann J, Becker B, Wedemeyer H, Manns MP, Steinmann J, Pietschmann T, Steinmann E (2010) How stable is the hepatitis C virus (HCV)? Environmental stability of HCV and its susceptibility to chemical biocides. J Infect Dis 201(12):1859–1866. doi:10.1086/652803

Clarke D, Griffin S, Beales L, Gelais CS, Burgess S, Harris M, Rowlands D (2006) Evidence for the formation of a heptameric ion channel complex by the hepatitis C virus p7 protein in vitro. J Biol Chem 281(48):37057–37068. doi:10.1074/jbc.M602434200

Cocquerel L, Meunier JC, Pillez A, Wychowski C, Dubuisson J (1998) A retention signal necessary and sufficient for endoplasmic reticulum localization maps to the transmembrane domain of hepatitis C virus glycoprotein E2. J Virol 72(3):2183–2191

Cocquerel L, Duvet S, Meunier JC, Pillez A, Cacan R, Wychowski C, Dubuisson J (1999) The transmembrane domain of hepatitis C virus glycoprotein E1 is a signal for static retention in the endoplasmic reticulum. J Virol 73(4):2641–2649

Cocquerel L, Wychowski C, Minner F, Penin F, Dubuisson J (2000) Charged residues in the transmembrane domains of hepatitis C virus glycoproteins play a major role in the processing, subcellular localization, and assembly of these envelope proteins. J Virol 74(8):3623–3633

Dao Thi VL, Granier C, Zeisel MB, Guerin M, Mancip J, Granio O, Penin F, Lavillette D, Bartenschlager R, Baumert TF, Cosset FL, Dreux M (2012) Characterization of hepatitis C virus particle subpopulations reveals multiple usage of the scavenger receptor BI for entry steps. J Biol Chem 287(37):31242–31257. doi:10.1074/jbc.M112.365924

de Beer MC, Durbin DM, Cai L, Jonas A, de Beer FC, van der Westhuyzen DR (2001) Apolipoprotein A-I conformation markedly influences HDL interaction with scavenger receptor BI. J Lipid Res 42(2):309–313

De Vos R, Verslype C, Depla E, Fevery J, Van Damme B, Desmet V, Roskams T (2002) Ultrastructural visualization of hepatitis C virus components in human and primate liver biopsies. J Hepatol 37(3):370–379

Deleersnyder V, Pillez A, Wychowski C, Blight K, Xu J, Hahn YS, Rice CM, Dubuisson J (1997) Formation of native hepatitis C virus glycoprotein complexes. J Virol 71(1):697–704

Deng L, Zhong L, Struble E, Duan H, Ma L, Harman C, Yan H, Virata-Theimer ML, Zhao Z, Feinstone S, Alter H, Zhang P (2013) Structural evidence for a bifurcated mode of action in the antibody-mediated neutralization of hepatitis C virus. Proc Natl Acad Sci U S A 110 (18):7418–7422. doi:10.1073/pnas.1305306110

Dreux M, Pietschmann T, Granier C, Voisset C, Ricard-Blum S, Mangeot PE, Keck Z, Foung S, Vu-Dac N, Dubuisson J, Bartenschlager R, Lavillette D, Cosset FL (2006) High density lipoprotein inhibits hepatitis C virus-neutralizing antibodies by stimulating cell entry via activation of the scavenger receptor BI. J Biol Chem 281(27):18285–18295. doi: M602706200 [pii]. 10.1074/jbc.M602706200

Dreux M, Boson B, Ricard-Blum S, Molle J, Lavillette D, Bartosch B, Pecheur EI, Cosset FL (2007) The exchangeable apolipoprotein ApoC-I promotes membrane fusion of hepatitis C virus. J Biol Chem 282(44):32357–32369. doi:10.1074/jbc.M705358200

Drummer HE, Poumbourios P (2004) Hepatitis C virus glycoprotein E2 contains a membrane-proximal heptad repeat sequence that is essential for E1E2 glycoprotein heterodimerization and viral entry. J Biol Chem 279(29):30066–30072. doi:10.1074/jbc.M405098200

Drummer HE, Maerz A, Poumbourios P (2003) Cell surface expression of functional hepatitis C virus E1 and E2 glycoproteins. FEBS Lett 546(2–3):385–390

Dubuisson J, Hsu HH, Cheung RC, Greenberg H, Russell DR, Rice CM (1994) Formation and intracellular localization of hepatitis C virus envelope glycoprotein complexes expressed by recombinant vaccinia and Sindbis viruses. J Virol 68:6147–6160

El Omari K, Iourin O, Harlos K, Grimes JM, Stuart DI (2013) Structure of a pestivirus envelope glycoprotein E2 clarifies its role in cell entry. Cell Rep 3(1):30–35. doi:10.1016/j.celrep.2012.12.001

Farci P, Alter HJ, Govindarajan S, Wong DC, Engle R, Lesniewski RR, Mushahwar IK, Desai SM, Miller RH, Ogata N et al (1992) Lack of protective immunity against reinfection with hepatitis C virus. Science 258(5079):135–140

Feinstone SM, Kapikian AZ, Purcell RH, Alter HJ, Holland PV (1975) Transfusion-associated hepatitis not due to viral hepatitis type A or B. N Engl J Med 292(15):767–770. doi:10.1056/NEJM197504102921502

Felmlee DJ, Sheridan DA, Bridge SH, Nielsen SU, Milne RW, Packard CJ, Caslake MJ, McLauchlan J, Toms GL, Neely RD, Bassendine MF (2010) Intravascular transfer contributes to postprandial increase in numbers of very-low-density hepatitis C virus particles. Gastroenterology 139(5):1774–1783, 1783 e1771-1776. doi:10.1053/j.gastro.2010.07.047

Fenouillet E, Lavillette D, Loureiro S, Krashias G, Maurin G, Cosset FL, Jones IM, Barbouche R (2008) Contribution of redox status to hepatitis C virus E2 envelope protein function and antigenicity. J Biol Chem 283(39):26340–26348. doi:10.1074/jbc.M805221200

Ferlenghi I, Clarke M, Ruttan T, Allison SL, Schalich J, Heinz FX, Harrison SC, Rey FA, Fuller SD (2001) Molecular organization of a recombinant subviral particle from tick-borne encephalitis virus. Mol Cell 7(3):593–602

Flint M, McKeating JA (1999) The C-terminal region of the hepatitis C virus E1 glycoprotein confers localization within the endoplasmic reticulum. J Gen Virol 80(Pt 8):1943–1947

Flint M, Logvinoff C, Rice CM, McKeating JA (2004) Characterization of infectious retroviral pseudotype particles bearing hepatitis C virus glycoproteins. J Virol 78(13):6875–6882. doi:10.1128/JVI.78.13.6875-6882.2004
Foster TL, Thompson GS, Kalverda AP, Kankanala J, Bentham M, Wetherill LF, Thompson J, Barker AM, Clarke D, Noerenberg M, Pearson AR, Rowlands DJ, Homans SW, Harris M, Foster R, Griffin S (2014) Structure-guided design affirms inhibitors of hepatitis C virus p7 as a viable class of antivirals targeting virion release. Hepatology 59(2):408–422. doi:10.1002/hep.26685
Gastaminza P, Kapadia SB, Chisari FV (2006) Differential biophysical properties of infectious intracellular and secreted hepatitis C virus particles. J Virol 80(22):11074–11081. doi:10.1128/JVI.01150-06
Gastaminza P, Cheng G, Wieland S, Zhong J, Liao W, Chisari FV (2008) Cellular determinants of hepatitis C virus assembly, maturation, degradation, and secretion. J Virol 82(5):2120–2129. doi:10.1128/JVI.02053-07
Gastaminza P, Dryden KA, Boyd B, Wood MR, Law M, Yeager M, Chisari FV (2010) Ultrastructural and biophysical characterization of hepatitis C virus particles produced in cell culture. J Virol 84(21):10999–11009. doi:10.1128/JVI.00526-10
Griffin SD, Beales LP, Clarke DS, Worsfold O, Evans SD, Jaeger J, Harris MP, Rowlands DJ (2003) The p7 protein of hepatitis C virus forms an ion channel that is blocked by the antiviral drug, Amantadine. FEBS Lett 535(1–3):34–38
Gusarova V, Brodsky JL, Fisher EA (2003) Apolipoprotein B100 exit from the endoplasmic reticulum (ER) is COPII-dependent, and its lipidation to very low density lipoprotein occurs post-ER. J Biol Chem 278(48):48051–48058. doi:10.1074/jbc.M306898200
Gusarova V, Seo J, Sullivan ML, Watkins SC, Brodsky JL, Fisher EA (2007) Golgi-associated maturation of very low density lipoproteins involves conformational changes in apolipoprotein B, but is not dependent on apolipoprotein E. J Biol Chem 282(27):19453–19462. doi:10.1074/jbc.M700475200
Hatters DM, Peters-Libeu CA, Weisgraber KH (2006) Apolipoprotein E structure: insights into function. Trends Biochem Sci 31(8):445–454. doi:10.1016/j.tibs.2006.06.008
He LF, Alling D, Popkin T, Shapiro M, Alter HJ, Purcell RH (1987) Determining the size of non-A, non-B hepatitis virus by filtration. J Infect Dis 156(4):636–640
Hebbachi AM, Brown AM, Gibbons GF (1999) Suppression of cytosolic triacylglycerol recruitment for very low density lipoprotein assembly by inactivation of microsomal triglyceride transfer protein results in a delayed removal of apoB-48 and apoB-100 from microsomal and Golgi membranes of primary rat hepatocytes. J Lipid Res 40(10):1758–1768
Helenius A, Aebi M (2001) Intracellular functions of N-linked glycans. Science 291(5512):2364–2369
Herker E, Harris C, Hernandez C, Carpentier A, Kaehlcke K, Rosenberg AR, Farese RV Jr, Ott M (2010) Efficient hepatitis C virus particle formation requires diacylglycerol acyltransferase-1. Nat Med 16(11):1295–1298. doi:10.1038/nm.2238
Hijikata M, Shimizu YK, Kato H, Iwamoto A, Shih JW, Alter HJ, Purcell RH, Yoshikura H (1993) Equilibrium centrifugation studies of hepatitis C virus: evidence for circulating immune complexes. J Virol 67(4):1953–1958
Hiramatsu N, Hayashi N, Haruna Y, Kasahara A, Fusamoto H, Mori C, Fuke I, Okayama H, Kamada T (1992) Immunohistochemical detection of hepatitis C virus-infected hepatocytes in chronic liver disease with monoclonal antibodies to core, envelope and NS3 regions of the hepatitis C virus genome. Hepatology 16(2):306–311
Holm L, Rosenstrom P (2010) Dali server: conservation mapping in 3D. Nucleic Acids Res 38 (Web Server issue):W545–W549. doi:10.1093/nar/gkq366
Huang H, Sun F, Owen DM, Li W, Chen Y, Gale M Jr, Ye J (2007) Hepatitis C virus production by human hepatocytes dependent on assembly and secretion of very low-density lipoproteins. Proc Natl Acad Sci U S A 104(14):5848–5853. doi:10.1073/pnas.0700760104

Jiang J, Luo G (2009) Apolipoprotein E but not B is required for the formation of infectious hepatitis C virus particles. J Virol 83(24):12680–12691. doi:10.1128/JVI.01476-09
Jiang J, Cun W, Wu X, Shi Q, Tang H, Luo G (2012) Hepatitis C virus attachment mediated by apolipoprotein E binding to cell surface heparan sulfate. J Virol 86(13):7256–7267. doi:10.1128/JVI.07222-11
Jonas APMC (2008) Lipoprotein structure. In: Cover biochemistry of lipids, lipoproteins and membranes, 5th edn. Elsevier Amsterdam, pp 485–506
Jones CT, Murray CL, Eastman DK, Tassello J, Rice CM (2007) Hepatitis C virus p7 and NS2 proteins are essential for production of infectious virus. J Virol 81(16):8374–8383. doi:10.1128/JVI.00690-07
Kaito M, Watanabe S, Tsukiyama-Kohara K, Yamaguchi K, Kobayashi Y, Konishi M, Yokoi M, Ishida S, Suzuki S, Kohara M (1994) Hepatitis C virus particle detected by immunoelectron microscopic study. J Gen Virol 75(Pt 7):1755–1760
Kanto T, Hayashi N, Takehara T, Hagiwara H, Mita E, Naito M, Kasahara A, Fusamoto H, Kamada T (1994) Buoyant density of hepatitis C virus recovered from infected hosts: two different features in sucrose equilibrium density-gradient centrifugation related to degree of liver inflammation. Hepatology 19(2):296–302
Karlsson H, Leanderson P, Tagesson C, Lindahl M (2005) Lipoproteomics I: mapping of proteins in low-density lipoprotein using two-dimensional gel electrophoresis and mass spectrometry. Proteomics 5(2):551–565. doi:10.1002/pmic.200300938
Keck ZY, Xia J, Wang Y, Wang W, Krey T, Prentoe J, Carlsen T, Li AY, Patel AH, Lemon SM, Bukh J, Rey FA, Foung SK (2012) Human monoclonal antibodies to a novel cluster of conformational epitopes on HCV E2 with resistance to neutralization escape in a genotype 2a isolate. PLoS Pathog 8(4):e1002653. doi:10.1371/journal.ppat.1002653
Khan AG, Whidby J, Miller MT, Scarborough H, Zatorski AV, Cygan A, Price AA, Yost SA, Bohannon CD, Jacob J, Grakoui A, Marcotrigiano J (2014) Structure of the core ectodomain of the hepatitis C virus envelope glycoprotein 2. Nature 509(7500):381–384. doi:10.1038/nature13117
Knott TJ, Pease RJ, Powell LM, Wallis SC, Rall SC Jr, Innerarity TL, Blackhart B, Taylor WH, Marcel Y, Milne R et al (1986) Complete protein sequence and identification of structural domains of human apolipoprotein B. Nature 323(6090):734–738. doi:10.1038/323734a0
Kolykhalov AA, Agapov EV, Blight KJ, Mihalik K, Feinstone SM, Rice CM (1997) Transmission of hepatitis C by intrahepatic inoculation with transcribed RNA. Science 277(5325):570–574
Kong L, Giang E, Nieusma T, Robbins JB, Deller MC, Stanfield RL, Wilson IA, Law M (2012a) Structure of hepatitis C virus envelope glycoprotein E2 antigenic site 412 to 423 in complex with antibody AP33. J Virol 86(23):13085–13088. doi:10.1128/JVI.01939-12
Kong L, Giang E, Robbins JB, Stanfield RL, Burton DR, Wilson IA, Law M (2012b) Structural basis of hepatitis C virus neutralization by broadly neutralizing antibody HCV1. Proc Natl Acad Sci U S A 109(24):9499–9504. doi:10.1073/pnas.1202924109
Kong L, Giang E, Nieusma T, Kadam RU, Cogburn KE, Hua Y, Dai X, Stanfield RL, Burton DR, Ward AB, Wilson IA, Law M (2013) Hepatitis C virus E2 envelope glycoprotein core structure. Science 342(6162):1090–1094. doi:10.1126/science.1243876
Kopp M, Murray CL, Jones CT, Rice CM (2010) Genetic analysis of the carboxy-terminal region of the hepatitis C virus core protein. J Virol 84(4):1666–1673. doi:10.1128/JVI.02043-09
Krey T, d'Alayer J, Kikuti CM, Saulnier A, Damier-Piolle L, Petitpas I, Johansson DX, Tawar RG, Baron B, Robert B, England P, Persson MAA, Martin A, Rey FA (2010) The disulfide bonds in glycoprotein E2 of hepatitis C virus reveal the tertiary organization of the molecule. PLoS Pathog 6(2):e1000762. doi:10.1371/journal.ppat.1000762
Kuhn RJ, Zhang W, Rossmann MG, Pletnev SV, Corver J, Lenches E, Jones CT, Mukhopadhyay S, Chipman PR, Strauss EG, Baker TS, Strauss JH (2002) Structure of dengue virus: implications for flavivirus organization, maturation, and fusion. Cell 108(5):717–725

Lamb RA, Parks GD (2007) Paramyxoviridae: the viruses and their replication. In: Knipe DM, Howley PM (eds) Fields virology, 5th edn. Lippincott Williams & Wilkins, Philadelphia, pp 1449–1496

Lazaro CA, Croager EJ, Mitchell C, Campbell JS, Yu C, Foraker J, Rhim JA, Yeoh GC, Fausto N (2003) Establishment, characterization, and long-term maintenance of cultures of human fetal hepatocytes. Hepatology 38(5):1095–1106. doi:10.1053/jhep.2003.50448

Le Sage V, Mouland AJ (2013) Viral subversion of the nuclear pore complex. Viruses 5 (8):2019–2042. doi:10.3390/v5082019

Lecluyse EL, Alexandre E (2010) Isolation and culture of primary hepatocytes from resected human liver tissue. Methods Mol Biol 640:57–82. doi:10.1007/978-1-60761-688-7_3

Li Y, Wang J, Kanai R, Modis Y (2013) Crystal structure of glycoprotein E2 from bovine viral diarrhea virus. Proc Natl Acad Sci U S A 110(17):6805–6810. doi:10.1073/pnas.1300524110

Liang Y, Shilagard T, Xiao SY, Snyder N, Lau D, Cicalese L, Weiss H, Vargas G, Lemon SM (2009) Visualizing hepatitis C virus infections in human liver by two-photon microscopy. Gastroenterology 137(4):1448–1458. doi:10.1053/j.gastro.2009.07.050

Libeu CP, Lund-Katz S, Phillips MC, Wehrli S, Hernaiz MJ, Capila I, Linhardt RJ, Raffai RL, Newhouse YM, Zhou F, Weisgraber KH (2001) New insights into the heparan sulfate proteoglycan-binding activity of apolipoprotein E. J Biol Chem 276(42):39138–39144. doi:10.1074/jbc.M104746200

Lindenbach BD, Evans MJ, Syder AJ, Wolk B, Tellinghuisen TL, Liu CC, Maruyama T, Hynes RO, Burton DR, McKeating JA, Rice CM (2005) Complete replication of hepatitis C virus in cell culture. Science 309(5734):623–626. doi:10.1126/science.1114016

Lindenbach BD, Meuleman P, Ploss A, Vanwolleghem T, Syder AJ, McKeating JA, Lanford RE, Feinstone SM, Major ME, Leroux-Roels G, Rice CM (2006) Cell culture-grown hepatitis C virus is infectious in vivo and can be recultured in vitro. Proc Natl Acad Sci U S A 103 (10):3805–3809. doi:10.1073/pnas.0511218103

Liu S, McCormick KD, Zhao W, Zhao T, Fan D, Wang T (2012) Human apolipoprotein E peptides inhibit hepatitis C virus entry by blocking virus binding. Hepatology 56(2):484–491. doi:10.1002/hep.25665

Lohmann V, Korner F, Koch JO, Herian U, Theilmann L, Bartenschlager R (1999) Replication of subgenomic hepatitis C virus RNAs in a hepatoma cell line. Science 285:110–113

Lussignol M, Kopp M, Molloy K, Vizcay-Barrena G, Fleck RA, Dorner M, Bell KL, Chait BT, Rice CM, Catanese MT (2016) Proteomics of HCV virions reveals an essential role for the nucleoporin Nup98 in virus morphogenesis. Proc Natl Acad Sci U S A 113(9):2484–2489. doi:10.1073/pnas.1518934113

Maillard P, Walic M, Meuleman P, Roohvand F, Huby T, Le Goff W, Leroux-Roels G, Pecheur EI, Budkowska A (2011) Lipoprotein lipase inhibits hepatitis C virus (HCV) infection by blocking virus cell entry. PLoS One 6(10):e26637. doi:10.1371/journal.pone.0026637

Mancone C, Amicone L, Fimia GM, Bravo E, Piacentini M, Tripodi M, Alonzi T (2007) Proteomic analysis of human very low-density lipoprotein by two-dimensional gel electrophoresis and MALDI-TOF/TOF. Proteomics 7(1):143–154. doi:10.1002/pmic.200600339

Mancone C, Steindler C, Santangelo L, Simonte G, Vlassi C, Longo MA, D'Offizi G, Di Giacomo C, Pucillo LP, Amicone L, Tripodi M, Alonzi T (2011) Hepatitis C virus production requires apolipoprotein A-I and affects its association with nascent low-density lipoproteins. Gut 60(3):378–386. doi:10.1136/gut.2010.211292

Mazumdar B, Banerjee A, Meyer K, Ray R (2011) Hepatitis C virus E1 envelope glycoprotein interacts with apolipoproteins in facilitating entry into hepatocytes. Hepatology 54 (4):1149–1156. doi:10.1002/hep.24523

McCaffrey K, Boo I, Tewierek K, Edmunds ML, Poumbourios P, Drummer HE (2012) Role of conserved cysteine residues in hepatitis C virus glycoprotein e2 folding and function. J Virol 86(7):3961–3974. doi:10.1128/JVI.05396-11

Mead JR, Irvine SA, Ramji DP (2002) Lipoprotein lipase: structure, function, regulation, and role in disease. J Mol Med (Berl) 80(12):753–769. doi:10.1007/s00109-002-0384-9

Mensenkamp AR, Jong MC, van Goor H, van Luyn MJ, Bloks V, Havinga R, Voshol PJ, Hofker MH, van Dijk KW, Havekes LM, Kuipers F (1999) Apolipoprotein E participates in the regulation of very low density lipoprotein-triglyceride secretion by the liver. J Biol Chem 274(50):35711–35718

Merz A, Long G, Hiet MS, Brugger B, Chlanda P, Andre P, Wieland F, Krijnse-Locker J, Bartenschlager R (2011) Biochemical and morphological properties of hepatitis C virus particles and determination of their lipidome. J Biol Chem 286(4):3018–3032. doi:10.1074/jbc.M110.175018

Meunier JC, Russell RS, Engle RE, Faulk KN, Purcell RH, Emerson SU (2008) Apolipoprotein c1 association with hepatitis C virus. J Virol 82(19):9647–9656. doi:10.1128/JVI.00914-08

Michalak JP, Wychowski C, Choukhi A, Meunier JC, Ung S, Rice CM, Dubuisson J (1997) Characterization of truncated forms of hepatitis C virus glycoproteins. J Gen Virol 78 (Pt 9):2299–2306

Miller RH, Purcell RH (1990) Hepatitis C virus shares amino acid sequence similarity with pestiviruses and flaviviruses as well as members of two plant virus supergroups. Proc Natl Acad Sci U S A 87(6):2057–2061

Miyamoto H, Okamoto H, Sato K, Tanaka T, Mishiro S (1992) Extraordinarily low density of hepatitis C virus estimated by sucrose density gradient centrifugation and the polymerase chain reaction. J Gen Virol 73(Pt 3):715–718

Miyanari Y, Atsuzawa K, Usuda N, Watashi K, Hishiki T, Zayas M, Bartenschlager R, Wakita T, Hijikata M, Shimotohno K (2007) The lipid droplet is an important organelle for hepatitis C virus production. Nat Cell Biol 9(9):1089–1097. doi:10.1038/ncb1631

Mukhopadhyay S, Kim BS, Chipman PR, Rossmann MG, Kuhn RJ (2003) Structure of West Nile virus. Science 302(5643):248. doi:10.1126/science.1089316

Nahmias Y, Goldwasser J, Casali M, van Poll D, Wakita T, Chung RT, Yarmush ML (2008) Apolipoprotein B-dependent hepatitis C virus secretion is inhibited by the grapefruit flavonoid naringenin. Hepatology 47(5):1437–1445. doi:10.1002/hep.22197

Neveu G, Barouch-Bentov R, Ziv-Av A, Gerber D, Jacob Y, Einav S (2012) Identification and targeting of an interaction between a tyrosine motif within hepatitis C virus core protein and AP2M1 essential for viral assembly. PLoS Pathog 8(8):e1002845. doi:10.1371/journal.ppat.1002845

Nielsen SU, Bassendine MF, Burt AD, Martin C, Pumeechockchai W, Toms GL (2006) Association between hepatitis C virus and very-low-density lipoprotein (VLDL)/LDL analyzed in iodixanol density gradients. J Virol 80(5):2418–2428. doi:10.1128/JVI.80.5.2418-2428.2006

Oehler V, Filipe A, Montserret R, da Costa D, Brown G, Penin F, McLauchlan J (2012) Structural analysis of hepatitis C virus core-E1 signal peptide and requirements for cleavage of the genotype 3a signal sequence by signal peptide peptidase. J Virol 86(15):7818–7828. doi:10.1128/JVI.00457-12

Op De Beeck A, Montserret R, Duvet S, Cocquerel L, Cacan R, Barberot B, Le Maire M, Penin F, Dubuisson J (2000) The transmembrane domains of hepatitis C virus envelope glycoproteins E1 and E2 play a major role in heterodimerization. J Biol Chem 275(40):31428–31437

Op De Beeck A, Voisset C, Bartosch B, Ciczora Y, Cocquerel L, Keck Z, Foung S, Cosset FL, Dubuisson J (2004) Characterization of functional hepatitis C virus envelope glycoproteins. J Virol 78(6):2994–3002

Owen DM, Huang H, Ye J, Gale M Jr (2009) Apolipoprotein E on hepatitis C virion facilitates infection through interaction with low-density lipoprotein receptor. Virology 394(1):99–108. doi:10.1016/j.virol.2009.08.037

Pan C, Kumar C, Bohl S, Klingmueller U, Mann M (2009) Comparative proteomic phenotyping of cell lines and primary cells to assess preservation of cell type-specific functions. Mol Cell Proteomics: MCP 8(3):443–450. doi:10.1074/mcp.M800258-MCP200

Parent R, Qu X, Petit MA, Beretta L (2009) The heat shock cognate protein 70 is associated with hepatitis C virus particles and modulates virus infectivity. Hepatology 49(6):1798–1809. doi:10.1002/hep.22852

Phillips ML, Schumaker VN (1989) Conformation of apolipoprotein B after lipid extraction of low density lipoproteins attached to an electron microscope grid. J Lipid Res 30(3):415–422
Podevin P, Carpentier A, Pene V, Aoudjehane L, Carriere M, Zaidi S, Hernandez C, Calle V, Meritet JF, Scatton O, Dreux M, Cosset FL, Wakita T, Bartenschlager R, Demignot S, Conti F, Rosenberg AR, Calmus Y (2010) Production of infectious hepatitis C virus in primary cultures of human adult hepatocytes. Gastroenterology 139(4):1355–1364. doi:10.1053/j.gastro.2010.06.058
Prince AM, Huima-Byron T, Parker TS, Levine DM (1996) Visualization of hepatitis C virions and putative defective interfering particles isolated from low-density lipoproteins. J Viral Hepat 3(1):11–17
Sakai A, Claire MS, Faulk K, Govindarajan S, Emerson SU, Purcell RH, Bukh J (2003) The p7 polypeptide of hepatitis C virus is critical for infectivity and contains functionally important genotype-specific sequences. Proc Natl Acad Sci U S A 100(20):11646–11651. doi:10.1073/pnas.1834545100
Scarselli E, Ansuini H, Cerino R, Roccasecca RM, Acali S, Filocamo G, Traboni C, Nicosia A, Cortese R, Vitelli A (2002) The human scavenger receptor class B type I is a novel candidate receptor for the hepatitis C virus. EMBO J 21(19):5017–5025
Schmidt WN, Stapleton JT, LaBrecque DR, Mitros FA, Kirby PA, Phillips MJ, Brashear DL (2000) Hepatitis C virus (HCV) infection and cryoglobulinemia: analysis of whole blood and plasma HCV-RNA concentrations and correlation with liver histology. Hepatology 31 (3):737–744. doi:10.1002/hep.510310326
Scholtes C, Ramiere C, Rainteau D, Perrin-Cocon L, Wolf C, Humbert L, Carreras M, Guironnet-Paquet A, Zoulim F, Bartenschlager R, Lotteau V, Andre P, Diaz O (2012) High plasma level of nucleocapsid-free envelope glycoprotein-positive lipoproteins in hepatitis C patients. Hepatology 56(1):39–48. doi:10.1002/hep.25628
Shelness GS, Sellers JA (2001) Very-low-density lipoprotein assembly and secretion. Curr Opin Lipidol 12(2):151–157
Shimizu Y, Hishiki T, Sugiyama K, Ogawa K, Funami K, Kato A, Ohsaki Y, Fujimoto T, Takaku H, Shimotohno K (2010) Lipoprotein lipase and hepatic triglyceride lipase reduce the infectivity of hepatitis C virus (HCV) through their catalytic activities on HCV-associated lipoproteins. Virology 407(1):152–159. doi:10.1016/j.virol.2010.08.011
Steinmann E, Penin F, Kallis S, Patel AH, Bartenschlager R, Pietschmann T (2007) Hepatitis C virus p7 protein is crucial for assembly and release of infectious virions. PLoS Pathog 3(7): e103. doi:10.1371/journal.ppat.0030103
Strauss JHSEG (2007) Viruses and human disease, 2nd edn. Elsevier Academic press Amsterdam, pp 35–62
Strauss JH, Strauss EG (2001) Virus evolution: how does an enveloped virus make a regular structure? Cell 105(1):5–8
Sumpter R Jr, Loo YM, Foy E, Li K, Yoneyama M, Fujita T, Lemon SM, Gale M Jr (2005) Regulating intracellular antiviral defense and permissiveness to hepatitis C virus RNA replication through a cellular RNA helicase, RIG-I. J Virol 79(5):2689–2699
Sun HY, Lin CC, Lee JC, Wang SW, Cheng PN, Wu IC, Chang TT, Lai MD, Shieh DB, Young KC (2013) Very low-density lipoprotein/lipo-viro particles reverse lipoprotein lipase-mediated inhibition of hepatitis C virus infection via apolipoprotein C-III. Gut 62(8):1193–1203. doi:10.1136/gutjnl-2011-301798
Swanson K, Wen X, Leser GP, Paterson RG, Lamb RA, Jardetzky TS (2010) Structure of the Newcastle disease virus F protein in the post-fusion conformation. Virology 402(2):372–379. doi:10.1016/j.virol.2010.03.050
Thomssen R, Bonk S (2002) Virolytic action of lipoprotein lipase on hepatitis C virus in human sera. Med Microbiol Immunol 191(1):17–24
Thomssen R, Bonk S, Propfe C, Heermann KH, Kochel HG, Uy A (1992) Association of hepatitis C virus in human sera with beta-lipoprotein. Med Microbiol Immunol 181(5):293–300

Thomssen R, Bonk S, Thiele A (1993) Density heterogeneities of hepatitis C virus in human sera due to the binding of beta-lipoproteins and immunoglobulins. Med Microbiol Immunol 182 (6):329–334

Trestard A, Bacq Y, Buzelay L, Dubois F, Barin F, Goudeau A, Roingeard P (1998) Ultrastructural and physicochemical characterization of the hepatitis C virus recovered from the serum of an agammaglobulinemic patient. Arch Virol 143(11):2241–2245

Tscherne DM, Jones CT, Evans MJ, Lindenbach BD, McKeating JA, Rice CM (2006) Time- and temperature-dependent activation of hepatitis C virus for low-pH-triggered entry. J Virol 80 (4):1734–1741. doi:10.1128/JVI.80.4.1734-1741.2006

Vaney MC, Rey FA (2011) Class II enveloped viruses. Cell Microbiol 13(10):1451–1459. doi:10.1111/j.1462-5822.2011.01653.x

Vieyres G, Thomas X, Descamps V, Duverlie G, Patel AH, Dubuisson J (2010) Characterization of the envelope glycoproteins associated with infectious hepatitis C virus. J Virol 84 (19):10159–10168. doi:10.1128/JVI.01180-10

Voisset C, Callens N, Blanchard E, Op De Beeck A, Dubuisson J, Vu-Dac N (2005) High density lipoproteins facilitate hepatitis C virus entry through the scavenger receptor class B type I. J Biol Chem 280(9):7793–7799. doi:10.1074/jbc.M411600200

Wahid A, Dubuisson J (2013) Virus-neutralizing antibodies to hepatitis C virus. J Viral Hepat 20 (6):369–376. doi:10.1111/jvh.12094

Wakita T, Pietschmann T, Kato T, Date T, Miyamoto M, Zhao Z, Murthy K, Habermann A, Krausslich HG, Mizokami M, Bartenschlager R, Liang TJ (2005) Production of infectious hepatitis C virus in tissue culture from a cloned viral genome. Nat Med 11(7):791–796. doi:10.1038/nm1268

White JM, Delos SE, Brecher M, Schornberg K (2008) Structures and mechanisms of viral membrane fusion proteins: multiple variations on a common theme. Crit Rev Biochem Mol Biol 43(3):189–219. doi:10.1080/10409230802058320

Williams DL, de La Llera-Moya M, Thuahnai ST, Lund-Katz S, Connelly MA, Azhar S, Anantharamaiah GM, Phillips MC (2000) Binding and cross-linking studies show that scavenger receptor BI interacts with multiple sites in apolipoprotein A-I and identify the class A amphipathic alpha-helix as a recognition motif. J Biol Chem 275(25):18897–18904. doi:10.1074/jbc.M002411200

Wozniak AL, Griffin S, Rowlands D, Harris M, Yi M, Lemon SM, Weinman SA (2010) Intracellular proton conductance of the hepatitis C virus p7 protein and its contribution to infectious virus production. PLoS Pathog 6(9):e1001087. doi:10.1371/journal.ppat.1001087

Yang CY, Chen SH, Gianturco SH, Bradley WA, Sparrow JT, Tanimura M, Li WH, Sparrow DA, DeLoof H, Rosseneu M et al (1986) Sequence, structure, receptor-binding domains and internal repeats of human apolipoprotein B-100. Nature 323(6090):738–742. doi:10.1038/323738a0

Yao Z, Zhou H, Figeys D, Wang Y, Sundaram M (2013) Microsome-associated lumenal lipid droplets in the regulation of lipoprotein secretion. Curr Opin Lipidol 24(2):160–170. doi:10.1097/MOL.0b013e32835aebe7

Yu X, Qiao M, Atanasov I, Hu Z, Kato T, Liang TJ, Zhou ZH (2007) Cryo-electron microscopy and three-dimensional reconstructions of hepatitis C virus particles. Virology 367(1):126–134. doi:10.1016/j.virol.2007.05.038

Zaiou M, Arnold KS, Newhouse YM, Innerarity TL, Weisgraber KH, Segall ML, Phillips MC, Lund-Katz S (2000) Apolipoprotein E;-low density lipoprotein receptor interaction. Influences of basic residue and amphipathic alpha-helix organization in the ligand. J Lipid Res 41 (7):1087–1095

Zeisel MB, Felmlee DJ, Baumert TF (2013) Hepatitis C virus entry. Curr Top Microbiol Immunol 369:87–112. doi:10.1007/978-3-642-27340-7_4

Zhang W, Chipman PR, Corver J, Johnson PR, Zhang Y, Mukhopadhyay S, Baker TS, Strauss JH, Rossmann MG, Kuhn RJ (2003a) Visualization of membrane protein domains by cryo-electron microscopy of dengue virus. Nat Struct Biol 10(11):907–912. doi:10.1038/nsb990

Zhang Y, Corver J, Chipman PR, Zhang W, Pletnev SV, Sedlak D, Baker TS, Strauss JH, Kuhn RJ, Rossmann MG (2003b) Structures of immature flavivirus particles. EMBO J 22 (11):2604–2613. doi:10.1093/emboj/cdg270
Zhang Y, Kostyuchenko VA, Rossmann MG (2007) Structural analysis of viral nucleocapsids by subtraction of partial projections. J Struct Biol 157(2):356–364. doi:10.1016/j.jsb.2006.09.002
Zhong J, Gastaminza P, Cheng G, Kapadia S, Kato T, Burton DR, Wieland SF, Uprichard SL, Wakita T, Chisari FV (2005) Robust hepatitis C virus infection in vitro. Proc Natl Acad Sci U S A 102(26):9294–9299. doi:10.1073/pnas.0503596102

Part II
The Viral Life Cycle

Cell Culture Systems for Propagation of HCV

Takaji Wakita

Abstract Hepatitis C virus (HCV) infection is seen worldwide and is a significant cause of severe chronic liver diseases. Recently, a large number of direct-acting antivirals (DAAs) have been developed against HCV infection, resulting in significant improvements in treatment efficacy. Rapid progress in HCV research has been largely dependent of the development of HCV culture systems and small animal infection models. In the development of HCV cell culture systems, the discovery of the JFH-1 clone, an HCV strain isolated from a fulminant hepatitis C patient, was a key finding. The JFH-1 strain was the first infectious HCV strain belonging to genotype 2a. JFH-1 replicated efficiently in cultured cell lines without acquiring adaptive mutations, providing the secretion of infectious viral particles into the culture medium. Other HCV strains, including JFH-2, that also were reported to be infectious in cultured cells. However, these other isolates proved to require adaptive mutations for replication. These infectious HCV systems has provided a powerful tool to study the viral life cycle, to construct anti-viral strategies, and to develop effective vaccines.

Keywords JFH-1 • Fulminant hepatitis • Subgenomic replicon • HCV culture • Viral life cycle

Abbreviations

DAAs	Direct-acting antivirals
HCV	Hepatitis C virus
LD	Lipid droplet
RT-PCR	Reverse transcription-polymerase chain reaction

T. Wakita, M.D., Ph.D. (✉)
National Institute of Infectious Diseases, Toyama 1-23-1, Shinjuku, Tokyo 162-8640, Japan
e-mail: wakita@nih.go.jp

T. Miyamura et al. (eds.), *Hepatitis C Virus I*, DOI 10.1007/978-4-431-56098-2_4

1 Introduction

HCV is a plus-stranded RNA virus that belongs to the family Flaviviridae and genus Hepacivirus. Most of the classical flaviviruses cause mosquito-borne infections; however, HCV causes blood-borne infection and chronic liver disease, and thus is classified as a hepatitis virus. HCV was first identified in 1989 as a principal agent in post-transfusion and sporadic acute hepatitis (Choo et al. 1989; Kuo et al. 1989). After this discovery, the frequency of new post-transfusion HCV infections was significantly decreased in many countries by implementation of HCV screening of blood products. However, the development of efficient anti-HCV treatments is still considered important, since HCV infection causes severe chronic liver disease and large numbers of HCV carriers exist worldwide. To this end, detailed analysis of the viral life cycle of HCV using virus culture has been critical (Fig. 1). However, the propagation of HCV in cultured cells has remained a challenge (Bartenschlager and Lohmann 2001). This limitation may be due to a number of factors, including a low replication capacity of the virus to establish persistent infection and tropism of the

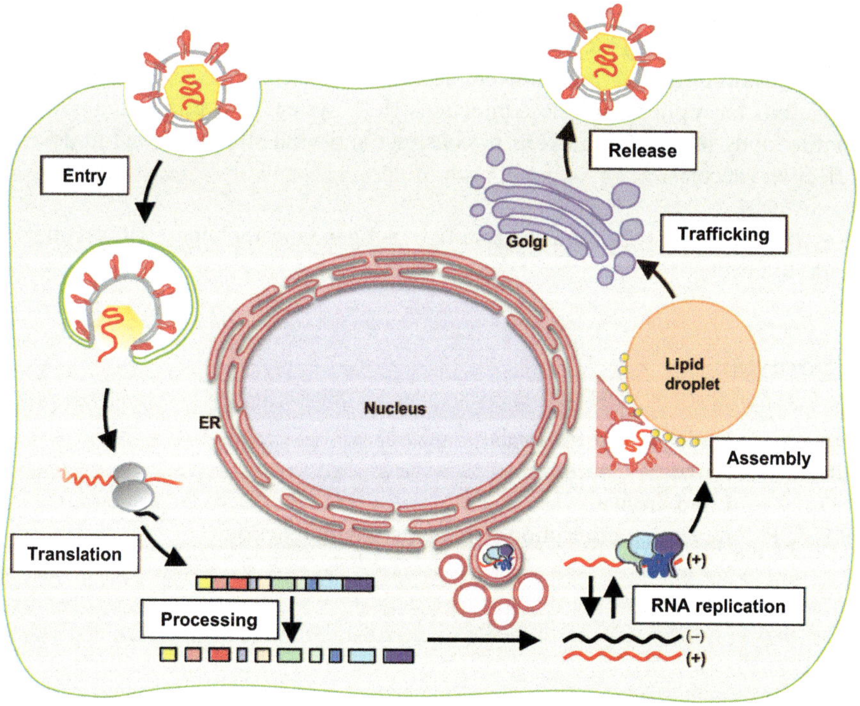

Fig. 1 HCV life cycle. Circulating virus particles bind and then invade the target cells by endocytosis. HCV genomic RNA is released in the cytoplasm and the viral polypeptide is translated. Ten viral proteins are processed from the precursor polypeptide by host and viral proteases. Viral proteins and RNA establish replication complexes that replicate the virus RNA. Replicated virus RNA and structural proteins assemble viral particles for secretion into circulation

virus for highly differentiated hepatocytes. Inoculation of patient sera or plasma into cultured cells results in only a limited level of HCV replication, as determined by nested reverse transcription-polymerase chain reaction (RT-PCR). In 1999, Lohmann et al. were the first to report efficient replication of a HCV subgenomic replicon by replacement of a HCV structural region with a neomycin-resistance gene (Lohmann et al. 1999). Transfection of replicon RNA into Huh7 hepatocellular carcinoma cells, followed by G418 selection culture for several weeks, permitted the establishment of replicons; robust replication of viral RNA was observed in these cells. Adaptive mutations were found in most replicon genomes; these lesions enhanced viral replication to different degrees, and some combinations of these adaptive mutations were observed to strongly enhance replication (Blight et al. 2000; Lohmann et al. 2001; Krieger et al. 2001). This replicon system proved to be a great advantage for the detailed analysis of viral RNA translation and replication mechanisms in cell culture (Fig. 1). Notably, the replicon system accelerated the development of anti-virals targeting the NS3/4a protease and the NS5b RNA polymerase. However, structural genes had been deleted in the construction of the subgenomic replicon genome, meaning that viral particles were not produced from the replicon cells. To establish a HCV infection system, genomic replicons containing a structural region with adaptive mutations in a nonstructural region also were generated; these constructs demonstrated efficient replication in transfected Huh7 cells (Ikeda et al. 2002; Pietschmann et al. 2002; Blight et al. 2003). However, these genomic replicons still did not produce viral particles. Furthermore, a full-length viral RNA genome (harboring adaptive mutations) that had been synthesized in vitro still was not infectious in chimpanzee, in contrast to the wild-type genome (Bukh et al. 2002).

In other work, Kato et al. isolated an infectious clinical HCV strain (JFH-1) from a patient with fulminant hepatitis (Kato et al. 2001). Intriguingly, the nucleotide sequence of JFH-1 was distinct from the sequences of chronic HCV isolates (Kato et al. 2001). Subgenomic replicons constructed using the JFH-1 cDNA were able to replicate with high colony formation efficiency in Huh7 cells, and in other cell lines, without adaptive mutations (Kato et al. 2003; Date et al. 2004). Thus, the original (unmutated) JFH-1 strain appeared to possess higher replication capacities in cultured cells than did other reported HCV strains. Therefore, we tested the JFH-1 strain for full-length viral RNA replication and infectious viral particle formation in cultured cells. When we transfected Huh7 cells with in-vitro-transcribed full-length JFH-1 HCV RNA, the JFH-1 RNA replicated efficiently (Fig. 2, Wakita et al. 2005). Furthermore, the transfected cells produced viral particles that were infectious for cultured cells and in chimpanzee (Wakita et al. 2005). Thus, a robust HCV culture system was established using the JFH-1 clone and permissive cells (Lindenbach et al. 2005; Zhong et al. 2005).

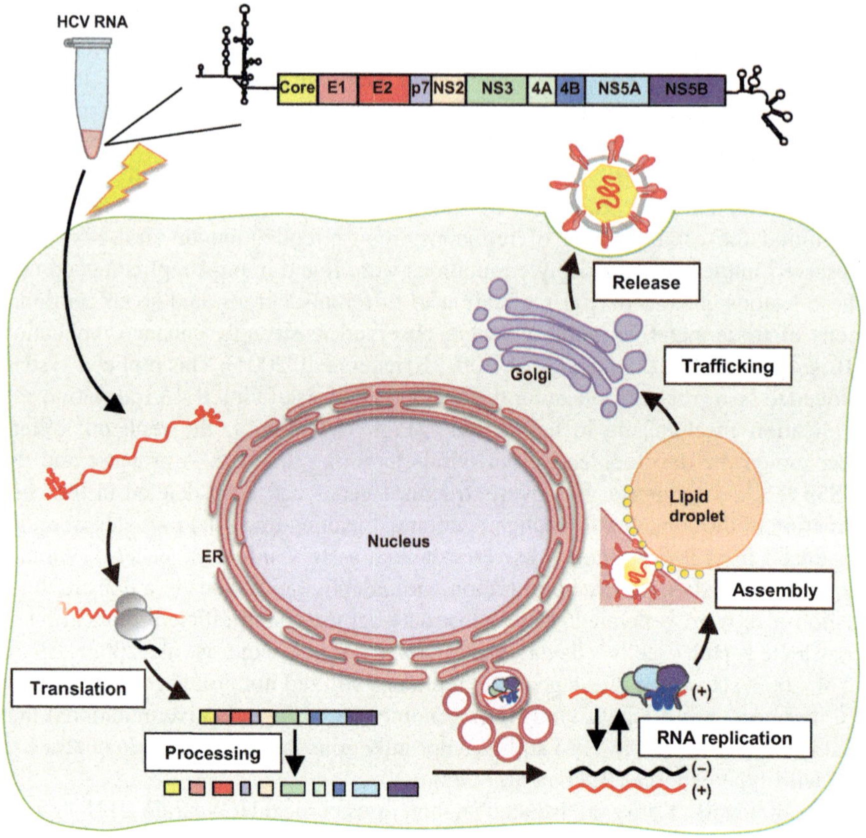

Fig. 2 Infectious HCV production by RNA transfection. In vitro-synthesized HCV RNA from template DNA is transfected into cultured cells. Transfected RNA serves as template RNA to drive HCV life cycle as described in Fig. 1

2 HCV Clones Isolated from Fulminant Hepatitis Patients and HCV Culture

The JFH-1 (Jikei or Japanese Fulminant Hepatitis patient) strain was isolated from a 32-year-old male patient (Kato et al. 2001); a distinct strain (JFH-2) was independently isolated from another 62-year-old male in the different hospitals (Date et al. 2012a). These patients were admitted with acute liver failure. Stage-II encephalopathy developed in both patients after admission, at which point both were diagnosed with fulminant hepatitis. HCV RNA was detected by RT-PCR using sera obtained during the acute phase. Both patients were seronegative for anti-HCV antibody at admission, but subsequently tested HCV seropositive. All viral markers indicating exposure to other hepatitis viruses were negative. These findings suggested that the fulminant hepatitis in each patient was in fact due to

Table 1 HCV replicon and infectious virus in cultured cells

	Replicon	Infectious virus
Genotype	Strain	
1a	H77	H77S, H77S2, H77S3, H77D, TN, HCV-1, H77C
1b	Con1, N, BK, J4, O, AH1, NC1	
2a	JFH-1, JFH-2	JFH-1, JFH-2, J6
2b		J8, DH8, DH10
3a	S310, S52	S310
4a	ED43	
5a	SA1	

HCV infection. To analyze the infecting strain of HCV, nested RT-PCR, as well as 5′-Rapid Amplification of cDNA Ends (5′-RACE) RT-PCR and 3′ RACE RT-PCR, were performed to cover the entire HCV genome. All of the PCR products were cloned and sequenced. Based on sequence analysis, both JFH-1 and JFH-2 belong to genotype 2a, with the nucleotide sequences of both deviating slightly from other genotype-2a clones isolated from patients with chronic hepatitis (Kato et al. 2001; Date et al. 2012b).

Subgenomic replicon and full-length constructs were assembled using cloned PCR fragments (Table 1, Kato et al. 2001; Date et al. 2012a). The colony formation efficiency of the JFH-1 replicon was much greater than that for the JFH-2 replicon. Furthermore, transient transfection of JFH-1 replicon RNA into Huh7 cells resulted in autonomous RNA replication (Kato et al. 2003, 2005a). Adaptive mutations were not necessary for efficient JFH-1 replicon replication in Huh7 cells. In contrast, adaptive mutations were detected in the JFH-2 replicon genome, and some of these mutations were demonstrated to enhance replicon replication. Furthermore, the JFH-1 replicon was observed to replicate in several other cell lines, including HepG2, IMY-N9, HeLa, and 293 human cells, as well as in mouse NIH3T3 fibroblast cells and mouse AML12, MMHD3, and MMH1-1 hepatocytes (Date et al. 2004; Kato et al. 2005b; Uprichard et al. 2006).

JFH-1 demonstrated marked replication efficiency compared to other reported HCV clones. To determine if the JFH-1 strain was capable of producing infectious virus, synthetic full-length JFH-1 RNA was transfected into Huh7 cells (Table 1, Wakita et al. 2005; Kato et al. 2006; Wakita 2009). Indeed, transfected full-length JFH-1 RNA replicated efficiently; secretion of infectious virus particles was confirmed by immunostaining of the infected cells for HCV proteins. In the original experiments, the infection efficiency was less than 0.5 % using standard Huh7 cells. However, this infection was specific, as demonstrated by showing that anti-CD81 or anti-E2 antibodies neutralized infection (Wakita et al. 2005; Zhong et al. 2005). Furthermore, the secreted virus particles were infectious in chimpanzee (Wakita et al. 2005; Kato et al. 2008). These results strongly suggested that the secreted JFH-1 virus particles were authentic HCV. The second fulminant hepatitis isolate, JFH-2, was evaluated in a similar fashion. In contrast to the case with JFH-1, transfection of Huh7 cells with the original (unmutated) full-length JFH-2 RNA did not yield viral replication or virus particle production. We then tested full-length JFH-2 RNA carrying the 2217AS mutation, which was the most efficient adaptive mutation in the replicon experiment. JFH-2 virion production was detected for at least 30 days after transfection. Secreted JFH-2 (2217AS) virus particles exhibited characteristics similar those of JFH-1 virions, with a notable exception: JFH-2 (2217AS) virions lacked in vivo infectivity in human liver-transplanted chimeric mice (Date et al. 2012b). In addition, further adaptive mutations were found in the virus genome of cell culture-adapted JFH-2 (2217AS) virus genome. Thus, the comparison between the JFH-1 and JFH-2 strains demonstrated that isolation of HCV from fulminant hepatitis cases was not the sole determinant of culture-competent phenotype.

In previous reports, adaptive mutations were shown to enhance viral RNA replication at the expense of virus particle formation efficiency (Pietschmann et al. 2002). Specifically, a highly cell culture-adapted Con1 strain replicated in cultured cells, but did not produce infectious virus particles. Interestingly, a highly adapted Con1 strain was not infectious for chimpanzee, while a moderately adapted Con1 strain was infectious. However, the virus recovered from the infected animal had the nucleotide sequence of the wild-type Con1 virus (Bukh et al. 2002). This result clearly suggested that HCV strains with lower replication efficiency were favored for in vivo infection. However, we must note that this "replication efficiency" was determined in cultured cells. In the case of JFH-2, full-length RNA with the most adaptive mutation, 2217AS, was able to replicate and produce virus particles. Secreted virus particles were infectious for naive Huh-7.5.1 cells, but additional adaptive mutations were found in the viral genomic RNA packaged in these virions. Interestingly, the 2695TI mutation in NS5B was shared among all of the adapted virus genomes, and isoleucine at NS5B amino acid residue 2695 also was found in the JFH-1 strain. However, the introduction of the 2695TI mutation alone into the JFH-2 virus genome was not sufficient to restore robust virus production. Based on the comparisons of cell culture-adapted viruses and their parental virus constructs, adaptive mutations are necessary to increase both viral genome replication and virus particle assembly/secretion efficiency.

Thus, a procedure for the production of cell culture-adapted HCV was established. The adaptive mutations identified in the subgenomic replicon assay were introduced into the full-length genomes, and the cells transfected with virus RNA were passaged repeatedly until virus particles were generated.

3 Permissive Cell Lines

Although the infection efficiency of JFH-1 virus in the original Huh7 cells maintained in our laboratory was quite low, efficient viral infection was achieved using "cured" cells, such as Huh7.5, Huh7.5.1, and Lunet/CD81 cells (Zhong et al. 2005; Blight et al. 2002; Koutsoudakis et al. 2007). These cell lines were "cured" by interferon treatment of cells carrying subgenomic replicons. Huh7.5

cells harbor a point mutation in the gene encoding RIG-I, resulting in defective intracellular interferon signaling in response to HCV RNA replication (Sumpter et al. 2005). Transfection of the JFH-1 genome or inoculation of infectious JFH-1 virus into Huh7.5 and Huh7.5.1 cells thus produces robust replication and HCV virus infection (Lindenbach et al. 2005; Zhong et al. 2005). Koutsoudakis et al. have reported that the level of CD81 cell surface expression is a key determinant of HCV infection (Koutsoudakis et al. 2007). In fact, we have observed that the Huh7 cells used in our initial infection assay were a mixture of cell clones with varying levels of CD81 expression and infectivity (Akazawa et al. 2007). Interestingly, we have isolated several Huh7 subclones that lack cell-surface CD81 expression. Among these subclones, Huh7-25 supported highly efficient subgenomic replicon replication. To test the role of CD81, we transfected a CD81 expression vector into these cell clones, yielding stable cells with a high level of ectopic CD81 expression. These cells supported a greater degree of infectivity of the JFH-1 virus than did the original Huh7 cells; infectivity also was elevated compared to the cured cell lines (Akazawa et al. 2007). Like the Huh7.5 cells, Huh7-25 cells were defective in the intracellular interferon induction pathway, but Huh7-25 cells additionally harbored decreased levels of TRIM25, a protein that functions as an ubiquitin ligase for RIG-I (Arnaud et al. 2010, 2011). Thus, high-level infectivity by HCV appears to require both interferon induction upon stimulation with double-strand RNA and cell-surface expression of CD81. Huh7 cells were first isolated more than 20 years ago (Nakabayashi et al. 1982), and since have been distributed worldwide. The phenotype of Huh7 cells, including permissivity for JFH-1 virus infectivity, may vary among the subclones maintained in different laboratories.

4 Other Infectious HCV Clones

It had been difficult to propagate HCV strains other than JFH-1 in cultured cells. A genotype-1a strain, H77S containing five adaptive mutations, has been reported to produce infectious virus after transfection of Huh7 cells with a synthesized RNA, albeit with limited efficiency (Table 1, Yi et al. 2006). H77S.2 and H77S.3, versions derived from the adapted H77S clone, exhibit greater infectivity than the original H77S adapted virus (Table 1, Shimakami et al. 2011). The JFH-1 strain, like J6 (the prototype HCV strain), belongs to genotype 2a. J6CF was established as an HCV clone that was infectious in chimpanzee (Yanagi et al. 1999); however, neither full-length nor subgenomic replicons of J6CF are able to replicate Huh7 cells (Murayama et al. 2007), despite similarity of nucleotide sequence over the entire HCV genome. To address the difference in replicative ability of these two strains, we performed chimeric replicon analysis. The swapping of segments between JFH-1 and J6CF revealed that the NS3 helicase-encoding region and the NS5B-encoding region (up to the 3′X site) were important for efficient replication and virus particle formation of the JFH-1 strain (Murayama et al. 2007). We found that a single point mutation (within the 561st codon of the NS5B-encoding region) was altered both for NS5B polymerase activity and for an RNA structure important for replicon replication (Murayama et al. 2010). Schmitt et al. also focused on structural and functional differences in NS5B between JFH-1 and J6 strains (Schmitt et al. 2011). Their research found that a point mutation at the 405th codon of the NS5B-encoding region was the most important in enhancing RNA synthesis. Recently, Li et al. reported the introduction of three mutations (LSG mutations; located in the NS3-, NS4A- and NS5B-encoding loci) permitted propagation of full-length J6 and J8 (genotype 2b) and adaptation, with infectivity titers comparable to those obtained in the JFH-1 system (Table 1, Li et al. 2012a). These researchers also developed highly infectious versions of TN, HCV-1, and H77C strains (genotype 1a) and of DH8 and DH10 strains (genotype 2b) based on the LSG substitutions (Table 1, Li et al. 2012b, 2015; Ramirez et al. 2014).

5 Genotype-3a Infection System

HCV is classified into seven major genotypes. Genotype 1b is most prevalent in Asian countries, followed by genotype 3a. A high incidence of hepatic steatosis is associated with clinical infection by genotype 3a. Combination therapy with interferon and ribavirin is more effective in treating patients infected with genotype 3a than those infected with genotype 1b (Hui et al. 2002); in contrast, protease inhibitors like telaprevir and boceprevir are less effective against genotype-3a infections (Gottwein et al. 2011). Thus, it is important to study the pathogenesis of genotype-3 HCV and to assess this genotype's susceptibility to different antiviral inhibitors. We first identified the S310 strain as a full-length genotype-3a HCV cDNA from a 71-year-old female patient suffering from post-transplantation recurrence of HCV infection. This patient had been diagnosed with HCV genotype-3a infection at the age of 59 years and received a liver transplant 4 years later due to liver cirrhosis. A subgenomic replicon of S310 yielded drug-resistant replicon colonies; sequencing of the resulting replicons revealed multiple independent adaptive mutations (T2186I; T2188A; R2198H; T2496I; R2895K; R2895G; and T2496I+R2895G) (Table 1, Saeed et al. 2013). Each of these mutations, as well as the S2210I mutation, was introduced into the full-length S310 HCV cDNA constructs. When each of the synthesized full-length S310 RNAs with adaptive mutations were transfected into Huh7.5.1 cells, the transfectants produced infectious virus in cell culture medium (Table 1, Kim et al. 2014). Virus particles secreted into the culture medium were found to be very similar to other cell culture-adapted viruses. We also compared anti-HCV drug susceptibility between cells infected with S310 and JFH-1. Notably, the protease inhibitor telaprevir appeared to be less effective against S310 infection compared to JFH-1 infection, consistent with results previously reported by replicon assay. In contrast, an NS5A inhibitor (PSI-6130), an NS5B inhibitor (BMS-790052), and host-targeting drugs exhibited similar levels of inhibition of infection by the two viral genotypes.

Genotype-3a HCV infections can result in hepatic steatosis (Hui et al. 2002). In addition, HCV core protein accumulates on the surface of lipid droplets (LDs) in infected cells, suggesting that the LD surface is an important site for HCV particle formation (Miyanari et al. 2007). The LD is an organelle used for the storage of neutral lipids. Interestingly, the lipid content in S310 virus-infected cells was higher than that of JFH-1 virus-infected cells and that of uninfected Huh7.5.1 cells, as assessed by LD staining area. This in-vitro observation was consistent with the in-vivo observation, in the original S310-infected patient, of microvesicular and macrovesicular steatosis both before and after liver transplantation. Hepatic steatosis might reflect the induction of lipid synthesis by activation (during infection by genotype-3a HCV) of SREBP-1 and PPARγ, or via the production of reactive oxygen species (Gavrilova et al. 2003; Videla et al. 2004; Ma et al. 2012). Alternatively, genotype-3a infection may impair lipid secretion and degradation by decreasing the activities of MTP and PPARα (Perlemuter et al. 2002; Mirandola et al. 2006).

Cells expressing genotype-3a core protein were used to study the genotype-3a HCV core protein association with steatosis (Jhaveri et al. 2008, 2009; Qiang and Jhaveri 2012). Expression of HCV genotype-3a core protein resulted in up-regulation of the promoter for the fatty acid synthase-encoding gene (Jackel-Cram et al. 2007). Domain 3 of the HCV core protein was sufficient for lipid accumulation, and specific polymorphisms in the genotype-3a HCV core protein increased cellular lipid levels, contributing to steatosis in cultured cells (Jhaveri et al. 2008). However, these studies used only core protein expression, so these experiments did not include the effects of other viral proteins or those of the viral life cycle. Furthermore, HCV core protein sublocalization on the LD surface is important for infectious virus particle formation (Miyanari et al. 2007). We therefore analyzed core protein and LD sublocalization using the genotype-3a infection system.

In cells expressing S310-derived core proteins, the core proteins stained with a punctate pattern, in contrast to the ring-like staining pattern seen when staining for core protein in JFH-1-infected cells. We further examined whether S310 infection resulted in the accumulation of LDs in a long term culture of infected cells. Interestingly, S310 virus–infected cells exhibited greater LD accumulation than uninfected or JFH-1 virus–infected Huh7.5.1 cells, suggesting that the S310-derived structural region was important for LD accumulation in S310-infected cells. Using real-time PCR, we assessed S310-infected cells for transcription of genes encoding proteins important for cellular lipid metabolism, including MTP, PPARα, and SREBP-1c. However, no differences in mRNA expression levels were found for these genes when comparing between infected and uninfected Huh7.5.1 cells.

5.1 Genotype-1b Infection System

As noted above (Table 1), subgenomic replicons are now available for HCV of genotypes 1a/b, 2a, 3a, 4a, and 5a, and virus culture systems are now available for HCV of genotypes 1a, 2a/b, and 3a. However, a robust virus culture system is still not available for genotype-1b HCV, although genotype 1b is the major subtype in Japan and many other countries. To address this shortcoming, we isolated a HCV cDNA (designated NC1) from a patient with acute severe hepatitis (Date et al. 2012b). A NC1 subgenomic replicon experiment identified several mutations that enhanced the colony formation efficiency of the NC1 replicon. Interestingly, the full-length NC1 genome with these adaptive mutations replicated in cultured cells and produced infectious virus particles. However, the viral infection efficiency was not sufficient for autonomous virus propagation in cultured cells, even though virus production efficiency was increased by the introduction of multiple mutations into the NC1 viral genome. In separate work, we also are attempting the establishment of a genotype-4a virus culture system; however, genotype-4a HCV thus far has shown only transient virus production after transfection with virus RNA. Clinically, both genotype-1b and -4a HCV infections are resistant to interferon therapy. We hypothesize that there may be some relation between interferon resistance and cell culture incompetence.

6 Viral Life Cycle

In the host, circulating HCV virions encounter and bind to receptor molecules on the surface of target cells, followed by invasion of the host cells by endocytosis. The endocytosed particle releases HCV genomic RNA in the cytoplasm, where the genome acts as a messenger RNA for viral protein translation. The resulting virus proteins combine with the RNA genome to establish a replication complex that replicates the viral RNA. The replicated viral RNA and structural proteins then form virus particles that are secreted from the surface of the infected cell. This series of steps is called the viral life cycle (Fig. 1). HCV RNA replication occurs at intracellular membrane structures, yielding an altered membrane structure ("membranous web") in the replicon cells when observed by electron microscopy (Gosert et al. 2003). This membranous web phenotype also is seen upon expression of the NS4B protein alone (Egger et al. 2002), and is observed in JFH-1 virus-infected cells along with increased LD volume (Rouillé et al. 2006). In a previous study, we found that transiently expressed core protein accumulated on the surface of the LDs in human osteosarcoma cells (Moradour et al. 1996), an observation confirmed by other HCV researchers (McLauchlan et al. 2002). In JFH-1 virus-infected Huh7 cells, LDs were surrounded by core proteins and further surrounded by NS proteins (Miyanari et al. 2007). Thus, each step of the viral life cycle of HCV has been the subject of detailed molecular analysis. Additionally, HCV has been shown to hijack a number of host factors for the reproduction of the virus in infected cells. It is thus possible that novel functions of HCV-related host factors will be uncovered from a detailed analysis of the HCV life cycle. These studies are expected to lead to development of novel anti-viral drugs and vaccines.

7 Conclusions

A robust HCV infection system was first reported in 2005; that isolate, the JFH-1 strain, remains the only HCV strain that can be propagated in cultured (Huh7) cells in the absence of adaptive mutations. This system has been useful for multiple purposes in the HCV research field, including the development of antivirals and vaccines, as well as improved characterization of the HCV viral life cycle. Infection systems for HCV of other genotypes have not yet progressed to this level. Future studies should be directed to the establishment of robust infectivity, especially for genotype-1b and -4 HCV clones.

Acknowledgement The author thanks Dr. Senko Tsukuda for her contribution for illustration. This work was supported by Grants-in-Aid for Scientific Research 24115003 from the Japan Society for the Promotion of Science, and by Grants-in-Aid for Scientific Research from the Ministry of Health, Labour and Welfare of Japan.

References

Akazawa D, Date T, Morikawa K, Murayama A, Miyamoto M, Kaga M, Barth H, Baumert TF, Dubuisson J, Wakita T (2007) CD81 expression is important for heterogeneous HCV permissiveness of Huh7 cell clones. J Virol 81(10):5036–5045

Arnaud N, Dabo S, Maillard P, Budkowska A, Kalliampakou KI, Mavromara P, Garcin D, Hugon J, Gatignol A, Akazawa D, Wakita T, Meurs EF (2010) Hepatitis C virus controls interferon production through PKR activation. PLoS One 5(5):e10575

Arnaud N, Dabo S, Akazawa D, Fukasawa M, Shinkai-Ouchi F, Hugon J, Wakita T, Meurs EF (2011) Hepatitis C virus reveals a novel early control in acute immune response. PLoS Pathog 7(10):e1002289

Bartenschlager R, Lohmann V (2001) Novel cell culture systems for the hepatitis C virus. Antiviral Res 52:1–17

Blight KJ, Kolykhalov AA, Rice CM (2000) Efficient initiation of HCV RNA replication in cell culture. Science 290:1972–1974

Blight KJ, McKeating JA, Rice CM (2002) Highly permissive cell lines for subgenomic and genomic hepatitis C virus RNA replication. J Virol 76:13001–13014

Blight KJ, McKeating JA, Marcotrigiano J, Rice CM (2003) Efficient replication of hepatitis C virus genotype 1a RNAs in cell culture. J Virol 77:3181–3190

Bukh J, Pietschmann T, Lohmann V, Krieger N, Faulk K, Engle RE, Govindarajan S, Shapiro M, St. Claire M, Bartenschlager R (2002) Mutations that permit efficient replication of hepatitis C virus RNA in Huh-7 cells prevent productive replication in chimpanzees. Proc Natl Acad Sci U S A 99:14416–14421

Choo QL, Kuo G, Weiner AJ, Overby LR, Bradley DW, Houghton M (1989) Isolation of a cDNA clone derived from a blood-borne non-A, non-B viral hepatitis genome. Science 244 (4902):359–362

Date T, Kato T, Miyamoto M, Zhao Z, Yasui K, Mizokami M, Wakita T (2004) Genotype 2a hepatitis C virus subgenomic replicon can replicate in HepG2 and IMY-N9 cells. J Biol Chem 279:22371–22376

Date T, Kato T, Kato J, Takahashi H, Morikawa K, Akazawa D, Murayama A, Tanaka-Kaneko K, Sata T, Tanaka Y, Mizokami M, Wakita T (2012a) Novel cell culture-adapted genotype 2a hepatitis C virus infectious clone. J Virol 86(19):10805–10820

Date T, Morikawa K, Tanaka Y, Tanaka-Kaneko K, Sata T, Mizokami M, Wakita T (2012b) Replication and infectivity of a novel genotype 1b hepatitis C virus clone. Microbiol Immunol 56(5):308–317

Egger D, Wölk B, Gosert R, Bianchi L, Blum HE, Moradpour D, Bienz K (2002) Expression of hepatitis C virus proteins induces distinct membrane alterations including a candidate viral replication complex. J Virol 76(12):5974–5984

Gavrilova O, Haluzik M, Matsusue K, Cutson JJ, Johnson L, Dietz KR, Nicol CJ et al (2003) Liver peroxisome proliferator-activated receptor gamma contributes to hepatic steatosis, triglyceride clearance, and regulation of body fat mass. J Biol Chem 278:34268–34276

Gosert R, Egger D, Lohmann V, Bartenschlager R, Blum HE, Bienz K, Moradpour D (2003) Identification of the hepatitis C virus RNA replication complex in Huh-7 cells harboring subgenomic replicons. J Virol 77(9):5487–5492

Gottwein JM, Scheel TK, Jensen TB, Ghanem L, Bukh J (2011) Differential efficacy of protease inhibitors against HCV genotypes 2a, 3a, 5a, and 6a NS3/4A protease recombinant viruses. Gastroenterology 141:1067–1079

Hui JM, Kench J, Farrell GC, Lin R, Samarasinghe D, Liddle C, Byth K et al (2002) Genotype-specific mechanisms for hepatic steatosis in chronic hepatitis C infection. J Gastroenterol Hepatol 17:873–881

Ikeda M, Yi M, Li K, Lemon SM (2002) Selectable subgenomic and genome-length dicistronic RNAs derived from an infectious molecular clone of the HCV-N strain of hepatitis C virus replicate efficiently in cultured Huh7 cells. J Virol 76(6):2997–3006

Jackel-Cram C, Babiuk LA, Liu Q (2007) Up-regulation of fatty acid synthase promoter by hepatitis C virus core protein: genotype-3a core has a stronger effect than genotype-1b core. J Hepatol 46:999–1008

Jhaveri R, McHutchison J, Patel K, Qiang G, Diehl AM (2008) Specific polymorphisms in hepatitis C virus genotype 3 core protein associated with intracellular lipid accumulation. J Infect Dis 197:283–291

Jhaveri R, Qiang G, Diehl AM (2009) Domain 3 of hepatitis C virus core protein is sufficient for intracellular lipid accumulation. J Infect Dis 200:1781–1788

Kato T, Furusaka A, Miyamoto M, Date T, Yasui K, Hiramoto J, Nagayama K, Tanaka T, Wakita T (2001) Sequence analysis of hepatitis C virus isolated from a fulminant hepatitis patient. J Med Virol 64:334–339

Kato T, Date T, Miyamoto M, Furusaka A, Tokushige K, Mizokami M, Wakita T (2003) Efficient replication of the genotype 2a hepatitis C virus subgenomic replicon. Gastroenterology 125:1808–1817

Kato T, Date T, Miyamoto M, Sugiyama M, Tanaka Y, Orito E, Ohno T, Sugihara K, Hasegawa I, Fujiwara K, Ito K, Ozasa A, Mizokami M, Wakita T (2005a) Detection of anti-hepatitis C virus effects of interferon and ribavirin by a sensitive replicon system. J Clin Microbiol 43:5679–5684

Kato T, Date T, Miyamoto M, Zhao Z, Mizokami M, Wakita T (2005b) Non-hepatic cell lines HeLa and 293 cells support efficient replication of hepatitis C virus genotype 2a subgenomic replicon. J Virol 79:592–596

Kato T, Date T, Murayama A, Morikawa K, Akazawa D, Wakita T (2006) Cell culture and infection system for hepatitis C virus. Nat Protoc 1(5):2334–2339

Kato T, Choi Y, Elmowalid G, Sapp RK, Barth H, Furusaka A, Mishiro S, Wakita T, Krawczynski K, Liang TJ (2008) Hepatitis C virus JFH-1 strain infection in chimpanzees is associated with low pathogenicity and emergence of an adaptive mutation. Hepatology 48 (3):732–740

Kim S, Date T, Yokokawa H, Kono T, Aizaki H, Maurel P, Gondeau C, Wakita T (2014) Development of hepatitis C virus genotype 3a cell culture system. Hepatology 60 (6):1838–1850

Koutsoudakis G, Herrmann E, Kallis S, Bartenschlager R, Pietschmann T (2007) The level of CD81 cell surface expression is a key determinant for productive entry of hepatitis C virus into host cells. J Virol 81:588–598

Krieger N, Lohmann V, Bartenschlager R (2001) Enhancement of hepatitis C virus RNA replication by cell culture-adaptive mutations. J Virol 75:4614–4624

Kuo G, Choo QL, Alter HJ, Gitnick GL, Redeker AG, Purcell RH, Miyamura T, Dienstag JL, Alter MJ, Stevens CE et al (1989) An assay for circulating antibodies to a major etiologic virus of human non-A, non-B hepatitis. Science 244(4902):362–364

Li YP, Ramirez S, Gottwein JM, Scheel TK, Mikkelsen L, Purcell RH, Bukh J (2012a) Robust full-length hepatitis C virus genotype 2a and 2b infectious cultures using mutations identified by a systematic approach applicable to patient strains. Proc Natl Acad Sci U S A 109(18):E1101–E1110

Li YP, Ramirez S, Jensen SB, Purcell RH, Gottwein JM, Bukh J (2012b) Highly efficient full-length hepatitis C virus genotype 1 (strain TN) infectious culture system. Proc Natl Acad Sci U S A 109(48):19757–19762

Li YP, Ramirez S, Mikkelsen L, Bukh J (2015) Efficient infectious cell culture systems of the hepatitis C virus (HCV) prototype strains HCV-1 and H77. J Virol 89(1):811–823

Lindenbach BD, Evans MJ, Syder AJ, Wolk B, Tellinghuisen TL, Liu CC, Maruyama T, Hynes RO, Burton DR, McKeating JA, Rice CM (2005) Complete replication of hepatitis C virus in cell culture. Science 309:623–626

Lohmann V, Körner F, Koch J, Herian U, Theilmann L, Bartenschlager R (1999) Replication of subgenomic hepatitis C virus RNAs in a hepatoma cell line. Science 285(5424):110–113

Lohmann V, Korner F, Dobierzewska A, Bartenschlager R (2001) Mutations in hepatitis C virus RNAs conferring cell culture adaptation. J Virol 75:1437–1449

Ma S, Yang D, Li D, Tan Y, Tang B, Yang Y (2012) Inhibition of uncoupling protein 2 with genipin exacerbates palmitate-induced hepatic steatosis. Lipids Health Dis 11:154

McLauchlan J, Lemberg MK, Hope G, Martoglio B (2002) Intramembrane proteolysis promotes trafficking of hepatitis C virus core protein to lipid droplets. EMBO J 21(15):3980–3988

Mirandola S, Realdon S, Iqbal J, Gerotto M, Dal Pero F, Bortoletto G, Marcolongo M et al (2006) Liver microsomal triglyceride transfer protein is involved in hepatitis C liver steatosis. Gastroenterology 130:1661–1669

Miyanari Y, Atsuzawa K, Usuda N, Watashi K, Hishiki T, Zayas M, Bartenschlager R, Wakita T, Hijikata M, Shimotohno K (2007) The lipid droplet is an important organelle for hepatitis C virus production. Nat Cell Biol 9(9):1089–1097

Moradpour D, Englert C, Wakita T, Wands JR (1996) Characterization of cell lines allowing tightly regulated expression of hepatitis C virus core protein. Virology 222(1):51–63

Murayama A, Date T, Morikawa K, Akazawa D, Miyamoto M, Kaga M, Ishii K, Suzuki T, Kato T, Mizokami M, Wakita T (2007) The NS3 helicase and NS5B-to-3′X regions are important for efficient hepatitis C virus strain JFH-1 replication in Huh7 cells. J Virol 81(15):8030–8040

Murayama A, Weng L, Date T, Akazawa D, Gao Y, Chen W, Suzuki T, Kato T, Mizokami M, Wakita T, Toyoda T (2010) Specific RNA structures and mutations implicated for HCV RNA replication and virus particle formation in cultured cells. PLoS Pathog 6(4):e1000885

Nakabayashi H, Taketa K, Miyano K, Yamane T, Sato J (1982) Growth of human hepatoma cells lines with differentiated functions in chemically defined medium. Cancer Res 42:3858–3863

Perlemuter G, Sabile A, Letteron P, Vona G, Topilco A, Chretien Y, Koike K et al (2002) Hepatitis C virus core protein inhibits microsomal triglyceride transfer protein activity and very low density lipoprotein secretion: a model of viral-related steatosis. FASEB J 16:185–194

Pietschmann T, Lohmann V, Kaul A, Krieger N, Rinck G, Rutter G, Strand D, Bartenschlager R (2002) Persistent and transient replication of full-length hepatitis C virus genomes in cell culture. J Virol 76(8):4008–4021

Qiang G, Jhaveri R (2012) Lipid droplet binding of hepatitis C virus core protein genotype 3. ISRN Gastroenterol 2012:176728

Ramirez S, Li YP, Jensen SB, Pedersen J, Gottwein JM, Bukh J (2014) Highly efficient infectious cell culture of three hepatitis C virus genotype 2b strains and sensitivity to lead protease, nonstructural protein 5A, and polymerase inhibitors. Hepatology 59(2):395–407

Rouillé Y, Helle F, Delgrange D, Roingeard P, Voisset C, Blanchard E, Belouzard S, McKeating J, Patel AH, Maertens G, Wakita T, Wychowski C, Dubuisson J (2006) Subcellular localization of hepatitis C virus structural proteins in a cell culture system that efficiently replicates the virus. J Virol 80(6):2832–2841

Saeed M, Gondeau C, Hmwe S, Yokokawa H, Date T, Suzuki T, Kato T, Maurel P, Wakita T (2013) Replication of hepatitis C virus genotype 3a in cultured cells. Gastroenterology 144 (1):56–58.e7

Schmitt M, Scrima N, Radujkovic D, Caillet-Saguy C, Simister PC, Friebe P, Wicht O, Klein R, Bartenschlager R, Lohmann V, Bressanelli S (2011) A comprehensive structure-function comparison of hepatitis C virus strain JFH1 and J6 polymerases reveals a key residue stimulating replication in cell culture across genotypes. J Virol 85(6):2565–2581

Shimakami T, Welsch C, Yamane D, McGivern DR, Yi M, Zeuzem S, Lemon SM (2011) Protease inhibitor-resistant hepatitis C virus mutants with reduced fitness from impaired production of infectious virus. Gastroenterology 140(2):667–675

Sumpter R Jr, Loo YM, Foy E, Li K, Yoneyama M, Fujita T, Lemon SM, Gale M Jr (2005) Regulating intracellular antiviral defense and permissiveness to hepatitis C virus RNA replication through a cellular RNA helicase. RIG-I J Virol 79:2689–2699

Uprichard SL, Chung J, Chisari FV, Wakita T (2006) Replication of a hepatitis C virus replicon clone in mouse cells. Virol J 3:89

Videla LA, Rodrigo R, Orellana M, Fernandez V, Tapia G, Quinones L, Varela N et al (2004) Oxidative stress-related parameters in the liver of non-alcoholic fatty liver disease patients. Clin Sci (Lond) 106:261–268

Wakita T (2009) Isolation of JFH-1 strain and development of an HCV infection system. Methods Mol Biol 510:305–327

Wakita T, Pietschmann T, Kato T, Date T, Miyamoto M, Zhao Z, Murthy K, Habermann A, Krausslich HG, Mizokami M, Bartenschlager R, Liang TJ (2005) Production of infectious hepatitis C virus in tissue culture from a cloned viral genome. Nat Med 11:791–796

Yanagi M, Purcell RH, Emerson SU, Bukh J (1999) Hepatitis C virus: an infectious molecular clone of a second major genotype (2a) and lack of viability of intertypic 1a and 2a chimeras. Virology 262(1):250–263

Yi M, Villanueva RA, Thomas DL, Wakita T, Lemon SM (2006) Production of infectious genotype 1a hepatitis C virus (Hutchinson strain) in cultured human hepatoma cells. Proc Natl Acad Sci U S A 103:2310–2315

Zhong J, Gastaminza P, Cheng G, Kapadia S, Kato T, Burton DR, Wieland SF, Uprichard SL, Wakita T, Chisari FV (2005) Robust hepatitis C virus infection in vitro. Proc Natl Acad Sci U S A 102:9294–9299

HCV Receptors and Virus Entry

Rajiv G. Tawar, Catherine Schuster, and Thomas F. Baumert

Abstract Hepatitis C virus (HCV) enters its target cells in a tightly regulated process that relies on several host factors. These host factors can be classified into three distinct categories depending on their role in the entry process, namely attachment factors, entry factors and facilitators. HCV entry is the target of host neutralizing anti-HCV antibody response. Furthermore, HCV entry is the first step of virus-host interaction and thus, in-part determines the hepato- and species tropism of the virus. Understanding HCV entry is, thus, not only crucial for development of novel entry-inhibitors and vaccine but also for developing new animal models to evaluate novel therapeutics and study disease pathogenesis. Research in the last 25 years has given unprecedented insights into the molecular aspects of HCV entry and its intricate association with the host. This review summarizes our current knowledge of the host factors involved in HCV entry and the molecular mechanisms determining the entry process.

Keywords HCV • Viral entry • Receptors • Co-receptor complex • Virus-host interactions

Abbreviations

Apo	Apolipoprotein
CD81	Cluster of differentiation 81

R.G. Tawar • C. Schuster
Inserm U1110, Institute de Recherche sur les Maladies Virales et Hépatiques, 3 rue Koeberlé, 67000 Strasbourg, France

Université de Strasbourg, Strasbourg, France

T.F. Baumert, M.D. (✉)
Inserm U1110, Institute de Recherche sur les Maladies Virales et Hépatiques, 3 rue Koeberlé, 67000 Strasbourg, France

Université de Strasbourg, Strasbourg, France

Pôle Hépato-digestif, Institut Hospitalo-Universitaire, Strasbourg University Hospitals, Strasbourg, France
e-mail: Thomas.Baumert@unistra.fr

T. Miyamura et al. (eds.), *Hepatitis C Virus I*, DOI 10.1007/978-4-431-56098-2_5

CLDN	Claudin
EGFR	Epidermal growth factor receptor
EphA2	Ephrin type-A receptor 2
HCV	Hepatitis C virus
HCVcc	Cell culture-derived HCV
HCV-LP	HCV-like particles
HCVpp	HCV pseudoparticles
HDL	High density lipoprotein
HRas	GTPase Harvey rat sarcoma viral oncogene homolog
HS	Heparan sulfate
HSPG	Heparan sulfate glycoprotein
HVRI	Hypervariable region I
IRES	Internal ribosomal entry site
LDL	Low density lipoprotein
LDL-R	Low density lipoprotein receptor
LEL	Large extracellular loop
LVP	Lipo-viral particle
NPC1L1	Niemann-Pick C1-like 1
OCLN	Occludin
PKA	Protein kinase A
SDC	Syndecan
SEL	Small extracellular loop
SR-BI	Scavenger receptor class B type I
TEM	Tetraspanin enriched microdomain
TfR1	Transferrin receptor protein 1
UTR	Untranslated region
VLDL	Very low density lipoprotein

1 Introduction

Hepatitis C virus (HCV) belongs to the *flaviviridae* family of viruses, further subclassified into the genus *hepacivirus*. HCV is a single-stranded positive strand enveloped RNA virus, which infects mainly the human hepatocytes. The ~9.6 Kb genome is flanked at 5′ and 3′ ends by highly conserved untranslated regions (UTR), which mediate translation of viral proteins and replication of the viral genome, respectively. HCV genome translation depends on an internal ribosomal entry site (IRES) and yields a ~3,000 amino acid polyprotein. The polyprotein is cleaved by host and viral proteases resulting in ten mature viral proteins. This includes three structural proteins (the capsid protein core and the envelop glycoproteins E1 and E2), a small integral transmembrane protein (p7) and six non-structural proteins (NS2, NS3, NS4A, NS4B, NS5A and NS5B).

HCV, like all viruses, relies on the host for its propagation and is thus required to enter the target cells. To facilitate entry, it has evolved a complex, highly orchestrated mechanism that relies on its envelope glycoproteins E1 and E2 and several host factors. A unique feature of HCV is its existence as lipoviral particles in the infected individual (Miyamoto et al. 1992; Thomssen et al. 1992, 1993; Andre et al. 2002). This association of virions with host lipoproteins not only allows the virus to evade host immune responses (Grove et al. 2008; Tao et al. 2009) but also plays a critical role in viral entry through the interaction of viral-bound host-derived apolipoproteins and host factors (see Sect. 2).

In hepatic cells, HCV persists by cell-free and cell-cell transmission, both depending on viral envelope glycoproteins E1 and E2 and several host cell entry factors, including CD81, scavenger receptor class B type 1 (SR-BI), claudin-1 (CLDN1), occludin (OCLN), epidermal growth factor receptor (EGFR) and GTPase Harvey rat sarcoma viral oncogene homolog (HRas) (Timpe et al. 2008; Brimacombe et al. 2011; Lupberger et al. 2011; Zona et al. 2013; Xiao et al. 2014). While the molecular mechanisms of early steps of cell-free entry have been extensively studied (Fig. 1; for a review see Zeisel et al. 2013), the late steps leading to fusion are poorly understood. The interaction of envelope glycoproteins E1 and E2 with specific cellular receptors lead to endocytosis of HCV particles via clathrin-coated vesicles and has been shown to involve dynamin (Blanchard et al. 2006; Meertens et al. 2006; Farquhar et al. 2012). Moreover, the role of specialized CD81 associated tetraspanin enriched microdomains (TEM) in HCV endocytosis has been described (Brazzoli et al. 2008; Farquhar et al. 2012; reviewed in Zona et al. 2014). HCV is subsequently delivered to early endosomes (Meertens et al. 2006; Coller et al. 2009) where low pH likely triggers conformational changes in the viral envelope glycoproteins leading to the fusion of viral and host membranes (Keck et al. 2005). That HCV fusion depends on low pH has been demonstrated using HCVpp (Bartosch et al. 2003a; Hsu et al. 2003; Codran et al. 2006; Kobayashi et al. 2006; Lavillette et al. 2006; Meertens et al. 2006) as well as HCVcc (Blanchard et al. 2006; Koutsoudakis et al. 2006; Haid et al. 2009). Enveloped viruses mediate fusion using their glycoproteins present on the surface of the viral particle. In case of HCV, both glycoproteins, E1 and E2, have been implicated in viral fusion (Takikawa et al. 2000; Drummer et al. 2007; Haid et al. 2009). Indeed, neutralizing antibody responses in HCV infected patients have been shown to target membrane fusion (Haberstroh et al. 2008). The recently reported high resolution crystal structure of HCV core E2 indicates that E2 is unlikely to have a role in membrane fusion (Kong et al. 2013; Khan et al. 2014). However, it is worth noting that the crystal structure lacks a significant portion of E2 ectodomain, which could potentially harbor a fusion peptide. Equally likely is the possibility that E1 or E1E2 are involved in the fusion process.

In addition to the cell-free HCV entry described above, the assembled viral particles can also be directly transmitted from an infected cell to an adjacent cell referred to as cell-to-cell transmission (Timpe et al. 2008; Brimacombe et al. 2011; Xiao et al. 2014). Viral persistence and dissemination in infected host is thought to be mainly mediated by cell-to-cell transmission. Indeed, recent reports showing the presence of clusters of infected hepatocytes in liver tissue from HCV infected individuals indicate that cell-to-cell spread is relevant in the human liver mode of

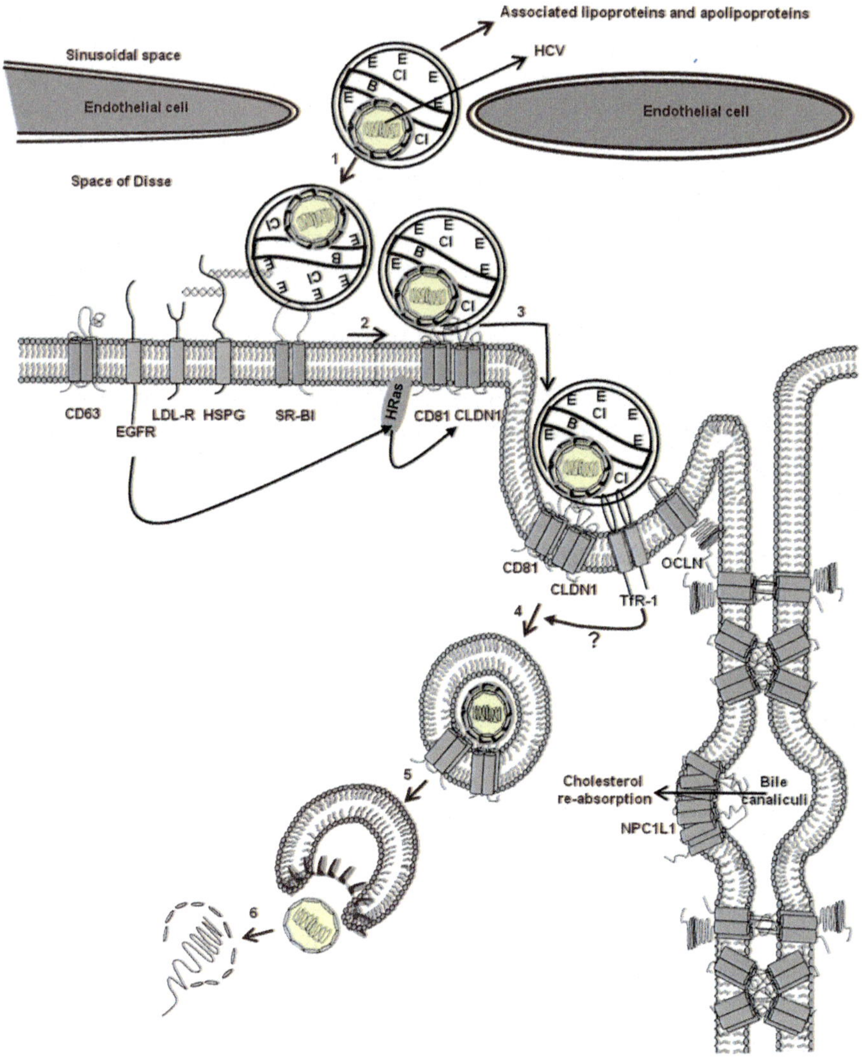

Fig. 1 Putative model of HCV receptors and their role in virus entry. In infected patients, HCV circulates as hybrid particles associated with lipoproteins e.g. VLDL and apolipoproteins including apoE (*E*), apoB (*B*) and apoCI (*CI*). These hybrid HCV particles are termed as lipoviral particles (*LVPs*). The LVPs access hepatocyte by entering the space of Dissé through the fenestrated endothelial cells lining the sinusoidal space. HCV enters the hepatocyte through a complex highly regulated process, which can be classified into three distinct steps: attachment, entry and fusion. Accordingly, the incoming virus particle first attaches to cell-surface expressed HSPG, LDL-R and SR-BI via apoE present on the LVP (to HSPG, LDL-R and SR-BI), thereby anchoring itself on the hepatocyte leading to concentration of the viral particles at the cell surface (step 1). This initial attachment primes HCV for entry into the hepatocyte by inducing conformational changes in the viral envelope glycoproteins and/or by bringing the LVP in close proximity to the host entry factors namely CD81 (step 2). Viral binding to CD81 activates EGFR signalling inducing the formation of CD81-CLDN1 co-receptor complex via GTPase HRas (step 2). In addition, engagement of CD81 by HCV activates Rho GTPases leading to lateral movement of the viral-receptor complex, thereby involving OCLN and TfR1 in the very late step of entry

transmission (Liang et al. 2009; Wieland et al. 2014). Although, E2 has been shown to be involved in cell-to-cell transmission of HCV, most of the patient-derived neutralizing anti-E2 antibodies reported so far are inefficient in blocking viral spread (Brimacombe et al. 2011). The underlying molecular mechanisms are unclear; however, it is possible that E2 adopts a different conformation during cell-to-cell transmission, which makes the epitope inaccessible.

The discovery of HCV host entry factors has paralleled advances in studying HCV life-cycle. The first model system to study HCV entry was based on use of recombinant soluble HCV E2 glycoprotein. This led to the identification of CD81 and SR-BI as host entry factors that could bind HCV at the surface of human hepatocellular carcinoma cells (Pileri et al. 1998; Scarselli et al. 2002). Subsequent development of surrogate HCV particle model system namely, hepatitis C-virus-like particles (HCV-LPs) identified cell surface heparan sulfate as an important host factor for mediating HCV binding to the target cells (Barth et al. 2003). Finally, the advent of HCVpp system consisting of ectopically expressed HCV envelope glycoproteins on retroviral particles (Bartosch et al. 2003a, b; Cormier et al. 2004; Zhang et al. 2004; Barth et al. 2006) and HCVcc (cell-culture derived HCV) system confirmed the role of CD81, SR-BI and heparin sulfate proteoglycans (HSPGs) in HCV entry (Heller et al. 2005; Lindenbach et al. 2005; Wakita et al. 2005; Zeisel et al. 2007; Dreux et al. 2009; Catanese et al. 2010; Jiang et al. 2013; Zahid et al. 2013; Lefevre et al. 2014) and helped uncover roles of several other host factors crucial for HCV entry namely, claudin 1, occludin, receptor tyrosine kinases – epidermal growth factor receptor (EGFR) and ephrin type-A receptor 2 (EphA2), Niemann-Pick C1-like 1 (NPC1L1), transferrin receptor 1 (TfR1), CD63 and Harvey rat sarcoma viral oncogene homolog (HRas) (Evans et al. 2007; Ploss et al. 2009; Lupberger et al. 2011; Sainz et al. 2012; Martin and Uprichard 2013; Park et al. 2013; Zona et al. 2013).

HCV receptors/entry factors can be broadly classified into three major categories based on their involvement in viral entry. This includes attachment factors, entry factors and facilitators. Thus several "attachment factors" help dock the virus on the cell surface, mostly through non-specific charge based interactions. Other host factors mediate specific interaction with viral glycoproteins either by themselves or by forming protein-protein complexes on the cell surface. These are referred to as

Fig. 1 (continued) (step 3). HCV along with CD81 and CLDN1 is subsequently internalized via clathrin-mediated endocytosis through events that are poorly understood (step 4). HCV is eventually delivered to early endosomes where low pH triggers fusion of virus and cell membrane releasing the viral capsid into the cytoplasm (step 5). In the cytosol, uncoating of the capsid liberates the HCV RNA for downstream steps of the viral life cycle (step 6). NPC1L1, present on the apical side of hepatocyte, regulates cholesterol homeostasis by absorbing excess cholesterol from the bile and is thus likely to play a role by modifying the cholesterol content of the membrane. TfR1 has been shown to directly interact with HCV and is likely to have a role in a post-entry step during viral endocytosis. Similarly, the tetraspanin CD63 has been shown to interact with HCV, however, the exact mechanism by which it modulates the entry process has not been uncovered

"entry factors". Finally, host proteins like host cell kinases, GTPase HRas and NPC1L1, although not directly implicated in the interaction with the virus, play a crucial role in supporting viral entry. These are described here as "facilitators".

2 Attachment Factors

Viral attachment is the first step for initiation of productive infection. In the liver, the space of Disse, space between the endothelial cells and the hepatocyte, is enriched in heparin sulfate (HS), a glycosaminoglycan. HS exhibit different patterns and degree of sulfation depending on the tissue and species of origin. HCV has been shown to particularly use highly sulfated heparan sulfate, which are especially found on the membrane of hepatocytes (Barth et al. 2003, 2006). This requirement of HCV might part-determine the hepatotropism of the virus. HCV uses several host factors to attach to the target cells via its envelope glycoproteins, particularly E2 and virus-associated apolipoproteins, particularly apolipoprotein E (apoE). Heparan sulfate (HS) is used as an initial cellular binding factor by many viruses including members of the *flaviviridae* family (Chen et al. 1997; Hulst et al. 2001; Mandl et al. 2001). Indeed, studies using HCV-LP demonstrated the role of cell surface heparan sulfate in binding and entry of HCV into the target cells (Barth et al. 2003, 2006). Subsequent studies using HCVpp and HCVcc system confirmed the role of HS in HCV entry (Barth et al. 2006; Jiang et al. 2013; Lefevre et al. 2014). While highly sulfated HS was identified as a key factor required for HCV docking on the cell surface, the nature of the core protein defining the HSPG remained unknown until recently. Silencing experiment combined with complementation assay, identified syndecan 1 (SDC1) and syndecan 4 (SDC4), members of syndecan family of HSPGs, as the host factors mediating initial attachment of HCV to the cell surface (Shi et al. 2013; Lefevre et al. 2014). SDC1 is highly expressed in epithelial cells such as hepatocytes, whereas SDC4 is widely expressed in range of cells (Kim et al. 1994). Given that SDC1 silencing was shown to activate expression of SDC4 and vice-versa, it is likely that both SDCs may play a role in HCV attachment and entry (Lefevre et al. 2014).

Both viral envelope glycoproteins, E1 and E2, have been shown to contribute to HSPG binding, E2 being a major determinant (Barth et al. 2003, 2006). Mapping studies using monoclonal antibodies have identified several regions of E2, particularly hyper-variable region I (HVRI), which was shown to be a major contributor to this binding (Barth et al. 2003). However, these experiments have been conducted with HCV-LPs, which do not reflect the physiological situation. Indeed, in the blood HCV infectious particles are associated with lipoproteins rich in apolipoproteins. Interestingly, apolipoprotein E (apoE), a host protein, has been recently shown to facilitate HCV's interaction with cell surface HSPG (Jiang et al. 2013; Lefevre et al. 2014). During the course of HCV life cycle, assembly and release of infectious viral particles require cytosolic lipid droplets and appear to be closely associated with very low density lipoprotein (VLDL) biogenesis and secretion, leading to incorporation of several apolipoproteins namely, apoE, apolipoprotien B (apoB), apolipoprotein AI (apoAI) and apolipoprotein CI (apoCI) on the newly synthesized viral particles (Meunier et al. 2005; Huang et al. 2007; Gastaminza et al. 2008; Catanese et al. 2013). Indeed, apoE was found to be present (200 molecules per virus particle) along with majority of E2-containing particles as studied by electron microscopy (Merz et al. 2011). The interaction of apoE with HSPGs has been mapped to the apoE HSPG-binding domain and consists of highly charged residues likely to participate in electrostatic interactions with negatively charged N- and O-sulfo groups of sulfated disaccharides (Jiang et al. 2013; Lefevre et al. 2014). Peptides mimicking apoE HSPG-binding domain as well as apoE antibodies inhibit HCV attachment to the cell surface. It is worth noting that differences, if any, between different HCV genotypes with regards to binding HSPGs have not been reported.

In addition to HSPGs, HCV has been shown to use low density lipoprotein – receptor (LDL-R) and SR-BI as other host factors to mediate its attachment to the target cells. The observation that in the infected patients HCV associates with VLDL led to the hypothesis that LDL-R can potentially act as an entry-receptor for HCV complexed VLDL. Indeed, using patient derived virus, it was shown that HCV is endocytosed via LDL-R into hepatoma cells and in lymphocyte derived cell line with upregulated LDL-R expression (Agnello et al. 1999). The role of LDL-R in HCV binding and internalization was further confirmed using purified lipo-viral particle (LVP) derived from patient and in an inhibition assay using purified VLDL and LDL (Andre et al. 2002). LDL-R does not bind HCV envelope glycoproteins but rather plays a role through interaction with its ligand apoE present on the surface of HCV LVP (Owen et al. 2009). However, since apoE binds several other cell-surface molecules, namely HSPGs and SR-BI that are implicated in HCV entry, the role of apoE-LDL-R interactions in productive HCV entry warrants further investigations. Indeed, the moderate effect of anti-LDL-R antibody on HCVcc entry and reduced entry despite increased binding of lipoprotein lipase treated HCVcc to LDL-R indicate that interaction with LDL-R likely leads to non-productive entry of HCV (Albecka et al. 2012). Interestingly, a recent report has shown that deletion of hypervariable region I (HVRI) of HCV E2 decreased dependency of three different HCV genotypes on LDL-R. This reduced dependency was shown to be apo-E independent using antibodies targeting three different regions on LDL-R; the antibody that potently blocked HCV entry was the one that did not affect LDL-R's association with apoE and apoB (Prentoe et al. 2014). Thus it seems that LDL-R plays a complex role in HCV entry that could go beyond its function as an attachment factor.

Unlike LDL-R, early identification of SR-B1 as a HCV receptor was based on its ability to directly bind HCV E2 (Scarselli et al. 2002, also see Sect. 3.2). However, it has been shown that SR-BI also plays a role in HCV attachment. SR-BI acts as a receptor for different classes of lipoproteins facilitating transfer of cholesterol ester into hepatocytes. HCV likely interacts with SR-BI through associated lipoproteins. Indeed, it has been shown that the attachment function of SR-BI cannot be abrogated with mutation in HVRI, anti-E2 antibodies or with small molecule inhibitor disrupting E2/SR-BI interaction. On the other hand apoE, LDL and VLDL potently blocked the binding of HCVcc to SR-BI ectopically expressed on CHO cells as well as on rat hepatoma cells (Maillard et al. 2006; Dao Thi et al. 2012). It is worth noting that the attachment function of SR-BI is not species dependent as both, human and mouse SR-BI have been shown to exhibit the same effect (Dao Thi et al. 2012).

Taken together, the present data indicate that HCV attachment is mediated by multiple host factors, acting alone or co-operatively. ApoE is an integral part of infectious HCV and seems to be the central player in HCV attachment as highlighted by its role in binding to HSPG, LDL-R and SR-BI. The role of apoE in HCV infection has been further demonstrated in a study where protection against persistent infection was shown to be associated with the presence of allele 2 that codes apoE isoform 2 (apoE2) (Price et al. 2006), likely because of its poor ability to bind LDL-R (reviewed in Hatters et al. 2006). Indeed, it has been subsequently shown that HCVcc particles produced in presence of apoE2 were less infectious than virus particles produced in presence of apoE3 and apoE4 isoforms (Hishiki et al. 2010). However, the in vitro effect of apoE2 was not reproduced in another study warranting further investigation into the role of apoE isoforms in HCV entry (Cun et al. 2010).

3 Entry Factors

3.1 CD81

Cluster of differentiation 81 (CD81), a 236 amino acid tetraspanin membrane protein, was the first host factor reported to interact with a soluble form of the HCV glycoprotein E2 (Pileri et al. 1998). Its role in HCV entry was subsequently confirmed using more sophisticated state of the art models namely, HCVpp and HCVcc (Bartosch et al. 2003a, b; Cormier et al. 2004; Zhang et al. 2004; Lavillette et al. 2005b; Lindenbach et al. 2005; Wakita et al. 2005). A soluble recombinant form of CD81 LEL and anti-CD81 antibodies have been shown to potently and pan-genotypically inhibit HCV entry and infectivity (Bartosch et al. 2003a; Hsu et al. 2003; Wakita et al. 2005; Zhong et al. 2005; Brimacombe et al. 2011). Similar effects have been observed upon silencing CD81 in hepatoma cells. (Zhang et al. 2004; Lavillette et al. 2005b). Moreover, HepG2 cells lacking CD81 have been shown to become susceptible to HCV infection after ectopic expression of CD81 (Bartosch et al. 2003b; McKeating et al. 2004; Zhang et al. 2004; Lindenbach et al. 2005). In line with these observations, it has been further demonstrated that efficiency of HCV entry is directly proportional to CD81 expression (Akazawa et al. 2007; Koutsoudakis et al. 2007).

CD81 belongs to the tetraspanin family of proteins consisting of four transmembrane domains, two extracellular loop (small extracellular loop (SEL) and large extracellular loop (LEL)) and intracellular N- and C-terminal domains (Boucheix and Rubinstein 2001; Levy and Shoham 2005b). Since the identification of CD81 as a HCV receptor, many studies have focused on defining the cellular and viral determinants of CD81-E2 interaction. Other tetraspanins e.g. CD9 and CD151 does not bind HCV E2, suggesting the interaction to be CD81 specific (Pileri et al. 1998; Flint et al. 1999; Petracca et al. 2000; Drummer et al. 2002). HCV binding has been shown to be exclusively mediated by the LEL with other regions e.g. the C-terminal region, transmembrane residues and post-translational modifications contributing to the binding (Masciopinto et al. 2001; Bertaux and Dragic 2006). Within the LEL, a specific secondary structure formed by the two disulphide bonds, is critical for its receptor function (Petracca et al. 2000). Several studies using recombinant soluble E2 and CD81 mutants have underscored the importance of residues L162, K171, I181, I182, N184, F186, D196 in binding (Higginbottom et al. 2000; Drummer et al. 2002). Some of these residues namely K171, I181, I182, F186 have been subsequently shown to be important in HCV entry using the HCVpp system (Bertaux and Dragic 2006; Flint et al. 2006)). Additionally, residues T149, E152, T153 of CD81 LEL take part in CLDN1 association to form CD81-CLDN1 co-receptor complexes (Davis et al. 2012; see Sect. 3.3).

The membrane lipid composition has been shown to affect expression of CD81 on the cell surface and in turn modulate HCV entry. In fact, enriching plasma membrane with ceramide induced internalization of CD81 and led to inhibition of HCVpp as well as HCVcc entry (Voisset et al. 2008). On the other hand depleting cholesterol from the plasma membrane has been shown to decrease HCVcc and HCVpp entry owing to reduced cell-surface expression of CD81 (Charrin et al. 2003; Kapadia et al. 2007). Furthermore, polarization of hepatoma cells has been shown to promote lateral diffusion of CD81, diffusion that has also been correlated with HCV entry (Harris et al. 2013). In addition, CD81 trafficking has been proposed to be intimately linked with HCV entry. Thus, while engagement of CD81, either by virus or by antibody induces clathrin-dependent internalization of CD81, a CD81-specific antibody has been shown to inhibit HCV entry after its internalization (Farquhar et al. 2012). These results highlight a delicate association of CD81 with the composition of plasma membrane, its fluidity and cell cytoskeleton (also see Sect. 4).

Taken together these data indicates that HCV initiates specific interaction with human CD81 and that modulation of CD81 expression on the cell surface for e.g. by its level of expression at the cell surface or changes in cell membrane composition directly affects HCV entry.

3.2 SR-BI

SR-BI is a 509 amino acid protein consisting of cytoplasmic N and C-terminus, two transmembrane domains and a large extracellular domain. After identification of CD81 as an entry factor of HCV, it was soon realized that HepG2 cells that lack CD81 efficiently bind soluble E2. Cross linking experiments with soluble E2 led to the identification of SR-BI (Scarselli et al. 2002). The role of SR-BI in HCV entry was further confirmed in HCVpp and HCVcc model systems by silencing SR-BI expression or by blocking viral entry using anti-SR-BI antibodies (Bartosch et al. 2003b, 2005; Zeisel et al. 2007; Catanese et al. 2010; Zahid et al. 2013). SR-BI has been shown to bind sE2, E1E2 heterodimer and HCVcc when overexpressed on CHO cells (Scarselli et al. 2002; Evans et al. 2007; Douam et al. 2014). The multifaceted role of SR-BI is slowly unraveling. HVR1 sequence of E2 has been shown to be the major viral determinant contributing to SR-BI binding. Deletion of HVR1 decreases infectivity of HCVpp and HCVcc in hepatoma cells because of reduced dependency on SR-BI (Prentoe et al. 2014). On the other hand, several regions of SR-BI protein have been shown to influence HCV entry. This includes the ectodomain of SR-BI (Scarselli et al. 2002; Catanese et al. 2010), the major determinant of HCV entry and the C-terminus cytoplasmic tail modulating the basal HCV entry process through intracellular trafficking and/or membrane relocalization (Dreux et al. 2009). The SR-BI ectodomain has been shown to influence HCV entry in two distinct manners namely, by direct interaction with HCV E2 and by promoting lipid transfer across the host membrane in a post-binding entry step (Scarselli et al. 2002; Catanese et al. 2010; Dao Thi et al. 2012; Zahid et al. 2013). The latter function increases the local cholesterol content on the plasma membrane that likely triggers formation of functional receptor complexes for HCV entry. Furthermore, SR-BI lipid transfer function has been shown to be important in cell-to-cell transmission that is independent of E2 binding (Zahid et al. 2013). In line with this, it has been shown that high density lipoproteins (HDLs) enhance HCV infection through their lipid transfer function and that this enhancement depends on HVRI (Bartosch et al. 2005; Voisset et al. 2005). Consistent with this it has been shown that enhanced entry of low density virus is HVRI dependent (Bankwitz et al. 2010; Prentoe et al. 2014). Interestingly, oxidized LDL (oxLDL) has been shown to be a potent inhibitor of HCV entry, likely through SR-BI-dependent mechanism and probably involving HVRI (von Hahn et al. 2006; Westhaus et al. 2013). Understanding the interplay of oxLDL with SR-BI and its HDL binding and lipid transfer function remains unsolved. The complex role of SR-BI in HCV entry has been further dissected in a recent study (Dao Thi et al. 2012). Accordingly, three different functions of SR-BI have been proposed namely, attachment (see Sect. 2), access (see Sect. 2) and enhancement (described in this section). Interestingly, it is only the enhancement function that was shown to rely on direct interaction with E2 in a post-binding step.

3.3 Claudins

The observation that several human and non-primate cell lines are non-permissive to HCV implicated additional host factors in viral entry (Bartosch et al. 2003a). This led to the discovery of two tight junction proteins namely CLDN1 and OCLN as HCV entry factors (Evans et al. 2007; Liu et al. 2009; Ploss et al. 2009).

CLDN1 was the first of the two tight junction proteins identified as critical for HCV entry (Evans et al. 2007). It is a 25 kDa protein highly expressed in the liver contributing to the formation of the tight junctions by homo and hetero-oligomerisation (reviewed in Gunzel and Yu 2013). While not a classical tetraspanin, its organization is like CD81 consisting of cytoplasmic N and C-termini, two extracellular loops, four transmembrane domain and a cytoplasmic loop. CLDN1 was identified as a co-receptor for HCV entry by transducing 293T with a cDNA library derived from permissive hepatoma cells Huh 7.5 (Evans et al. 2007). Although CLDN1 has not been shown to interact with HCV sE2 or with HCVcc when overexpressed on CHO cells (Evans et al. 2007; Krieger et al. 2010), a recent report demonstrated that E1E2 can interact with extracellular loop 1 of CLDN1 highlighting the role of E1 in modulating receptor binding (Douam et al. 2014). CLDN1 is used for entry by all the major genotypes as anti-CLDN1 antibodies have been shown to pan-genotypically inhibit HCV entry (Fofana et al. 2010; Hotzel et al. 2011). On the surface of the cells CLDN1 interacts with CD81 to form the CD81-CLDN1 co-receptor complex, which is central to HCV entry (Harris et al. 2010). The importance of CD81-CLDN1 co-receptor complex is further highlighted by the fact that anti-CLDN1 antibodies that break this complex potently neutralize HCV entry (Krieger et al. 2010). Determinants of HCV entry have been mapped to the extracellular loop 1, which not only participate in cell-cell contact but also mediates association with CD81 LEL (Evans et al. 2007; Cukierman et al. 2009; Davis et al. 2012). In addition to tight junctions, CLDN1 also localizes at the basolateral membrane in polarized HepG2 cells and in the human liver (Reynolds et al. 2008; Mee et al. 2009). CLDN1 mutants that lead to enrichment of CLDN1 at cell-cell contacts promote HCV entry as against mutants that causes enrichment in the cytoplasm and at plasma membrane in non-hepatic 293T cells (Yang et al. 2008; Cukierman et al. 2009). However, in human hepatoblastoma HepG2 cells polarization was shown to restrict HCV entry, which correlated with increased CLDN1 localization at the tight junction and undetectable CLDN1-CD81 co-receptor complex (Mee et al. 2009). Furthermore, in HCV infected liver increased CLDN1 staining with more prominent CLDN1/CD81 co-localization in the basolateral region was observed compared to normal liver (Reynolds et al. 2008). While the different experimental approaches could explain the above results, it likely reflects the dynamic nature of HCV entry. Indeed it has been shown that initial interaction of HCV with CD81 at the basolateral membrane activates EGFR/HRas signaling, thereby promoting lateral diffusion of CD81 and/or virus-receptor complex facilitating CD81-CLDN1 co-receptor complex formation that is necessary for HCV entry (Brazzoli et al. 2008; Diao et al. 2012; Zona et al. 2013).

Several recent reports have implicated other CLDNs notably CLDN6 and CLDN9 in HCV entry. Using cell culture model systems and 293T cells expressing different claudins, it was demonstrated that HCV can efficiently use CLDN6 and 9 for its entry, which could be blocked by respective anti-CLDN antibody (Meertens et al. 2008; Fofana et al. 2013b). The role of CLDN6 was further investigated using Huh6 hepatoma cells that predominantly express CLDN6

(Haid et al. 2014). While CLDN1 conferred permissivity to all the genotypes studied, CLDN6 did so to only some viral strains. However, in primary human hepatocytes (PHH) and Huh 7.5.1 cells, anti-CLDN6 and CLDN9 antibodies were shown to have no effect on HCVpp entry (Fofana et al. 2013b). It is worth noting that of the three claudins – CLDN1, CLDN6 and CLDN9, only CLDN1 is predominantly expressed in primary human hepatocytes (PHH). Thus the in vivo relevance, if any, for a role of other CLDNs, particularly CLDN6, remains to be further evaluated.

3.4 Occludin

Unlike CLDN1, the role of OCLN in HCV entry is poorly understood. Silencing of OCLN was shown to inhibit HCVpp entry into Huh 7.5.1 cells (Liu et al. 2009). Subsequently, screening a cDNA library derived from Huh 7.5.1 cells identified OCLN as a human specific HCV entry factor that rendered mouse hepatoma cell lines permissive to HCV entry (Ploss et al. 2009). In fact expression of human CD81 and human OCLN was shown to recapitulate early steps of HCV entry in fully immunocompetent inbred mice (Dorner et al. 2011). OCLN is a 60 kDa four transmembrane domain protein regulating tight junction formation. Determinants of HCV entry have been mapped to the extracellular loop 2 (Ploss et al. 2009). However, a recent report using OCLN mutants with FLAG insertion has also implicated regions of extracellular loop 1 in HCV entry (Sourisseau et al. 2013). The viral determinants of OCLN interaction have been mapped to HCV E2, as shown by immunoprecipitation experiments (Liu et al. 2009). This is further supported by the finding that OCLN usage seems to be isolate-specific indicating that HCV glycoproteins directly interact with OCLN (Benedicto et al. 2008; Sourisseau et al. 2013). However, residues involved in this interaction remains to be mapped.

3.5 Others

Using an in silico integrative-genomic approach, the tetraspanin CD63 was recently reported as an entry factor for HCV. The role of CD63 in HCV entry was validated by silencing (in HCVcc and HCVpp) and antibody mediated blocking experiments (in HCVcc). Interestingly, like CD81, extracellular loop 2 of CD63 was shown to inhibit HCVcc entry indicating that HCV directly interacts with CD63 (Park et al. 2013). CD63 and CD81 have been shown to be involved in formation of tetraspanin enriched microdomain (TEM) (Levy and Shoham 2005a); however, whether they interact with each other in hepatocytes is not known. Besides its role in HCV entry, CD63 could facilitate HCV uptake through its role in clathrin-dependent endocytosis and vesicle trafficking (Rous et al. 2002; Janvier and Bonifacino 2005).

HCV infection has been shown to increase hepatic iron load and alter TfR1 expression in patients. This clinical observation subsequently led to the identification of TfR1 as a host factor facilitating HCV entry (Martin and Uprichard 2013). TfR1 is a major receptor for cellular iron uptake. The available evidences indicate that TfR1 likely plays a role during a late step of HCV entry. Indeed, silencing TfR1 trafficking protein (TTP), a cargo protein required for internalization of TfR1, has also been shown to inhibit HCVpp and HCVcc entry indicating that TfR1 might play a role in HCV endocytosis. Interestingly, HCVcc particles have been shown to bind TfR1 when overexpressed on CHO cells indicating that neither HCV-associated lipoproteins interfere with the interaction nor a pre-engagement with other receptors is required to prime the binding.

4 Facilitators

4.1 Receptor Tyrosine Kinases

Many viruses activate signaling pathways after attachment and receptor clustering on the surface of their target cells to facilitate their entry and subsequent infection (reviewed in Yamauchi and Helenius 2013). In case of HCV, genome-wide host kinase RNAi screen identified receptor tyrosine kinases, namely EGFR and EphA2, as crucial regulators of HCV entry and infection (Lupberger et al. 2011). EGFR mediates its pro-HCV effect by promoting the formation of CD81-CLDN1 co-receptor complex. Indeed, silencing EGFR expression or inhibiting its activity by erlotinib reduces CD81-CLDN1 association at the cell surface (Lupberger et al. 2011). In human hepatoma cells and in PHH activation of EGFR was shown to predominantly transduce through MAPK signaling. (Zona et al. 2013). In line with this, blocking of upstream EGFR-signal transducers (e.g. EGFR, Ras, Raf) was shown to have a greater inhibitory effect on HCV entry compared to downstream signaling components (e.g. MEK1/2 and ERK1/2). Differential stable isotope labeling by amino acids in cell culture (SILAC) approach along with mass spectrometry led to the identification of GTPase HRas as a molecular switch transducing receptor tyrosine kinase (RTK) signals required for HCV entry. HRas facilitates formation and stabilization of HCV-receptor complex by modulating lateral membrane diffusion of CD81 (Zona et al. 2013). Furthermore, it has been demonstrated that virus engagement by CD81 leads to virus internalization via activation of PI3/AKT and EGFR signaling pathways (Diao et al. 2012; Liu et al. 2012). In addition to RTKs, protein kinase A (PKA) and phosphatidylinositol 4-kinase type III-alpha/beta (PI4KIIIalpha/beta) have been shown to modulate HCV entry. The former was shown to regulate CLDN1 cellular localization and

CD81-CLDN1 co-receptor formation (Farquhar et al. 2008), while for the latter, a clear mechanism has not been proposed (Trotard et al. 2009).

4.2 NPC1L1

In addition to host signaling, HCV uses other cellular mechanism to facilitate its entry into hepatocyte. The recently identified cholesterol transporter NPC1L1 is such a host factor exploited by HCV (Sainz et al. 2012; Martin and Uprichard 2013). NPC1L1 is a cholesterol absorption receptor present on the apical canalicular surface of polarized hepatocytes. It reabsorbs free cholesterol secreted into bile and thus is important in maintaining whole body cholesterol homeostasis. The role of NPC1L1 in HCV entry was demonstrated by three different approaches namely by silencing NPC1L1, by blocking antibodies and by approved pharmacological inhibitors (Sainz et al. 2012). Interestingly, silencing or inhibiting NPC1L1 only blocked HCVcc infection but not HCVpp entry indicating cholesterol-dependent role of NPC1L1 in HCV entry. Given that NPC1L1 is present at the apical canalicular surface in polarized cells, it probably facilitates HCV entry by modulating cholesterol levels. Indeed, the extracellular N-terminal domain, which is involved in cholesterol binding and uptake, has been implicated in HCV entry (Sainz et al. 2012). Nevertheless, the exact molecular mechanism of interaction between HCV and NPC1L1 remains to be determined.

5 Conclusion

Tremendous progress has been made in our understanding of the HCV entry process, aided by the development of different model systems to study HCV life-cycle. Several host factors have been identified and the functional role of most of them has been validated in animal model system (Meuleman et al. 2008; Lupberger et al. 2011; Lacek et al. 2012; Sainz et al. 2012). In fact, identification of these host receptors and co-receptors has enabled development of a genetically humanized immunocompetent mouse model to study HCV entry in vivo (Dorner et al. 2011; Chen et al. 2014). Kinetic studies of HCV entry using antibodies targeting host factors have given useful insight into the step-wise process of viral entry (Fig. 1). Thus, we now know that attachment of the viral particle to the cell surface is the first step of the entry process and that it involves HSPGs, LDL-R and E2-independent SR-BI function. Attachment of viral particles likely leads to clustering of receptor proteins CD81 and SR-BI on the cell surface in a post-attachment entry step. It is worth noting that SR-BI has been identified as a component of CD81 TEM indicating that attachment to SR-BI might bring the virus near CD81 (Zona et al. 2013). Binding of viral particles to CD81 activates EGFR signaling leading to the formation of CD81-CLDN1 co-receptor complex facilitated by GTPase HRas

(Diao et al. 2012; Zona et al. 2013). This sequence of events is supported by kinetic studies demonstrating that CD81, SR-BI and CLDN1 acts at similar time-points during HCV entry (Zeisel et al. 2007; Krieger et al. 2010). The virus-receptor multi–complex then interacts with OCLN (Sourisseau et al. 2013) and TfR1 (Martin and Uprichard 2013) in a very late step of HCV entry; interaction being modulated by CD81 mediated lateral diffusion of the complex (Harris et al. 2013). Whether this happens at the basolateral membrane or at the tight junction is controversial. While CD81 and CLDN1 have been shown to be present in the endocytic vesicle associated with the virus, other host factors e.g. SR-BI and OCLN have not been shown to be concomitantly endocytosed (Coller et al. 2009; Farquhar et al. 2012). The step-wise usage of other co-factors namely NPC1L1 and CD63 remains to be dissected.

The wide array of host factors implicated in HCV entry could also serve as targets for developing anti-viral therapy. Indeed small molecules targeting SR-BI and EGFR are in phase 1 clinical trial (Sulkowski et al. 2014, ClinicalTrials.gov: NCT01835938). Targeting host factors involved in HCV entry offers high genetic barrier to resistance and thus presents an attractive option that could complement existing direct acting anti-virals (DAA). Furthermore, as HCV persists in the host; primarily through cell-to-cell transmission that involves most of the entry host factors, host-entry inhibitors offer an added advantage. Indeed, cell-to-cell transmission has been shown to be resistant to most of the neutralizing antibodies reported so far, whereas anti-receptor antibodies have been shown to block both, cell-free and cell-cell, modes of transmission (Fofana et al. 2010, 2013a; Lacek et al. 2012; Zahid et al. 2013). Besides attractive option for drug development, HCV entry is a major target of host humoral immune response. Early appearance of neutralizing antibodies has been associated with spontaneous HCV clearance (Lavillette et al. 2005a; Pestka et al. 2007; Osburn et al. 2014). Intriguingly, all neutralizing antibodies reported so far have been shown to target CD81 binding regions on HCV E2 (Owsianka et al. 2001; Keck et al. 2008; Mancini et al. 2009; Kong et al. 2013). This indicates that E2-CD81 interaction is critical for HCV entry, a feature that could be exploited for vaccine development. Development of a global prophylactic vaccine is a high priority since the huge cost of current therapies is unlikely to benefit a big number of patients living in resource limited countries. Building on our current understanding of HCV entry and translating it into a successful vaccine is a challenge for the future.

References

Agnello V, Abel G, Elfahal M, Knight GB, Zhang QX (1999) Hepatitis C virus and other flaviviridae viruses enter cells via low density lipoprotein receptor. Proc Natl Acad Sci U S A 96:12766–12771

Akazawa D, Date T, Morikawa K, Murayama A, Miyamoto M, Kaga M, Barth H, Baumert TF, Dubuisson J, Wakita T (2007) CD81 expression is important for the permissiveness of Huh7 cell clones for heterogeneous hepatitis C virus infection. J Virol 81:5036–5045

Albecka A, Belouzard S, Op De Beeck A, Descamps V, Goueslain L, Bertrand-Michel J, Terce F, Duverlie G, Rouille Y, Dubuisson J (2012) Role of low-density lipoprotein receptor in the hepatitis C virus life cycle. Hepatology 55:998–1007

Andre P, Komurian-Pradel F, Deforges S, Perret M, Berland JL, Sodoyer M, Pol S, Brechot C, Paranhos-Baccala G, Lotteau V (2002) Characterization of low- and very-low-density hepatitis C virus RNA-containing particles. J Virol 76:6919–6928

Bankwitz D, Steinmann E, Bitzegeio J, Ciesek S, Friesland M, Herrmann E, Zeisel MB, Baumert TF, Keck ZY, Foung SK, Pecheur EI, Pietschmann T (2010) Hepatitis C virus hypervariable region 1 modulates receptor interactions, conceals the CD81 binding site, and protects conserved neutralizing epitopes. J Virol 84:5751–5763

Barth H, Schafer C, Adah MI, Zhang F, Linhardt RJ, Toyoda H, Kinoshita-Toyoda A, Toida T, Van Kuppevelt TH, Depla E, Von Weizsacker F, Blum HE, Baumert TF (2003) Cellular binding of hepatitis C virus envelope glycoprotein E2 requires cell surface heparan sulfate. J Biol Chem 278:41003–41012

Barth H, Schnober EK, Zhang F, Linhardt RJ, Depla E, Boson B, Cosset FL, Patel AH, Blum HE, Baumert TF (2006) Viral and cellular determinants of the hepatitis C virus envelope-heparan sulfate interaction. J Virol 80:10579–10590

Bartosch B, Dubuisson J, Cosset FL (2003a) Infectious hepatitis C virus pseudo-particles containing functional E1-E2 envelope protein complexes. J Exp Med 197:633–642

Bartosch B, Vitelli A, Granier C, Goujon C, Dubuisson J, Pascale S, Scarselli E, Cortese R, Nicosia A, Cosset FL (2003b) Cell entry of hepatitis C virus requires a set of co-receptors that include the CD81 tetraspanin and the SR-B1 scavenger receptor. J Biol Chem 278:41624–41630

Bartosch B, Verney G, Dreux M, Donot P, Morice Y, Penin F, Pawlotsky JM, Lavillette D, Cosset FL (2005) An interplay between hypervariable region 1 of the hepatitis C virus E2 glycoprotein, the scavenger receptor BI, and high-density lipoprotein promotes both enhancement of infection and protection against neutralizing antibodies. J Virol 79:8217–8229

Benedicto I, Molina-Jimenez F, Barreiro O, Maldonado-Rodriguez A, Prieto J, Moreno-Otero R, Aldabe R, Lopez-Cabrera M, Majano PL (2008) Hepatitis C virus envelope components alter localization of hepatocyte tight junction-associated proteins and promote occludin retention in the endoplasmic reticulum. Hepatology 48:1044–1053

Bertaux C, Dragic T (2006) Different domains of CD81 mediate distinct stages of hepatitis C virus pseudoparticle entry. J Virol 80:4940–4948

Blanchard E, Belouzard S, Goueslain L, Wakita T, Dubuisson J, Wychowski C, Rouille Y (2006) Hepatitis C virus entry depends on clathrin-mediated endocytosis. J Virol 80:6964–6972

Boucheix C, Rubinstein E (2001) Tetraspanins. Cell Mol Life Sci: CMLS 58:1189–1205

Brazzoli M, Bianchi A, Filippini S, Weiner A, Zhu Q, Pizza M, Crotta S (2008) CD81 is a central regulator of cellular events required for hepatitis C virus infection of human hepatocytes. J Virol 82:8316–8329

Brimacombe CL, Grove J, Meredith LW, Hu K, Syder AJ, Flores MV, Timpe JM, Krieger SE, Baumert TF, Tellinghuisen TL, Wong-Staal F, Balfe P, Mckeating JA (2011) Neutralizing antibody-resistant hepatitis C virus cell-to-cell transmission. J Virol 85:596–605

Catanese MT, Ansuini H, Graziani R, Huby T, Moreau M, Ball JK, Paonessa G, Rice CM, Cortese R, Vitelli A, Nicosia A (2010) Role of scavenger receptor class B type I in hepatitis C virus entry: kinetics and molecular determinants. J Virol 84:34–43

Catanese MT, Uryu K, Kopp M, Edwards TJ, Andrus L, Rice WJ, Silvestry M, Kuhn RJ, Rice CM (2013) Ultrastructural analysis of hepatitis C virus particles. Proc Natl Acad Sci U S A 110:9505–9510

Charrin S, Manie S, Thiele C, Billard M, Gerlier D, Boucheix C, Rubinstein E (2003) A physical and functional link between cholesterol and tetraspanins. Eur J Immunol 33:2479–2489

Chen Y, Maguire T, Hileman RE, Fromm JR, Esko JD, Linhardt RJ, Marks RM (1997) Dengue virus infectivity depends on envelope protein binding to target cell heparan sulfate. Nat Med 3:866–871

Chen J, Zhao Y, Zhang C, Chen H, Feng J, Chi X, Pan Y, Du J, Guo M, Cao H, Wang Z, Pei R, Wang Q, Pan L, Niu J, Chen X, Tang H (2014) Persistent hepatitis C virus infections and hepatopathological manifestations in immune-competent humanized mice. Cell Res 24:1050–1066

Codran A, Royer C, Jaeck D, Bastien-Valle M, Baumert TF, Kieny MP, Pereira CA, Martin JP (2006) Entry of hepatitis C virus pseudotypes into primary human hepatocytes by clathrin-dependent endocytosis. J Gen Virol 87:2583–2593

Coller KE, Berger KL, Heaton NS, Cooper JD, Yoon R, Randall G (2009) RNA interference and single particle tracking analysis of hepatitis C virus endocytosis. PLoS Pathog 5:e1000702

Cormier EG, Tsamis F, Kajumo F, Durso RJ, Gardner JP, Dragic T (2004) CD81 is an entry coreceptor for hepatitis C virus. Proc Natl Acad Sci U S A 101:7270–7274

Cukierman L, Meertens L, Bertaux C, Kajumo F, Dragic T (2009) Residues in a highly conserved claudin-1 motif are required for hepatitis C virus entry and mediate the formation of cell-cell contacts. J Virol 83:5477–5484

Cun W, Jiang J, Luo G (2010) The C-terminal alpha-helix domain of apolipoprotein E is required for interaction with nonstructural protein 5A and assembly of hepatitis C virus. J Virol 84:11532–11541

Dao Thi VL, Granier C, Zeisel MB, Guerin M, Mancip J, Granio O, Penin F, Lavillette D, Bartenschlager R, Baumert TF, Cosset FL, Dreux M (2012) Characterization of hepatitis C virus particle subpopulations reveals multiple usage of the scavenger receptor BI for entry steps. J Biol Chem 287:31242–31257

Davis C, Harris HJ, Hu K, Drummer HE, Mckeating JA, Mullins JG, Balfe P (2012) In silico directed mutagenesis identifies the CD81/claudin-1 hepatitis C virus receptor interface. Cell Microbiol 14:1892–1903

Diao J, Pantua H, Ngu H, Komuves L, Diehl L, Schaefer G, Kapadia SB (2012) Hepatitis C virus induces epidermal growth factor receptor activation via CD81 binding for viral internalization and entry. J Virol 86:10935–10949

Dorner M, Horwitz JA, Robbins JB, Barry WT, Feng Q, Mu K, Jones CT, Schoggins JW, Catanese MT, Burton DR, Law M, Rice CM, Ploss A (2011) A genetically humanized mouse model for hepatitis C virus infection. Nature 474:208–211

Douam F, Dao Thi VL, Maurin G, Fresquet J, Mompelat D, Zeisel MB, Baumert TF, Cosset FL, Lavillette D (2014) Critical interaction between E1 and E2 glycoproteins determines binding and fusion properties of hepatitis C virus during cell entry. Hepatology 59:776–788

Dreux M, Dao Thi VL, Fresquet J, Guerin M, Julia Z, Verney G, Durantel D, Zoulim F, Lavillette D, Cosset FL, Bartosch B (2009) Receptor complementation and mutagenesis reveal SR BI as an essential HCV entry factor and functionally imply its intra and extra cellular domains. PLoS Pathog 5:e1000310

Drummer HE, Wilson KA, Poumbourios P (2002) Identification of the hepatitis C virus E2 glycoprotein binding site on the large extracellular loop of CD81. J Virol 76:11143–11147

Drummer HE, Boo I, Poumbourios P (2007) Mutagenesis of a conserved fusion peptide-like motif and membrane-proximal heptad-repeat region of hepatitis C virus glycoprotein E1. J Gen Virol 88:1144–1148

Evans MJ, Von Hahn T, Tscherne DM, Syder AJ, Panis M, Wolk B, Hatziioannou T, Mckeating JA, Bieniasz PD, Rice CM (2007) Claudin-1 is a hepatitis C virus co-receptor required for a late step in entry. Nature 446:801–805

Farquhar MJ, Harris HJ, Diskar M, Jones S, Mee CJ, Nielsen SU, Brimacombe CL, Molina S, Toms GL, Maurel P, Howl J, Herberg FW, Van Ijzendoorn SC, Balfe P, Mckeating JA (2008) Protein kinase A-dependent step(s) in hepatitis C virus entry and infectivity. J Virol 82:8797–8811

Farquhar MJ, Hu K, Harris HJ, Davis C, Brimacombe CL, Fletcher SJ, Baumert TF, Rappoport JZ, Balfe P, Mckeating JA (2012) Hepatitis C virus induces CD81 and claudin-1 endocytosis. J Virol 86:4305–4316

Flint M, Maidens C, Loomis-Price LD, Shotton C, Dubuisson J, Monk P, Higginbottom A, Levy S, Mckeating JA (1999) Characterization of hepatitis C virus E2 glycoprotein interaction with a putative cellular receptor, CD81. J Virol 73:6235–6244

Flint M, Von Hahn T, Zhang J, Farquhar M, Jones CT, Balfe P, Rice CM, Mckeating JA (2006) Diverse CD81 proteins support hepatitis C virus infection. J Virol 80:11331–11342

Fofana I, Krieger SE, Grunert F, Glauben S, Xiao F, Fafi-Kremer S, Soulier E, Royer C, Thumann C, Mee CJ, Mckeating JA, Dragic T, Pessaux P, Stoll-Keller F, Schuster C, Thompson J, Baumert TF (2010) Monoclonal anti-claudin 1 antibodies prevent hepatitis C virus infection of primary human hepatocytes. Gastroenterology 139:953–964, 964 e1-4

Fofana I, Xiao F, Thumann C, Turek M, Zona L, Tawar RG, Grunert F, Thompson J, Zeisel MB, Baumert TF (2013a) A novel monoclonal anti-CD81 antibody produced by genetic immunization efficiently inhibits Hepatitis C virus cell-cell transmission. PLoS One 8:e64221

Fofana I, Zona L, Thumann C, Heydmann L, Durand SC, Lupberger J, Blum HE, Pessaux P, Gondeau C, Reynolds GM, Mckeating JA, Grunert F, Thompson J, Zeisel MB, Baumert TF (2013b) Functional analysis of claudin-6 and claudin-9 as entry factors for hepatitis C virus infection of human hepatocytes by using monoclonal antibodies. J Virol 87:10405–10410

Gastaminza P, Cheng G, Wieland S, Zhong J, Liao W, Chisari FV (2008) Cellular determinants of hepatitis C virus assembly, maturation, degradation, and secretion. J Virol 82:2120–2129

Grove J, Nielsen S, Zhong J, Bassendine MF, Drummer HE, Balfe P, Mckeating JA (2008) Identification of a residue in hepatitis C virus E2 glycoprotein that determines scavenger receptor BI and CD81 receptor dependency and sensitivity to neutralizing antibodies. J Virol 82:12020–12029

Gunzel D, Yu AS (2013) Claudins and the modulation of tight junction permeability. Physiol Rev 93:525–569

Haberstroh A, Schnober EK, Zeisel MB, Carolla P, Barth H, Blum HE, Cosset FL, Koutsoudakis G, Bartenschlager R, Union A, Depla E, Owsianka A, Patel AH, Schuster C, Stoll-Keller F, Doffoel M, Dreux M, Baumert TF (2008) Neutralizing host responses in hepatitis C virus infection target viral entry at postbinding steps and membrane fusion. Gastroenterology 135(1719–1728):e1

Haid S, Pietschmann T, Pecheur EI (2009) Low pH-dependent hepatitis C virus membrane fusion depends on E2 integrity, target lipid composition, and density of virus particles. J Biol Chem 284:17657–17667

Haid S, Grethe C, Dill MT, Heim M, Kaderali L, Pietschmann T (2014) Isolate-dependent use of claudins for cell entry by hepatitis C virus. Hepatology 59:24–34

Harris HJ, Davis C, Mullins JG, Hu K, Goodall M, Farquhar MJ, Mee CJ, Mccaffrey K, Young S, Drummer H, Balfe P, Mckeating JA (2010) Claudin association with CD81 defines hepatitis C virus entry. J Biol Chem 285:21092–21102

Harris HJ, Clerte C, Farquhar MJ, Goodall M, Hu K, Rassam P, Dosset P, Wilson GK, Balfe P, Ijzendoorn SC, Milhiet PE, Mckeating JA (2013) Hepatoma polarization limits CD81 and hepatitis C virus dynamics. Cell Microbiol 15:430–445

Hatters DM, Peters-Libeu CA, Weisgraber KH (2006) Apolipoprotein E structure: insights into function. Trends Biochem Sci 31:445–454

Heller T, Saito S, Auerbach J, Williams T, Moreen TR, Jazwinski A, Cruz B, Jeurkar N, Sapp R, Luo G, Liang TJ (2005) An in vitro model of hepatitis C virion production. Proc Natl Acad Sci U S A 102:2579–2583

Higginbottom A, Quinn ER, Kuo CC, Flint M, Wilson LH, Bianchi E, Nicosia A, Monk PN, Mckeating JA, Levy S (2000) Identification of amino acid residues in CD81 critical for interaction with hepatitis C virus envelope glycoprotein E2. J Virol 74:3642–3649

Hishiki T, Shimizu Y, Tobita R, Sugiyama K, Ogawa K, Funami K, Ohsaki Y, Fujimoto T, Takaku H, Wakita T, Baumert TF, Miyanari Y, Shimotohno K (2010) Infectivity of hepatitis C virus is influenced by association with apolipoprotein E isoforms. J Virol 84:12048–12057

Hotzel I, Chiang V, Diao J, Pantua H, Maun HR, Kapadia SB (2011) Efficient production of antibodies against a mammalian integral membrane protein by phage display. Protein Eng Des Sel: PEDS 24:679–689

Hsu M, Zhang J, Flint M, Logvinoff C, Cheng-Mayer C, Rice CM, Mckeating JA (2003) Hepatitis C virus glycoproteins mediate pH-dependent cell entry of pseudotyped retroviral particles. Proc Natl Acad Sci U S A 100:7271–7276

Huang H, Sun F, Owen DM, Li W, Chen Y, Gale M Jr, Ye J (2007) Hepatitis C virus production by human hepatocytes dependent on assembly and secretion of very low-density lipoproteins. Proc Natl Acad Sci U S A 104:5848–5853

Hulst MM, Van Gennip HG, Vlot AC, Schooten E, De Smit AJ, Moormann RJ (2001) Interaction of classical swine fever virus with membrane-associated heparan sulfate: role for virus replication in vivo and virulence. J Virol 75:9585–9595

Janvier K, Bonifacino JS (2005) Role of the endocytic machinery in the sorting of lysosome-associated membrane proteins. Mol Biol Cell 16:4231–4242

Jiang J, Wu X, Tang H, Luo G (2013) Apolipoprotein E mediates attachment of clinical hepatitis C virus to hepatocytes by binding to cell surface heparan sulfate proteoglycan receptors. PLoS One 8:e67982

Kapadia SB, Barth H, Baumert T, Mckeating JA, Chisari FV (2007) Initiation of hepatitis C virus infection is dependent on cholesterol and cooperativity between CD81 and scavenger receptor B type I. J Virol 81:374–383

Keck ZY, Li TK, Xia J, Bartosch B, Cosset FL, Dubuisson J, Foung SK (2005) Analysis of a highly flexible conformational immunogenic domain a in hepatitis C virus E2. J Virol 79:13199–13208

Keck ZY, Li TK, Xia J, Gal-Tanamy M, Olson O, Li SH, Patel AH, Ball JK, Lemon SM, Foung SK (2008) Definition of a conserved immunodominant domain on hepatitis C virus E2 glycoprotein by neutralizing human monoclonal antibodies. J Virol 82:6061–6066

Khan AG, Whidby J, Miller MT, Scarborough H, Zatorski AV, Cygan A, Price AA, Yost SA, Bohannon CD, Jacob J, Grakoui A, Marcotrigiano J (2014) Structure of the core ectodomain of the hepatitis C virus envelope glycoprotein 2. Nature 509:381–384

Kim CW, Goldberger OA, Gallo RL, Bernfield M (1994) Members of the syndecan family of heparan sulfate proteoglycans are expressed in distinct cell-, tissue-, and development-specific patterns. Mol Biol Cell 5:797–805

Kobayashi M, Bennett MC, Bercot T, Singh IR (2006) Functional analysis of hepatitis C virus envelope proteins, using a cell-cell fusion assay. J Virol 80:1817–1825

Kong L, Giang E, Nieusma T, Kadam RU, Cogburn KE, Hua Y, Dai X, Stanfield RL, Burton DR, Ward AB, Wilson IA, Law M (2013) Hepatitis C virus E2 envelope glycoprotein core structure. Science 342:1090–1094

Koutsoudakis G, Kaul A, Steinmann E, Kallis S, Lohmann V, Pietschmann T, Bartenschlager R (2006) Characterization of the early steps of hepatitis C virus infection by using luciferase reporter viruses. J Virol 80:5308–5320

Koutsoudakis G, Herrmann E, Kallis S, Bartenschlager R, Pietschmann T (2007) The level of CD81 cell surface expression is a key determinant for productive entry of hepatitis C virus into host cells. J Virol 81:588–598

Krieger SE, Zeisel MB, Davis C, Thumann C, Harris HJ, Schnober EK, Mee C, Soulier E, Royer C, Lambotin M, Grunert F, Dao Thi VL, Dreux M, Cosset FL, Mckeating JA, Schuster C, Baumert TF (2010) Inhibition of hepatitis C virus infection by anti-claudin-1 antibodies is mediated by neutralization of E2-CD81-claudin-1 associations. Hepatology 51:1144–1157

Lacek K, Vercauteren K, Grzyb K, Naddeo M, Verhoye L, Slowikowski MP, Fafi-Kremer S, Patel AH, Baumert TF, Folgori A, Leroux-Roels G, Cortese R, Meuleman P, Nicosia A (2012) Novel human SR-BI antibodies prevent infection and dissemination of HCV in vitro and in humanized mice. J Hepatol 57:17–23

Lavillette D, Morice Y, Germanidis G, Donot P, Soulier A, Pagkalos E, Sakellariou G, Intrator L, Bartosch B, Pawlotsky JM, Cosset FL (2005a) Human serum facilitates hepatitis C virus infection, and neutralizing responses inversely correlate with viral replication kinetics at the acute phase of hepatitis C virus infection. J Virol 79:6023–6034

Lavillette D, Tarr AW, Voisset C, Donot P, Bartosch B, Bain C, Patel AH, Dubuisson J, Ball JK, Cosset FL (2005b) Characterization of host-range and cell entry properties of the major genotypes and subtypes of hepatitis C virus. Hepatology 41:265–274

Lavillette D, Bartosch B, Nourrisson D, Verney G, Cosset FL, Penin F, Pecheur EI (2006) Hepatitis C virus glycoproteins mediate low pH-dependent membrane fusion with liposomes. J Biol Chem 281:3909–3917

Lefevre M, Felmlee DJ, Parnot M, Baumert TF, Schuster C (2014) Syndecan 4 is involved in mediating HCV entry through interaction with lipoviral particle-associated apolipoprotein E. PLoS One 9:e95550

Levy S, Shoham T (2005a) Protein-protein interactions in the tetraspanin web. Physiology 20:218–224

Levy S, Shoham T (2005b) The tetraspanin web modulates immune-signalling complexes. Nat Rev Immunol 5:136–148

Liang Y, Shilagard T, Xiao SY, Snyder N, Lau D, Cicalese L, Weiss H, Vargas G, Lemon SM (2009) Visualizing hepatitis C virus infections in human liver by two-photon microscopy. Gastroenterology 137:1448–1458

Lindenbach BD, Evans MJ, Syder AJ, Wolk B, Tellinghuisen TL, Liu CC, Maruyama T, Hynes RO, Burton DR, Mckeating JA, Rice CM (2005) Complete replication of hepatitis C virus in cell culture. Science 309:623–626

Liu S, Yang W, Shen L, Turner JR, Coyne CB, Wang T (2009) Tight junction proteins claudin-1 and occludin control hepatitis C virus entry and are downregulated during infection to prevent superinfection. J Virol 83:2011–2014

Liu Z, Tian Y, Machida K, Lai MM, Luo G, Foung SK, Ou JH (2012) Transient activation of the PI3K-AKT pathway by hepatitis C virus to enhance viral entry. J Biol Chem 287:41922–41930

Lupberger J, Zeisel MB, Xiao F, Thumann C, Fofana I, Zona L, Davis C, Mee CJ, Turek M, Gorke S, Royer C, Fischer B, Zahid MN, Lavillette D, Fresquet J, Cosset FL, Rothenberg SM, Pietschmann T, Patel AH, Pessaux P, Doffoel M, Raffelsberger W, Poch O, Mckeating JA, Brino L, Baumert TF (2011) EGFR and EphA2 are host factors for hepatitis C virus entry and possible targets for antiviral therapy. Nat Med 17:589–595

Maillard P, Huby T, Andreo U, Moreau M, Chapman J, Budkowska A (2006) The interaction of natural hepatitis C virus with human scavenger receptor SR-BI/Cla1 is mediated by ApoB-containing lipoproteins. FASEB J 20:735–737

Mancini N, Diotti RA, Perotti M, Sautto G, Clementi N, Nitti G, Patel AH, Ball JK, Clementi M, Burioni R (2009) Hepatitis C virus (HCV) infection may elicit neutralizing antibodies targeting epitopes conserved in all viral genotypes. PLoS One 4:e8254

Mandl CW, Kroschewski H, Allison SL, Kofler R, Holzmann H, Meixner T, Heinz FX (2001) Adaptation of tick-borne encephalitis virus to BHK-21 cells results in the formation of multiple heparan sulfate binding sites in the envelope protein and attenuation in vivo. J Virol 75:5627–5637

Martin DN, Uprichard SL (2013) Identification of transferrin receptor 1 as a hepatitis C virus entry factor. Proc Natl Acad Sci U S A 110:10777–10782

Masciopinto F, Campagnoli S, Abrignani S, Uematsu Y, Pileri P (2001) The small extracellular loop of CD81 is necessary for optimal surface expression of the large loop, a putative HCV receptor. Virus Res 80:1–10

Mckeating JA, Zhang LQ, Logvinoff C, Flint M, Zhang J, Yu J, Butera D, Ho DD, Dustin LB, Rice CM, Balfe P (2004) Diverse hepatitis C virus glycoproteins mediate viral infection in a CD81-dependent manner. J Virol 78:8496–8505

Mee CJ, Harris HJ, Farquhar MJ, Wilson G, Reynolds G, Davis C, Van ISC, Balfe P, Mckeating JA (2009) Polarization restricts hepatitis C virus entry into HepG2 hepatoma cells. J Virol 83:6211–6221

Meertens L, Bertaux C, Dragic T (2006) Hepatitis C virus entry requires a critical postinternalization step and delivery to early endosomes via clathrin-coated vesicles. J Virol 80:11571–11578

Meertens L, Bertaux C, Cukierman L, Cormier E, Lavillette D, Cosset FL, Dragic T (2008) The tight junction proteins claudin-1, -6, and -9 are entry cofactors for hepatitis C virus. J Virol 82:3555–3560

Merz A, Long G, Hiet MS, Brugger B, Chlanda P, Andre P, Wieland F, Krijnse-Locker J, Bartenschlager R (2011) Biochemical and morphological properties of hepatitis C virus particles and determination of their lipidome. J Biol Chem 286:3018–3032

Meuleman P, Hesselgesser J, Paulson M, Vanwolleghem T, Desombere I, Reiser H, Leroux-Roels G (2008) Anti-CD81 antibodies can prevent a hepatitis C virus infection in vivo. Hepatology 48:1761–1768

Meunier JC, Engle RE, Faulk K, Zhao M, Bartosch B, Alter H, Emerson SU, Cosset FL, Purcell RH, Bukh J (2005) Evidence for cross-genotype neutralization of hepatitis C virus pseudo-particles and enhancement of infectivity by apolipoprotein C1. Proc Natl Acad Sci U S A 102:4560–4565

Miyamoto H, Okamoto H, Sato K, Tanaka T, Mishiro S (1992) Extraordinarily low density of hepatitis C virus estimated by sucrose density gradient centrifugation and the polymerase chain reaction. J Gen Virol 73(Pt 3):715–718

Osburn WO, Snider AE, Wells BL, Latanich R, Bailey JR, Thomas DL, Cox AL, Ray SC (2014) Clearance of hepatitis C infection is associated with the early appearance of broad neutralizing antibody responses. Hepatology 59:2140–2151

Owen DM, Huang H, Ye J, Gale M Jr (2009) Apolipoprotein E on hepatitis C virion facilitates infection through interaction with low-density lipoprotein receptor. Virology 394:99–108

Owsianka A, Clayton RF, Loomis-Price LD, Mckeating JA, Patel AH (2001) Functional analysis of hepatitis C virus E2 glycoproteins and virus-like particles reveals structural dissimilarities between different forms of E2. J Gen Virol 82:1877–1883

Park JH, Park S, Yang JS, Kwon OS, Kim S, Jang SK (2013) Discovery of cellular proteins required for the early steps of HCV infection using integrative genomics. PLoS One 8:e60333

Pestka JM, Zeisel MB, Blaser E, Schurmann P, Bartosch B, Cosset FL, Patel AH, Meisel H, Baumert J, Viazov S, Rispeter K, Blum HE, Roggendorf M, Baumert TF (2007) Rapid induction of virus-neutralizing antibodies and viral clearance in a single-source outbreak of hepatitis C. Proc Natl Acad Sci U S A 104:6025–6030

Petracca R, Falugi F, Galli G, Norais N, Rosa D, Campagnoli S, Burgio V, Di Stasio E, Giardina B, Houghton M, Abrignani S, Grandi G (2000) Structure-function analysis of hepatitis C virus envelope-CD81 binding. J Virol 74:4824–4830

Pileri P, Uematsu Y, Campagnoli S, Galli G, Falugi F, Petracca R, Weiner AJ, Houghton M, Rosa D, Grandi G, Abrignani S (1998) Binding of hepatitis C virus to CD81. Science 282:938–941

Ploss A, Evans MJ, Gaysinskaya VA, Panis M, You H, De Jong YP, Rice CM (2009) Human occludin is a hepatitis C virus entry factor required for infection of mouse cells. Nature 457:882–886

Prentoe J, Serre SB, Ramirez S, Nicosia A, Gottwein JM, Bukh J (2014) Hypervariable region 1 deletion and required adaptive envelope mutations confer decreased dependency on scavenger receptor class B type I and low-density lipoprotein receptor for hepatitis C virus. J Virol 88:1725–1739

Price DA, Bassendine MF, Norris SM, Golding C, Toms GL, Schmid ML, Morris CM, Burt AD, Donaldson PT (2006) Apolipoprotein epsilon3 allele is associated with persistent hepatitis C virus infection. Gut 55:715–718

Reynolds GM, Harris HJ, Jennings A, Hu K, Grove J, Lalor PF, Adams DH, Balfe P, Hubscher SG, Mckeating JA (2008) Hepatitis C virus receptor expression in normal and diseased liver tissue. Hepatology 47:418–427

Rous BA, Reaves BJ, Ihrke G, Briggs JA, Gray SR, Stephens DJ, Banting G, Luzio JP (2002) Role of adaptor complex AP-3 in targeting wild-type and mutated CD63 to lysosomes. Mol Biol Cell 13:1071–1082

Sainz B Jr, Barretto N, Martin DN, Hiraga N, Imamura M, Hussain S, Marsh KA, Yu X, Chayama K, Alrefai WA, Uprichard SL (2012) Identification of the Niemann-Pick C1-like 1 cholesterol absorption receptor as a new hepatitis C virus entry factor. Nat Med 18:281–285

Scarselli E, Ansuini H, Cerino R, Roccasecca RM, Acali S, Filocamo G, Traboni C, Nicosia A, Cortese R, Vitelli A (2002) The human scavenger receptor class B type I is a novel candidate receptor for the hepatitis C virus. EMBO J 21:5017–5025

Shi Q, Jiang J, Luo G (2013) Syndecan-1 serves as the major receptor for attachment of hepatitis C virus to the surfaces of hepatocytes. J Virol 87:6866–6875

Sourisseau M, Michta ML, Zony C, Israelow B, Hopcraft SE, Narbus CM, Parra Martin A, Evans MJ (2013) Temporal analysis of hepatitis C virus cell entry with occludin directed blocking antibodies. PLoS Pathog 9:e1003244

Sulkowski MS, Kang M, Matining R, Wyles D, Johnson VA, Morse GD, Amorosa V, Bhattacharya D, Coughlin K, Wong-Staal F, Glesby MJ (2014) Safety and antiviral activity of the HCV entry inhibitor ITX5061 in treatment-naive HCV-infected adults: a randomized, double-blind, phase 1b study. J Infect Dis 209:658–667

Takikawa S, Ishii K, Aizaki H, Suzuki T, Asakura H, Matsuura Y, Miyamura T (2000) Cell fusion activity of hepatitis C virus envelope proteins. J Virol 74:5066–5074

Tao W, Xu C, Ding Q, Li R, Xiang Y, Chung J, Zhong J (2009) A single point mutation in E2 enhances hepatitis C virus infectivity and alters lipoprotein association of viral particles. Virology 395:67–76

Thomssen R, Bonk S, Propfe C, Heermann KH, Kochel HG, Uy A (1992) Association of hepatitis C virus in human sera with beta-lipoprotein. Med Microbiol Immunol 181:293–300

Thomssen R, Bonk S, Thiele A (1993) Density heterogeneities of hepatitis C virus in human sera due to the binding of beta-lipoproteins and immunoglobulins. Med Microbiol Immunol 182:329–334

Timpe JM, Stamataki Z, Jennings A, Hu K, Farquhar MJ, Harris HJ, Schwarz A, Desombere I, Roels GL, Balfe P, Mckeating JA (2008) Hepatitis C virus cell-cell transmission in hepatoma cells in the presence of neutralizing antibodies. Hepatology 47:17–24

Trotard M, Lepere-Douard C, Regeard M, Piquet-Pellorce C, Lavillette D, Cosset FL, Gripon P, Le Seyec J (2009) Kinases required in hepatitis C virus entry and replication highlighted by small interference RNA screening. FASEB J 23:3780–3789

Voisset C, Callens N, Blanchard E, Op De Beeck A, Dubuisson J, Vu-Dac N (2005) High density lipoproteins facilitate hepatitis C virus entry through the scavenger receptor class B type I. J Biol Chem 280:7793–7799

Voisset C, Lavie M, Helle F, Op De Beeck A, Bilheu A, Bertrand-Michel J, Terce F, Cocquerel L, Wychowski C, Vu-Dac N, Dubuisson J (2008) Ceramide enrichment of the plasma membrane induces CD81 internalization and inhibits hepatitis C virus entry. Cell Microbiol 10:606–617

Von Hahn T, Lindenbach BD, Boullier A, Quehenberger O, Paulson M, Rice CM, Mckeating JA (2006) Oxidized low-density lipoprotein inhibits hepatitis C virus cell entry in human hepatoma cells. Hepatology 43:932–942

Wakita T, Pietschmann T, Kato T, Date T, Miyamoto M, Zhao Z, Murthy K, Habermann A, Krausslich HG, Mizokami M, Bartenschlager R, Liang TJ (2005) Production of infectious hepatitis C virus in tissue culture from a cloned viral genome. Nat Med 11:791–796

Westhaus S, Bankwitz D, Ernst S, Rohrmann K, Wappler I, Agne C, Luchtefeld M, Schieffer B, Sarrazin C, Manns MP, Pietschmann T, Ciesek S, Von Hahn T (2013) Characterization of the inhibition of hepatitis C virus entry by in vitro-generated and patient-derived oxidized low-density lipoprotein. Hepatology 57:1716–1724

Wieland S, Makowska Z, Campana B, Calabrese D, Dill MT, Chung J, Chisari FV, Heim MH (2014) Simultaneous detection of hepatitis C virus and interferon stimulated gene expression in infected human liver. Hepatology 59:2121–2130
Xiao F, Fofana I, Heydmann L, Barth H, Soulier E, Habersetzer F, Doffoel M, Bukh J, Patel AH, Zeisel MB, Baumert TF (2014) Hepatitis C virus cell-cell transmission and resistance to direct-acting antiviral agents. PLoS Pathog 10:e1004128
Yamauchi Y, Helenius A (2013) Virus entry at a glance. J Cell Sci 126:1289–1295
Yang W, Qiu C, Biswas N, Jin J, Watkins SC, Montelaro RC, Coyne CB, Wang T (2008) Correlation of the tight junction-like distribution of Claudin-1 to the cellular tropism of hepatitis C virus. J Biol Chem 283:8643–8653
Zahid MN, Turek M, Xiao F, Thi VL, Guerin M, Fofana I, Bachellier P, Thompson J, Delang L, Neyts J, Bankwitz D, Pietschmann T, Dreux M, Cosset FL, Grunert F, Baumert TF, Zeisel MB (2013) The postbinding activity of scavenger receptor class B type I mediates initiation of hepatitis C virus infection and viral dissemination. Hepatology 57:492–504
Zeisel MB, Koutsoudakis G, Schnober EK, Haberstroh A, Blum HE, Cosset FL, Wakita T, Jaeck D, Doffoel M, Royer C, Soulier E, Schvoerer E, Schuster C, Stoll-Keller F, Bartenschlager R, Pietschmann T, Barth H, Baumert TF (2007) Scavenger receptor class B type I is a key host factor for hepatitis C virus infection required for an entry step closely linked to CD81. Hepatology 46:1722–1731
Zeisel MB, Felmlee DJ, Baumert TF (2013) Hepatitis C virus entry. Curr Top Microbiol Immunol 369:87–112
Zhang J, Randall G, Higginbottom A, Monk P, Rice CM, Mckeating JA (2004) CD81 is required for hepatitis C virus glycoprotein-mediated viral infection. J Virol 78:1448–1455
Zhong J, Gastaminza P, Cheng G, Kapadia S, Kato T, Burton DR, Wieland SF, Uprichard SL, Wakita T, Chisari FV (2005) Robust hepatitis C virus infection in vitro. Proc Natl Acad Sci U S A 102:9294–9299
Zona L, Lupberger J, Sidahmed-Adrar N, Thumann C, Harris HJ, Barnes A, Florentin J, Tawar RG, Xiao F, Turek M, Durand SC, Duong FH, Heim MH, Cosset FL, Hirsch I, Samuel D, Brino L, Zeisel MB, Le Naour F, Mckeating JA, Baumert TF (2013) HRas signal transduction promotes hepatitis C virus cell entry by triggering assembly of the host tetraspanin receptor complex. Cell Host Microbe 13:302–313
Zona L, Tawar RG, Zeisel MB, Xiao F, Schuster C, Lupberger J, Baumert TF (2014) CD81-receptor associations – impact for hepatitis C virus entry and antiviral therapies. Viruses 6:875–892

Structural Proteins of HCV and Biological Functions

Kohji Moriishi and Yoshiharu Matsuura

Abstract Hepatitis C virus (HCV) is a major causative agent of liver disorders and a major risk factor for hepatocellular carcinoma. The induction of hepatocellular carcinoma by HCV is thought to involve not only chronic inflammation, but also the biological activity of HCV components. Structural proteins of HCV are composed of the core protein and two envelope proteins, E1 and E2. The HCV core protein has been reported to exhibit multiple biological functions involved in lipid synthesis, iron metabolism, insulin response, oxidative stress and cell growth, and to thereby contribute to the development of carcinogenesis and metabolic disorders. Moreover, several reports suggest that envelope proteins also play an important role in viral entry as well as HCV-related pathogenic events. However, the mechanism by which the structural proteins induce hepatitis C-related disorders has not been fully understood. This review focuses on the current status of biological responses mediated by HCV structural proteins.

Keywords HCV • Structural proteins • Core protein • Envelope protein • Oxidative stress • Insulin resistance • Lipid metabolism • Mitochondria

Hepatitis C virus (HCV) possesses a genome consisting of a single positive strand RNA with a nucleotide length of 9.6 kb, which encodes a single polyprotein. This polyprotein is matured by processing dependent on host and viral proteases, resulting in structural and nonstructural proteins (Grakoui et al. 1993a, b; Harada et al. 1991; Hijikata et al. 1991). Structural proteins consisting of the core protein and two envelope proteins E1 and E2 occupy one third of the N-terminal region of the polyprotein, while the remaining viral proteins consist of the viroporin p7 and nonstructural proteins which form a replication complex with host factors (Grakoui et al. 1993c). The structural proteins and host lipid components are employed for

K. Moriishi, D.V.M.,Ph.D.
Department of Microbiology, Division of Medicine, Graduate School of Medicine and Engineering, University of Yamanashi, 409-3898 Chuo-shi, Yamanashi, Japan

Y. Matsuura, D.V.M., Ph.D. (✉)
Department of Molecular Virology, Research Institute for Microbial Diseases, Osaka University, 565-0871 Osaka, Japan
e-mail: matsuura@biken.osaka-u.ac.jp

T. Miyamura et al. (eds.), *Hepatitis C Virus I*, DOI 10.1007/978-4-431-56098-2_6

formation of the viral particle (for review see (Moriishi and Matsuura 2012). The nucleocapsid consisting of mature core proteins and a viral genome is surrounded by an envelope composed of host lipids and viral envelope proteins.

The HCV core protein and envelope proteins are released from the viral polyprotein by host proteases. HCV structural proteins may provide a host for severe liver disorders over several decades of persistent infection. The HCV core protein is involved in formation of the viral particle as well as the induction of liver disorders, including metabolic diseases. In addition, it is more important that the core protein could induce hepatocellular carcinoma in mice regardless of other HCV viral proteins (Moriya et al. 1998). Accumulating evidence supports the notion that envelope proteins induce a stress response during persistent infection to lead to liver disorders. This review summarizes the biological functions of HCV structural proteins in the development of HCV-related disorders.

1 Maturation of HCV Structural Proteins for Assembly of Viral Particles

1.1 Processing and Modification of HCV Core Protein

Hepatitis C virus (HCV) belongs to the genus Hepacivirus of the *Flaviviridae* family. The *Flaviviridae* family is composed of four genera, Flavivirus, Pestivirus, Pegivirus and Hepacivirus. The viral genomic structures and the composition of viral proteins differ among these genera. The capsid, or core, protein is encoded in the 5′-regions of the viral genomes of three of the four genera, with the exception being Pegivirus. Pegivirus does not have a capsid protein, suggesting that unknown viral or host proteins may be involved in formation of the viral particle. The structure and processing of the capsid protein are variable in genera of the *Flaviviridae* family. The structural proteins of HCV were detected as processed proteins at the first time in mammal and insect cells (Matsuura et al. 1992), contributing to identification of their cleavage sites. The capsid proteins of HCV and GBV-B, which are classified into the genus Hepacivirus, are cleaved by signal peptide peptidase (SPP), following signal peptidase-dependent processing (McLauchlan et al. 2002; Targett-Adams et al. 2006), while the capsid protein of classical swine fever virus (CSFV), which belongs to the genus Pestivirus, is cleaved by SPP (Heimann et al. 2006). The C-terminal end of the mature HCV core protein expressed in insect cells was reported to be Phe177 or Leu179 (Hussy et al. 1996; Ogino et al. 2004). The C-terminal residue of the mature HCV core protein that was expressed in a human cell line was identified as Phe177 by mass spectrometry (Okamoto et al. 2008). Non-primate hepaciviruses have recently been identified in dogs, horses, rodents and bats (Burbelo et al. 2012; Drexler et al. 2013; Kapoor et al. 2011, 2013; Lyons et al. 2012; Tanaka et al. 2014). The C-terminal hydrophobic membrane-anchoring region of HCV core protein shows high homology to the core protein of equine hepacivirus, which is the most closely related homologue of HCV among non-primate hepaciviruses (Burbelo et al. 2012; Kapoor et al. 2013; Lyons et al. 2012). We recently reported that the core protein of equine hepacivirus was cleaved by SPP and then localized on the lipid droplets and partially on lipid-raft like membranes in a manner similar to HCV core protein (Tanaka et al. 2014). The secondary structures and cis-acting elements of the equine hepacivirus genome also exhibit characteristics similar to those of the HCV genome (Tanaka et al. 2014). The mechanism of the viral propagation may thus be conserved between equine hepacivirus and HCV.

A hydrophocity/hydrophilicity plot suggests that the core protein consists of three domains, domain 1 (2–118), 2 (119–174), and 3 (175–191) (Hope and McLauchlan 2000; McLauchlan 2000). The helix-loop-helix structure located in domain 2 is critical for association of the core protein with lipid droplets and shares common features with the core proteins of GBV-B (Hope et al. 2002). Three hydrophobic amino acid residues, Leu139, Val140, and Leu144, in domain 2 exhibit hydrophobic peaks within domain 2 and are responsible for SPP-dependent cleavage, membrane anchoring and virus production (Okamoto et al. 2004, 2008). Furthermore, comparative analysis between the JFH1 and Jc1 strains suggests that the efficiency of virus assembly is determined by the binding ability of domain 2 to lipid droplets (Shavinskaya et al. 2007). Cysteine residue 172 of HCV core protein is palmitoylated. Palmitoylation of the core protein is responsible for the virus production but not for SPP-dependent processing or LD localization of the core protein (Majeau et al. 2009). These results suggest that the hydrophobicity of domains 2 and 3 is critical for intracellular localization and SPP cleavage of the core protein and viral production.

Recently, herpesviruses and other pathogens have been reported to employ SPP for their life cycles and pathogenesis. Human cytomegalovirus protein US2 promoted dislocation of the class I major histocompatibility complex (MHC) heavy chain from the endoplasmic reticulum (ER) by direct interaction with SPP, resulting in proteasome-dependent degradation of the MHC class I heavy chain (Loureiro et al. 2006). Herpes simplex virus-1 exploited SPP by binding to the viral glycoprotein gK for its own replication (Allen et al. 2014). The human malaria *Plasmodium falciparum* expresses its own SPP on the cell surface. The malaria SPP recognizes Band3 in the red blood cells for invasion (Li et al. 2008). SPP inhibitors, L-685,458, NITD731 and LY411.575, were shown to block the growth of *P. falciparum* and the rodent malarial parasite *P. berghei* (Li et al. 2009c; Harbut et al. 2012; Parvanova et al. 2009). SPP or SPP-like proteases may be employed by other pathogens for their propagation and will be target molecules for the development of therapeutic compounds against several pathogens.

1.2 Structure of Envelope Proteins for the Viral Entry and Assembly

HCV envelope proteins E1 and E2 are cleaved from the polyprotein by signal peptidase (Hijikata et al. 1991). Both E1 and E2, each of which consists of a large ectodomain with a C-terminal transmembrane region, are classified into a group of type I membrane proteins and are reported to form non-covalent heterodimers (Deleersnyder et al. 1997). The envelope proteins are highly modified post-translationally at 6 and 11 potential sites for N-glycosylation (Goffard and Dubuisson 2003; Zhang et al. 2004), some of which are responsible for infectivity (Goffard et al. 2005). The core domain (the ectodomain E2 lacking HVR1) of the HCV E2 protein shares some basic characteristics with other class II fusion proteins, such as an immunoglobulin-like fold consisting of a β-sheet structure (Kong et al. 2013; Khan et al. 2014). However, the precise function of E1 and E2 in membrane fusion has not yet been fully clarified. Two hydrophobic regions spanning from 504 to 522 and from 604 to 624 in E2 are predicted to be potential fusion peptides (Khan et al. 2014; Lavillette et al. 2007; Krey et al. 2010), while the region spanning from 262 to 290 in E1 is reported to be important for membrane fusion (Li et al. 2009b). E2 should be responsible for the HCV entry step in cooperation with E1, but the mechanism underlying this step remains unclear.

The synthesized viral genome is wrapped with the core proteins to form a nucleocapsid on lipid droplets close to the ER, on which the viral genome is synthesized (Miyanari et al. 2007). A nucleocapsid egresses with envelope proteins into the ER membrane in close proximity to the lipid droplets. HCV particles in the patients' sera have been reported to exhibit densities of 1.03–1.25 g/ml (Thomssen et al. 1992, 1993). HCV particles with a density of lower than 1.06 g/ml are infectious to chimpanzees, while those with a higher density exhibit lower infectivity (Bradley et al. 1991; Hijikata et al. 1993). HCV particles interacting with lipoproteins in the sera of patients (Andre et al. 2002) were prepared from the fractions with very low to low buoyant densities (1.03–1.25 g/ml), and have been designated lipo-viro-particles (LVP) (Andre et al. 2002; Nielsen et al. 2006). LVP are composed of HCV particle components and very low-density lipoproteins (VLDL), including apolipoprotein B (ApoB) and apolipoprotein E (ApoE) (Andre et al. 2002). The HCV entry process on the surface of hepatocytes has been reported to be carried out by using entry factors including LDLR, CD81, scavenger receptor class B type I (SR-BI), and the tight junction proteins claudin-1 and occludin (Bartosch et al. 2003; Evans et al. 2007; Pileri et al. 1998; Ploss et al. 2009). Lectin receptors including DC-SIGN, L-SIGN, and langerin may be responsible for the invasive step from sinusoidal endothelial cells (Lozach et al. 2003; Pohlmann et al. 2003; Gardner et al. 2003; Chen et al. 2014). Envelope proteins with Man8/9 N-glycans exhibit higher binding to lectin receptors (DC-SIGN, L-SIGN and langerin) than to non-lectin receptors (CD81, SRBI, claudin-1, and occludin) in the presence of calcium ions, while HCV envelope proteins with Man5 N-glycans bound to non-lectin receptors at a higher affinity than lectin receptors (Chen et al. 2014). The HCV viral particle may be captured by lectin receptors on sinusoidal endothelial cells at a high affinity followed by infection to hepatocytes via non-lectin receptors. HCV envelope proteins interact with ApoE and ApoB in ER (Boyer et al. 2014). Intracellular and extracellular infectious particles also associate with ApoE and ApoB (Boyer et al. 2014). ApoE, but not E2, on the surface of LVP mediates the SR-BI-dependent entry step via the lipid transfer activity of SR-BI, although the HVR1 of E2 affects this step (Dao Thi et al. 2012). These results suggest that E2 HVR1 enhances the SR-BI-ApoE interaction for HCV entry. Further study will be required to clarify the mechanism by which HCV utilizes SR-BI for its entry step.

2 Biological Functions of Structural Proteins

2.1 Modulation of Lipid Metabolism by HCV Core Protein

Liver steatosis is frequently found in persistent HCV infection and results in accumulation of triglyceride and fatty acids in hepatocytes (see Negro 2010). However, the involvement of HCV infection in the development of fatty liver has not yet been clarified completely. Several reports support the notion that HCV core protein contributes to the accumulation of lipid droplets and hepatic steatosis in transgenic mice and cultural cells (Barba et al. 1997; Hope and McLauchlan 2000; Moriya et al. 1997). The lipid profiling of a core transgenic mouse of genotype 1b showed a similar composition to that of a hepatitis C patient (Koike et al. 2010; Miyoshi et al. 2011). Syntheses of triglycerides and fatty acids are transcriptionally regulated by the sterol regulatory element-binding proteins (SREBPs) (Horton et al. 2002). An HCV cell culture system derived from the genotype 3a strain showed that lipid accumulation was enhanced in cells infected with HCV genotype 3a compared to those infected with the genotype 2a strain JFH-1 (Kim et al. 2014). Patients infected with HCV genotype 3a exhibited progression of steatosis at a significantly higher rate than those with genotype 1a or 1b (Adinolfi et al. 2011; Mihm et al. 1997). Expression of the HCV core protein derived from genotype 3a induced lipid accumulation in lipid-free cultured cells at a higher level than expression of other genotype core proteins (Abid et al. 2005). The HCV core protein of genotype 3a stimulated activity of the fatty acid synthetase promoter at a significantly higher level than that of genotype 1b (Abid et al. 2005).

HCV infection or expression of the genotype 3a core protein was found to enhance the cleavage of SREBPs, leading to posttranslational activation of SREBPs (Waris et al. 2007). The recent report by Bose et al. suggested that the forkhead box transcription factor FoxO1 was activated by the HCV core protein or infection followed by activation of srebp-1c promoter activity, leading to the accumulation of lipids (Bose et al. 2014). However, controversial results were reported from hepatitis C patients. McPherson reported that SREBP-1c was not involved in

HCV-related steatosis (McPherson et al. 2008), whereas Lima-Cabello et al. reported that LXRα, SREBP-1c and -2, and fatty acid synthetase were overexpressed in the livers of HCV patients with steatosis (Lima-Cabello et al. 2011), suggesting that LXRα transcriptionally upregulates SREBP-1c expression followed by fatty acid synthetase expression. It has been reported that most of the genes under the control of SREBPs were upregulated during the early stage of HCV infection in the livers of chimpanzees (Bigger et al. 2004). Our previous data indicated that the core protein potentiates the binding ability of the LXRα-RXRα complex to the *srebp-1c* promoter in cultured cells and in the livers of core-transgenic mice (Moriishi et al. 2007). Upregulation of *srebp-1c* promoter activity may be associated with direct interaction between the core protein and RXRα (Tsutsumi et al. 2002b). Cholesterol and ApoB were significantly reduced in patients with severe hepatitis C or core-transgenic mice (Perlemuter et al. 2002). The microsomal triglyceride transfer protein (MTP) positively regulates the formation and secretion of very low-density lipoproteins. In core-transgenic mice, MTP-specific activity is significantly decreased (Perlemuter et al. 2002), resulting in accumulation of lipids in the liver. The gene related to the synthesis and secretion of lipids may be regulated by HCV infection or the core protein at a transcriptional and/or post-translational step.

Peroxisome proliferator activated receptors are nuclear receptors that transcriptionally regulate metabolic signaling (Halilbasic et al. 2013). PPARα regulates the genes encoding enzymes associated with peroxisomal microsomal and mitochondrial γ oxidation (Halilbasic et al. 2013). PPARα is expressed in the liver and downregulated in the HCV-infected liver and the core-expressing HepG2 cells (Dharancy et al. 2005). PPARα was decreased in mice infected with adenovirus expressing the HCV core protein (Yamaguchi et al. 2005). In an earlier study, severe liver steatosis was induced in core-transgenic mice (Moriya et al. 1997). HCV replication transcriptionally induced the expression of miR-27 in cell culture and an in vivo mouse model (Singaravelu et al. 2014). Both the HCV core protein and NS4B promote the expression of miR-27 through a PI3-K-dependent pathway. Transfection of miR-27 enhances the size and volume of lipid droplets in cultured cells and also impairs PPARα signaling. PPARα transcriptionally increases the genes regulating mitochondrial and peroxisomal fatty acid oxidation (Desvergne and Wahli 1999). An increase in miR-27 in infected cells also downregulates ANGPTL3, which is an inhibitor of lipoprotein lipase responsible for fatty acid uptake (Mattijssen and Kersten 2012). These data suggest that induction of miR-27 by HCV infection downregulates fatty acid oxidation via impairment of PPARα signaling and up-regulates fatty acid uptake via inhibition of ANGPTL3 expression, leading to development of liver steatosis. Unexpectedly, PPARα-knockout core-transgenic mice did not show steatosis (Tanaka et al. 2008). Furthermore, PPARα expression was required for induction of hepatocellular carcinoma by HCV core protein (Tanaka et al. 2008). Therefore, the HCV core protein may require a small amount of PPARα for the development of liver disorders and may maintain PPARα at a steady level.

PPARγ is involved in adipocyte differentiation and energy storage by adipocytes mediating an anabolic energy state (Halilbasic et al. 2013). In addition, PPARγ plays an important role in the development of liver steatosis (Gavrilova et al. 2003; Yu et al. 2003). The HCV core protein has been shown to potentiate PPARγ activity and transcriptionally upregulate SREBP1 activity, resulting in lipid accumulation. Furthermore, HCV core protein expression induced leptin receptor activation in hepatic stellate cells and contributed to transcriptional upregulation of MMP-1, PAPRγ and SREBP-1c, leading to promotion of hepatic fibrogenesis (Wu et al. 2013). PPARγ may thus be involved in HCV core-induced liver steatosis, in cooperation with PPARα.

2.2 *Regulation of Iron Metabolism by HCV Core Protein*

Iron overload has been reported as a common hallmark of chronic hepatitis C infection (Bonkovsky 2002; Boucher et al. 1997; Di Bisceglie et al. 1992). Accumulation of iron in the liver by HCV infection promotes liver inflammation and interferon resistance due to inhibition of the JAK-STAT pathway by oxidative stress (Olynyk et al. 1995; Bassett et al. 1999; Nishina et al. 2008; Fujita et al. 2007). Iron is involved in induction of reactive oxygen species (ROS). Iron Fe^{2+} reacts with hydrogen peroxide (H_2O_2) to yield Fe^{3+}, hydroxyl radical ($^{\cdot}OH$), and hydroxide ion ($OH-$) (Fenton reaction) (Graf et al. 1984). Hydroxyl radical reacts with lipids, resulting in lipid peroxidation (Okada 1996). Iron concentration in the liver is regulated by an import protein transferrin receptor and an export protein ferroportin (Pantopoulos et al. 2012). Imported iron atoms are enclosed with ferritin in cells and stored as iron-ferritin complexes (Ganz and Nemeth 2012; Liu and Theil 2005). Another iron-regulating protein, hepcidin, which is encoded on the gene HAMP, is a short peptide inducing internalization and degradation of ferroportin and regulates plasma iron concentration and iron metabolism in the liver (Ganz and Nemeth 2012; Nemeth et al. 2004, 2006). Expression of hepcidin is stimulated by iron overload and inflammation, and is suppressed by anemia and hypoxia (Nemeth and Ganz 2006). BMP6 is produced and secreted by various cell types and is a main regulator of hepcidin expression (Andriopoulos et al. 2009). BMP6 binds and stimulates dimers of BMP-RI/II to cooperate with the coreceptor hemojuvelin (Andriopoulos et al. 2009; Meynard et al. 2009; Xia et al. 2008), leading to downstream signaling including dimerization of Smad4 with Smad1/5/8 (Wang et al. 2005). Smad dimers transcriptionally induce expression of hepcidin. Screening of a whole genome using an siRNA library revealed that hepcidin-knockdown reduced HCV replication significantly (Tai et al. 2009), suggesting that hepcidin expression is required for HCV replication and control of iron metabolism. Hepcidin was shown to be transcriptionally enhanced by the HCV core protein through Smad4, STAT3 and CK2 (Foka et al. 2014). Knockdown of hepcidin impaired HCV replication in a replicon cell line (Bartolomei et al. 2011). Iron upregulates HCV replication by enhancement of IRES-dependent translation and expression of eIF3 and La (Cho et al. 2008; Theurl et al. 2004; Wang et al. 2012). However, knockdown of hepcidin suppressed IRES- as well as CAP-dependent translation (Tai et al. 2009). Hepcidin may contribute to translational regulations of the viral proteins and host proteins by accumulation of iron in HCV-infected cells.

2.3 Quality Control of HCV Structural Proteins

The core protein is modified with ubiquitin by host enzymes. The host E3 ligase E6AP catalyzes ubiquitination of the core protein to suppress viral production (Shirakura et al. 2007). The poly-ubiquitinated core protein is degraded in the cytosol in a proteasome-dependent manner. HCV core protein is also degraded in a ubiquitin-independent PA28γ-dependent pathway, leading to upregulation of the viral production by suppression of cytosolic ubiquitin-dependent degradation of the core protein (Moriishi et al. 2003, 2007). The host mechanisms of protein degradation may regulate HCV production and control the quality of the core protein for viral propagation. Although qualitative limitations of HCV envelope proteins have been regulated by an ER-associated degradation (ERAD) system (Saeed et al. 2011), inhibition of ER enhanced the viral production (Saeed et al. 2011), suggesting that unfolded envelope proteins positively regulate the HCV production.

The unfolded protein response (UPR) is carried out by three pathways, an IRE1α, a PERK and an ATF6-dependent pathway (Gardner et al. 2013). The luminal domains of the PERK, ATF6 and IRE1α proteins interact with the ER resident chaperone BiP (Bertolotti et al. 2000). BiP renders PERK, ATF6 and IRE1α inactive without accumulation of unfolded protein (Bertolotti et al. 2000), while the accumulation of unfolded proteins stimulates release of BiP from PERK, ATF6 or IRE1α, leading to the induction of genes related to protein folding, cell survival, autophagy and so on (Bertolotti et al. 2000). UPR stimulates expression of both MAP1LC3B and ATG5 by ATF4 and CHOP, which are induced by activation of PARK and ATF6 (Rouschop et al. 2010; Wang et al. 2014). ATF4 also activates transcription of CHOP (Kojima et al. 2003). The HCV core protein was recently shown to activate both the PERK and ATP6 pathways, but not the IRE1α pathway, to stimulate expression of MAP1LC3B, ATG12 and ATG5 (Wang et al. 2014), suggesting that autophagy is induced by the upregulation of ATG proteins through the UPR of HCV core protein. Expression of HCV envelop proteins induced the expression of CHOP through PERK and IRE1α pathways (Chan and Egan 2005). CHOP stimulates IP3R through Ero1α activation, followed by accumulation of Ca $^{2+}$ in mitochondria (Li et al. 2009a). UPR-induced accumulation of Ca^{2+} in mitochondria may be associated with ROS production in HCV infected cells, as described later.

2.4 Effect of HCV Infection on Mitochondria

The HCV core protein can enhance the production of ROS by damaging the mitochondrial electron transport system, and thereby contribute to the emergence of hepatocellular carcinoma (Moriya et al. 2001; Nunez et al. 2004; Okuda et al. 2002), suggesting that accumulation of lipids advances the occurrence of hepatocellular carcinoma by enhancing ROS production. Expression of HCV polyproteins in cultured sarcoma cells promoted the production of ROS and nitrogen species and inhibited complex I activity, resulting in activation of mitochondrial calcium uptake (Piccoli et al. 2007). The HCV core protein is localized in the lipid droplets, ER and mitochondria (Okuda et al. 2002) and could induce ROS, leading to accumulation of lipid peroxidation products and enhancement of antioxidant gene expression (Okuda et al. 2002). Upregulation of lipid peroxidation was observed in core-transgenic mice but not in wild type mice following treatment with CCl_4 (Okuda et al. 2002). The mitochondria of transgenic mice expressing HCV polyprotein exhibited enhancement of glutathione oxidation, decrease in NADPH contents, impairment of complex I activity and promotion of ROS production (Korenaga et al. 2005). Glutathione oxidation and ROS upregulation were also found in isolated mitochondria in the presence of recombinant core protein (Korenaga et al. 2005). Ca^{2+} uptake was increased by the recombinant core protein in isolated mitochondria (Korenaga et al. 2005). HCV core protein induced ER stress via an unfolded protein response and then potentiated production of ER chaperone proteins and release of Ca^{2+} from the ER store (Benali-Furet et al. 2005; Bergqvist et al. 2003). In addition, the HCV core protein was found to enhance mitochondrial Ca^{2+} uptake via the Ca^{2+} uniporter, which is localized in the mitochondrial inner membrane (Li et al. 2007). Furthermore, HCV core protein interacted with the mitochondria chaperone prohibitin to upregulate prohibitin stability in cultured cells and the transgenic mouse liver (Tsutsumi et al. 2009). HCV core protein inhibited the interaction between prohibitin and COX, resulting in the impairment of COX activity (Tsutsumi et al. 2009). These reports suggest that HCV core protein induces ER stress and Ca^{2+} release from the ER and then stimulates mitochondrial Ca^{2+} uptake to upregulate ROS production. In addition, the HCV core protein may impair COX activity by both sequestering prohibitin and decreasing glutathione, leading to further enhancement of ROS production.

2.5 Insulin Resistance

Epidemiological studies have clearly established an association between type 2 diabetes mellitus and HCV infection (Cavaghan et al. 2000; Kahn 1998). Type 2 diabetes is a complex disease characterized by the high-level production of hepatic glucose due to insulin resistance, resulting in glucose tolerance hyperglycemia (Cavaghan et al. 2000; Kahn 1998). Insulin is ordinarily produced at a sufficient level in type 2 diabetes mellitus patients; however, the glucose level cannot be decreased due to a disorder in insulin signaling. Insulin receptor is a tyrosine kinase composed of two subunits (Draznin 2006; Youngren 2007). Binding of insulin activates insulin receptor, which triggers Tyr phosphorylation of insulin receptor substrate 1 (IRS1) (Draznin 2006; Youngren 2007). The phosphorylated IRS1 and IRS2 positively regulate PI3K, which phosphorylates phosphatidylinositol 4, 5-bisphophate into phosphatidylinositol-3,4,5-triphosphate (PIP3). PDK1 and PDK2 are recruited by the resulting PIP3 with Akt and then phosphorylate Akt at Thr308 and Ser 473 (Burgering and Coffer 1995; Taniguchi et al. 2006; Alessi et al. 1996; Manning and Cantley 2007), resulting in activation of Akt. Phosphorylated Akt itself phosphorylates a glucose transporter, GLUT-4, contributing to translocation of GLUT-4 to the plasma membrane for upregulation of glucose uptake (Taniguchi et al. 2006; Thirone et al. 2006).

Elevation of TNFα production is one of the risk factors for insulin resistance (Gurav 2012). TNFα can indirectly mediate phosphorylation of IRS1at multiple sites through the activation of several Ser kinases, including JNK, IKKβ and ERK (Gao et al. 2003; Solinas and Karin 2010). TNFα stimulates activation of MEKK1, ASK1 and TAK1, which phosphorylate MKK7 for activation (Nakajima et al. 2006). In the same study, phosphorylated MKK7 was able to phosphorylate JNK (Nakajima et al. 2006). JNK1 has been shown to interact with IRS1 through the region spanning from the residues 555–898 (Aguirre et al. 2000). Phosphorylation of IRS1 Ser307 was detected in cultured cells treated with a JNK agonist, resulting in a decrease in Tyr phosphorylation of IRS1(Aguirre et al. 2000). Rui et al. suggested that a TNFα-dependent, JNK-independent mechanism may also be associated with phosphorylation of IRS1 Ser307 (Rui et al. 2001). Insulin was shown to stimulate the PI3K pathway to enhance phosphorylation of IRS1 Ser307 (Aguirre et al. 2002; Rui et al. 2001). Insulin-stimulated JNK phosphorylates IRS1 S307 may be a negative feedback pathway of insulin signaling (Lee et al. 2003). Phosphorylation of IRS1 Ser307 by JNK impairs the binding ability of IRS1 to insulin receptor (Aguirre et al. 2002), while phosphorylation of IRS1 Ser302 by JNK may also be involved in the negative feedback of insulin signaling (Werner et al. 2004). An increase in TNFα and a decrease in Tyr phosphorylation of IRS1 were observed in both the livers of HCV core gene transgenic mice and hepatitis C patients (Shintani et al. 2004; Miyamoto et al. 2007). Furthermore, JNK and its downstream factor AP-1 have been shown to be activated in core gene transgenic mice (Tsutsumi et al. 2002a). The HCV core protein also activated JNK and enhanced phosphorylation of IRS1 Ser312, leading to a decrease in Tyr phosphorylation of IRS1 and inhibition of insulin signaling (Banerjee et al. 2008). HCV core protein may activate JNK through upregulation of TNFα production, leading to insulin resistance through Ser phosphorylation of IRS1.

Evidence of the involvement of suppressor of cytokine signaling (SOCS) proteins in HCV-associated insulin resistance has been accumulating. SOCS-1 and SOCS-3 show relatively high homology and share similar functions. Both SOCS-1 and SOCS-3 can bind to insulin receptors irrespective of their phosphorylation status and impair Try phosphorylation of IRS-1 (Ueki et al. 2004). Further, SOCS-1 and SOCS-3 were shown to promote the degradation of IRS-1/2 via the ubiquitin-proteasome pathway (Rui et al. 2002). SOCS-1, but not SOCS-3, was decreased in the livers of core gene transgenic mice and hepatitis C patients, as well as in HepG2 cells expressing HCV core protein (Miyoshi et al. 2005). Expression of SOCS-3 was promoted after IFN treatment in HCV-infected chimpanzees, whereas the human liver showed variable responses to different treatments (Huang et al. 2007). SOCS-3 expression was significantly promoted in peripheral lymphocytes prepared from genotype 1b-infected IFN-non-responders (Persico et al. 2007). HCV core protein stimulated the expression of SOCS-3 and then enhanced ubiquitination of IRS1 and IRS2, leading to a decrease in IRS1/2 in a proteasome-dependent pathway (Kawaguchi et al. 2004). Induction of SOCS3 expression by the HCV core protein may be associated with the core protein mutations of Met70 and Leu91 (Funaoka et al. 2011), which are statistical predictors of low response to IFN/ribavirin therapy (Akuta et al. 2005, 2010). The regulation of IRS1 by HCV core protein may be accomplished by a genotype-specific pathway. Ubiquitin-dependent degradation of IRS1 was observed by the HCV core protein of genotypes 1b and 3a (Pazienza et al. 2007). In addition, IRS1 was decreased transcriptionally by downregulation of PPARγ and post-translationally by upregulation of SOCS-7 (Pazienza et al. 2007), while the core protein of genotype 1b activated mTOR, which suppresses IRS1 by Ser/Thr phosphorylation (Pazienza et al. 2007). E2 transcriptionally promoted the expression of SOCS3 and ubiquitination-dependent downregulation of IRS1, resulting in the impairment of Akt and GSK3 (Hsieh et al. 2012). The HCV core protein may regulate SOCS proteins cooperating with E2 under infectious conditions.

2.6 *Involvement of Envelope Proteins in Biological Functions*

Envelope proteins may control tight junction and facilitate secondary invasion of HCV after primary infection. Occludin, claudin-1 and ZO-1, which are tight junction proteins, are localized in the baso-lateral membrane position of Huh7, while these tight junction proteins were defused in Huh7 harboring a full-genomic but not a sub-genomic HCV replicon (Benedicto et al. 2008). Exogenous expression of HCV structural proteins, but not core alone, resulted in the translocation of tight junction proteins irrespective of the viral replication (Benedicto et al. 2008). HCV envelope proteins may facilitate subsequent virus infection by disruption of tight junction.

HCV envelope proteins may regulate ROS production and cell death. HCV infection was shown to stimulate production of ROS and NO and to reduce mitochondrial transmembrane potential (Machida et al. 2006), leading to double-stranded DNA breaks and apoptosis. Although expression of core, E1, or NS3 could induce ROS production in cultured cells (Machida et al. 2006), regulation of apoptosis by E2 is controversial. Chiou et al. reported that E2 induces apoptosis by cytoplasmic release of cytochrome c, upregulation of Bax and downregulation of Bcl-2 followed by activations of caspases-3, 8, and 9 (Chiou et al. 2006), whereas apoptosis induced by the death ligand TRAIL was suppressed by the expression of E2 (Lee et al. 2005). The expression of E2 may be capable of supporting the HCV replication by inhibiting apoptosis (Lee et al. 2005).

Phosphorylation and activation of STAT1 were enhanced by the expression of both HCV E2 and HIV gp120 (Balasubramanian et al. 2006). Lyn kinase, p38MAP kinase and protein kinase C δ are responsible for STAT1 phosphorylation (Balasubramanian et al. 2006). An increase in STAT1 might contribute to apoptosis in the hepatocytes of patients co-infected with HCV and HIV. HCV infection down-regulated the amounts of miR181c, which targets homeobox A1 (HOXA1), by modulating C/EBP-β (Mukherjee et al. 2014). In the same study, HOXA1 expression was potentiated in HCV-infected cells (Mukherjee et al. 2014). In addition, miR-181c was shown to bind directly to the E1 or NS5A gene (Mukherjee et al. 2014). Finally, HOXA1 promotes cell growth through upregulation of STAT3 and STAT5 (Mohankumar et al. 2007). These data suggest that the transcriptional and posttranscriptional down-regulation of miR-181c by HCV infection might contribute to activation of HOXA1 followed by upregulation of STA3 and STAT5.

3 Conclusions

The structural proteins of HCV are basically employed for formation of a viral particle like structural proteins of other enveloped viruses. The HCV core proteins is processed by host proteases, and then associated with lipid droplets and intracellular compartments for formation of nucleocapsid, while HCV glycoproteins, E1 and E2, are localized in ER membrane in close proximity to the lipid droplets. Both envelope proteins are classified into a group of type I membrane proteins, and are reported to form non-covalent heterodimers. The recent report of structural analysis revealed that HCV E2 protein is classified into the family of class II fusion protein. The envelope proteins play an important role in an entry step cooperating with several host entry factors, lectin and lipoproteins. In this text, we also summarized the biological functions of HCV structural proteins (Fig. 1). To date, it has not been fully clarified how HCV can cause hepatocellular carcinoma in humans. Persistent inflammation over a long period of time is expected to be associated with the development of hepatocellular carcinoma, due to both genomic alterations and biological functions of the HCV proteins. HCV core protein upregulates uptake of free fatty acids and the transcriptional activities of SREBPs, and down-regulates MTP function and β-oxidation, leading to liver steatosis. In addition, the structural proteins induce accumulation of mitochondrial Ca^{2+} and iron via ER stress and functions of hepcidin and Ca^{2+} uniporter, resulting in an increase in ROS. Oxidative stress induced by HCV infection may be one of the causative agents related to the

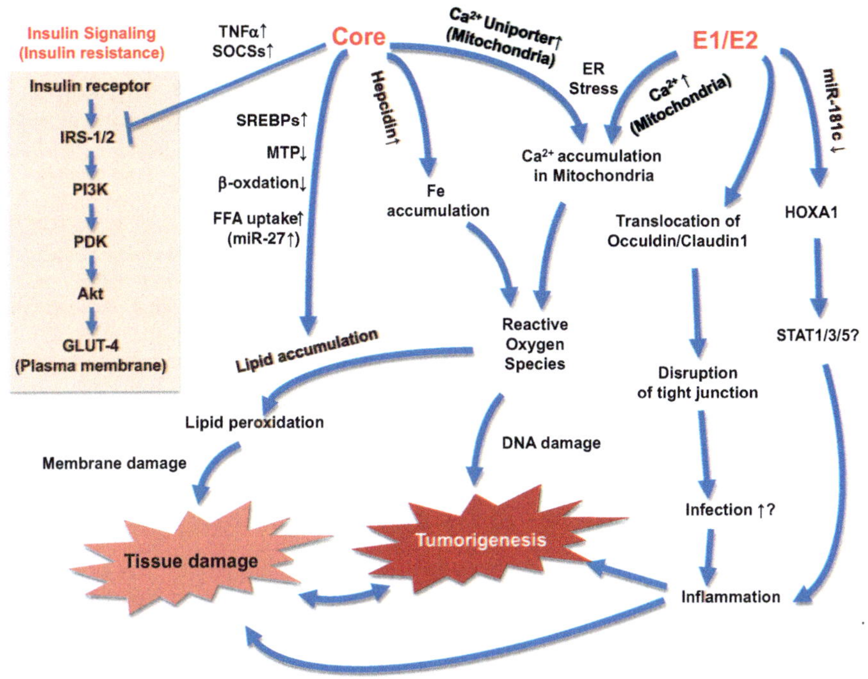

Fig. 1 Schematic diagram of the biological functions of HCV structural proteins. The biological and pathological actions induced by HCV structural proteins are summarized following the text

genetic alterations, including DNA double-strand breaks. In addition, retrotransposition targeting specific genes is predicted to be one of the potential causative agents of hepatocellular carcinoma in hepatitis B and C patients (Shukla et al. 2013). However, the mechanism by which HCV-related retrotransposition is induced has not been fully understood. Further study will be required to understand how carcinogenesis is related to hepatitis viruses and to develop antiviral agents for the eradication of these viruses in humans.

References

Abid K, Pazienza V, de Gottardi A, Rubbia-Brandt L, Conne B, Pugnale P, Rossi C, Mangia A, Negro F (2005) An in vitro model of hepatitis C virus genotype 3a-associated triglycerides accumulation. J Hepatol 42(5):744–751

Adinolfi LE, Restivo L, Zampino R, Lonardo A, Loria P (2011) Metabolic alterations and chronic hepatitis C: treatment strategies. Expert Opin Pharmacother 12(14):2215–2234

Aguirre V, Uchida T, Yenush L, Davis R, White MF (2000) The c-Jun NH(2)-terminal kinase promotes insulin resistance during association with insulin receptor substrate-1 and phosphorylation of Ser(307). J Biol Chem 275(12):9047–9054

Aguirre V, Werner ED, Giraud J, Lee YH, Shoelson SE, White MF (2002) Phosphorylation of Ser307 in insulin receptor substrate-1 blocks interactions with the insulin receptor and inhibits insulin action. J Biol Chem 277(2):1531–1537

Akuta N, Suzuki F, Sezaki H, Suzuki Y, Hosaka T, Someya T, Kobayashi M, Saitoh S, Watahiki S, Sato J, Matsuda M, Kobayashi M, Arase Y, Ikeda K, Kumada H (2005) Association of amino acid substitution pattern in core protein of hepatitis C virus genotype 1b high viral load and non-virological response to interferon-ribavirin combination therapy. Intervirology 48 (6):372–380

Akuta N, Suzuki F, Hirakawa M, Kawamura Y, Yatsuji H, Sezaki H, Suzuki Y, Hosaka T, Kobayashi M, Kobayashi M, Saitoh S, Arase Y, Ikeda K, Chayama K, Nakamura Y, Kumada H (2010) Amino acid substitution in hepatitis C virus core region and genetic variation near the interleukin 28B gene predict viral response to telaprevir with peginterferon and ribavirin. Hepatology 52(2):421–429

Alessi DR, Andjelkovic M, Caudwell B, Cron P, Morrice N, Cohen P, Hemmings BA (1996) Mechanism of activation of protein kinase B by insulin and IGF-1. EMBO J 15(23):6541–6551

Allen SJ, Mott KR, Matsuura Y, Moriishi K, Kousoulas KG, Ghiasi H (2014) Binding of HSV-1 glycoprotein K (gK) to signal peptide peptidase (SPP) is required for virus infectivity. PLoS One 9(1):e85360

Andre P, Komurian-Pradel F, Deforges S, Perret M, Berland JL, Sodoyer M, Pol S, Brechot C, Paranhos-Baccala G, Lotteau V (2002) Characterization of low- and very-low-density hepatitis C virus RNA-containing particles. J Virol 76(14):6919–6928

Andriopoulos B Jr, Corradini E, Xia Y, Faasse SA, Chen S, Grgurevic L, Knutson MD, Pietrangelo A, Vukicevic S, Lin HY, Babitt JL (2009) BMP6 is a key endogenous regulator of hepcidin expression and iron metabolism. Nat Genet 41(4):482–487

Balasubramanian A, Ganju RK, Groopman JE (2006) Signal transducer and activator of transcription factor 1 mediates apoptosis induced by hepatitis C virus and HIV envelope proteins in hepatocytes. J Infect Dis 194(5):670–681

Banerjee S, Saito K, Ait-Goughoulte M, Meyer K, Ray RB, Ray R (2008) Hepatitis C virus core protein upregulates serine phosphorylation of insulin receptor substrate-1 and impairs the downstream akt/protein kinase B signaling pathway for insulin resistance. J Virol 82 (6):2606–2612

Barba G, Harper F, Harada T, Kohara M, Goulinet S, Matsuura Y, Eder G, Schaff Z, Chapman MJ, Miyamura T, Brechot C (1997) Hepatitis C virus core protein shows a cytoplasmic localization and associates to cellular lipid storage droplets. Proc Natl Acad Sci U S A 94(4):1200–1205

Bartolomei G, Cevik RE, Marcello A (2011) Modulation of hepatitis C virus replication by iron and hepcidin in Huh7 hepatocytes. J Gen Virol 92(Pt 9):2072–2081

Bartosch B, Vitelli A, Granier C, Goujon C, Dubuisson J, Pascale S, Scarselli E, Cortese R, Nicosia A, Cosset FL (2003) Cell entry of hepatitis C virus requires a set of co-receptors that include the CD81 tetraspanin and the SR-B1 scavenger receptor. J Biol Chem 278 (43):41624–41630

Bassett SE, Di Bisceglie AM, Bacon BR, Sharp RM, Govindarajan S, Hubbard GB, Brasky KM, Lanford RE (1999) Effects of iron loading on pathogenicity in hepatitis C virus-infected chimpanzees. Hepatology 29(6):1884–1892

Benali-Furet NL, Chami M, Houel L, De Giorgi F, Vernejoul F, Lagorce D, Buscail L, Bartenschlager R, Ichas F, Rizzuto R, Paterlini-Brechot P (2005) Hepatitis C virus core triggers apoptosis in liver cells by inducing ER stress and ER calcium depletion. Oncogene 24 (31):4921–4933

Benedicto I, Molina-Jimenez F, Barreiro O, Maldonado-Rodriguez A, Prieto J, Moreno-Otero R, Aldabe R, Lopez-Cabrera M, Majano PL (2008) Hepatitis C virus envelope components alter localization of hepatocyte tight junction-associated proteins and promote occludin retention in the endoplasmic reticulum. Hepatology 48(4):1044–1053

Bergqvist A, Sundstrom S, Dimberg LY, Gylfe E, Masucci MG (2003) The hepatitis C virus core protein modulates T cell responses by inducing spontaneous and altering T-cell receptor-triggered Ca2+ oscillations. J Biol Chem 278(21):18877–18883
Bertolotti A, Zhang Y, Hendershot LM, Harding HP, Ron D (2000) Dynamic interaction of BiP and ER stress transducers in the unfolded-protein response. Nat Cell Biol 2(6):326–332
Bigger CB, Guerra B, Brasky KM, Hubbard G, Beard MR, Luxon BA, Lemon SM, Lanford RE (2004) Intrahepatic gene expression during chronic hepatitis C virus infection in chimpanzees. J Virol 78(24):13779–13792
Bonkovsky HL (2002) Iron as a comorbid factor in chronic viral hepatitis. Am J Gastroenterol 97 (1):1–4
Bose SK, Kim H, Meyer K, Wolins N, Davidson NO, Ray R (2014) Forkhead box transcription factor regulation and lipid accumulation by hepatitis C virus. J Virol 88(8):4195–4203
Boucher E, Bourienne A, Adams P, Turlin B, Brissot P, Deugnier Y (1997) Liver iron concentration and distribution in chronic hepatitis C before and after interferon treatment. Gut 41 (1):115–120
Boyer A, Dumans A, Beaumont E, Etienne L, Roingeard P, Meunier JC (2014) The association of hepatitis C virus glycoproteins with apolipoproteins E and B early in assembly is conserved in lipoviral particles. J Biol Chem 289(27):18904–18913
Bradley D, McCaustland K, Krawczynski K, Spelbring J, Humphrey C, Cook EH (1991) Hepatitis C virus: buoyant density of the factor VIII-derived isolate in sucrose. J Med Virol 34 (3):206–208
Burbelo PD, Dubovi EJ, Simmonds P, Medina JL, Henriquez JA, Mishra N, Wagner J, Tokarz R, Cullen JM, Iadarola MJ, Rice CM, Lipkin WI, Kapoor A (2012) Serology-enabled discovery of genetically diverse hepaciviruses in a new host. J Virol 86(11):6171–6178
Burgering BM, Coffer PJ (1995) Protein kinase B (c-Akt) in phosphatidylinositol-3-OH kinase signal transduction. Nature 376(6541):599–602
Cavaghan MK, Ehrmann DA, Polonsky KS (2000) Interactions between insulin resistance and insulin secretion in the development of glucose intolerance. J Clin Invest 106(3):329–333
Chan SW, Egan PA (2005) Hepatitis C virus envelope proteins regulate CHOP via induction of the unfolded protein response. FASEB J Off Publ Fed Am Soc Exp Biol 19(11):1510–1512
Chen PC, Chuang PK, Chen CH, Chan YT, Chen JR, Lin SW, Ma C, Hsu TL, Wong CH (2014) Role of N-linked glycans in the interactions of recombinant HCV envelope glycoproteins with cellular receptors. ACS Chem Biol 9(7):1437–1443
Chiou HL, Hsieh YS, Hsieh MR, Chen TY (2006) HCV E2 may induce apoptosis of Huh-7 cells via a mitochondrial-related caspase pathway. Biochem Biophys Res Commun 345(1):453–458
Cho H, Lee HC, Jang SK, Kim YK (2008) Iron increases translation initiation directed by internal ribosome entry site of hepatitis C virus. Virus Genes 37(2):154–160
Dao Thi VL, Granier C, Zeisel MB, Guerin M, Mancip J, Granio O, Penin F, Lavillette D, Bartenschlager R, Baumert TF, Cosset FL, Dreux M (2012) Characterization of hepatitis C virus particle subpopulations reveals multiple usage of the scavenger receptor BI for entry steps. J Biol Chem 287(37):31242–31257
Deleersnyder V, Pillez A, Wychowski C, Blight K, Xu J, Hahn YS, Rice CM, Dubuisson J (1997) Formation of native hepatitis C virus glycoprotein complexes. J Virol 71(1):697–704
Desvergne B, Wahli W (1999) Peroxisome proliferator-activated receptors: nuclear control of metabolism. Endocr Rev 20(5):649–688
Dharancy S, Malapel M, Perlemuter G, Roskams T, Cheng Y, Dubuquoy L, Podevin P, Conti F, Canva V, Philippe D, Gambiez L, Mathurin P, Paris JC, Schoonjans K, Calmus Y, Pol S, Auwerx J, Desreumaux P (2005) Impaired expression of the peroxisome proliferator-activated receptor alpha during hepatitis C virus infection. Gastroenterology 128(2):334–342
Di Bisceglie AM, Axiotis CA, Hoofnagle JH, Bacon BR (1992) Measurements of iron status in patients with chronic hepatitis. Gastroenterology 102(6):2108–2113

Draznin B (2006) Molecular mechanisms of insulin resistance: serine phosphorylation of insulin receptor substrate-1 and increased expression of p85alpha: the two sides of a coin. Diabetes 55 (8):2392–2397

Drexler JF, Corman VM, Muller MA, Lukashev AN, Gmyl A, Coutard B, Adam A, Ritz D, Leijten LM, van Riel D, Kallies R, Klose SM, Gloza-Rausch F, Binger T, Annan A, Adu-Sarkodie Y, Oppong S, Bourgarel M, Rupp D, Hoffmann B, Schlegel M, Kummerer BM, Kruger DH, Schmidt-Chanasit J, Setien AA, Cottontail VM, Hemachudha T, Wacharapluesadee S, Osterrieder K, Bartenschlager R, Matthee S, Beer M, Kuiken T, Reusken C, Leroy EM, Ulrich RG, Drosten C (2013) Evidence for novel hepaciviruses in rodents. PLoS Pathog 9(6): e1003438

Evans MJ, von Hahn T, Tscherne DM, Syder AJ, Panis M, Wolk B, Hatziioannou T, McKeating JA, Bieniasz PD, Rice CM (2007) Claudin-1 is a hepatitis C virus co-receptor required for a late step in entry. Nature 446(7137):801–805

Foka P, Dimitriadis A, Kyratzopoulou E, Giannimaras DA, Sarno S, Simos G, Georgopoulou U, Mamalaki A (2014) A complex signaling network involving protein kinase CK2 is required for hepatitis C virus core protein-mediated modulation of the iron-regulatory hepcidin gene expression. Cell Mol Life Sci CMLS 71(21):4243–4258

Fujita N, Sugimoto R, Urawa N, Araki J, Mifuji R, Yamamoto M, Horiike S, Tanaka H, Iwasa M, Kobayashi Y, Adachi Y, Kaito M (2007) Hepatic iron accumulation is associated with disease progression and resistance to interferon/ribavirin combination therapy in chronic hepatitis C. J Gastroenterol Hepatol 22(11):1886–1893

Funaoka Y, Sakamoto N, Suda G, Itsui Y, Nakagawa M, Kakinuma S, Watanabe T, Mishima K, Ueyama M, Onozuka I, Nitta S, Kitazume A, Kiyohashi K, Murakawa M, Azuma S, Tsuchiya K, Watanabe M (2011) Analysis of interferon signaling by infectious hepatitis C virus clones with substitutions of core amino acids 70 and 91. J Virol 85(12):5986–5994

Ganz T, Nemeth E (2012) Hepcidin and iron homeostasis. Biochim Biophys Acta 1823 (9):1434–1443

Gao Z, Zuberi A, Quon MJ, Dong Z, Ye J (2003) Aspirin inhibits serine phosphorylation of insulin receptor substrate 1 in tumor necrosis factor-treated cells through targeting multiple serine kinases. J Biol Chem 278(27):24944–24950

Gardner JP, Durso RJ, Arrigale RR, Donovan GP, Maddon PJ, Dragic T, Olson WC (2003) L-SIGN (CD 209L) is a liver-specific capture receptor for hepatitis C virus. Proc Natl Acad Sci U S A 100(8):4498–4503

Gardner BM, Pincus D, Gotthardt K, Gallagher CM, Walter P (2013) Endoplasmic reticulum stress sensing in the unfolded protein response. Cold Spring Harb Perspect Biol 5(3):a013169

Gavrilova O, Haluzik M, Matsusue K, Cutson JJ, Johnson L, Dietz KR, Nicol CJ, Vinson C, Gonzalez FJ, Reitman ML (2003) Liver peroxisome proliferator-activated receptor gamma contributes to hepatic steatosis, triglyceride clearance, and regulation of body fat mass. J Biol Chem 278(36):34268–34276

Goffard A, Dubuisson J (2003) Glycosylation of hepatitis C virus envelope proteins. Biochimie 85 (3–4):295–301

Goffard A, Callens N, Bartosch B, Wychowski C, Cosset FL, Montpellier C, Dubuisson J (2005) Role of N-linked glycans in the functions of hepatitis C virus envelope glycoproteins. J Virol 79(13):8400–8409

Graf E, Mahoney JR, Bryant RG, Eaton JW (1984) Iron-catalyzed hydroxyl radical formation. Stringent requirement for free iron coordination site. J Biol Chem 259(6):3620–3624

Grakoui A, McCourt DW, Wychowski C, Feinstone SM, Rice CM (1993a) Characterization of the hepatitis C virus-encoded serine proteinase: determination of proteinase-dependent polyprotein cleavage sites. J Virol 67(5):2832–2843

Grakoui A, McCourt DW, Wychowski C, Feinstone SM, Rice CM (1993b) A second hepatitis C virus-encoded proteinase. Proc Natl Acad Sci U S A 90(22):10583–10587

Grakoui A, Wychowski C, Lin C, Feinstone SM, Rice CM (1993c) Expression and identification of hepatitis C virus polyprotein cleavage products. J Virol 67(3):1385–1395

Gurav AN (2012) Periodontitis and insulin resistance: casual or causal relationship? Diabetes Metab J 36(6):404–411
Halilbasic E, Baghdasaryan A, Trauner M (2013) Nuclear receptors as drug targets in cholestatic liver diseases. Clin Liver Dis 17(2):161–189
Harada S, Watanabe Y, Takeuchi K, Suzuki T, Katayama T, Takebe Y, Saito I, Miyamura T (1991) Expression of processed core protein of hepatitis C virus in mammalian cells. J Virol 65 (6):3015–3021
Harbut MB, Patel BA, Yeung BK, McNamara CW, Bright AT, Ballard J, Supek F, Golde TE, Winzeler EA, Diagana TT, Greenbaum DC (2012) Targeting the ERAD pathway via inhibition of signal peptide peptidase for antiparasitic therapeutic design. Proc Natl Acad Sci U S A 109 (52):21486–21491
Heimann M, Roman-Sosa G, Martoglio B, Thiel HJ, Rumenapf T (2006) Core protein of pestiviruses is processed at the C terminus by signal peptide peptidase. J Virol 80 (4):1915–1921
Hijikata M, Kato N, Ootsuyama Y, Nakagawa M, Shimotohno K (1991) Gene mapping of the putative structural region of the hepatitis C virus genome by in vitro processing analysis. Proc Natl Acad Sci U S A 88(13):5547–5551
Hijikata M, Shimizu YK, Kato H, Iwamoto A, Shih JW, Alter HJ, Purcell RH, Yoshikura H (1993) Equilibrium centrifugation studies of hepatitis C virus: evidence for circulating immune complexes. J Virol 67(4):1953–1958
Hope RG, McLauchlan J (2000) Sequence motifs required for lipid droplet association and protein stability are unique to the hepatitis C virus core protein. J Gen Virol 81(Pt 8):1913–1925
Hope RG, Murphy DJ, McLauchlan J (2002) The domains required to direct core proteins of hepatitis C virus and GB virus-B to lipid droplets share common features with plant oleosin proteins. J Biol Chem 277(6):4261–4270
Horton JD, Goldstein JL, Brown MS (2002) SREBPs: activators of the complete program of cholesterol and fatty acid synthesis in the liver. J Clin Invest 109(9):1125–1131
Hsieh MJ, Lan KP, Liu HY, Zhang XZ, Lin YF, Chen TY, Chiou HL (2012) Hepatitis C virus E2 protein involve in insulin resistance through an impairment of Akt/PKB and GSK3beta signaling in hepatocytes. BMC Gastroenterol 12:74
Huang Y, Feld JJ, Sapp RK, Nanda S, Lin JH, Blatt LM, Fried MW, Murthy K, Liang TJ (2007) Defective hepatic response to interferon and activation of suppressor of cytokine signaling 3 in chronic hepatitis C. Gastroenterology 132(2):733–744
Hussy P, Langen H, Mous J, Jacobsen H (1996) Hepatitis C virus core protein: carboxy-terminal boundaries of two processed species suggest cleavage by a signal peptide peptidase. Virology 224(1):93–104
Kahn BB (1998) Type 2 diabetes: when insulin secretion fails to compensate for insulin resistance. Cell 92(5):593–596
Kapoor A, Simmonds P, Gerold G, Qaisar N, Jain K, Henriquez JA, Firth C, Hirschberg DL, Rice CM, Shields S, Lipkin WI (2011) Characterization of a canine homolog of hepatitis C virus. Proc Natl Acad Sci U S A 108(28):11608–11613
Kapoor A, Simmonds P, Scheel TK, Hjelle B, Cullen JM, Burbelo PD, Chauhan LV, Duraisamy R, Sanchez Leon M, Jain K, Vandegrift KJ, Calisher CH, Rice CM, Lipkin WI (2013) Identification of rodent homologs of hepatitis C virus and pegiviruses. mBiol 4(2):e00216-00213
Kawaguchi T, Yoshida T, Harada M, Hisamoto T, Nagao Y, Ide T, Taniguchi E, Kumemura H, Hanada S, Maeyama M, Baba S, Koga H, Kumashiro R, Ueno T, Ogata H, Yoshimura A, Sata M (2004) Hepatitis C virus down-regulates insulin receptor substrates 1 and 2 through up-regulation of suppressor of cytokine signaling 3. Am J Pathol 165(5):1499–1508
Khan AG, Whidby J, Miller MT, Scarborough H, Zatorski AV, Cygan A, Price AA, Yost SA, Bohannon CD, Jacob J, Grakoui A, Marcotrigiano J (2014) Structure of the core ectodomain of the hepatitis C virus envelope glycoprotein 2. Nature 509(7500):381–384

Kim S, Date T, Yokokawa H, Kono T, Aizaki H, Maurel P, Gondeau C, Wakita T (2014) Development of hepatitis C virus genotype 3a cell culture system. Hepatology in press. doi:10.1002/hep.27197

Koike K, Tsutsumi T, Yotsuyanagi H, Moriya K (2010) Lipid metabolism and liver disease in hepatitis C viral infection. Oncology 78(Suppl 1):24–30

Kojima E, Takeuchi A, Haneda M, Yagi A, Hasegawa T, Yamaki K, Takeda K, Akira S, Shimokata K, Isobe K (2003) The function of GADD34 is a recovery from a shutoff of protein synthesis induced by ER stress: elucidation by GADD34-deficient mice. FASEB J Off Publ Fed Am Soc Exp Biol 17(11):1573–1575

Kong L, Giang E, Nieusma T, Kadam RU, Cogburn KE, Hua Y, Dai X, Stanfield RL, Burton DR, Ward AB, Wilson IA, Law M (2013) Hepatitis C virus E2 envelope glycoprotein core structure. Science 342(6162):1090–1094

Korenaga M, Wang T, Li Y, Showalter LA, Chan T, Sun J, Weinman SA (2005) Hepatitis C virus core protein inhibits mitochondrial electron transport and increases reactive oxygen species (ROS) production. J Biol Chem 280(45):37481–37488

Krey T, d'Alayer J, Kikuti CM, Saulnier A, Damier-Piolle L, Petitpas I, Johansson DX, Tawar RG, Baron B, Robert B, England P, Persson MA, Martin A, Rey FA (2010) The disulfide bonds in glycoprotein E2 of hepatitis C virus reveal the tertiary organization of the molecule. PLoS Pathog 6(2):e1000762

Lavillette D, Pecheur EI, Donot P, Fresquet J, Molle J, Corbau R, Dreux M, Penin F, Cosset FL (2007) Characterization of fusion determinants points to the involvement of three discrete regions of both E1 and E2 glycoproteins in the membrane fusion process of hepatitis C virus. J Virol 81(16):8752–8765

Lee YH, Giraud J, Davis RJ, White MF (2003) c-Jun N-terminal kinase (JNK) mediates feedback inhibition of the insulin signaling cascade. J Biol Chem 278(5):2896–2902

Lee SH, Kim YK, Kim CS, Seol SK, Kim J, Cho S, Song YL, Bartenschlager R, Jang SK (2005) E2 of hepatitis C virus inhibits apoptosis. J Immunol 175(12):8226–8235

Li Y, Boehning DF, Qian T, Popov VL, Weinman SA (2007) Hepatitis C virus core protein increases mitochondrial ROS production by stimulation of Ca2+ uniporter activity. FASEB J Off Publ Fed Am Soc Exp Biol 21(10):2474–2485

Li X, Chen H, Oh SS, Chishti AH (2008) A Presenilin-like protease associated with Plasmodium falciparum micronemes is involved in erythrocyte invasion. Mol Biochem Parasitol 158 (1):22–31

Li G, Mongillo M, Chin KT, Harding H, Ron D, Marks AR, Tabas I (2009a) Role of ERO1-alpha-mediated stimulation of inositol 1,4,5-triphosphate receptor activity in endoplasmic reticulum stress-induced apoptosis. J Cell Biol 186(6):783–792

Li HF, Huang CH, Ai LS, Chuang CK, Chen SS (2009b) Mutagenesis of the fusion peptide-like domain of hepatitis C virus E1 glycoprotein: involvement in cell fusion and virus entry. J Biomed Sci 16:89

Li X, Chen H, Bahamontes-Rosa N, Kun JF, Traore B, Crompton PD, Chishti AH (2009c) Plasmodium falciparum signal peptide peptidase is a promising drug target against blood stage malaria. Biochem Biophys Res Commun 380(3):454–459

Lima-Cabello E, Garcia-Mediavilla MV, Miquilena-Colina ME, Vargas-Castrillon J, Lozano-Rodriguez T, Fernandez-Bermejo M, Olcoz JL, Gonzalez-Gallego J, Garcia-Monzon C, Sanchez-Campos S (2011) Enhanced expression of pro-inflammatory mediators and liver X-receptor-regulated lipogenic genes in non-alcoholic fatty liver disease and hepatitis C. Clin Sci (Lond) 120(6):239–250

Liu X, Theil EC (2005) Ferritins: dynamic management of biological iron and oxygen chemistry. Acc Chem Res 38(3):167–175

Loureiro J, Lilley BN, Spooner E, Noriega V, Tortorella D, Ploegh HL (2006) Signal peptide peptidase is required for dislocation from the endoplasmic reticulum. Nature 441 (7095):894–897

Lozach PY, Lortat-Jacob H, de Lacroix de Lavalette A, Staropoli I, Foung S, Amara A, Houles C, Fieschi F, Schwartz O, Virelizier JL, Arenzana-Seisdedos F, Altmeyer R (2003) DC-SIGN and L-SIGN are high affinity binding receptors for hepatitis C virus glycoprotein E2. J Biol Chem 278(22):20358–20366

Lyons S, Kapoor A, Sharp C, Schneider BS, Wolfe ND, Culshaw G, Corcoran B, McGorum BC, Simmonds P (2012) Nonprimate hepaciviruses in domestic horses, United kingdom. Emerg Infect Dis 18(12):1976–1982

Machida K, Cheng KT, Lai CK, Jeng KS, Sung VM, Lai MM (2006) Hepatitis C virus triggers mitochondrial permeability transition with production of reactive oxygen species, leading to DNA damage and STAT3 activation. J Virol 80(14):7199–7207

Majeau N, Fromentin R, Savard C, Duval M, Tremblay MJ, Leclerc D (2009) Palmitoylation of hepatitis C virus core protein is important for virion production. J Biol Chem 284 (49):33915–33925

Manning BD, Cantley LC (2007) AKT/PKB signaling: navigating downstream. Cell 129 (7):1261–1274

Matsuura Y, Harada S, Suzuki R, Watanabe Y, Inoue Y, Saito I, Miyamura T (1992) Expression of processed envelope protein of hepatitis C virus in mammalian and insect cells. J Virol 66 (3):1425–1431

Mattijssen F, Kersten S (2012) Regulation of triglyceride metabolism by Angiopoietin-like proteins. Biochim Biophys Acta 1821(5):782–789

McLauchlan J (2000) Properties of the hepatitis C virus core protein: a structural protein that modulates cellular processes. J Viral Hepat 7(1):2–14

McLauchlan J, Lemberg MK, Hope G, Martoglio B (2002) Intramembrane proteolysis promotes trafficking of hepatitis C virus core protein to lipid droplets. EMBO J 21(15):3980–3988

McPherson S, Jonsson JR, Barrie HD, O'Rourke P, Clouston AD, Powell EE (2008) Investigation of the role of SREBP-1c in the pathogenesis of HCV-related steatosis. J Hepatol 49 (6):1046–1054

Meynard D, Kautz L, Darnaud V, Canonne-Hergaux F, Coppin H, Roth MP (2009) Lack of the bone morphogenetic protein BMP6 induces massive iron overload. Nat Genet 41(4):478–481

Mihm S, Fayyazi A, Hartmann H, Ramadori G (1997) Analysis of histopathological manifestations of chronic hepatitis C virus infection with respect to virus genotype. Hepatology 25 (3):735–739

Miyamoto H, Moriishi K, Moriya K, Murata S, Tanaka K, Suzuki T, Miyamura T, Koike K, Matsuura Y (2007) Involvement of the PA28gamma-dependent pathway in insulin resistance induced by hepatitis C virus core protein. J Virol 81(4):1727–1735

Miyanari Y, Atsuzawa K, Usuda N, Watashi K, Hishiki T, Zayas M, Bartenschlager R, Wakita T, Hijikata M, Shimotohno K (2007) The lipid droplet is an important organelle for hepatitis C virus production. Nat Cell Biol 9(9):1089–1097

Miyoshi H, Fujie H, Shintani Y, Tsutsumi T, Shinzawa S, Makuuchi M, Kokudo N, Matsuura Y, Suzuki T, Miyamura T, Moriya K, Koike K (2005) Hepatitis C virus core protein exerts an inhibitory effect on suppressor of cytokine signaling (SOCS)-1 gene expression. J Hepatol 43 (5):757–763

Miyoshi H, Moriya K, Tsutsumi T, Shinzawa S, Fujie H, Shintani Y, Fujinaga H, Goto K, Todoroki T, Suzuki T, Miyamura T, Matsuura Y, Yotsuyanagi H, Koike K (2011) Pathogenesis of lipid metabolism disorder in hepatitis C: polyunsaturated fatty acids counteract lipid alterations induced by the core protein. J Hepatol 54(3):432–438

Mohankumar KM, Xu XQ, Zhu T, Kannan N, Miller LD, Liu ET, Gluckman PD, Sukumar S, Emerald BS, Lobie PE (2007) HOXA1-stimulated oncogenicity is mediated by selective upregulation of components of the p44/42 MAP kinase pathway in human mammary carcinoma cells. Oncogene 26(27):3998–4008

Moriishi K, Matsuura Y (2012) Exploitation of lipid components by viral and host proteins for hepatitis C virus infection. Front Microbiol 3:54

Moriishi K, Okabayashi T, Nakai K, Moriya K, Koike K, Murata S, Chiba T, Tanaka K, Suzuki R, Suzuki T, Miyamura T, Matsuura Y (2003) Proteasome activator PA28gamma-dependent nuclear retention and degradation of hepatitis C virus core protein. J Virol 77 (19):10237–10249

Moriishi K, Mochizuki R, Moriya K, Miyamoto H, Mori Y, Abe T, Murata S, Tanaka K, Miyamura T, Suzuki T, Koike K, Matsuura Y (2007) Critical role of PA28gamma in hepatitis C virus-associated steatogenesis and hepatocarcinogenesis. Proc Natl Acad Sci U S A 104 (5):1661–1666

Moriya K, Yotsuyanagi H, Shintani Y, Fujie H, Ishibashi K, Matsuura Y, Miyamura T, Koike K (1997) Hepatitis C virus core protein induces hepatic steatosis in transgenic mice. J Gen Virol 78(Pt 7):1527–1531

Moriya K, Fujie H, Shintani Y, Yotsuyanagi H, Tsutsumi T, Ishibashi K, Matsuura Y, Kimura S, Miyamura T, Koike K (1998) The core protein of hepatitis C virus induces hepatocellular carcinoma in transgenic mice. Nat Med 4(9):1065–1067

Moriya K, Nakagawa K, Santa T, Shintani Y, Fujie H, Miyoshi H, Tsutsumi T, Miyazawa T, Ishibashi K, Horie T, Imai K, Todoroki T, Kimura S, Koike K (2001) Oxidative stress in the absence of inflammation in a mouse model for hepatitis C virus-associated hepatocarcinogenesis. Cancer Res 61(11):4365–4670

Mukherjee A, Shrivastava S, Bhanja Chowdhury J, Ray R, Ray RB (2014) Transcriptional suppression of miR-181c by hepatitis C virus enhances homeobox A1 expression. J Virol 88 (14):7929–7940

Nakajima A, Komazawa-Sakon S, Takekawa M, Sasazuki T, Yeh WC, Yagita H, Okumura K, Nakano H (2006) An antiapoptotic protein, c-FLIPL, directly binds to MKK7 and inhibits the JNK pathway. EMBO J 25(23):5549–5559

Negro F (2010) Hepatitis C virus-induced steatosis: an overview. Dig Dis 28(1):294–299

Nemeth E, Ganz T (2006) Regulation of iron metabolism by hepcidin. Annu Rev Nutr 26:323–342

Nemeth E, Tuttle MS, Powelson J, Vaughn MB, Donovan A, Ward DM, Ganz T, Kaplan J (2004) Hepcidin regulates cellular iron efflux by binding to ferroportin and inducing its internalization. Science 306(5704):2090–2093

Nemeth E, Preza GC, Jung CL, Kaplan J, Waring AJ, Ganz T (2006) The N-terminus of hepcidin is essential for its interaction with ferroportin: structure-function study. Blood 107(1):328–333

Nielsen SU, Bassendine MF, Burt AD, Martin C, Pumeechockchai W, Toms GL (2006) Association between hepatitis C virus and very-low-density lipoprotein (VLDL)/LDL analyzed in iodixanol density gradients. J Virol 80(5):2418–2428

Nishina S, Hino K, Korenaga M, Vecchi C, Pietrangelo A, Mizukami Y, Furutani T, Sakai A, Okuda M, Hidaka I, Okita K, Sakaida I (2008) Hepatitis C virus-induced reactive oxygen species raise hepatic iron level in mice by reducing hepcidin transcription. Gastroenterology 134(1):226–238

Nunez O, Fernandez-Martinez A, Majano PL, Apolinario A, Gomez-Gonzalo M, Benedicto I, Lopez-Cabrera M, Bosca L, Clemente G, Garcia-Monzon C, Martin-Sanz P (2004) Increased intrahepatic cyclooxygenase 2, matrix metalloproteinase 2, and matrix metalloproteinase 9 expression is associated with progressive liver disease in chronic hepatitis C virus infection: role of viral core and NS5A proteins. Gut 53(11):1665–1672

Ogino T, Fukuda H, Imajoh-Ohmi S, Kohara M, Nomoto A (2004) Membrane binding properties and terminal residues of the mature hepatitis C virus capsid protein in insect cells. J Virol 78 (21):11766–11777

Okada S (1996) Iron-induced tissue damage and cancer: the role of reactive oxygen species-free radicals. Pathol Int 46(5):311–332

Okamoto K, Moriishi K, Miyamura T, Matsuura Y (2004) Intramembrane proteolysis and endoplasmic reticulum retention of hepatitis C virus core protein. J Virol 78(12):6370–6380

Okamoto K, Mori Y, Komoda Y, Okamoto T, Okochi M, Takeda M, Suzuki T, Moriishi K, Matsuura Y (2008) Intramembrane processing by signal peptide peptidase regulates the

membrane localization of hepatitis C virus core protein and viral propagation. J Virol 82 (17):8349–8361
Okuda M, Li K, Beard MR, Showalter LA, Scholle F, Lemon SM, Weinman SA (2002) Mitochondrial injury, oxidative stress, and antioxidant gene expression are induced by hepatitis C virus core protein. Gastroenterology 122(2):366–375
Olynyk JK, Reddy KR, Di Bisceglie AM, Jeffers LJ, Parker TI, Radick JL, Schiff ER, Bacon BR (1995) Hepatic iron concentration as a predictor of response to interferon alfa therapy in chronic hepatitis C. Gastroenterology 108(4):1104–1109
Pantopoulos K, Porwal SK, Tartakoff A, Devireddy L (2012) Mechanisms of mammalian iron homeostasis. Biochemistry 51(29):5705–5724
Parvanova I, Epiphanio S, Fauq A, Golde TE, Prudencio M, Mota MM (2009) A small molecule inhibitor of signal peptide peptidase inhibits Plasmodium development in the liver and decreases malaria severity. PLoS One 4(4):e5078
Pazienza V, Clement S, Pugnale P, Conzelman S, Foti M, Mangia A, Negro F (2007) The hepatitis C virus core protein of genotypes 3a and 1b downregulates insulin receptor substrate 1 through genotype-specific mechanisms. Hepatology 45(5):1164–1171
Perlemuter G, Sabile A, Letteron P, Vona G, Topilco A, Chretien Y, Koike K, Pessayre D, Chapman J, Barba G, Brechot C (2002) Hepatitis C virus core protein inhibits microsomal triglyceride transfer protein activity and very low density lipoprotein secretion: a model of viral-related steatosis. FASEB J 16(2):185–194
Persico M, Capasso M, Persico E, Svelto M, Russo R, Spano D, Croce L, La Mura V, Moschella F, Masutti F, Torella R, Tiribelli C, Iolascon A (2007) Suppressor of cytokine signaling 3 (SOCS3) expression and hepatitis C virus-related chronic hepatitis: insulin resistance and response to antiviral therapy. Hepatology 46(4):1009–1015
Piccoli C, Scrima R, Quarato G, D'Aprile A, Ripoli M, Lecce L, Boffoli D, Moradpour D, Capitanio N (2007) Hepatitis C virus protein expression causes calcium-mediated mitochondrial bioenergetic dysfunction and nitro-oxidative stress. Hepatology 46(1):58–65
Pileri P, Uematsu Y, Campagnoli S, Galli G, Falugi F, Petracca R, Weiner AJ, Houghton M, Rosa D, Grandi G, Abrignani S (1998) Binding of hepatitis C virus to CD81. Science 282 (5390):938–941
Ploss A, Evans MJ, Gaysinskaya VA, Panis M, You H, de Jong YP, Rice CM (2009) Human occludin is a hepatitis C virus entry factor required for infection of mouse cells. Nature 457 (7231):882–886
Pohlmann S, Zhang J, Baribaud F, Chen Z, Leslie GJ, Lin G, Granelli-Piperno A, Doms RW, Rice CM, McKeating JA (2003) Hepatitis C virus glycoproteins interact with DC-SIGN and DC-SIGNR. J Virol 77(7):4070–4080
Rouschop KM, van den Beucken T, Dubois L, Niessen H, Bussink J, Savelkouls K, Keulers T, Mujcic H, Landuyt W, Voncken JW, Lambin P, van der Kogel AJ, Koritzinsky M, Wouters BG (2010) The unfolded protein response protects human tumor cells during hypoxia through regulation of the autophagy genes MAP1LC3B and ATG5. J Clin Invest 120(1):127–141
Rui L, Aguirre V, Kim JK, Shulman GI, Lee A, Corbould A, Dunaif A, White MF (2001) Insulin/IGF-1 and TNF-alpha stimulate phosphorylation of IRS-1 at inhibitory Ser307 via distinct pathways. J Clin Invest 107(2):181–189
Rui L, Yuan M, Frantz D, Shoelson S, White MF (2002) SOCS-1 and SOCS-3 block insulin signaling by ubiquitin-mediated degradation of IRS1 and IRS2. J Biol Chem 277 (44):42394–42398
Saeed M, Suzuki R, Watanabe N, Masaki T, Tomonaga M, Muhammad A, Kato T, Matsuura Y, Watanabe H, Wakita T, Suzuki T (2011) Role of the endoplasmic reticulum-associated degradation (ERAD) pathway in degradation of hepatitis C virus envelope proteins and production of virus particles. J Biol Chem 286(43):37264–37273
Shavinskaya A, Boulant S, Penin F, McLauchlan J, Bartenschlager R (2007) The lipid droplet binding domain of hepatitis C virus core protein is a major determinant for efficient virus assembly. J Biol Chem 282(51):37158–37169

Shintani Y, Fujie H, Miyoshi H, Tsutsumi T, Tsukamoto K, Kimura S, Moriya K, Koike K (2004) Hepatitis C virus infection and diabetes: direct involvement of the virus in the development of insulin resistance. Gastroenterology 126(3):840–848

Shirakura M, Murakami K, Ichimura T, Suzuki R, Shimoji T, Fukuda K, Abe K, Sato S, Fukasawa M, Yamakawa Y, Nishijima M, Moriishi K, Matsuura Y, Wakita T, Suzuki T, Howley PM, Miyamura T, Shoji I (2007) E6AP ubiquitin ligase mediates ubiquitylation and degradation of hepatitis C virus core protein. J Virol 81(3):1174–1185

Shukla R, Upton KR, Munoz-Lopez M, Gerhardt DJ, Fisher ME, Nguyen T, Brennan PM, Baillie JK, Collino A, Ghisletti S, Sinha S, Iannelli F, Radaelli E, Dos Santos A, Rapoud D, Guettier C, Samuel D, Natoli G, Carninci P, Ciccarelli FD, Garcia-Perez JL, Faivre J, Faulkner GJ (2013) Endogenous retrotransposition activates oncogenic pathways in hepatocellular carcinoma. Cell 153(1):101–111

Singaravelu R, Chen R, Lyn RK, Jones DM, O'Hara S, Rouleau Y, Cheng J, Srinivasan P, Nasheri N, Russell RS, Tyrrell DL, Pezacki JP (2014) Hepatitis C virus induced up-regulation of microRNA-27: a novel mechanism for hepatic steatosis. Hepatology 59 (1):98–108

Solinas G, Karin M (2010) JNK1 and IKKbeta: molecular links between obesity and metabolic dysfunction. FASEB J Off Publ Fed Am Soc Exp Biol 24(8):2596–2611

Tai AW, Benita Y, Peng LF, Kim SS, Sakamoto N, Xavier RJ, Chung RT (2009) A functional genomic screen identifies cellular cofactors of hepatitis C virus replication. Cell Host Microbe 5(3):298–307

Tanaka N, Moriya K, Kiyosawa K, Koike K, Gonzalez FJ, Aoyama T (2008) PPARalpha activation is essential for HCV core protein-induced hepatic steatosis and hepatocellular carcinoma in mice. J Clin Invest 118(2):683–694

Tanaka T, Kasai H, Yamashita A, Okuyama-Dobashi K, Yasumoto J, Maekawa S, Enomoto N, Okamoto T, Matsuura Y, Morimatsu M, Manabe N, Ochiai K, Yamashita K, Moriishi K (2014) Hallmarks of hepatitis C virus in equine hepacivirus. J Virol in press. doi:10.1128/JVI.02280-02214

Taniguchi CM, Emanuelli B, Kahn CR (2006) Critical nodes in signalling pathways: insights into insulin action. Nat Rev Mol Cell Biol 7(2):85–96

Targett-Adams P, Schaller T, Hope G, Lanford RE, Lemon SM, Martin A, McLauchlan J (2006) Signal peptide peptidase cleavage of GB virus B core protein is required for productive infection in vivo. J Biol Chem 281(39):29221–29227

Theurl I, Zoller H, Obrist P, Datz C, Bachmann F, Elliott RM, Weiss G (2004) Iron regulates hepatitis C virus translation via stimulation of expression of translation initiation factor 3. J Infect Dis 190(4):819–825

Thirone AC, Huang C, Klip A (2006) Tissue-specific roles of IRS proteins in insulin signaling and glucose transport. Trends Endocrinol Metab TEM 17(2):72–78

Thomssen R, Bonk S, Propfe C, Heermann KH, Kochel HG, Uy A (1992) Association of hepatitis C virus in human sera with beta-lipoprotein. Med Microbiol Immunol 181(5):293–300

Thomssen R, Bonk S, Thiele A (1993) Density heterogeneities of hepatitis C virus in human sera due to the binding of beta-lipoproteins and immunoglobulins. Med Microbiol Immunol 182 (6):329–334

Tsutsumi T, Suzuki T, Moriya K, Yotsuyanagi H, Shintani Y, Fujie H, Matsuura Y, Kimura S, Koike K, Miyamura T (2002a) Alteration of intrahepatic cytokine expression and AP-1 activation in transgenic mice expressing hepatitis C virus core protein. Virology 304 (2):415–424

Tsutsumi T, Suzuki T, Shimoike T, Suzuki R, Moriya K, Shintani Y, Fujie H, Matsuura Y, Koike K, Miyamura T (2002b) Interaction of hepatitis C virus core protein with retinoid X receptor alpha modulates its transcriptional activity. Hepatology 35(4):937–946

Tsutsumi T, Matsuda M, Aizaki H, Moriya K, Miyoshi H, Fujie H, Shintani Y, Yotsuyanagi H, Miyamura T, Suzuki T, Koike K (2009) Proteomics analysis of mitochondrial proteins reveals

overexpression of a mitochondrial protein chaperon, prohibitin, in cells expressing hepatitis C virus core protein. Hepatology 50(2):378–386
Ueki K, Kondo T, Kahn CR (2004) Suppressor of cytokine signaling 1 (SOCS-1) and SOCS-3 cause insulin resistance through inhibition of tyrosine phosphorylation of insulin receptor substrate proteins by discrete mechanisms. Mol Cell Biol 24(12):5434–5446
Wang RH, Li C, Xu X, Zheng Y, Xiao C, Zerfas P, Cooperman S, Eckhaus M, Rouault T, Mishra L, Deng CX (2005) A role of SMAD4 in iron metabolism through the positive regulation of hepcidin expression. Cell Metab 2(6):399–409
Wang Q, Liu Y, An D, Diao H, Xu W, He X, Sun R, Wei L, Li L (2012) Regulation of hepatitis C virus translation initiation by iron: role of eIF3 and La protein. Virus Res 167(2):302–309
Wang J, Kang R, Huang H, Xi X, Wang B, Wang J, Zhao Z (2014) Hepatitis C virus core protein activates autophagy through EIF2AK3 and ATF6 UPR pathway-mediated MAP1LC3B and ATG12 expression. Autophagy 10(5):766–784
Waris G, Felmlee DJ, Negro F, Siddiqui A (2007) Hepatitis C virus induces proteolytic cleavage of sterol regulatory element binding proteins and stimulates their phosphorylation via oxidative stress. J Virol 81(15):8122–8130
Werner ED, Lee J, Hansen L, Yuan M, Shoelson SE (2004) Insulin resistance due to phosphorylation of insulin receptor substrate-1 at serine 302. J Biol Chem 279(34):35298–35305
Wu CF, Lin YL, Huang YT (2013) Hepatitis C virus core protein stimulates fibrogenesis in hepatic stellate cells involving the obese receptor. J Cell Biochem 114(3):541–550
Xia Y, Babitt JL, Sidis Y, Chung RT, Lin HY (2008) Hemojuvelin regulates hepcidin expression via a selective subset of BMP ligands and receptors independently of neogenin. Blood 111 (10):5195–5204
Yamaguchi A, Tazuma S, Nishioka T, Ohishi W, Hyogo H, Nomura S, Chayama K (2005) Hepatitis C virus core protein modulates fatty acid metabolism and thereby causes lipid accumulation in the liver. Dig Dis Sci 50(7):1361–1371
Youngren JF (2007) Regulation of insulin receptor function. Cell Mol Life Sci: CMLS 64 (7–8):873–891
Yu S, Matsusue K, Kashireddy P, Cao WQ, Yeldandi V, Yeldandi AV, Rao MS, Gonzalez FJ, Reddy JK (2003) Adipocyte-specific gene expression and adipogenic steatosis in the mouse liver due to peroxisome proliferator-activated receptor gamma1 (PPARgamma1) overexpression. J Biol Chem 278(1):498–505
Zhang M, Gaschen B, Blay W, Foley B, Haigwood N, Kuiken C, Korber B (2004) Tracking global patterns of N-linked glycosylation site variation in highly variable viral glycoproteins: HIV, SIV, and HCV envelopes and influenza hemagglutinin. Glycobiology 14(12):1229–1246

Role of Nonstructural Proteins in HCV Replication

Tetsuro Suzuki and Ryosuke Suzuki

Abstract The open reading frame of the HCV RNA genome is translated at the rough endoplasmic reticulum, yielding a polyprotein precursor that is co- and posttranslationally processed by cellular and viral proteases. The carboxy-terminal two-thirds of the polyprotein encodes various nonstructural (NS) proteins (NS2, NS3, NS4A, NS4B, NS5A, and NS5B), which, it is thought, are not incorporated into virus particles but instead coordinate the intracellular aspects of viral replication. Together with host factors, NS3, NS4A, NS4B, NS5A, and NS5B form a membrane-associated replication complex. Cleavage of the precursor at the NS2/3 junction is accomplished by a cysteine protease encoded by NS2 and the N-terminal portion of NS3. The NS3-4A heterodimeric complex contains a serine protease domain and an RNA helicase/nucleoside triphosphatase domain. The protease activity is responsible for downstream cleavage events in the NS region. NS4B and NS5A have no enzymatic activity. NS4B is thought to serve as a scaffold for the viral replication complex and to induce substantial rearrangements of membrane vesicles. NS5A exists in hypo- and hyperphosphorylated forms and interacts with a large number of host proteins that are important for assembly and function of the replication complex. NS5B RNA-dependent RNA polymerase, which presumably mediates de novo RNA synthesis, is the catalytic core of the HCV replication machinery. Evidence clarifying the roles of several NS proteins is accumulating. Here, we present an overview of the mechanisms by which HCV NS proteins contribute to the viral replication cycle.

Keywords NS2 • NS3 • NS4A • NS4B • NS5A • NS5B • Replication

T. Suzuki, Ph.D. (✉)
Department of Infectious Diseases, Hamamatsu University School of Medicine, Shizuoka 431-3192, Japan
e-mail: tesuzuki@hama-med.ac.jp

R. Suzuki, Ph.D.
Department of Virology II, National Institute of Infectious Diseases, Tokyo 162-8640, Japan
e-mail: ryosuke@nih.go.jp

T. Miyamura et al. (eds.), *Hepatitis C Virus I*, DOI 10.1007/978-4-431-56098-2_7

1 Introduction

HCV open reading frame of the RNA genome is translated at the rough endoplasmic reticulum (ER), yielding a polyprotein precursor of about 3000 amino acids (aa) that is co- and post-translationally processed by cellular- and viral proteases. Nonstructural (NS) proteins are processed by two viral proteases: cleavage between NS2 and NS3 is a rapid intramolecular reaction mediated by NS2-3 protease, while the remaining four junctions are processed by NS3-4A protease. It is considered that the NS3-NS5B region is processed by sequential cleavage: NS3/4A→NS5A/5B→NS4A/4B→NS4B/5A (Tanji et al. 1994; Failla et al. 1995; Lin et al. 1994; Bartenschlager et al. 1994). NS3, NS4A, NS4B, NS5A and NS5B are the viral components of functional replicase complexes required for genome RNA replication. Understanding of molecular structures of individual NS proteins and their roles in the viral lifecycle has increased substantially over the last decades. Progress in the virus research as well as development of powerful model systems stimulated efforts for drug development. Recently, direct antivirals targeting NS3, NS5A and NS5B have been approved successively, as described in detail in *Volume II*.

2 NS2

The N-terminus of NS2 is cleaved by host signal peptidase from the p7 protein. NS2 is a 23-kDa transmembrane protein, with a highly hydrophobic 96-residue N-terminus. The N-terminal domain consists of three transmembrane helices that insert into the ER membrane (Jirasko et al. 2008; Phan et al. 2009). The C-terminal domain of NS2 resides in the cytoplasm, together with one-third of the N-terminal domain of NS3, enabling the NS2-3 protease activity that is necessary for cleavage of the NS2/3 junction. Although NS2 is not essential for the replication of subgenomic RNA replicons (Lohmann et al. 1999), NS2/3 cleavage is required for formation of the viral replication complex (RC), or replicase. Accumulating evidence from in vitro studies of HCV infection suggests that NS2 is required for virus assembly, independent of protease activity (Jones et al. 2007; Jirasko et al. 2008).

The NS2-3 protease was initially thought to be a metalloprotease because its enzymatic activity is stimulated by zinc and inhibited by EDTA (Hijikata et al. 1993). However, a later study revealed that the zinc-binding domain is located in the NS3 region, a domain that is important for NS2 protease activity, as discussed below (Schregel et al. 2009). NS2-3 is currently thought to function as a cysteine protease (Lackner et al. 2004). Mutational analyses showed that the highly conserved His-143, Glu-163, and Cys-184 residues are essential for NS2-3 protease activity (Grakoui et al. 1993; Welbourn et al. 2005; Hijikata et al. 1993). Analysis of the crystal structure of the C-terminal domain of NS2 revealed that the region can homodimerize, creating two composite active sites (Lorenz et al. 2006). The catalytic His-143 and Glu-163 residues are contributed by one monomer, and Cys-184 is contributed by the other monomer, forming the catalytic triad (Lorenz et al. 2006). NS2/3 processing is critical for genome replication (Welbourn et al. 2005); however, some studies have demonstrated replication of the subgenomic replicon that encodes NS3-5B (Lohmann et al. 1999), indicating that NS2 is not essential for replication of the viral genome in vitro.

The involvement of NS2 in the production of infectious virus independent of its protease activity was initially demonstrated using a dicistronic expression system with the viral genome containing the internal ribosomal entry site from encephalomyocarditis virus between NS2 and NS3 (Jones et al. 2007; Jirasko et al. 2008; Dentzer et al. 2009). Several adaptive mutations in NS2 that enhance virus production have also been reported (Kato et al. 2008; Popescu et al. 2011; Russell et al. 2008; Yi et al. 2007; Jensen et al. 2008; Scheel et al. 2008). Furthermore, there is increasing evidence for genetic and biochemical interactions between NS2 and other HCV proteins, including E1, E2, p7, NS3-4A, and NS5A (Phan et al. 2009; Popescu et al. 2011; Ma et al. 2011; Jirasko et al. 2010; Stapleford and Lindenbach 2011; Yi et al. 2009; Selby et al. 1994; Kiiver et al. 2006; Counihan et al. 2011). Thus, it is now thought that NS2 functions as a scaffold, coordinating interactions between structural and NS proteins during viral assembly. Recently, signal peptidase complex subunit 1 (SPCS1), a component of the microsomal signal peptidase complex that is responsible for cleavage of the signal peptides of many secreted and membrane-associated proteins, was identified as an NS2-binding host factor (Suzuki et al. 2013). SPCS1 was shown to interact with both NS2 and E2, forming an E2-SPCS1-NS2 complex. SPCS1 thus plays a critical role in the assembly of infectious HCV particles (Suzuki et al. 2013), as interaction between E2 and NS2 is important for infectious viral particle assembly (Popescu et al. 2011; Stapleford and Lindenbach 2011; Jirasko et al. 2010).

Cyclosporin A (CsA) is known to inhibit replication of the HCV RNA (Watashi et al. 2003) as described below. Besides inhibiting the replication of the subgenomic replicon containing NS3 to NS5B genes, it has been shown that its inhibitory effect was remarkably enhanced in cultures of full-length virus (Ciesek et al. 2009). The CsA's inhibitory activity is dependent on NS2 as well as on the peptidyl-prolyl cis-trans isomerase (PPI) activity of cyclophilin (Cyp) A (Ciesek et al. 2009).

3 NS3-4A

NS3 is a 70-kDa protein containing a serine protease motif in the N-terminal domain and nucleoside triphosphatase (NTPase) and RNA helicase motifs in the C-terminal domain. NS4A is a 54-aa 8-kDa protein that acts as a cofactor for NS3 serine protease (Bartenschlager et al. 1994; Failla et al. 1994). The NS3-4A complex associates with the ER membrane through the transmembrane α-helix of the NS4A N-terminus (Brass et al. 2008). The NS3-4A complex is essential for viral polyprotein processing and replication of the RNA genome.

Numerous studies utilizing various expression systems have revealed that NS3-4A cleaves at the NS3/4A, NS4A/4B, NS4B/5A, and NS5A/5B sites (Hijikata et al. 1993; Bartenschlager et al. 1993; Eckart et al. 1993; Tomei et al. 1993; D'Souza et al. 1994; Manabe et al. 1994; Shoji et al. 1995; Suzuki et al. 1995). The cleavage of NS3/4A is an autocatalytic and cotranslational event that is mediated in cis, whereas cleavage at the NS4A/4B, NS4B/5A, and NS5A/5B sites occurs in trans (Bartenschlager et al. 1994; Tanji et al. 1994; Lin et al. 1994; Failla et al. 1994). The catalytic triad of NS3-4A protease, which is a member of the trypsin/chymotrypsin superfamily of serine proteases, is formed by His-57, Asp-81, and Ser-139. Detailed mutagenesis experiments have indicated that the N-terminal domain of NS3 and the central domain of NS4A are important for stable interaction between NS3 and NS4A (Koch et al. 1996; Shimizu et al. 1996). Analyses of the NS3 serine protease crystal structure have revealed that NS3 has a trypsin-like fold and a structural zinc-binding site (Love et al. 1996). Zinc is tetrahedrally coordinated by Cys-97, Cys-99, Cys-145, and His-149. This domain is also important for cleavage of the NS2/3 junction by NS2 protease (Schregel et al. 2009). Direct interaction between NS3 and synthetic NS4A peptide has also been confirmed by X-ray crystallography (Kim et al. 1996). Sequence alignment of NS3-4A substrate among various HCV isolates indicates that a consensus sequence would include the Asp or Glu residue at the P6 position and Cys for Thr residue at the P1 position, and Ala or Ser residue at the P1′ position (D/E-X-X-X-X-C/T|A/S-X-X-X) (Bartenschlager et al. 1995).

RNA helicase and NTPase activities in the C-terminal domain of NS3 have also been demonstrated. The RNA helicase/NTPase domain is presumably responsible for unwinding replicative double-stranded (ds)RNA intermediates in order to remove RNA secondary structures or separate the genome from RNA-binding proteins during RNA replication. Recombinant NS3 expressed in *E. coli* has been shown to bind to the poly(U) sequence located at the 3′ end of the HCV RNA (Kanai et al. 1995). Furthermore, the NTPase activity is enhanced by adding poly (U) to the reaction mixture. NS3 helicase, a DExH/D-box helicase, potentially unwinds dsRNA in the 3–5′ direction (Tai et al. 1996). Crystal structure analyses revealed that NS3 helicase has distinct NTPase and RNA-binding domains. This structure supports a mechanism of helicase activity involving initial recognition of a 3′ single-stranded region on the nucleic acid by an Arg-rich region of the RNA-binding domain (Yao et al. 1997). These findings suggest that NS3 binds to the 3′ end of the HCV genome and regulates the replication of viral RNA.

Adenosine 5′-triphosphate (ATP), the major energy currency of cells, is involved in a variety of processes, including viral replication. ATP-dependent reactions essential for replication of viral genomes are catalyzed by various virus-encoded enzymes, including helicases and polymerases. High-energy phosphoryl groups are required for NS3 RNA helicase and NS5B RNA polymerase activities (see below). Previous research demonstrated that creatine kinase B (CKB), which is critical for the maintenance and regulation of cellular energy stores, accumulates in particular subcellular sites where the viral genome potentially replicates in infected cells. It is likely that CKB is directed to the HCV replication machinery through interaction with NS4A and that the enzyme functions as a positive regulator of the viral replicase by providing ATP (Hara et al. 2009).

It is now apparent that HCV NS3 is also involved in viral morphogenesis (Ma et al. 2008; Han et al. 2009; Suzuki et al. 2012), although its precise role and the underlying molecular mechanism(s) have not fully been elucidated. Three cell-culture adaptive mutations involved in HCV assembly have been identified in the *NS3* gene. The Q221L mutation in helicase subdomain 1 results in approximately 30-fold higher production of HCV without affecting the NS3 enzymatic activity (Ma et al. 2008). The M260K adaptive mutation is also located in subdomain 1 of the NS3 helicase (Han et al. 2009). The N556D mutation, located in subdomain 3 of the helicase, enhances the assembly of infectious viral particles by increasing specific infectivity without affecting the efficiency of viral RNA replication (Suzuki et al. 2012). Considering the possibility that NS3 plays a role in linking the viral replicase and assembly sites (Jones et al. 2011), it is likely that the NS3 helicase mediates interactions with structural proteins.

4 NS4B

NS4B is a 27-kDa integral membrane protein consisting of a 69-residue N-terminal domain, a central transmembrane domain (aa residues 70–190), and a C-terminal domain (aa residues 191–261). NS4B is thought to play an essential role in viral replication through remodeling of the intracellular membrane by self-oligomerization, providing a platform for the viral RC (Egger et al. 2002; Lundin et al. 2003; Gouttenoire et al. 2010b; Paul et al. 2011). NS4B induces the formation of a complicated multivesiculated cytoplasmic structure known as the membranous web. This specialized membrane structure, derived at least in part from the ER, is rich in HCV NS proteins and viral RNA and likely contains single-, double-, and multi-membrane vesicles. It was recently reported that double-membrane vesicles are possible sites of viral genome replication (Romero-Brey et al. 2012; Ferraris et al. 2010; Gouttenoire et al. 2010a; Paul et al. 2013). The N-terminal domain of NS4B contains two amphipathic helices (designated amphiphathic α-helix [AH]1 and AH2) and reportedly has a dual-membrane topology (Lundin et al. 2003, 2006). Although the N-terminal domain is predicted to reside in the cytoplasmic face, NS4B AH2 was shown to translocate at least partially across the membrane into the ER lumen, presumably through a posttranslational mechanism (Gouttenoire et al. 2009; Lundin et al. 2006). Such a topology change might contribute to the induction of membranous vesicle formation. AH2 is also a major determinant of NS4B oligomerization (Gouttenoire et al. 2010b). The central domain of NS4B contains four predicted transmembrane segments. The C-terminal domain, which comprises two α-helices, is believed to be on the cytosolic side of the ER membrane and includes arginine residues that may be important in binding between RNA and

NS4B (Einav et al. 2008). Two conserved dimerization motifs (GXXXG and S/T clusters) have been identified within the transmembrane domains, and mutations in each of the putative motifs result in a reduction in viral replication (Han et al. 2011). In addition, a nucleotide-binding motif (Thompson et al. 2009) and two palmitoylation sites (Yu et al. 2006) were indicated in the region. Heterotypic interaction between the highly conserved cytoplasmic C-terminal domain and the NS4B N-terminus was shown to be important for induction of functional membrane vesicle formation (Paul et al. 2011). Specific charged residues in the N- and C-termini of NS4B play a role in formation of the viral RC (Blight 2011). Further studies on the impact of mutations in the C-terminal domain on the HCV infection cycle demonstrated that this NS4B domain is not only essential for viral genome replication but contributes to the assembly of infectious particles, possibly in the process of genome RNA encapsidation (Paul et al. 2011; Han et al. 2013).

5 NS5A

NS5A is an approximately 450-residue hydrophilic phosphoprotein that functions as a key regulator of viral genome replication and virion assembly. NS5A is predicted to contain three domains that are separated by repetitive low-complexity segments (LCS1 and 2). The N-terminal domain (D1; aa residues 28–213) immediately follows the membrane-anchoring α-helix (Brass et al. 2002; Penin et al. 2004; Tellinghuisen et al. 2004) and forms a dimeric structure with an unconventional zinc-coordinating motif (Tellinghuisen et al. 2005). Two different crystal structures for D1 have been reported. Although the dimeric D1 interface differs markedly in each structure, both models suggest that the dimeric structure associates with the lipid bilayer in membrane-bound HCV RCs (Love et al. 2009; Tellinghuisen et al. 2005). The model for dimer formation at the membrane surface led to the proposal that an RNA-binding groove is exposed to the cytosol (Tellinghuisen et al. 2005); binding of RNA by the protein has been demonstrated (Huang et al. 2005).

NS5A interacts with uridylate- and guanylate-rich RNAs as well as the 3′ ends of positive- and negative-stranded HCV RNAs (Huang et al. 2005). In the native state, domains II (D2; aa residues 250–342) and III (D3; residues aa 356-) are unfolded monomers (Hanoulle et al. 2009; Verdegem et al. 2011), but no other structural information has been reported for these domains. Such structural flexibility might enable this enigmatic protein to interact with a variety of host-derived proteins. D3 exhibits genotypic length polymorphisms. D1 and part of D2 are required for genome RNA replication. Most of D3 is unessential for this process; however, D3 is important for assembly of the infectious particles (Appel et al. 2008; Masaki et al. 2008; Hughes et al. 2009; Kim et al. 2011; Tellinghuisen et al. 2008).

Although NS5A has no known enzymatic activity, it has a variety of other functions, foremost as an essential component of the viral RC. A large number of cell culture-adaptive mutations mapped to NS5A gene have been shown to enhance

RNA replication of HCV replicons in human hepatoma HuH-7 cells (Blight et al. 2000; Krieger et al. 2001; Lohmann et al. 2001). Unlike the other NS proteins, NS5A can be trans-complemented in viral replicons containing mutated NS5A that does not support viral RNA replication (Appel et al. 2005a). Association of HCV NS proteins with the membrane is a key feature of RC formation in the specialized membrane compartment. Mapping and structural studies have shown that the N-terminal amphipathic α-helix of NS5A, which is well conserved among HCV isolates, serves as a membrane anchor, allowing association with ER-derived membrane, in which NS5A embeds in the plane of the cytosolic leaflet of the membrane bilayer (Penin et al. 2004). It has been proposed that the N-terminal domain of NS5A forms a platform that is involved in specific protein-protein interactions essential for the assembly of the RC (Penin et al. 2004).

Interaction between NS5A and RNA and other NS proteins is thought to be a crucial aspect of this protein's function in RNA replication. NS5A also interacts with a variety of cellular proteins that are potentially involved in viral replication. Cyps and FK506-binding proteins (FKBPs) are classified as immunophilins, proteins capable of binding to the immunosuppressants cyclosporine and FK506, respectively. Cyps exhibit PPI activity, catalyzing the cis-trans isomerization of peptide bonds preceding proline residues. Cyps also mediate de novo protein folding and isomerization of native proteins. CypA is a major host factor required for HCV replication (Kaul et al. 2009), and its PPI activity is thought to play a role in viral replication (Liu et al. 2009; Kaul et al. 2009). Increasing evidence supports a direct interaction between NS5A and CypA; D2 and LCS2 of NS5A contain sites shown to be important for binding to CypA (Hanoulle et al. 2009; Grise et al. 2012; Fernandes et al. 2010; Ross-Thriepland et al. 2013). How the CypA-NS5A interaction influences HCV replication remains to be elucidated, but it can be speculated that CypA-catalyzed isomerization of prolines in NS5A is required for proper binding of NS5A to viral RNA and/or for high polymerase activity of NS5B molecules (see below) directly associated with NS5A.

The immunophilin FKBP8 also plays a role in HCV genome replication. FKBP8 interacts with D1 of NS5A and recruits the molecular chaperone heat shock protein 90 (Hsp90) to the HCV RC through interaction of the carboxylate clump structure of FKBP8 with the C-terminal MEEVD motif of Hsp90 (Okamoto et al. 2008, 2006). Research suggests that Hsp90 and cochaperones are involved in formation of the HCV RC through interaction with NS5A or other HCV proteins (Taguwa et al. 2008, 2009; Ujino et al. 2009; Nakagawa et al. 2007). Human butyrate-induced transcript 1, which has significant homology with cochaperone p23, is also thought to be involved in HCV replication through interactions with NS5A and Hsp90 (Taguwa et al. 2008).

Small-interfering RNA screening studies identified phosphatidylinositol-4-phosphate kinase III α (PI4KIIIα) as a critical factor for HCV replication. The lipid kinase PI4KIIIα belongs to the family of type III phosphatidylinositol-4 kinases, which catalyze the conversion of phosphatidylinositol to phosphatidylinositol-4-phosphate (PI4P). PI4P is the most abundant monophosphorylated inositol phospholipid in mammalian cells and is thought to play roles in intracellular signaling and membrane trafficking pathways. PI4KIIIα was shown to interact with NS5A, leading to activation of lipid kinase activity and increased PI4P levels (Bianco et al. 2012; Tai and Salloum 2011; Berger et al. 2011; Reiss et al. 2011). Another study demonstrated that mutations within the PI4KIIIα-binding site in the C-terminal portion of NS5A D1 result in reduced HCV RNA replication and alterations in the morphology of possible replication sites; the phenotypes are the same as those observed in gene silencing or treatment with inhibitors of PI4KIIIα (Reiss et al. 2013). It is of interest that the level of basally phosphorylated NS5A increases with overexpression of PI4KIIIα, indicating that the phosphorylation status of NS5A is regulated by PI4KIIIα (Reiss et al. 2013).

Vesicle-associated membrane protein (VAMP)-associated protein (VAP) subtypes A (VAP-A) and B (VAP-B) also interact with NS5A. VAP-A and VAP-B localize primarily at the cytoplasmic face of the ER and Golgi apparatus and are involved in maintaining ER homeostasis and vesicular trafficking. In HCV-replicating cells, VAP-A can be detected in the detergent-resistant membrane fraction that contains the viral RC. Interaction between VAP-A and NS5A is required for the efficient replication of HCV genomic RNA (Gao et al. 2004). VAP-B may also be involved in viral replication through the formation of homodimers and/or heterodimers with VAP-A (Hamamoto et al. 2005). VAP-A and VAP-B form either hetero- or homodimers through their transmembrane regions and interact with NS5A and NS5B. A recent study showed that VAP-A and VAP-B are enriched in the cytoplasmic membrane compartment, the formation of which is induced by HCV RNA replication (Paul et al. 2013). VAP-C, a third VAP subtype that is an alternatively spliced isoform of VAP-B, has been shown to act as a negative regulator of HCV replication. It is likely that VAP-C is involved at least in part in the determination of virus tissue tropism (Kukihara et al. 2009). Other host-derived factors, such as F-box, leucine-rich repeat protein 2 (FBL2) (Kapadia and Chisari 2005; Wang et al. 2005), and tail-interacting protein 47 (TIP47) (Vogt et al. 2013) may play roles in viral replication via interaction with NS5A.

NS5A is phosphorylated on multiple serine and threonine residues and exists in basally phosphorylated (p56) and hyperphosphorylated (p58) forms (Reed et al. 1997; Tanji et al. 1995; Kaneko et al. 1994). Regions around the center and near the C-terminus of NS5A appear to be required for basal phosphorylation, whereas hyperphosphorylation primarily targets a number of serine residues located within LCS1 (Tanji et al. 1995; Blight et al. 2000; Appel et al. 2005b; Evans et al. 2004). Phosphorylation of NS5A occurs primarily on serine residues, with minor phosphorylation on threonine residues. No phosphorylation on tyrosine residues in NS5A has been shown. It is thought that phosphorylation of NS5A modulates its function. It has been suggested that HCV replication could be regulated by controlling the p58:p56 ratio. Some adaptive mutations that reduce the level of NS5A hyperphosphorylation enhance HCV replicon replication (Appel et al. 2005b; Blight et al. 2000). Similarly, suppression of NS5A hyperphosphorylation through either the use of kinase inhibitors or mutagenesis enhances RNA replication in non-culture-adapted replicons (Neddermann et al. 2004; Appel et al. 2005b). In contrast, HCV RNA replication is inhibited in cells carrying adapted replicons after treatment with the same kinase inhibitor (Neddermann et al. 2004). Thus, viral RNA replication might be regulated through a delicate balance between basal- and hyperphosphorylation of NS5A.

The C-terminal region of NS5A D3 is essential for the production of infectious virus, presumably because this region mediates the interaction between NS5A and core proteins (Appel et al. 2008; Masaki et al. 2008; Hughes et al. 2009; Kim et al. 2011; Tellinghuisen et al. 2008). In infected cells, a proportion of the NS5A molecules present in the cell localizes with core in close proximity to lipid droplets, the site of virion assembly. Alanine substitutions in the serine cluster in D3 impair phosphorylation of NS5A, leading to a decrease in NS5A-core interaction, perturbation of the subcellular distribution of NS5A, and disruption of virion production (Masaki et al. 2008; Tellinghuisen et al. 2008; Appel et al. 2008). It has been proposed, therefore, that NS5A acts as a transport vehicle, carrying newly synthesized viral genomes from replication sites to encapsidation sites (Masaki et al. 2008). Other genetic evidence suggests that NS5A, in association with NS2, functions at a later step in virus production, following virion assembly but prior to virus release (Yi et al. 2009).

Many studies have attempted to identify the cellular protein kinases responsible for NS5A phosphorylation, and several serine/threonine protein kinases have been identified as potential candidates (Chen et al. 2010; Kim et al. 1999; Coito et al. 2004; Ide et al. 1997; Quintavalle et al. 2007; Tellinghuisen et al. 2008). The α isoform of casein kinase I (CKI-α) was shown to be involved in replication of the subgenomic replicon derived from genotype 1b, with moderate inhibition of viral RNA replication resulting from CKI-α knockdown (Quintavalle et al. 2006). A study utilizing inhibitors of NS5A hyperphosphorylation coupled with inhibitor affinity chromatography identified NS5A as a direct substrate of CKI-α (Quintavalle et al. 2007). Recently, a kinome-wide in vitro high-throughput binding screening combined with phosphorylation analysis identified CKI-α as a major NS5A-associated kinase that is critical for NS5A hyperphosphorylation and the production of infectious virus (Masaki et al. 2014). A phosphoproteomic analysis of NS5A with and without CKI-α depletion in HCV-replicating cells identified an LCS1-derived peptide fragment that contains eight highly conserved serine residues that are important for CKI-α–mediated hyperphosphorylation. Subsequent mutagenesis analyses demonstrated that the NS5A (genotype 2a) residues serine-225 and serine-232 might be important for NS5A hyperphosphorylation and hyperphosphorylation-dependent regulation of virion production (Masaki et al. 2014). Casein kinase II (CKII) is also known to be involved in NS5A phosphorylation. In vitro, CKII binds to and phosphorylates the C-terminal domain of NS5A (Kim et al. 1999). Mapping studies have suggested that a specific serine residue in a consensus CKII phosphorylation motif within NS5A D3 is important for regulation of infectious HCV production (Tellinghuisen et al. 2008). Polo-like kinase 1 (Plk1) was identified as a candidate for NS5A phosphorylation through siRNA screening. Plk1 interacts with NS5A and plays a role in viral replication through hyperphosphorylation of NS5A (Chen et al. 2010). Other serine/threonine kinases, including AKT, p70S6K, MEK1, and MKK6 have also been shown to phosphorylate NS5A in vitro (Chen et al. 2010). Considering the complexity of NS5A phosphorylation, it is likely that the differential phosphorylation of NS5A by multiple kinases coordinates the specific protein-protein interactions that regulate NS5A activity during HCV replication.

6 NS5B

NS5B is a 65-kDa tail-anchored protein that has characteristics of viral RNA-dependent RNA polymerase (RdRp). NS5B is the catalytic core of the viral RC and plays a key role in the synthesis of complementary negative-strand RNA during viral replication, using the genome as a template and subsequently the genomic RNA from the negative-strand RNA template. The fidelity of NS5B RdRp is approximately one million times lower than that of typical prokaryotic and eukaryotic DNA polymerases, presumably because it exhibits no exonuclease or proofreading activities. A 21-residue C-terminal region forms an α-helical transmembrane domain that is dispensable for in vitro polymerase activity but is necessary for posttranslational targeting to the cytoplasmic side of the ER in cells (Moradpour et al. 2004; Schmidt-Mende et al. 2001).

The structure of NS5B RdRp resembles a typical "right hand" polymerase shape and contains finger, palm, and thumb subdomains, similar to other template-dependent polymerases (Ago et al. 1999; Bressanelli et al. 2002; Lesburg et al. 1999). The finger and thumb subdomains serve important functions as sites for interacting with the nucleic acid substrate. The palm domain contains conserved active-site motifs, such as the GDD motif, in which Asp residues coordinate divalent cations (Mg^{2+} and/or Mn^{2+}). It has been shown that Mn^{2+} ions stabilize the conformation of the active site (Ranjith-Kumar et al. 2002). An unusual feature of this RdRp is that the finger and thumb subdomains interact extensively, leading to the formation of a fully enclosed, or encircled, active site in which nucleotide binding is controlled and de novo RNA synthesis is initiated (Moradpour and Penin 2013). Because such a closed conformation appears insufficient to accommodate the partially duplexed RNA that must form during RNA synthesis, it is highly likely that the RdRp must transition to a more open conformation after initiation of RNA synthesis. It has been suggested that the closed conformation is almost suitable for de novo initiation, with enough room to bind a single strand of template RNA and priming nucleotides (Schmitt et al. 2011). Both partially open as well as closed forms have been identified by X-ray crystallography of genotype 2a-derived RdRp, indicating that the crucial difference between the two conformations is the relative orientation of the thumb domain in relation to the finger and palm domains (Biswal et al. 2005). HCV RdRp also has an unusual 12-residue hairpin loop that protrudes into the active site. The loop structure purportedly interferes with binding to dsRNA and helps position the 3′ end of the RNA template for proper initiation of RNA synthesis (Butcher et al. 2001; Hong et al. 2001).

It is highly likely that the initiation of HCV RNA synthesis at the 3′ end of the positive- and negative-strand RNAs in infected cells begins de novo with a

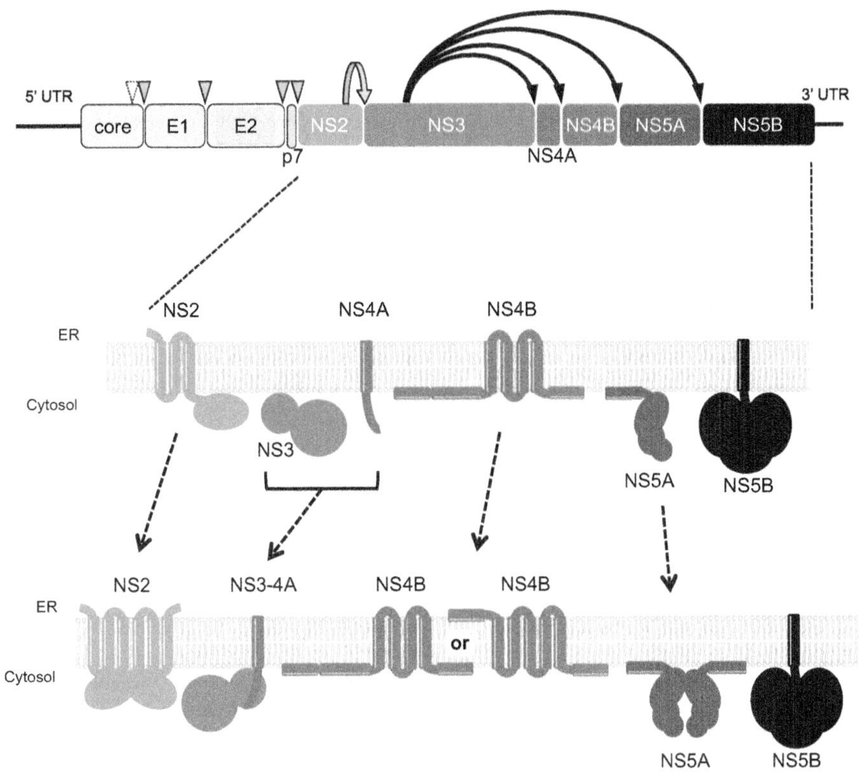

Fig. 1 Gene organization of HCV, processing of viral proteins and maturation of NS proteins (*top*). The HCV RNA genome contains a large open reading frame encoding a precursor polyprotein. Untranslated regions of 5′ and 3′ (5′ UTR, 3′ UTR) are shown on the left and the right side of the genome structure, respectively. Cleavages at the core/E1, E1/E2, E2/p7, and p7/NS2 junctions are mediated by host signal peptidases indicated by *gray triangles*. The *open triangle* indicates cleavage by host signal peptide peptidase. Processing at the NS2/NS3 junction mediated by the NS2–3 protease is indicated by a *gray arrow*. NS3-4A serine protease cleavage sites are indicated by *black arrows* (*middle*). Membrane topologies of HCV NS proteins (*bottom*). NS2 and NS5A are shown as a homodimer. NS3 associates ER membrane through interacting with NS4A. The two alternative membrane topologies of NS4B are shown

1-nucleotide primer, as copy-back or self-priming RNA synthesis cannot produce authentic viral RNA with precise 5′ and 3′ ends. HCV RdRp tends to selectively utilize purine nucleotides for initiation of RNA replication, and de novo initiation in vitro occurs most efficiently with guanine (Zhong et al. 2000; Luo et al. 2000). A study to determine the initiating nucleotides in HCV-replicating cells demonstrated that the replication of positive-strand RNA is preferentially initiated with purine nucleotides, whereas negative-strand RNA replication is invariably initiated with ATP (Cai et al. 2004). Coordination of the catalytic Asp residues with divalent metals may facilitate the formation of a phosphodiester bond between the initiating and second nucleotides (O’Farrell et al. 2003; Ferrari et al. 1999).

References

Ago H, Adachi T, Yoshida A, Yamamoto M, Habuka N, Yatsunami K, Miyano M (1999) Crystal structure of the RNA-dependent RNA polymerase of hepatitis C virus. Structure 7 (11):1417–1426

Appel N, Herian U, Bartenschlager R (2005a) Efficient rescue of hepatitis C virus RNA replication by trans-complementation with nonstructural protein 5A. J Virol 79(2):896–909. doi:10.1128/JVI.79.2.896-909.2005

Appel N, Pietschmann T, Bartenschlager R (2005b) Mutational analysis of hepatitis C virus nonstructural protein 5A: potential role of differential phosphorylation in RNA replication and identification of a genetically flexible domain. J Virol 79(5):3187–3194. doi:10.1128/JVI.79.5.3187-3194.2005

Appel N, Zayas M, Miller S, Krijnse-Locker J, Schaller T, Friebe P, Kallis S, Engel U, Bartenschlager R (2008) Essential role of domain III of nonstructural protein 5A for hepatitis C virus infectious particle assembly. PLoS Pathog 4(3):e1000035. doi:10.1371/journal.ppat.1000035

Bartenschlager R, Ahlborn-Laake L, Mous J, Jacobsen H (1993) Nonstructural protein 3 of the hepatitis C virus encodes a serine-type proteinase required for cleavage at the NS3/4 and NS4/5 junctions. J Virol 67(7):3835–3844

Bartenschlager R, Ahlborn-Laake L, Mous J, Jacobsen H (1994) Kinetic and structural analyses of hepatitis C virus polyprotein processing. J Virol 68(8):5045–5055

Bartenschlager R, Ahlborn-Laake L, Yasargil K, Mous J, Jacobsen H (1995) Substrate determinants for cleavage in cis and in trans by the hepatitis C virus NS3 proteinase. J Virol 69 (1):198–205

Berger KL, Kelly SM, Jordan TX, Tartell MA, Randall G (2011) Hepatitis C virus stimulates the phosphatidylinositol 4-kinase III alpha-dependent phosphatidylinositol 4-phosphate production that is essential for its replication. J Virol 85(17):8870–8883. doi:10.1128/JVI.00059-11

Bianco A, Reghellin V, Donnici L, Fenu S, Alvarez R, Baruffa C, Peri F, Pagani M, Abrignani S, Neddermann P, De Francesco R (2012) Metabolism of phosphatidylinositol 4-kinase IIIalpha-dependent PI4P Is subverted by HCV and is targeted by a 4-anilino quinazoline with antiviral activity. PLoS Pathog 8(3):e1002576. doi:10.1371/journal.ppat.1002576

Biswal BK, Cherney MM, Wang M, Chan L, Yannopoulos CG, Bilimoria D, Nicolas O, Bedard J, James MN (2005) Crystal structures of the RNA-dependent RNA polymerase genotype 2a of hepatitis C virus reveal two conformations and suggest mechanisms of inhibition by non-nucleoside inhibitors. J Biol Chem 280(18):18202–18210. doi:10.1074/jbc.M413410200

Blight KJ (2011) Charged residues in hepatitis C virus NS4B are critical for multiple NS4B functions in RNA replication. J Virol 85(16):8158–8171. doi:10.1128/JVI.00858-11

Blight KJ, Kolykhalov AA, Rice CM (2000) Efficient initiation of HCV RNA replication in cell culture. Science 290(5498):1972–1974

Brass V, Bieck E, Montserret R, Wolk B, Hellings JA, Blum HE, Penin F, Moradpour D (2002) An amino-terminal amphipathic alpha-helix mediates membrane association of the hepatitis C virus nonstructural protein 5A. J Biol Chem 277(10):8130–8139. doi:10.1074/jbc.M111289200

Brass V, Berke JM, Montserret R, Blum HE, Penin F, Moradpour D (2008) Structural determinants for membrane association and dynamic organization of the hepatitis C virus NS3-4A complex. Proc Natl Acad Sci U S A 105(38):14545–14550. doi:10.1073/pnas.0807298105

Bressanelli S, Tomei L, Rey FA, De Francesco R (2002) Structural analysis of the hepatitis C virus RNA polymerase in complex with ribonucleotides. J Virol 76(7):3482–3492

Butcher SJ, Grimes JM, Makeyev EV, Bamford DH, Stuart DI (2001) A mechanism for initiating RNA-dependent RNA polymerization. Nature 410(6825):235–240. doi:10.1038/35065653

Cai Z, Liang TJ, Luo G (2004) Effects of mutations of the initiation nucleotides on hepatitis C virus RNA replication in the cell. J Virol 78(7):3633–3643

Chen YC, Su WC, Huang JY, Chao TC, Jeng KS, Machida K, Lai MM (2010) Polo-like kinase 1 is involved in hepatitis C virus replication by hyperphosphorylating NS5A. J Virol 84 (16):7983–7993. doi:10.1128/JVI.00068-10

Ciesek S, Steinmann E, Wedemeyer H, Manns MP, Neyts J, Tautz N, Madan V, Bartenschlager R, von Hahn T, Pietschmann T (2009) Cyclosporine A inhibits hepatitis C virus nonstructural protein 2 through cyclophilin A. Hepatology 50(5):1638–1645. doi:10.1002/hep.23281

Coito C, Diamond DL, Neddermann P, Korth MJ, Katze MG (2004) High-throughput screening of the yeast kinome: identification of human serine/threonine protein kinases that phosphorylate the hepatitis C virus NS5A protein. J Virol 78(7):3502–3513

Counihan NA, Rawlinson SM, Lindenbach BD (2011) Trafficking of hepatitis C virus core protein during virus particle assembly. PLoS Pathog 7(10):e1002302. doi:10.1371/journal.ppat.1002302 PPATHOGENS-D-11-00418 [pii]

Dentzer TG, Lorenz IC, Evans MJ, Rice CM (2009) Determinants of the hepatitis C virus nonstructural protein 2 protease domain required for production of infectious virus. J Virol 83(24):12702–12713. doi:JVI.01184-09 [pii] 10.1128/JVI.01184-09

D'Souza ED, O'Sullivan E, Amphlett EM, Rowlands DJ, Sangar DV, Clarke BE (1994) Analysis of NS3-mediated processing of the hepatitis C virus non-structural region in vitro. J Gen Virol 75(Pt 12):3469–3476

Eckart MR, Selby M, Masiarz F, Lee C, Berger K, Crawford K, Kuo C, Kuo G, Houghton M, Choo QL (1993) The hepatitis C virus encodes a serine protease involved in processing of the putative nonstructural proteins from the viral polyprotein precursor. Biochem Biophys Res Commun 192(2):399–406. doi:10.1006/bbrc.1993.1429

Egger D, Wolk B, Gosert R, Bianchi L, Blum HE, Moradpour D, Bienz K (2002) Expression of hepatitis C virus proteins induces distinct membrane alterations including a candidate viral replication complex. J Virol 76(12):5974–5984

Einav S, Gerber D, Bryson PD, Sklan EH, Elazar M, Maerkl SJ, Glenn JS, Quake SR (2008) Discovery of a hepatitis C target and its pharmacological inhibitors by microfluidic affinity analysis. Nat Biotechnol 26(9):1019–1027. doi:10.1038/nbt.1490

Evans MJ, Rice CM, Goff SP (2004) Phosphorylation of hepatitis C virus nonstructural protein 5A modulates its protein interactions and viral RNA replication. Proc Natl Acad Sci U S A 101 (35):13038–13043. doi:10.1073/pnas.0405152101

Failla C, Tomei L, De Francesco R (1994) Both NS3 and NS4A are required for proteolytic processing of hepatitis C virus nonstructural proteins. J Virol 68(6):3753–3760

Failla C, Tomei L, De Francesco R (1995) An amino-terminal domain of the hepatitis C virus NS3 protease is essential for interaction with NS4A. J Virol 69(3):1769–1777

Fernandes F, Ansari IU, Striker R (2010) Cyclosporine inhibits a direct interaction between cyclophilins and hepatitis C NS5A. PLoS ONE 5(3):e9815. doi:10.1371/journal.pone.0009815

Ferrari E, Wright-Minogue J, Fang JW, Baroudy BM, Lau JY, Hong Z (1999) Characterization of soluble hepatitis C virus RNA-dependent RNA polymerase expressed in Escherichia coli. J Virol 73(2):1649–1654

Ferraris P, Blanchard E, Roingeard P (2010) Ultrastructural and biochemical analyses of hepatitis C virus-associated host cell membranes. J Gen Virol 91(Pt 9):2230–2237. doi:10.1099/vir.0.022186-0

Gao L, Aizaki H, He JW, Lai MM (2004) Interactions between viral nonstructural proteins and host protein hVAP-33 mediate the formation of hepatitis C virus RNA replication complex on lipid raft. J Virol 78(7):3480–3488

Gouttenoire J, Castet V, Montserret R, Arora N, Raussens V, Ruysschaert JM, Diesis E, Blum HE, Penin F, Moradpour D (2009) Identification of a novel determinant for membrane association in hepatitis C virus nonstructural protein 4B. J Virol 83(12):6257–6268. doi:10.1128/JVI.02663-08

Gouttenoire J, Penin F, Moradpour D (2010a) Hepatitis C virus nonstructural protein 4B: a journey into unexplored territory. Rev Med Virol 20(2):117–129. doi:10.1002/rmv.640

Gouttenoire J, Roingeard P, Penin F, Moradpour D (2010b) Amphipathic alpha-helix AH2 is a major determinant for the oligomerization of hepatitis C virus nonstructural protein 4B. J Virol 84(24):12529–12537. doi:10.1128/JVI.01798-10
Grakoui A, McCourt DW, Wychowski C, Feinstone SM, Rice CM (1993) A second hepatitis C virus-encoded proteinase. Proc Natl Acad Sci U S A 90(22):10583–10587
Grise H, Frausto S, Logan T, Tang H (2012) A conserved tandem cyclophilin-binding site in hepatitis C virus nonstructural protein 5A regulates Alisporivir susceptibility. J Virol 86 (9):4811–4822. doi:10.1128/JVI.06641-11
Hamamoto I, Nishimura Y, Okamoto T, Aizaki H, Liu M, Mori Y, Abe T, Suzuki T, Lai MM, Miyamura T, Moriishi K, Matsuura Y (2005) Human VAP-B is involved in hepatitis C virus replication through interaction with NS5A and NS5B. J Virol 79(21):13473–13482. doi:79/21/ 13473 [pii] 10.1128/JVI.79.21.13473-13482.2005
Han Q, Xu C, Wu C, Zhu W, Yang R, Chen X (2009) Compensatory mutations in NS3 and NS5A proteins enhance the virus production capability of hepatitis C reporter virus. Virus Res 145 (1):63–73
Han Q, Aligo J, Manna D, Belton K, Chintapalli SV, Hong Y, Patterson RL, van Rossum DB, Konan KV (2011) Conserved GXXXG- and S/T-like motifs in the transmembrane domains of NS4B protein are required for hepatitis C virus replication. J Virol 85(13):6464–6479. doi:10. 1128/JVI.02298-10
Han Q, Manna D, Belton K, Cole R, Konan KV (2013) Modulation of hepatitis C virus genome encapsidation by nonstructural protein 4B. J Virol 87(13):7409–7422. doi:10.1128/JVI.03523-12
Hanoulle X, Badillo A, Wieruszeski JM, Verdegem D, Landrieu I, Bartenschlager R, Penin F, Lippens G (2009) Hepatitis C virus NS5A protein is a substrate for the peptidyl-prolyl cis/trans isomerase activity of cyclophilins A and B. J Biol Chem 284(20):13589–13601. doi:10.1074/ jbc.M809244200
Hara H, Aizaki H, Matsuda M, Shinkai-Ouchi F, Inoue Y, Murakami K, Shoji I, Kawakami H, Matsuura Y, Lai MM, Miyamura T, Wakita T, Suzuki T (2009) Involvement of creatine kinase B in hepatitis C virus genome replication through interaction with the viral NS4A protein. J Virol 83(10):5137–5147. doi:10.1128/jvi.02179-08
Hijikata M, Mizushima H, Akagi T, Mori S, Kakiuchi N, Kato N, Tanaka T, Kimura K, Shimotohno K (1993) Two distinct proteinase activities required for the processing of a putative nonstructural precursor protein of hepatitis C virus. J Virol 67(8):4665–4675
Hong Z, Cameron CE, Walker MP, Castro C, Yao N, Lau JY, Zhong W (2001) A novel mechanism to ensure terminal initiation by hepatitis C virus NS5B polymerase. Virology 285(1):6–11. doi:10.1006/viro.2001.0948
Huang L, Hwang J, Sharma SD, Hargittai MR, Chen Y, Arnold JJ, Raney KD, Cameron CE (2005) Hepatitis C virus nonstructural protein 5A (NS5A) is an RNA-binding protein. J Biol Chem 280(43):36417–36428. doi:10.1074/jbc.M508175200
Hughes M, Griffin S, Harris M (2009) Domain III of NS5A contributes to both RNA replication and assembly of hepatitis C virus particles. J Gen Virol 90(Pt 6):1329–1334. doi:10.1099/vir.0. 009332-0
Ide Y, Tanimoto A, Sasaguri Y, Padmanabhan R (1997) Hepatitis C virus NS5A protein is phosphorylated in vitro by a stably bound protein kinase from HeLa cells and by cAMP-dependent protein kinase A-alpha catalytic subunit. Gene 201(1–2):151–158
Jensen TB, Gottwein JM, Scheel TK, Hoegh AM, Eugen-Olsen J, Bukh J (2008) Highly efficient JFH1-based cell-culture system for hepatitis C virus genotype 5a: failure of homologous neutralizing-antibody treatment to control infection. J Infect Dis 198(12):1756–1765. doi:10. 1086/593021
Jirasko V, Montserret R, Appel N, Janvier A, Eustachi L, Brohm C, Steinmann E, Pietschmann T, Penin F, Bartenschlager R (2008) Structural and functional characterization of nonstructural protein 2 for its role in hepatitis C virus assembly. J Biol Chem 283(42):28546–28562

Jirasko V, Montserret R, Lee JY, Gouttenoire J, Moradpour D, Penin F, Bartenschlager R (2010) Structural and functional studies of nonstructural protein 2 of the hepatitis C virus reveal its key role as organizer of virion assembly. PLoS Pathog 6(12):e1001233. doi:10.1371/journal.ppat.1001233

Jones CT, Murray CL, Eastman DK, Tassello J, Rice CM (2007) Hepatitis C virus p7 and NS2 proteins are essential for production of infectious virus. J Virol 81(16):8374–8383. doi:JVI.00690-07 [pii] 10.1128/JVI.00690-07

Jones DM, Atoom AM, Zhang X, Kottilil S, Russell RS (2011) A genetic interaction between the core and NS3 proteins of hepatitis C virus is essential for production of infectious virus. J Virol 85(23):12351–12361. doi:JVI.05313-11 [pii] 10.1128/JVI.05313-11

Kanai A, Tanabe K, Kohara M (1995) Poly(U) binding activity of hepatitis C virus NS3 protein, a putative RNA helicase. FEBS Lett 376(3):221–224

Kaneko T, Tanji Y, Satoh S, Hijikata M, Asabe S, Kimura K, Shimotohno K (1994) Production of two phosphoproteins from the NS5A region of the hepatitis C viral genome. Biochem Biophys Res Commun 205(1):320–326. doi:10.1006/bbrc.1994.2667

Kapadia SB, Chisari FV (2005) Hepatitis C virus RNA replication is regulated by host geranylgeranylation and fatty acids. Proc Natl Acad Sci U S A 102(7):2561–2566. doi:10.1073/pnas.0409834102

Kato T, Choi Y, Elmowalid G, Sapp RK, Barth H, Furusaka A, Mishiro S, Wakita T, Krawczynski K, Liang TJ (2008) Hepatitis C virus JFH-1 strain infection in chimpanzees is associated with low pathogenicity and emergence of an adaptive mutation. Hepatology 48(3):732–740. doi:10.1002/hep.22422

Kaul A, Stauffer S, Berger C, Pertel T, Schmitt J, Kallis S, Zayas M, Lohmann V, Luban J, Bartenschlager R (2009) Essential role of cyclophilin A for hepatitis C virus replication and virus production and possible link to polyprotein cleavage kinetics. PLoS Pathog 5(8):e1000546. doi:10.1371/journal.ppat.1000546

Kiiver K, Merits A, Ustav M, Zusinaite E (2006) Complex formation between hepatitis C virus NS2 and NS3 proteins. Virus Res 117(2):264–272. doi:S0168-1702(05)00336-9 [pii] 10.1016/j.virusres.2005.10.021

Kim JL, Morgenstern KA, Lin C, Fox T, Dwyer MD, Landro JA, Chambers SP, Markland W, Lepre CA, O'Malley ET, Harbeson SL, Rice CM, Murcko MA, Caron PR, Thomson JA (1996) Crystal structure of the hepatitis C virus NS3 protease domain complexed with a synthetic NS4A cofactor peptide. Cell 87(2):343–355

Kim J, Lee D, Choe J (1999) Hepatitis C virus NS5A protein is phosphorylated by casein kinase II. Biochem Biophys Res Commun 257(3):777–781. doi:10.1006/bbrc.1999.0460

Kim S, Welsch C, Yi M, Lemon SM (2011) Regulation of the production of infectious genotype 1a hepatitis C virus by NS5A domain III. J Virol 85(13):6645–6656. doi:10.1128/JVI.02156-10

Koch JO, Lohmann V, Herian U, Bartenschlager R (1996) In vitro studies on the activation of the hepatitis C virus NS3 proteinase by the NS4A cofactor. Virology 221(1):54–66. doi:10.1006/viro.1996.0352

Krieger N, Lohmann V, Bartenschlager R (2001) Enhancement of hepatitis C virus RNA replication by cell culture-adaptive mutations. J Virol 75(10):4614–4624. doi:10.1128/JVI.75.10.4614-4624.2001

Kukihara H, Moriishi K, Taguwa S, Tani H, Abe T, Mori Y, Suzuki T, Fukuhara T, Taketomi A, Maehara Y, Matsuura Y (2009) Human VAP-C negatively regulates hepatitis C virus propagation. J Virol 83(16):7959–7969. doi:10.1128/JVI.00889-09

Lackner T, Muller A, Pankraz A, Becher P, Thiel HJ, Gorbalenya AE, Tautz N (2004) Temporal modulation of an autoprotease is crucial for replication and pathogenicity of an RNA virus. J Virol 78(19):10765–10775. doi:10.1128/jvi.78.19.10765-10775.2004

Lesburg CA, Cable MB, Ferrari E, Hong Z, Mannarino AF, Weber PC (1999) Crystal structure of the RNA-dependent RNA polymerase from hepatitis C virus reveals a fully encircled active site. Nat Struct Biol 6(10):937–943. doi:10.1038/13305

Lin C, Pragai BM, Grakoui A, Xu J, Rice CM (1994) Hepatitis C virus NS3 serine proteinase: trans-cleavage requirements and processing kinetics. J Virol 68(12):8147–8157

Liu Z, Yang F, Robotham JM, Tang H (2009) Critical role of cyclophilin A and its prolyl-peptidyl isomerase activity in the structure and function of the hepatitis C virus replication complex. J Virol 83(13):6554–6565. doi:10.1128/JVI.02550-08

Lohmann V, Korner F, Koch J, Herian U, Theilmann L, Bartenschlager R (1999) Replication of subgenomic hepatitis C virus RNAs in a hepatoma cell line. Science 285(5424):110–113. doi:7638 [pii]

Lohmann V, Korner F, Dobierzewska A, Bartenschlager R (2001) Mutations in hepatitis C virus RNAs conferring cell culture adaptation. J Virol 75(3):1437–1449. doi:10.1128/JVI.75.3.1437-1449.2001

Lorenz IC, Marcotrigiano J, Dentzer TG, Rice CM (2006) Structure of the catalytic domain of the hepatitis C virus NS2-3 protease. Nature 442(7104):831–835. doi:nature04975 [pii] 10.1038/nature04975

Love RA, Parge HE, Wickersham JA, Hostomsky Z, Habuka N, Moomaw EW, Adachi T, Hostomska Z (1996) The crystal structure of hepatitis C virus NS3 proteinase reveals a trypsin-like fold and a structural zinc binding site. Cell 87(2):331–342

Love RA, Brodsky O, Hickey MJ, Wells PA, Cronin CN (2009) Crystal structure of a novel dimeric form of NS5A domain I protein from hepatitis C virus. J Virol 83(9):4395–4403. doi:10.1128/JVI.02352-08

Lundin M, Monne M, Widell A, Von Heijne G, Persson MA (2003) Topology of the membrane-associated hepatitis C virus protein NS4B. J Virol 77(9):5428–5438

Lundin M, Lindstrom H, Gronwall C, Persson MA (2006) Dual topology of the processed hepatitis C virus protein NS4B is influenced by the NS5A protein. J Gen Virol 87(Pt 11):3263–3272. doi:10.1099/vir.0.82211-0

Luo G, Hamatake RK, Mathis DM, Racela J, Rigat KL, Lemm J, Colonno RJ (2000) De novo initiation of RNA synthesis by the RNA-dependent RNA polymerase (NS5B) of hepatitis C virus. J Virol 74(2):851–863

Ma Y, Yates J, Liang Y, Lemon SM, Yi M (2008) NS3 helicase domains involved in infectious intracellular hepatitis C virus particle assembly. J Virol 82(15):7624–7639

Ma Y, Anantpadma M, Timpe JM, Shanmugam S, Singh SM, Lemon SM, Yi M (2011) Hepatitis C virus NS2 protein serves as a scaffold for virus assembly by interacting with both structural and nonstructural proteins. J Virol 85(1):86–97. doi:JVI.01070-10 [pii] 10.1128/JVI.01070-10

Manabe S, Fuke I, Tanishita O, Kaji C, Gomi Y, Yoshida S, Mori C, Takamizawa A, Yosida I, Okayama H (1994) Production of nonstructural proteins of hepatitis C virus requires a putative viral protease encoded by NS3. Virology 198(2):636–644. doi:10.1006/viro.1994.1075

Masaki T, Suzuki R, Murakami K, Aizaki H, Ishii K, Murayama A, Date T, Matsuura Y, Miyamura T, Wakita T, Suzuki T (2008) Interaction of hepatitis C virus nonstructural protein 5A with core protein is critical for the production of infectious virus particles. J Virol 82 (16):7964–7976. doi:10.1128/JVI.00826-08

Masaki T, Matsunaga S, Takahashi H, Nakashima K, Kimura Y, Ito M, Matsuda M, Murayama A, Kato T, Hirano H, Endo Y, Lemon SM, Wakita T, Sawasaki T, Suzuki T (2014) Involvement of hepatitis C virus NS5A hyperphosphorylation mediated by casein kinase I-alpha in infectious virus production. J Virol 88(13):7541–7555. doi:10.1128/jvi.03170-13

Moradpour D, Penin F (2013) Hepatitis C virus proteins: from structure to function. Curr Top Microbiol Immunol 369:113–142. doi:10.1007/978-3-642-27340-7_5

Moradpour D, Brass V, Bieck E, Friebe P, Gosert R, Blum HE, Bartenschlager R, Penin F, Lohmann V (2004) Membrane association of the RNA-dependent RNA polymerase is essential for hepatitis C virus RNA replication. J Virol 78(23):13278–13284. doi:10.1128/JVI.78.23.13278-13284.2004

Nakagawa S, Umehara T, Matsuda C, Kuge S, Sudoh M, Kohara M (2007) Hsp90 inhibitors suppress HCV replication in replicon cells and humanized liver mice. Biochem Biophys Res Commun 353(4):882–888. doi:10.1016/j.bbrc.2006.12.117

Neddermann P, Quintavalle M, Di Pietro C, Clementi A, Cerretani M, Altamura S, Bartholomew L, De Francesco R (2004) Reduction of hepatitis C virus NS5A hyperphosphorylation by selective inhibition of cellular kinases activates viral RNA replication in cell culture. J Virol 78(23):13306–13314. doi:10.1128/JVI.78.23.13306-13314.2004

O'Farrell D, Trowbridge R, Rowlands D, Jager J (2003) Substrate complexes of hepatitis C virus RNA polymerase (HC-J4): structural evidence for nucleotide import and de-novo initiation. J Mol Biol 326(4):1025–1035

Okamoto T, Nishimura Y, Ichimura T, Suzuki K, Miyamura T, Suzuki T, Moriishi K, Matsuura Y (2006) Hepatitis C virus RNA replication is regulated by FKBP8 and Hsp90. EMBO J 25 (20):5015–5025. doi:10.1038/sj.emboj.7601367

Okamoto T, Omori H, Kaname Y, Abe T, Nishimura Y, Suzuki T, Miyamura T, Yoshimori T, Moriishi K, Matsuura Y (2008) A single-amino-acid mutation in hepatitis C virus NS5A disrupting FKBP8 interaction impairs viral replication. J Virol 82(7):3480–3489. doi:10.1128/JVI.02253-07

Paul D, Romero-Brey I, Gouttenoire J, Stoitsova S, Krijnse-Locker J, Moradpour D, Bartenschlager R (2011) NS4B self-interaction through conserved C-terminal elements is required for the establishment of functional hepatitis C virus replication complexes. J Virol 85(14):6963–6976. doi:10.1128/JVI.00502-11

Paul D, Hoppe S, Saher G, Krijnse-Locker J, Bartenschlager R (2013) Morphological and biochemical characterization of the membranous hepatitis C virus replication compartment. J Virol 87(19):10612–10627. doi:10.1128/JVI.01370-13

Penin F, Brass V, Appel N, Ramboarina S, Montserret R, Ficheux D, Blum HE, Bartenschlager R, Moradpour D (2004) Structure and function of the membrane anchor domain of hepatitis C virus nonstructural protein 5A. J Biol Chem 279(39):40835–40843. doi:10.1074/jbc.M404761200

Phan T, Beran RK, Peters C, Lorenz IC, Lindenbach BD (2009) Hepatitis C virus NS2 protein contributes to virus particle assembly via opposing epistatic interactions with the E1-E2 glycoprotein and NS3-NS4A enzyme complexes. J Virol 83(17):8379–8395. doi:JVI.00891-09 [pii] 10.1128/JVI.00891-09

Popescu CI, Callens N, Trinel D, Roingeard P, Moradpour D, Descamps V, Duverlie G, Penin F, Heliot L, Rouille Y, Dubuisson J (2011) NS2 protein of hepatitis C virus interacts with structural and non-structural proteins towards virus assembly. PLoS Pathog 7(2):e1001278

Quintavalle M, Sambucini S, Di Pietro C, De Francesco R, Neddermann P (2006) The alpha isoform of protein kinase CKI is responsible for hepatitis C virus NS5A hyperphosphorylation. J Virol 80(22):11305–11312. doi:10.1128/JVI.01465-06

Quintavalle M, Sambucini S, Summa V, Orsatti L, Talamo F, De Francesco R, Neddermann P (2007) Hepatitis C virus NS5A is a direct substrate of casein kinase I-alpha, a cellular kinase identified by inhibitor affinity chromatography using specific NS5A hyperphosphorylation inhibitors. J Biol Chem 282(8):5536–5544. doi:10.1074/jbc.M610486200

Ranjith-Kumar CT, Kim YC, Gutshall L, Silverman C, Khandekar S, Sarisky RT, Kao CC (2002) Mechanism of de novo initiation by the hepatitis C virus RNA-dependent RNA polymerase: role of divalent metals. J Virol 76(24):12513–12525

Reed KE, Xu J, Rice CM (1997) Phosphorylation of the hepatitis C virus NS5A protein in vitro and in vivo: properties of the NS5A-associated kinase. J Virol 71(10):7187–7197

Reiss S, Rebhan I, Backes P, Romero-Brey I, Erfle H, Matula P, Kaderali L, Poenisch M, Blankenburg H, Hiet MS, Longerich T, Diehl S, Ramirez F, Balla T, Rohr K, Kaul A, Buhler S, Pepperkok R, Lengauer T, Albrecht M, Eils R, Schirmacher P, Lohmann V, Bartenschlager R (2011) Recruitment and activation of a lipid kinase by hepatitis C virus NS5A is essential for integrity of the membranous replication compartment. Cell Host Microbe 9(1):32–45. doi:10.1016/j.chom.2010.12.002

Reiss S, Harak C, Romero-Brey I, Radujkovic D, Klein R, Ruggieri A, Rebhan I, Bartenschlager R, Lohmann V (2013) The lipid kinase phosphatidylinositol-4 kinase III alpha regulates the phosphorylation status of hepatitis C virus NS5A. PLoS Pathog 9(5): e1003359. doi:10.1371/journal.ppat.1003359

Romero-Brey I, Merz A, Chiramel A, Lee JY, Chlanda P, Haselman U, Santarella-Mellwig R, Habermann A, Hoppe S, Kallis S, Walther P, Antony C, Krijnse-Locker J, Bartenschlager R (2012) Three-dimensional architecture and biogenesis of membrane structures associated with hepatitis C virus replication. PLoS Pathog 8(12):e1003056. doi:10.1371/journal.ppat.1003056

Ross-Thriepland D, Amako Y, Harris M (2013) The C terminus of NS5A domain II is a key determinant of hepatitis C virus genome replication, but is not required for virion assembly and release. J Gen Virol 94(Pt 5):1009–1018. doi:10.1099/vir.0.050633-0

Russell RS, Meunier JC, Takikawa S, Faulk K, Engle RE, Bukh J, Purcell RH, Emerson SU (2008) Advantages of a single-cycle production assay to study cell culture-adaptive mutations of hepatitis C virus. Proc Natl Acad Sci U S A 105(11):4370–4375

Scheel TK, Gottwein JM, Jensen TB, Prentoe JC, Hoegh AM, Alter HJ, Eugen-Olsen J, Bukh J (2008) Development of JFH1-based cell culture systems for hepatitis C virus genotype 4a and evidence for cross-genotype neutralization. Proc Natl Acad Sci USA 105(3):997–1002. doi:0711044105 [pii] 10.1073/pnas.0711044105

Schmidt-Mende J, Bieck E, Hugle T, Penin F, Rice CM, Blum HE, Moradpour D (2001) Determinants for membrane association of the hepatitis C virus RNA-dependent RNA polymerase. J Biol Chem 276(47):44052–44063. doi:10.1074/jbc.M103358200

Schmitt M, Scrima N, Radujkovic D, Caillet-Saguy C, Simister PC, Friebe P, Wicht O, Klein R, Bartenschlager R, Lohmann V, Bressanelli S (2011) A comprehensive structure-function comparison of hepatitis C virus strain JFH1 and J6 polymerases reveals a key residue stimulating replication in cell culture across genotypes. J Virol 85(6):2565–2581. doi:10.1128/JVI.02177-10

Schregel V, Jacobi S, Penin F, Tautz N (2009) Hepatitis C virus NS2 is a protease stimulated by cofactor domains in NS3. Proc Natl Acad Sci U S A 106(13):5342–5347. doi:10.1073/pnas.0810950106

Selby MJ, Glazer E, Masiarz F, Houghton M (1994) Complex processing and protein: protein interactions in the E2:NS2 region of HCV. Virology 204(1):114–122. doi:S0042-6822(84)71515-7 [pii] 10.1006/viro.1994.1515

Shimizu Y, Yamaji K, Masuho Y, Yokota T, Inoue H, Sudo K, Satoh S, Shimotohno K (1996) Identification of the sequence on NS4A required for enhanced cleavage of the NS5A/5B site by hepatitis C virus NS3 protease. J Virol 70(1):127–132

Shoji I, Suzuki T, Chieda S, Sato M, Harada T, Chiba T, Matsuura Y, Miyamura T (1995) Proteolytic activity of NS3 serine proteinase of hepatitis C virus efficiently expressed in Escherichia coli. Hepatology 22(6):1648–1655

Stapleford KA, Lindenbach BD (2011) Hepatitis C virus NS2 coordinates virus particle assembly through physical interactions with the E1-E2 glycoprotein and NS3-NS4A enzyme complexes. J Virol 85(4):1706–1717

Suzuki T, Sato M, Chieda S, Shoji I, Harada T, Yamakawa Y, Watabe S, Matsuura Y, Miyamura T (1995) In vivo and in vitro trans-cleavage activity of hepatitis C virus serine proteinase expressed by recombinant baculoviruses. J Gen Virol 76(Pt 12):3021–3029

Suzuki R, Saito K, Kato T, Shirakura M, Akazawa D, Ishii K, Aizaki H, Kanegae Y, Matsuura Y, Saito I, Wakita T, Suzuki T (2012) Trans-complemented hepatitis C virus particles as a versatile tool for study of virus assembly and infection. Virology 432:29–38. doi:S0042-6822(12)00284-X [pii] 10.1016/j.virol.2012.05.033

Suzuki R, Matsuda M, Watashi K, Aizaki H, Matsuura Y, Wakita T, Suzuki T (2013) Signal peptidase complex subunit 1 participates in the assembly of hepatitis C virus through an interaction with E2 and NS2. PLoS Pathog 9(8):e1003589. doi:10.1371/journal.ppat.1003589

Taguwa S, Okamoto T, Abe T, Mori Y, Suzuki T, Moriishi K, Matsuura Y (2008) Human butyrate-induced transcript 1 interacts with hepatitis C virus NS5A and regulates viral replication. J Virol 82(6):2631–2641. doi:10.1128/JVI.02153-07

Taguwa S, Kambara H, Omori H, Tani H, Abe T, Mori Y, Suzuki T, Yoshimori T, Moriishi K, Matsuura Y (2009) Cochaperone activity of human butyrate-induced transcript 1 facilitates hepatitis C virus replication through an Hsp90-dependent pathway. J Virol 83 (20):10427–10436. doi:10.1128/JVI.01035-09

Tai AW, Salloum S (2011) The role of the phosphatidylinositol 4-kinase PI4KA in hepatitis C virus-induced host membrane rearrangement. PLoS ONE 6(10):e26300. doi:10.1371/journal.pone.0026300

Tai CL, Chi WK, Chen DS, Hwang LH (1996) The helicase activity associated with hepatitis C virus nonstructural protein 3 (NS3). J Virol 70(12):8477–8484

Tanji Y, Hijikata M, Hirowatari Y, Shimotohno K (1994) Hepatitis C virus polyprotein processing: kinetics and mutagenic analysis of serine proteinase-dependent cleavage. J Virol 68 (12):8418–8422

Tanji Y, Kaneko T, Satoh S, Shimotohno K (1995) Phosphorylation of hepatitis C virus-encoded nonstructural protein NS5A. J Virol 69(7):3980–3986

Tellinghuisen TL, Marcotrigiano J, Gorbalenya AE, Rice CM (2004) The NS5A protein of hepatitis C virus is a zinc metalloprotein. J Biol Chem 279(47):48576–48587. doi:10.1074/jbc.M407787200

Tellinghuisen TL, Marcotrigiano J, Rice CM (2005) Structure of the zinc-binding domain of an essential component of the hepatitis C virus replicase. Nature 435(7040):374–379. doi:10.1038/nature03580

Tellinghuisen TL, Foss KL, Treadaway J (2008) Regulation of hepatitis C virion production via phosphorylation of the NS5A protein. PLoS Pathog 4(3):e1000032. doi:10.1371/journal.ppat.1000032

Thompson AA, Zou A, Yan J, Duggal R, Hao W, Molina D, Cronin CN, Wells PA (2009) Biochemical characterization of recombinant hepatitis C virus nonstructural protein 4B: evidence for ATP/GTP hydrolysis and adenylate kinase activity. Biochemistry 48 (5):906–916. doi:10.1021/bi801747p

Tomei L, Failla C, Santolini E, De Francesco R, La Monica N (1993) NS3 is a serine protease required for processing of hepatitis C virus polyprotein. J Virol 67(7):4017–4026

Ujino S, Yamaguchi S, Shimotohno K, Takaku H (2009) Heat-shock protein 90 is essential for stabilization of the hepatitis C virus nonstructural protein NS3. J Biol Chem 284 (11):6841–6846. doi:10.1074/jbc.M806452200

Verdegem D, Badillo A, Wieruszeski JM, Landrieu I, Leroy A, Bartenschlager R, Penin F, Lippens G, Hanoulle X (2011) Domain 3 of NS5A protein from the hepatitis C virus has intrinsic alpha-helical propensity and is a substrate of cyclophilin A. J Biol Chem 286 (23):20441–20454. doi:10.1074/jbc.M110.182436

Vogt DA, Camus G, Herker E, Webster BR, Tsou CL, Greene WC, Yen TS, Ott M (2013) Lipid droplet-binding protein TIP47 regulates hepatitis C Virus RNA replication through interaction with the viral NS5A protein. PLoS Pathog 9(4):e1003302. doi:10.1371/journal.ppat.1003302

Wang C, Gale M Jr, Keller BC, Huang H, Brown MS, Goldstein JL, Ye J (2005) Identification of FBL2 as a geranylgeranylated cellular protein required for hepatitis C virus RNA replication. Mol Cell 18(4):425–434. doi:10.1016/j.molcel.2005.04.004

Watashi K, Hijikata M, Hosaka M, Yamaji M, Shimotohno K (2003) Cyclosporin A suppresses replication of hepatitis C virus genome in cultured hepatocytes. Hepatology 38(5):1282–1288. doi:10.1053/jhep.2003.50449

Welbourn S, Green R, Gamache I, Dandache S, Lohmann V, Bartenschlager R, Meerovitch K, Pause A (2005) Hepatitis C virus NS2/3 processing is required for NS3 stability and viral RNA replication. J Biol Chem 280(33):29604–29611. doi:10.1074/jbc.M505019200

Yao N, Hesson T, Cable M, Hong Z, Kwong AD, Le HV, Weber PC (1997) Structure of the hepatitis C virus RNA helicase domain. Nat Struct Biol 4(6):463–467

Yi M, Ma Y, Yates J, Lemon SM (2007) Compensatory mutations in E1, p7, NS2, and NS3 enhance yields of cell culture-infectious intergenotypic chimeric hepatitis C virus. J Virol 81 (2):629–638. doi:JVI.01890-06 [pii] 10.1128/JVI.01890-06

Yi M, Ma Y, Yates J, Lemon SM (2009) Trans-complementation of an NS2 defect in a late step in hepatitis C virus (HCV) particle assembly and maturation. PLoS Pathog 5(5):e1000403. doi:10.1371/journal.ppat.1000403

Yu GY, Lee KJ, Gao L, Lai MM (2006) Palmitoylation and polymerization of hepatitis C virus NS4B protein. J Virol 80(12):6013–6023. doi:10.1128/JVI.00053-06

Zhong W, Ferrari E, Lesburg CA, Maag D, Ghosh SK, Cameron CE, Lau JY, Hong Z (2000) Template/primer requirements and single nucleotide incorporation by hepatitis C virus nonstructural protein 5B polymerase. J Virol 74(19):9134–9143

The HCV Replicase Complex and Viral RNA Synthesis

Inés Romero-Brey and Volker Lohmann

Abstract Replication of hepatitis C virus (HCV) is tightly linked to membrane alterations designated the membranous web, harboring the viral replicase complex. In this chapter we describe the morphology and 3D architecture of the HCV-induced replication organelles, mainly consisting of double membrane vesicles, which are generated by a concerted action of the nonstructural proteins NS3 to NS5B. Recent studies have furthermore identified a number of host cell proteins and lipids contributing to the biogenesis of the membranous web, which are discussed in this chapter. Viral RNA synthesis is tightly associated with these membrane alterations and mainly driven by the viral RNA dependent RNA polymerase NS5B. We summarize our current knowledge of the structure and function of NS5B, the role of *cis*-acting replication elements at the termini of the genome in regulating RNA synthesis and the contribution of additional viral and host factors to viral RNA synthesis, which is still ill defined.

Keywords Replication organelles/factories/complexes • Membranous web • Membrane rearrangements • Double membrane vesicles • Replicase • Nonstructural proteins • Virus-cell proteins/lipids interactions • RdRp • Polymerase • Cis-acting replication elements • Host factor • RNA synthesis

Abbreviations

CRE	*cis*-acting replication element
DENV	Dengue virus
DMV	Double membrane vesicle
DRM	Detergent resistant membranes
EM	Electron microscopy
ER	Endoplasmic reticulum
ET	Electron tomography

I. Romero-Brey • V. Lohmann (✉)
Department of Infectious Diseases, Molecular Virology, University of Heidelberg, Im Neuenheimer Feld, 345, 69120 Heidelberg, Germany
e-mail: Volker_lohmann@med.uni-heidelberg.de

T. Miyamura et al. (eds.), *Hepatitis C Virus I*, DOI 10.1007/978-4-431-56098-2_8

FAPP2	Four-phosphate-adaptor protein 2
GFP	Green fluorescence protein
HCV	Hepatitis C virus
IRES	Internal ribosome entry site
JFH1	Japanese fulminant hepatitis 1
LD	Lipid droplet
MMV	Multimembrane vesicle
MW	Membranous web
NS	Non structural
OSBP	Oxysterol-binding protein
PI4KIIIα	Phosphatidylinositol 4-kinase IIIα
PI4P	Phosphatidylinositol 4-phosphate
RdRp	RNA dependent RNA polymerase
SMV	Single membrane vesicle

1 Introduction

The genome of HCV encompasses a single ~9,600 nts long RNA molecule containing one large open reading frame (ORF) that is flanked by non-translated regions (NTRs), important for viral translation and replication. The viral genome is not capped and the 5′NTR contains an internal ribosome entry site (IRES), enabling viral translation. Upon release into the cytoplasm of an infected cell, the genome is translated into a polyprotein, which is co- and posttranslationally cleaved into ten functional subunits by cellular and viral proteases: core, envelope glycoproteins E1 and E2, p7 and the nonstructural proteins (NS) NS2, NS3, NS4A, NS4B, NS5A and NS5B. Core to NS2 are primarily involved in the formation of infectious virus (reviewed in Lindenbach and Rice 2013), whereas the NS proteins NS3 to NS5B are necessary and sufficient for viral RNA replication, which will be the focus of this chapter. After translation and processing, the nonstructural proteins (NS3 to NS5B) induce intracellular membrane alterations designated the membranous web (MW) and harboring the viral replicase (Fig. 1). The formation of these replication factories is a common step during the replication of positive- strand RNA viruses (reviewed in Romero-Brey and Bartenschlager 2014). Although their function is not fully clear yet, it is generally assumed that they facilitate RNA synthesis by concentrating viral and host proteins involved in replication and that they might shield replication intermediates from detection by intracellular pattern recognition receptors to avoid activation of innate immune responses.

RNA synthesis is associated with these replication organelles, first generating a negative strand RNA genome, probably as part of a double stranded replication intermediate (dsRNA). Negative strand RNA is then the template for progeny positive strand RNA, which is produced in five- to tenfold excess. The newly synthesized positive strand RNA either re-enters a new translation/replication

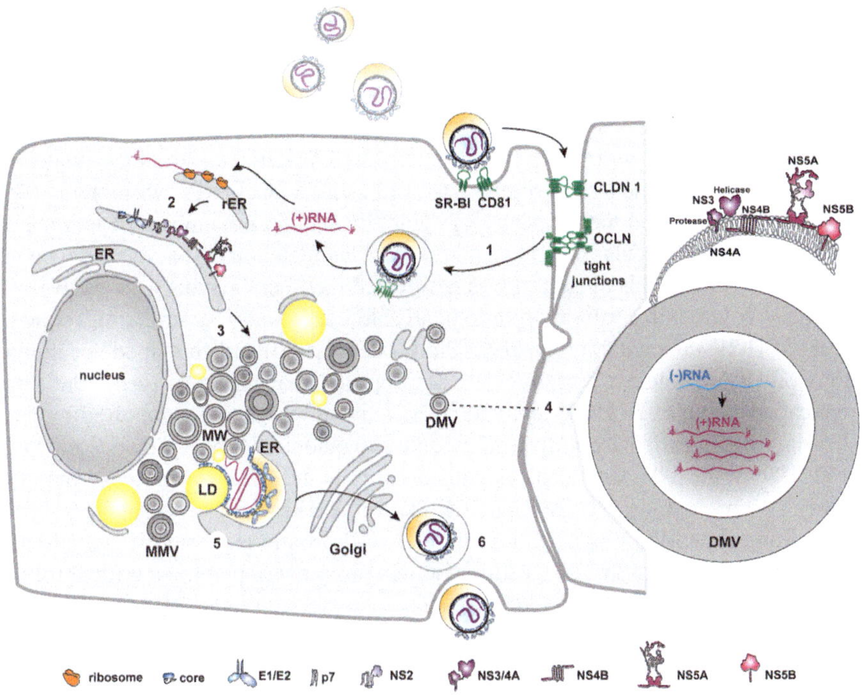

Fig. 1 Scheme of the HCV replication cycle (adapted from Bartenschlager et al. 2013). Following initial binding of HCV lipoviroparticle to scavenger receptor class B member 1 (SRB1) and CD81 and further interactions with the tight junction proteins claudin 1 (*CLDN1*) and occludin (*OCLN*), the virus enters the cell via receptor-mediated endocytosis (step 1, reviewed in Lindenbach and Rice 2013). The positive-sense single-stranded viral RNA genome is released into the cytoplasm and translated in an IRES-dependent manner at the rough ER. The resulting polyprotein is cleaved into ten mature proteins (step 2). Viral proteins, in conjunction with host cell factors, induce the formation of the membranous web (*MW*) mainly composed of double- and multi-membrane vesicles (DMVs and MMVs, respectively), usually in close proximity to lipid droplets (*LDs*) (step 3). RNA replication is thought to take place at DMVs and proceeds via a negative-sense copy ((−)RNA) that serves as a template for the production of excess amounts of positive-sense progeny RNAs ((+)RNA) (step 4, highlighted to the *right*). Assembly of HCV particles is supposed to take place in close proximity to the ER and LDs, where core protein and viral RNA accumulate (reviewed in Bartenschlager et al. 2011). The viral envelope is acquired by budding through the ER membrane in a process that is linked to lipoprotein synthesis (step 5). HCV particles are thought to be released via the constitutive secretory pathway (step 6)

cycle or is packaged into virions (Fig. 1). This chapter summarizes our current knowledge of HCV induced membrane alterations, providing a detailed view of viral and host cell proteins as well as lipids engaged in their biogenesis. We furthermore provide an overview of the role of the NS proteins and *cis*-acting elements in distinct steps of RNA synthesis.

2 Architecture of the HCV Replication Factories

The first visualization of HCV-induced intracellular changes in cell culture by transmission electron microscopy (TEM) was performed by Egger and coworkers (Egger et al. 2002) in U2-OS human osteosarcoma-derived cell lines inducibly expressing the entire HCV polyprotein, as well as individual HCV NS proteins. The main alterations found were vesicles of approximately 85 nm embedded in a membranous matrix of circular or very tightly undulating membranes that formed a rather compact structure. This alteration was designated the 'membranous web' (MW).

The MW structures were morphologically similar to web-like structures found in livers of HCV-infected chimpanzees (Pfeifer et al. 1980) and contained accumulations of HCV proteins, therefore it was proposed that the MW forms the viral replication complex in HCV-infected cells. This view was later supported by data showing that in cells harboring a bicistronic subgenomic HCV replicon (Lohmann et al. 1999a) the MW structures contained not only the HCV NS proteins, but also newly synthesized viral RNA (Gosert et al. 2003). These initial studies found that formation of the MW was triggered by the NS4B protein expressed in the absence of any other HCV protein, suggesting that this protein was playing a critical role in the formation of the MW. In addition, the sole expression of NS3/4A also induced vesicular membrane alterations, while this could not be observed for any of the other NS proteins. Additional HCV specific membrane alterations were observed, so called contiguous vesicles with an irregular shape that often surrounded the web, and vacuoles of dilated rER of various sizes, however, both structures were less frequent and contained less viral antigens (Table 1).

Ten years later, the use of cutting edge approaches like correlative light electron microscopy (CLEM) and electron tomography (ET), as well as the cryo-fixation of cells by means of high pressure freezing (HPF) have allowed us to gain further insights into the morphology and the 3D architecture of these HCV-induced membranous modifications with higher resolution (Romero-Brey et al. 2012). In this manner we found that HCV-infected cells show high abundance of double membrane vesicles (DMVs), having two lipid bilayers attached to each other (Fig. 2a). Furthermore, TEM analysis of infected Huh7.5 cells at different times post-infection revealed that the abundance of DMVs correlated with the amount of HCV RNA being synthesized, indicating that they might play a role in the replication of the HCV genome. Their diameter also increases over the course of infection from an average diameter of 125 nm 16 hpi to around 185 nm 48 hpi. Analysis by ET revealed also that approximately half of these double-layered vesicles remained connected through neck-like structures to their donor membranes at the ER. In addition, late in infection (from 36 hpi onwards), multi-membrane vesicles (MMVs) were observed, having an average diameter of 340 nm (Fig. 2b). These much more complicated structures might be generated from pre-existing DMVs that undergo secondary invaginations or from late synthesized large DMTs (double membrane tubules) that enwrap pre-existing DMVs. In our study, the abundance of DMVs correlated best with RNA synthesis, suggesting that DMVs might indeed

Table 1 Summary of HCV-induced structures described in ultrastructural analyses reported in the literature

Report	System/viral strain	Cell type	Fixation method	Structure	Size
Egger et al. 2002	Expression of the polyprotein/genotype 1a	U-2 OS human osteosarcoma-derived	2.5 % GA	Membranous web (MW)	85 nm
				Contiguous vesicles (CVs)	Not-provided
				Dilated ER vesicles	Various sizes (not provided)
Gosert et al. 2003	Subgenomic replicon/9–13 (genotype 1b)	Huh7	2.5 % GA	Membranous web (MW)	80–100 nm
Ferraris et al. 2010	Sugenomic replicon/JFH1 (genotype 2a)	Huh7.5	4 % PFA and 1 % GA	DMVs	200–500 nm
				MMVs	150–200 nm
Romero-Brey et al. 2012	Infection/Jc1 (genotype 2a)	Huh7.5	4 % PFA and 0.1 % GA + HPF	DMVs	~170 nm
	Subgenomic replicon/JFH1 (genotype 2a)	Huh7	HPF	MMVs	340 nm
	Expression of polyprotein/JFH1 (genotype 2a)	Huh7-Lunet T7	HPF		
Ferraris et al. 2013	Infection/JFH1 (genotype 2a)	Huh7.5	4 % PFA and 1 % GA	Vesicles in clusters (ViCs)	100–200 nm
				Contiguous vesicles (CVs)	~100 nm
				DMVs	150–1,000 nm

GA glutaraldehyde, *PFA* paraformaldehyde, *HPF* high-pressure freezing, *DMVs* Double Membrane Vesicles, *MMVs* Multi-membrane Vesicles

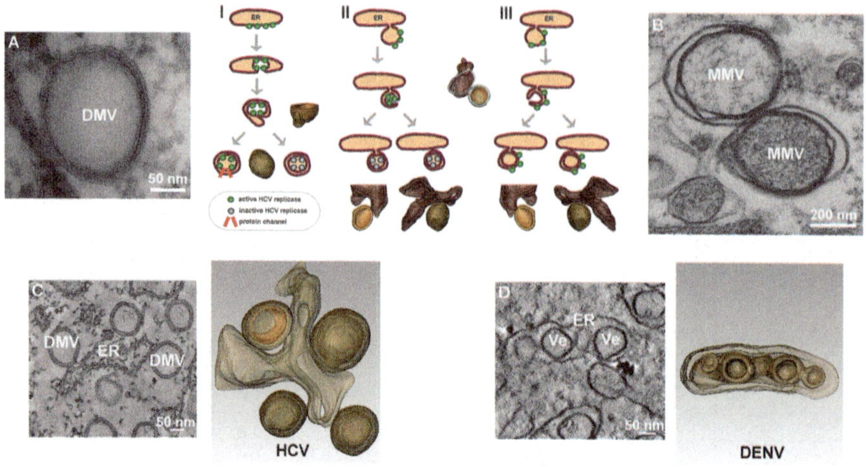

Fig. 2 3D architecture of HCV-induced structures. (**a**) DMVs, double membrane vesicles, are the main hallmark of HCV infection: they are composed of two lipid bilayers tightly opposed to each other. I–III represent different ways in which DMVs and active replicase might originate from the ER (Romero-Brey et al. 2012). (**b**) MMVs, multi-membrane vesicles, are found late in infection and consist of several circular concentric membranes. (**c**, **d**) Slices through electron tomograms (on the *left*) and 3D representation (on the *right*) of the HCV and DENV replication vesicles (reproduced with permission from Romero-Brey and Bartenschlager, 2014). (**c**) Some of the DMVs originated upon HCV infection remain connected to the ER via their outer membrane, indicating that they might be originated by evagination of ER membranes. (**d**) The DENV replication sites are formed as invaginations towards the ER lumen, generating several vesicles (*Ve*) inside the ER lumen

represent the sites of RNA replication. This view is supported by EM and biochemical data, showing that affinity-purified membranes from HCV replicon cells mainly consist of DMVs, that are able to support RNA synthesis in vitro (Paul et al. 2013).

DMVs and MMVs were also described previously in Huh 7.5 cells harboring a subgenomic replicon (JFH1) (Ferraris et al. 2010). Furthermore, both structures were also observed in purified membranes associated with HCV that were found to contain NS3, NS5A and HCV RNA. In a more recent study Ferraris et al. identified two additional types of membrane alterations using JFH1-infected cells: contiguous vesicles (CVs), small single-membrane vesicles, present in large numbers and widely distributed throughout the cytoplasm, with a more homogeneous size (around 100 nm). They were tightly associated with each other and tended to form a collar around lipid droplets (LDs). Vesicles in clusters (ViCs), small single-membrane vesicles of variable size (100–200 nm), were also found in infected cells, grouped together in well-delimited areas (Table 1, Ferraris et al. 2013). A 3D reconstruction of a complete HCV-infected cell revealed that all three membrane structures were tightly connected and closely associated with LD clusters (Ferraris et al. 2013). In this study, the number of CVs correlated with intracellular HCV RNA levels, arguing for a possible role of CVs in the early stages of viral replication. Alternatively, CVs might constitute the membranous platform for viral assembly. In fact, the core protein is present in these structures (16 %) as well as on the LD surface (81 %).

However, thus far visualization of virus assembly sites and virus particles in infected cells has not been possible, making this hypothesis difficult to prove. While most of the dsRNA signal was located within DMVs or at DMV membranes, ViCs were free of viral components and RNA and these structures as well as CVs were very rarely observed in cells with a subgenomic JFH1 replicon (Ferraris et al. 2010), or absent in cells infected with a JFH1 variant designated Jc1 (Romero-Brey et al. 2012), having the same replicase, but producing far higher amounts of virus (Pietschmann et al. 2006). Altogether, these data suggest that CVs and ViCs might arise from accumulations of structural proteins and are probably linked to assembly rather than being associated with RNA synthesis.

The differences observed for HCV-induced membrane alterations can most likely be attributed to variations in the cell culture models and fixation techniques. Still, recent studies agree that the predominant membranous structures detected in HCV-infected cells and in cells harboring subgenomic replicons are DMVs (Table 1). However it is not proven that these structures represent *bona fide* active replication factories. A functional link between DMVs and RNA replication is suggested by the fact that dsRNA, the presumptive RNA intermediate, localizes to the lumen of these vesicles (Ferraris et al. 2010), that purified DMVs contain HCV RNA (Ferraris et al. 2010), and also retain the ability to synthesize RNA in vitro (Paul et al. 2013). The most important argument against their role as active replication sites is the apparent absence of a pore or an opening connecting them with the surrounding cytosol, which is likely to be required to supply nucleotides and to deliver newly synthesized RNA to the cytoplasm. Only in ~10 % of DMVs has an opening towards the cytosol been observed, possibly indicating that only a minority of the DMVs are actively engaged in replication at a given time (Romero-Brey et al. 2012). Alternatively, many pores might be beyond the resolution limits of the methods used so far to visualize them. Thus it could be that proteinaceous (viral or cellular) or lipid channels might allow the transport of newly synthesized RNA towards the cytosol and the uptake of nucleotides from the cytosol (see below, Role of cell proteins and lipids in the biogenesis of the HCV replication sites). However, it still cannot be excluded that DMVs represent late stages of inactive replication organelles that originate from less prominent and less abundant active replication vesicles, e.g. from CVs (Ferraris et al. 2013), due to the lack of time-resolved data showing newly synthesized RNA in these structures.

3 Biogenesis of the Membranous Web and Comparison to Replication Organelles Induced by Different Members of the Family *Flaviviridae* and by Other Positive- Strand RNA Viruses

HCV, as the prototype virus of the genus *Hepacivirus*, seems to induce the formation of membrane-bound RNA factories that are distinct from those observed in cells infected with other members of the family *Flaviviridae*. Analysis by ET revealed that DMVs are evaginations from the ER (Fig. 2c). Still it is not clear how they are generated and several alternative models might be envisaged (Fig. 2a, I–III) (Romero-Brey et al. 2012): (I) It could be that HCV proteins induce invaginations of the ER membrane, and the second membrane is acquired by a local contraction of the ER lumen. In this case enzymatically active HCV replicase (green dots) resides in the lumen of the invagination and remains active as long as the vesicle is linked to the cytosol. DMVs might be connected to the cytosol via protein channels formed by viral or host factors. Alternatively, such structures might be formed by autophagy, which has been proposed to play an essential role in MW formation (Sir et al. 2012). (II) HCV proteins or host cell machinery recruited by viral proteins might induce a single-membrane evagination from the ER membrane, which then undergoes a secondary invagination. These DMVs might initially contain an opening to the cytoplasm, which could later be closed to render the replicase inactive. The connection of the outer DMV membrane with the ER membrane could be pinched off by host factors. (III) A third possibility is that induction of DMVs follows the same pathway as described above, but the active replicase resides on the outside of the DMV. In this case, no connection of the DMV to the cytoplasm is required, explaining the low frequency of pores observed. However, this model seems unlikely, since viral replicase activity and RNA is highly resistant to proteases and nucleases, respectively, in replication complexes purified from replicon cells (Quinkert et al. 2005; Paul et al. 2013; Miyanari et al. 2003). This protection is detergent sensitive, arguing for the active replicase being shielded by membranes.

Interestingly, no prominent membrane alterations associated with RNA replication have been found in cells infected with the closely related pestiviruses, other than vesicles in multi-vesicular bodies (MVBs) containing dsRNA (Schmeiser et al. 2014).

Surprisingly, replication sites produced by viruses of the genus *Flavivirus* are entirely different from the HCV MW. Generally, flavivirus replication vesicles appear as single membrane invaginations in the lumen of the ER (Fig. 2d). These vesicles have been observed for instance in Dengue Virus (DENV)-infected cells (Welsch et al. 2009) having a diameter of 80–90 nm, as well as in Kunjin Virus -the Australian variant of West Nile Virus- (WNV_{KUN})- (Westaway et al. 1997b; Mackenzie et al. 2001; Gillespie et al. 2010) and Tick-Borne Encephalitis Virus (TBEV)-infected cells (Overby et al. 2010; Miorin et al. 2013). ET analysis of cells infected with either one of these three viruses revealed that such vesicles are frequently found in groups filling up the ER lumen and, therefore, they have been termed vesicle packets (VPs). Another aspect that distinguishes these vesicles from those seen in HCV-infected cells is that they remain connected to the cytosol via pores of ~10 nm diameter, which allows the export of the newly synthesized RNA to the flaviviral assembly sites. In fact, virions were found in close proximity to the pores of the replication vesicles, suggesting that replication factories could represent a continuous membrane network that provides a platform for the transport of viral proteins and genomes between sites of RNA replication, ribosome-containing compartments (RNA translation) and virus assembly sites (Welsch et al. 2009). In addition to these invaginations, convoluted membranes (CM) have been found in DENV (Welsch et al. 2009) and WNV_{KUN} (Westaway et al. 1997b; Mackenzie et al. 1996) infected cells. Morphologically, CMs resemble smooth ER membranes, lack ribosomes and in the case of DENV are induced by the sole expression of NS4A (Roosendaal et al. 2006; Miller et al. 2007). Based on the localization of several NS proteins to these structures, it has been suggested that they may represent the site of RNA translation/polyprotein processing (Westaway et al. 1997a, b), or might represent a storage site for proteins and lipids involved in viral replication that can be recruited to vesicles upon demand (Welsch et al. 2009). However their exact role in the flaviviral life cycle needs still to be determined.

These findings indicate that all members of the genus *Flavivirus* studied to date, as well as HCV (genus *Hepacivirus*) utilize the ER as a source of membranes for the formation of their replication factories, although they seem to re-shape these membranes in a fundamentally different fashion: while members of the genus *Flavivirus* induce invaginations of the host membranes, HCV seems to form evaginations of these membranes, generating DMVs that look like protusions of the ER (Fig. 2c, d).

However, HCV is not the only positive- strand RNA virus inducing the formation of DMVs, which involves the formation of positive curvature membranes. Structures with two lipid bilayers have been also found in cells infected with positive-strand viruses classified within the families *Picornaviridae*, *Coronaviridae*, *Arteriviridae* and *Nodaviridae* (reviewed in Romero-Brey and Bartenschlager 2014). In contrast, alphaviruses and Rubella Virus are known to induce the formation of negatively curved membranes, initiated by invaginations of the pre-existing membrane bilayer and giving rise to vesicles (also called spherules or vacuoles) towards the lumen of the targeted cell organelle, as described above for members of the genus *Flavivirus* (reviewed in Romero-Brey and Bartenschlager 2014). The conservation of these two types of induced organelles in distantly related viruses supports the possibility of an evolutionary conserved mechanism, which might involve distinct cellular machineries.

4 Role of the Viral Proteins in the Formation of DMVs

As already discussed in this chapter, cells harboring subgenomic replicons contain DMVs as those found following HCV infection (Ferraris et al. 2010; Romero-Brey et al. 2012), indicating that the structural proteins are not only dispensable for replication (Lohmann et al. 1999a), but also for the formation of DMVs (Fig. 3a). This finding also provided an indirect indication that the function of these structures is related to the viral replication of HCV. Interestingly the replication efficiency of replicons correlates with the amount of DMVs that are found in the cytosol of cells transfected with replicons: replicon NS2-5B produces lower amount of DMVs than replicon NS3-5B (Romero-Brey et al. 2012). This is again an indirect evidence of the role of DMVs in replication.

The very same membrane alterations identified in replicon cells were also found upon expression of a NS3-5B polyprotein fragment demonstrating that formation of DMVs is solely induced by viral proteins independent from HCV RNA replication

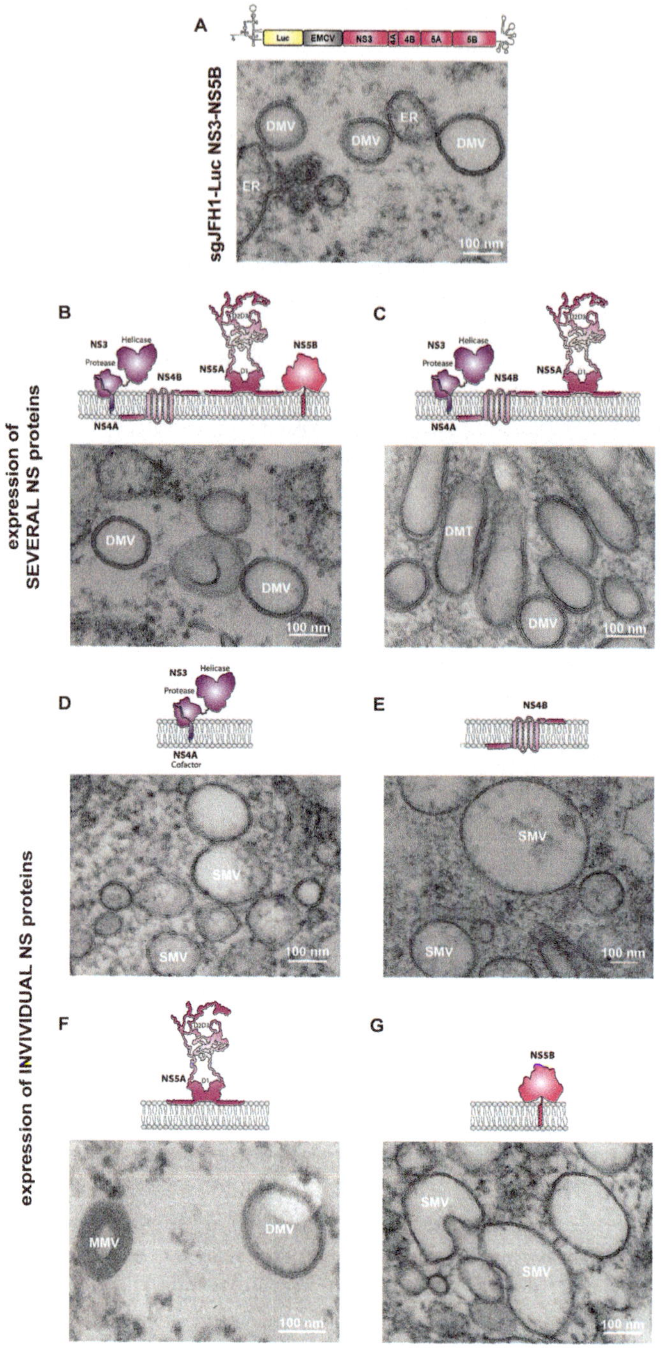

Fig. 3 Analysis of the formation of DMVs in cells harboring HCV subgenomic replicons (**a**) and upon expression of several NS proteins (**b**, **c**) and single proteins (**d–g**). The NS proteins were expressed in Huh7-Lunet T7 cells and membrane alterations were assessed by EM in cell sections after epon embedding. (**b**) Expression of all the NS replicase proteins (NS3-5B) induces the

(Romero-Brey et al. 2012) and also in the absence of the NTRs (Berger et al. 2014) (Fig. 3b). Therefore, expression systems can be used to assess the contribution of the replicase NS proteins to the formation of DMVs. In addition, the use of expression models allows studying the impact of inhibitors of viral replication and of mutations interfering with RNA synthesis on MW morphology (Romero-Brey et al. 2012).

Earlier studies already showed the potential of individual HCV proteins to induce membrane alterations: the NS3/4A complex induced the formation of large amounts of smooth SMVs and expression of NS4B was able to generate the MW (Egger et al. 2002), which was defined as compact clusters of ~85 nm vesicles. We obtained similar results following expression of NS3/4A and NS4B, but in addition also found specific membrane alterations upon expression of NS5A and NS5B (Fig. 3d–g, respectively). Importantly, expression of NS5A led to the formation of DMVs and MMVs, albeit with low abundance and different size as compared to NS3-NS5B expression, whereas NS5B induced enlargements of the ER occasionally containing invaginations. Therefore, all the replicase components were capable of inducing membrane vesiculation with NS5A having the highest potential to trigger membrane curvature. Some of these NS5A-induced structures were DMV-like vesicles, pointing to NS5A as an important contributor to DMV biogenesis. Still, the individual expression of the NS replicase components NS3/4A, NS4B, NS5A and NS5B did not lead to the formation of DMVs with the same abundance, morphology and heterogeneity as expression of NS3-5B (Romero-Brey et al. 2012).

Although our results showed that NS4B is clearly not able to induce the formation of DMVs and the complexity of the MW, it still plays an important role in triggering rearrangements of intracellular membranes. A recent study has demonstrated that NS4B oligomerizes through multiple conserved determinants and that oligomerization appears to be required for MW induction (Gouttenoire et al. 2010). Indeed, mutations affecting the highly conserved C-terminal domain

Fig. 3 (continued) formation of DMVs, morphologically identical to those found in infected cells. (**c**) Expression of an NS3 to 5A polyprotein fragment lacking NS5B (RdRp) induced the formation of small clusters of DMVs and elongated double membrane tubules (*DMTs*) having a highly variable diameter (166 ± 92 nm). (**d–g**) Analysis of the contribution of individual HCV proteins to the formation of the DMVs: NS3/4A, NS4B, NS5A or NS5B. (**c**) Note that NS3 and NS4A were not expressed individually, because both proteins form a stable membrane-associated complex and only this complex is physiologically relevant (Wolk et al. 2000). Expression of the NS3/4A complex led to the formation of swollen ER sheets and single membrane vesicles (*SMVs*) of variable diameter (203 ± 93 nm). (**d**) In cells expressing NS4B SMVs were also found having an average size of 325 nm (±168 nm). (**e**) Cells expressing NS5A alone showed vesicles containing several lipid bilayers in a concentric manner (MMVs, multimembrane vesicles) with an average diameter of 125 nm (±35 nm). Some of these vesicles displayed only two lipid bilayers and their morphology was indistinguishable from DMVs observed in HCV-infected cells or cells containing a subgenomic replicon. (**f**) Expression of NS5B induced enlarged ER sacs with an average diameter of 370 nm (±150 nm) and occasional curvature (The EM micrographs were taken from Romero-Brey et al. 2012, with permission)

impairing NS4B self-interaction resulted in the formation of aberrant DMVs, arguing for a central role of NS4B in the formation of functional replication compartments (Paul et al. 2011).

Although the molecular mechanisms orchestrating the formation of these complex structures are still poorly understood, it is now clear that all the replicase NS proteins (NS3/4A, NS4B, NS5A and NS5B) work in a concerted action to generate DMVs and the overall complexity of HCV replication sites. Even expression of NS3-5A, lacking the RNA-dependent RNA polymerase (RdRp), NS5B, generated different MW structures mainly consisting of elongated DMVs (so called DMTs, double membrane tubules, Fig. 3c), indicating that NS5B is not a mandatory building block for the formation of DMVs but still influences their morphology.

5 Inhibition of MW Formation by Direct Acting Antivirals

In agreement with the view that MW formation requires a complex interplay of all replicase proteins, recent studies brought up evidence that existing classes of replication inhibitors indeed act by interfering with MW biogenesis. NS5A inhibitors have been identified in high throughput screening with replicon cells and are characterized by a very high potency against HCV in vitro and in vivo (Gao et al. 2010). Among them the clinical lead compound Daclatasvir (DCV) has been approved for treatment of patients. This class of compounds indeed blocks the replication of HCV RNA already at the stage of the DMV formation (Berger et al. 2014). CLEM on cells expressing the NS3-5B polyprotein revealed unequivocally that co-treatment (addition of drug directly before DNA transfection) with NS5A inhibitors had no severe effect on the subcellular distribution of NS5A, but completely prevented biogenesis of DMVs. This complete abrogation of membranous vesicle formation was not found upon expression of an inhibitor-resistant NS5A mutant when the drug was applied after transfection. Therefore, highly potent DCV-like NS5A inhibitors disrupt DMV formation independent of RNA replication at a very early stage of the viral replication cycle.

Inhibition of NS4B is another way to interfere with MW formation. A recent study has shown that resistance mutations to intravenously administered Silibinin (Legalon-SIL [SIL]), which has been successfully used for HCV therapy, indeed map to NS4B (Esser-Nobis et al. 2013). Ultrastructural analyses revealed changes in the morphology of viral membrane alterations upon SIL treatment of a susceptible genotype 1b isolate, but not of a resistant NS4B mutant or genotype 2a. Most of the vesicles observed were MMVs and no DMVs, suggesting that SIL indeed modulated the morphology of viral replication sites. Other HCV inhibitors targeting NS4B might also act by interfering with MW morphogenesis, however, formal proof is lacking (Dvory-Sobol et al. 2010; Dufner-Beattie et al. 2014).

In addition to inhibitors directly targeting viral proteins and thereby interfering with MW formation, a number of drugs targeting host factors have been shown to change the morphology of the HCV replication sites. This includes inhibitors of

PI4KIIIα (Bianco et al. 2012), OSBP (Wang et al. 2014) and cyclophilins (Madan et al. 2014) and will be discussed in more detail in the next section. Generally, a surprisingly high number of drugs targeting various viral and cellular proteins seem to inhibit viral replication by disturbing the morphogenesis of the MW. Targeting viral replication sites might therefore be a promising strategy to develop inhibitors also to other positive- strand RNA viruses.

6 Role of Cellular Proteins and Lipids in the Biogenesis of the HCV Replication Sites

Literally hundreds of cellular proteins and lipids linked to viral RNA synthesis have been identified in numerous high content screenings (e.g. Li et al. 2009; Tai et al. 2009) and a comprehensive overview goes beyond the scope of this review, particularly since still little is known about their mechanisms of action. However, since several of those factors have recently been shown to be involved in the formation of the HCV replication sites, we will focus on the following here: PI4KIIIα, OSBP, VAP-A and VAP-B, cyclophilins, GBF1-Arf1-COPI, Nups, the autophagy machinery, Rabs, PSTPIP2 and PLA2G4C as summarized in Fig. 4.

6.1 Role of Cellular Lipids and Proteins Involved in Lipid Synthesis and Transport in the Biogenesis of the MW

The ability to shape a target membrane into a RNA replicating vesicle is not exclusively dependent on the ability of viral and cellular proteins to bend membranes. Indeed growing evidence shows that lipids are also key players in membrane-bound viral RNA synthesis (reviewed in Alvisi et al. 2011). Multiple reports have indicated that HCV modulates lipid metabolism (e.g. cholesterol and fatty acid biosynthesis) to promote viral replication (Kapadia and Chisari 2005; Su et al. 2002; Takano et al. 2011; Waris et al. 2007; Diamond et al. 2010). This modulation results in *de novo* lipid formation in order to increase membrane surface area, which is required for the formation of viral replication factories. A cellular cofactor involved in lipogenesis during HCV infection is the sterol regulatory element-binding protein (SREBP) (Waris et al. 2007). SREBPs are major regulators of lipid metabolism and major transcription factors for expression of genes required for lipid biosynthesis. Over-expression of NS4B has been shown to activate SREBP, leading to elevated levels of transcripts involved in lipogenesis such as fatty acid synthase (FASN) (Waris et al. 2007; Park et al. 2009). However, by using S1P-specific inhibitors SREBP-mediated lipogenesis was found to be dispensable for HCV RNA replication but required for assembly and release of progeny virus (Olmstead et al. 2012). Similarly modulation of triglyceride and cholesterol ester

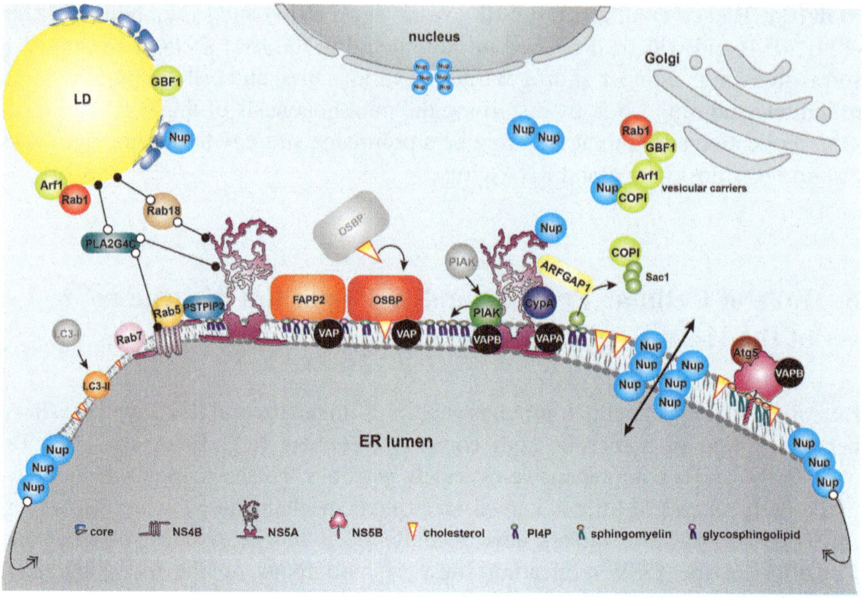

Fig. 4 Schematic representation of an ER membrane, depicting a map of interactions between the HCV NS proteins and several cellular cofactors that might be directly or indirectly involved in the formation and/or functionality of DMVs. For an understating of the different interactions please refer to the text

synthesis –lipids stored in LDs- seems to be also related to HCV assembly (Liefhebber et al. 2014).

Sphingolipid synthesis is also stimulated by HCV, resulting in globally and locally increased sphingomyelin levels that are required for replication and might contribute to detergent resistance of HCV replication sites (Hirata et al. 2012). In addition, sphingomyelin interacts with the NS5B polymerase in a genotype-specific manner, enhancing HCV RNA synthesis (Weng et al. 2010) (Fig. 4).

A very recent report stressed not only the importance of *de novo* synthesis of lipids, but also the need to limit the oxidative degradation of lipids, designated lipid peroxidation, to achieve robust replication of HCV in cell culture (Yamane et al. 2014). Thus lipid peroxidation, regulated in part through sphingosine kinase-2, severely restricts HCV replication of all genotypes in Huh-7 cells and primary human hepatoblasts and reduces the HCV-induced MW abundance. The genotype 2a JFH1 strain is an exception and is resistant to lipid peroxidation (Yamane et al. 2014). For further details, readers are referred to the chapter by Yamane and Lemon in this volume.

Besides generally enhancing cellular lipid content, HCV also exerts a more direct influence on the lipid composition of its replication sites. The HCV replication complex has been reported to be associated with lipid rafts (Aizaki et al. 2004; Shi et al. 2003). These membrane lipid microdomains are enriched for cholesterol, sphingolipids and certain proteins and form nanoscale-ordered protein-lipid assemblies (reviewed in Simons and Sampaio 2011): they generally contain three- to fivefold the amount of cholesterol found in the surrounding bilayer (Anchisi et al. 2012). Consistent with this, analysis of the molecular composition of the membranous replication compartment revealed that HCV-remodeled membranes are highly enriched in cholesterol (Paul et al. 2013). Higher concentrations of this cone-shaped lipid on membranes enhances rigidity and stability, likely serving as an important structural component of HCV-remodeled membranes (Fig. 4). The accumulation of these lipids within HCV-induced membranes might help the bending of these membranes into DMVs, via the formation of lipid-enriched microdomains that favour the positive contouring of these membranes. Several mechanisms have been unraveled concerning how HCV modulates the lipid environment of its replication sites, including the activation of the lipid kinase PI4KIIIα to generate enhanced levels of PI4P at the replication sites, which in turn attracts lipid transport proteins delivering cholesterol and glycosphingolipids.

6.1.1 PI4KIIIα

One essential host factor implicated in HCV RNA replication is the phosphatidylinositol 4-kinase IIIα (PI4KIIIα), an ER-resident enzyme that catalyzes the synthesis of phosphatidylinositol 4-phosphate (PI4P). In mammalian cells, the family of PI4-kinases comprises two types with two isoforms each (PI4KIIα, PI4KIIβ, PI4KIIIα and PI4KIIIβ) differing in subcellular localization and being responsible for the synthesis of distinct PI4P pools (reviewed in Balla 2006). PI4KIIIα is present in the ER, the plasma membrane (Balla et al. 2005) and parts of the Golgi (Bianco et al. 2012). PI4P accumulates in HCV replicating cells due to recruitment and activation of PI4KIIIα (Reiss et al. 2011; Berger et al. 2011; Tai and Salloum 2011; Bianco et al. 2012; Hsu et al. 2010) by NS5A and NS5B (Reiss et al. 2013) (Fig. 4). The interaction site of PI4KIIIα with NS5A has been mapped to the carboxyterminal end of domain I, encompassing seven amino acids, which are essential for viral RNA replication and PI4KIIIα activation (Reiss et al. 2013). However, only the C-terminal half of PI4KIIIα is required for HCV replication (Harak et al. 2014). In addition to PI4KIIIα, several other studies also implicated a role of the Golgi resident PI4KIIIβ isoform, particularly in the replication of genotype 1 (Borawski et al. 2009; Hsu et al. 2010). This indeed suggests that PI4P metabolism has an essential role in HCV replication.

PI4P localizes mainly to the Golgi and the inner leaflet of the plasma membrane, where it fulfills important functions by providing "signatures" to distinct membrane compartments and by recruiting multiple factors involved in vesicle budding and lipid biosynthesis (Hammond et al. 2012). Activation of PI4KIIIα by HCV results in strongly elevated intracellular PI4P levels with an altered intracellular distribution, partially co-localizing with HCV replication sites (Reiss et al. 2011). PI4KIIIα thereby influences the morphogenesis of the HCV replication compartments, most likely involving altered levels of PI4P. Depletion of PI4KIIIα by RNA interference or inhibitors, or by abrogating NS5A interaction, results in more homogenous MW structures consisting solely of DMVs with reduced diameter and organized in huge clusters (Reiss et al. 2011, 2013).

In addition to its role in MW morphogenesis, PI4KIIIα might also impact HCV replication in a PI4P-independent fashion, influencing regulation of NS5A phosphorylation, thereby facilitating synthesis of the p56 form of this NS protein (Reiss et al. 2013). Alternatively, PI4P may recruit cellular and/or viral proteins required for replication (Berger et al. 2009; Bishe et al. 2012).

6.1.2 Lipid Transporters: OSBP, FAPP2

In addition to a potential direct role in the morphogenesis of the MW, PI4P indirectly affects the lipid composition of membranes by attracting lipid transfer proteins, thereby orchestrating changes in lipid composition at HCV replication sites. Lipid transfer proteins (LTPs) are implicated in the nonvesicular transport of lipids within the cell. Several of these proteins contain an N-terminal pleckstrin homology (PH) domain that mediates PI4P binding and a C-terminal lipid transfer domain that interacts with a second lipid that needs to be transferred between membranes. The direction of transport is typically driven by PI4P concentration gradients. The ER membrane is depleted of PI4P by the phosphoinositide phosphatase Sac1 (the enzyme that catalyzing PI4P dephosphorylation), whereas oxysterol-binding protein (OSBP) regularly transfers PI4P from Golgi to ER and transports back cholesterol from ER to Golgi in a process assisted by VAP-A and -B (Mesmin et al. 2013) (described below).

OSBP was first shown to be important for HCV virion secretion (Amako et al. 2009, 2011). However, it was recently shown that OSBP is also essential for HCV RNA replication (Wang et al. 2014). Colocalization of OSBP and NS5A is lost when PI4KIIIα is silenced, suggesting that OSBP is recruited to the replication compartment via PI4P (Fig. 4). Using fluorescent cholesterol analogs, the authors confirmed that cholesterol is enriched in NS5A-positive structures (Wang et al. 2014), corroborating previous findings with purified DMVs (Paul et al. 2013). Furthermore, inhibition of OSBP caused clustering and reduction in the diameter of the membranous vesicles, mimicking the phenotype observed upon silencing of PI4KIIIα and suggesting that this phenotype in fact is mediated by cholesterol depletion. In summary, these results suggest that OSBP is a PI4P effector required for HCV replication membrane integrity and cholesterol trafficking. Activation of PI4KIIIα and enrichment of PI4P thereby generates a concentration gradient to deliver cholesterol to replication sites. Interestingly, OSBP is also required for poliovirus infection, suggesting that DMV-inducing viruses might generally require PI4P- and cholesterol- enriched membranes in order to form their replication compartments.

Recently, an additional PI4P-specific lipid transfer protein, the four-phosphate adaptor protein 2 (FAPP2), was shown to be important for HCV replication (Khan et al. 2014). FAPP2 exchanges PI4P with glycosphingolipids, which are thereby enriched in the MW, again due to PI4P concentration gradient generated by activation of PI4KIIIα. Depletion of FAPP2 could be partially rescued by providing exogenous glycosphingolipids, arguing for a direct role of the lipid rather than the protein (Khan et al. 2014). Overall, HCV-mediated induction of PI4KIIIα has a fundamental impact on the lipid composition of the viral replication compartment beyond PI4P and might also substantially contribute to the lipid-raft properties of these membranes (Aizaki et al. 2004) and provide lipids required for virion formation (Aizaki et al. 2008).

6.2 *Role of Vesicular Trafficking in the Biogenesis of the MW*

6.2.1 VAPs

Two additional well-described host factors are the vesicle-associated membrane protein–associated proteins A and B (VAP-A and VAP-B) (Evans et al. 2004; Gao et al. 2004; Hamamoto et al. 2005) that are involved in ER homeostasis and vesicular trafficking (Wyles et al. 2002; Amarilio et al. 2005). In a recent study, Paul and colleagues found that these HCV–co-opted host proteins VAP-A and VAP-B are enriched in purified DMVs (Paul et al. 2013). These findings together with the fact that VAP-A interacts with OSBP to modify export from the ER (Amarilio et al. 2005), where the DMVs are being formed, suggests that an interplay of both proteins might help to re-shape ER membranes (Fig. 4). In addition, VAP-A and -B might be involved in the recruitment of lipid transport proteins like OSBP and FAPP2 and thereby help to shape the lipid composition of the HCV replication sites (Mesmin et al. 2013). Furthermore, VAP-A and VAP-B proteins have been shown to interact with NS5A in a phosphorylation-dependent manner (Evans et al. 2004; Gao et al. 2004; Hamamoto et al. 2005) and with NS5B (Hamamoto et al. 2005), highlighting again the central role of NS5A in the formation of the membranous replication factories (Fig. 4).

6.2.2 Rabs

The Ras superfamily of small GTPases is involved in the regulation of vesicle budding, transport and fusion with target membranes. They cycle between active (GTP-bound) and inactive (GDP-bound) forms, requiring cognate GTPase-activating proteins (GAPs) for catalysis and guanine nucleotide exchange factors (GEFs) for GTP binding. Rabs are associated with specific host membranes, for instance, Rab5 is bound to the early endosome (EE) where it regulates endocytosis and EE fusion, Rab7 is found in the late endosome (LE) and facilitates transport from LE to lysosomes, Rab6 is mostly associated with EE recycling to the Golgi as well as intra-Golgi traffic and Rab1 and 2 are mostly found in the Golgi and regulate traffic between ER and the Golgi apparatus (reviewed in Mizuno-Yamasaki et al. 2012).

Rab5 has been shown to be involved in HCV genome replication (Stone et al. 2007; Berger et al. 2009) and the NS4B C-terminal domain might play a role in Rab5 recruitment to NS4B foci (Aligo et al. 2009) (Fig. 4), suggesting that the early endosome contributes to the formation of HCV replication sites. Another study demonstrated that Rab5 and 7 strongly co-localize with NS4B positive foci and Rab2, 5 and 7 are required for HCV RNA replication (Manna et al. 2010) (Fig. 4).

Rab1b, a regulator of membrane dynamics in the early secretory pathway (Sklan et al. 2007) and its negative regulator TBC1D20 (Sklan et al. 2007; Nevo-Yassaf et al. 2012) have been reported to be involved in the replication of HCV as well.

Rab18 is believed to promote physical interaction between LDs and ER membranes (Ozeki et al. 2005). Salloum and colleagues (2013) have shown that Rab18 promotes the physical association of the other replicase components with LDs through direct association between NS5A and the active, GTP-bound form of Rab18 (Fig. 4). Rab18 might, therefore, help to establish a physical coupling between the HCV replication and virion production. Consistent with this hypothesis, a recent report has shown that Rab18 is required for viral assembly (Dansako et al. 2014).

Altogether these findings suggest an involvement of several Rabs in the formation of HCV replication sites, which are likely dependent on a mechanism more complex than a simple remodeling of ER membranes.

6.2.3 GBF1-Arf1-COPI

Several independent studies point to a role for GBF1-ArfI-COPI, a protein complex involved in retrograde transport from *cis*-Golgi to the ER (reviewed in Bonifacino and Glick 2004; Donaldson and Jackson 2011), in the biogenesis of the MW. GBF1 (Golgi complex-specific BFA resistance factor 1) is a guanine nucleotide exchange factor (GEF), which is recruited to the membrane of the *cis*-Golgi and activates the small GTPase-protein Arf1 (ADP-ribosylation factor 1). Once activated by the binding of GTP, Arf1 in turn recruits different effectors, including the coat protein complex COPI, which then forms vesicular carriers. Recent reports indicate that the replication of HCV depends on the GBF1-Arf1-COPI complex (Tai et al. 2009; Goueslain et al. 2010; Matto et al. 2011; Zhang et al. 2012; Farhat et al. 2013), however the precise role of this pathway in HCV replication is still elusive.

Pharmacological inhibition of Arf1 by the fungal metabolite brefeldin A (BFA), has been shown to decrease HCV RNA replication (Tai et al. 2009; Goueslain et al. 2010; Matto et al. 2011). The inhibition of HCV replication by BFA is much stronger at the beginning of the infection than when the infection is already established (Goueslain et al. 2010), suggesting a crucial role at the onset of replication. EM analyses indicated that BFA does not block the formation of MW-like structures induced by expression of HCV proteins in a non-replicative context, suggesting that GBF1 is probably not involved in the formation of HCV replication complexes but, rather, in their activity (Goueslain et al. 2010). HCV replication is almost as sensitive to BFA in BFA-resistant hepatoma-derived cells as in Huh-7 cells, suggesting that GBF1 might fulfill another function, in addition to regulation of the secretory pathway (Farhat et al. 2013). Goueslain and coworkers hypothesized that GBF1-associated mechanisms function to deliver proteins or lipids to the HCV replication sites (Goueslain et al. 2010). Indeed Arf1 and GBF1 seem to be involved in generating a PI4P-enriched environment supportive of HCV replication (Zhang et al. 2012). Along the same lines, it has been recently shown that ARFGAP1 is hijacked by NS5A to remove COPI cargo Sac1 from the site of HCV replication (Li et al. 2014). Since Sac1 catalyzes the PI4P dephosphorylation, its removal by ARFGAP1 might thereby help to maintain high levels of PI4P at HCV replication sites (Fig. 4).

6.3 Cyclophilins

Cyclophilins (Cyps) are ubiquitous molecular chaperones catalyzing the *cis-trans* isomerization of proline residues and hence are called peptidyl-prolyl *cis-trans*-isomerases (PPIases). Their isomerase activity is thought to be important for proper folding of certain proteins. At present, 16 Cyp members have been identified with seven major members found in humans (CypA-E, Cyp40, CypNK; (Wang and Heitman 2005)).

The role of Cyps in HCV replication was identified by the inhibition of HCV replication by Cyclosporin A (Watashi et al. 2003). Initially, CypB was supposed to be the key player in HCV replication (Watashi et al. 2005; Nakagawa et al. 2004), however, more recent evidence rather points to CypA (Yang et al. 2008; Kaul et al. 2009; Chatterji et al. 2009). For both CypA (Yang et al. 2008) and CypB (Watashi et al. 2005; Nakagawa et al. 2005; Heck et al. 2009), direct interactions with NS5B were reported, but more recently, binding sites in NS5A domain II and III have been identified and characterized by NMR as well (Hanoulle et al. 2009; Verdegem et al. 2011). Indeed, mutations conferring resistance to CypA inhibitors have been found in NS5A (Yang et al. 2010; Grise et al. 2012) as well as NS5B (Liu et al. 2009a). Initially, CypA was regarded as a host factor directly regulating RNA synthesis by *cis-trans* isomerization of particular proline residues in NS5A and NS5B, thereby modulating polyprotein processing kinetics (Kaul et al. 2009), RNA binding by NS5B (Watashi et al. 2005; Liu et al. 2009a), or recruitment of NS5B to replication sites (Liu et al. 2009b). A more recent study additionally suggests that CypA and NS5B share a binding site in NS5A, which might regulate replication (Rosnoblet et al. 2012). However, Madan and colleagues have reported that cyclosporine also blocks the *de novo* formation of DMVs, while having little effect on established membranous replication factories (Madan et al. 2014). Furthermore, this block was prevented by cyclosporine resistance mutations in NS5A. These data suggest that CypA-dependent modification of NS5A is required for the biogenesis of the HCV replication compartment (Fig. 4).

6.4 Nups

Several components of the nuclear transport machinery have been shown to be involved in the life cycle of HCV (Kim et al. 1999; Suzuki et al. 2005; Yamanaka et al. 2002; Ide et al. 1996; Isoyama et al. 2002; Chung et al. 2000; de Chassey et al. 2008), including soluble nuclear transport factors (NTFs), many of which are members of a family of proteins termed karyopherins (Kaps). Kaps bind nuclear localization/import signal (NLS) or nuclear export signal (NES) containing molecules in the cytoplasm or nucleus and escort these cargos across the nuclear envelope (NE) through passageways formed by large macromolecular structures termed nuclear pore complexes (NPCs) (reviewed in (Wente and Rout 2010). Each NPC is comprised of ~30 distinct proteins, called nucleoporins (Nups), which form a cylindrical channel lined by Nups that facilitate movement of the NTF across the nuclear envelope.

Following HCV infection Neufeldt and colleagues (Neufeldt et al. 2013), observed an increase in cytoplasmic levels of various Nups and their recruitment to regions of cytoplasm containing HCV replication or assembly complexes. Consistent with these observations, they also demonstrated an association between various HCV proteins and specific Nups, as well as the NTFs Kap β3/IPO5 and Kap α. These interactions appear to play a role in the viral life cycle, as depletion of specific Nups or Kap β3/IPO5 inhibits HCV replication. Interaction of HCV proteins with Nups and Kaps could therefore potentially alter host cell nucleocytoplasmic transport to facilitate HCV replication. Alternatively, Nups may be recruited to the MW, in part, to usurp their functions in contouring of membranes (Neufeldt et al. 2013) (Fig. 4). The curvature of membrane domains at sites of viral particle budding into the ER lumen is indeed topologically similar to the nuclear pore membrane (Bartenschlager et al. 2010; Shimizu et al. 2011). Finally, Nups could act as a permeability barrier allowing NLS-containing molecules to access regions within the MW (Fig. 4). NLS-like sequences or NTF binding domains have been identified in core (Ide et al. 1996), NS5A (Suzuki et al. 2005), and NS3 (Kim et al. 1999). Moreover, previous studies have found that core and NS5A proteins interact with Kap β3/IPO5 and Kap α (Isoyama et al. 2002; Chung et al. 2000; de Chassey et al. 2008). NLS sequences have also been detected in several host-cell factors associated with the MW (Lee et al. 2011; Isken et al. 2007).

6.5 Autophagy Machinery

The morphological similarity of DMVs and autophagosomes argues in favour of the autophagy machinery as a player during their biogenesis. The autophagy machinery is implicated in the degradation and recycling of cellular material, being essential to maintain the cell homeostasis. It has been shown to play a major role for the replication of several other positive-strand RNA viruses, including poliovirus, coronaviruses and dengue virus (reviewed in Paul and Bartenschlager 2013).

In the case of HCV, the role of autophagy is still a matter of controversy. Ferraris et al. found that the fractions enriched in HCV-induced DMVs contain microtubule-associated protein 1 light chain 3-II (LC3-II), the lipidated species of LC3-I, a hallmark of autophagy induction (Fig. 4) (Ferraris et al. 2010). However, there are conflicting results concerning the stage of the HCV life cycle requiring the autophagy machinery. Hence, it has been proposed that autophagy is involved in HCV RNA translation (Dreux et al. 2009), initiation of RNA replication (Sir et al. 2008a, b; Guevin et al. 2010), production of infectious virus particles (Tanida et al. 2009) or suppression of the innate antiviral defense leading to viral persistence and chronic hepatitis (Ke and Chen 2011; Shrivastava et al. 2011; Granato et al. 2014). A recent study even suggests that autophagosomes indeed represent the actual HCV replication sites (Sir et al. 2012). However, other studies observed only limited co-localization of autophagy markers with HCV proteins and failed to detect autophagosome precursors early after infection which might develop into DMVs (Romero-Brey et al. 2012).

It might well be possible that individual factors of the autophagy pathway, rather than the complete machinery, are involved in the formation of the MW. Indeed it has been recently reported that NS4B forms a complex with Rab5 and Vps34 and induces autophagy (Su et al. 2011). Moreover, NS5B appears to interact via its thumb domain with the autophagy protein 5 (Atg5), which regularly initiates the formation of autophagic DMVs (Guevin et al. 2010).

Autophagy might also play a role in the formation of MMVs late in HCV infection (Fig. 2). MMVs could originate from autophagosomes engulfing DMVs as part of a cellular stress response induced by massive virus-induced membrane alterations, to degrade or recycle these structures after RNA synthesis has finished (Romero-Brey et al. 2012).

6.6 PSTPIP2

Proline-serine-threonine phosphatase interacting protein 2 (PSTPIP2) is a protein with membrane-deforming activity critical for MW formation in HCV replication (Chao et al. 2012). Immunoprecipitation results indicated that PSTPIP2 has the potential to interact with NS4B and NS5A directly. PSTPIP2 was predominantly localized in detergent-resistant membranes (DRMs) where HCV replication occurs

(Shi et al. 2003). Importantly, PSTPIP2 knockdown caused a significant reduction in the formation of HCV- and NS4B-induced MWs, whereas its over-expression induced cytoplasmic tubular membranes, highlighting its ability to induce positive membrane curvature. Furthermore, a PSTPIP2 mutant defective in inducing membrane curvature failed to support HCV replication, confirming that the membrane-deforming ability of PSTPIP2 is essential for HCV replication.

Consistent with these results, the F-BAR domain of PSTPIP2 can bind to phosphatidylinositide (PI) lipids (Hu et al. 2009; Tsujita et al. 2006) and may thereby target them to intracellular membranes to induce the formation of membrane curvature, thus initiating and/or stabilizing the MW (Fig. 4).

6.7 PLA2G4C

The phospholipase A2 gamma group IVC (PLA2G4C) gene was identified as a host gene that is upregulated in expression upon HCV infection (Xu et al. 2012). While PLA2G4C was barely detectable in the hepatoma cell lines Huh7.5.1 and Lunet, its expression was enhanced after HCV infection, contributing to HCV replication and assembly and colocalizing with NS4B and NS5A. PLA2G4C is a membrane-bound, calcium-independent, cytosolic phospholipase (Stewart et al. 2002; Underwood et al. 1998) that hydrolyzes fatty acid from the sn-2 and sn-1 positions of phosphatidylcholine generating lipid signaling molecules such as arachidonic acid. EM analysis demonstrated that the MW formation was defective after PLA2G4C knockdown in HCV replicon-containing cells, suggesting that this protein is required for the biogenesis of the MW. In addition, PLA2G4C overexpression relocates the NS4B protein to LDs, where virion assembly occurs. Thus, PLA2G4C may bridge the steps of RNA replication and HCV assembly by translocation of replication complexes to LDs.

As mentioned above, NS5A is a key player in the formation of DMVs (Romero-Brey et al. 2012). Indeed, most of the cellular proteins that are known to have an impact in the morphology of DMVs are interaction partners of this multifunctional NS protein, including PI4KIIIα, VAP-A, VAP-B, cyclophilins, GBF1-Arf1-COPI, Nups, some members of the autophagy machinery, Rab18, PSTPIP2 and PLA2G4C (Fig. 4). However, the precise functions of host cell factors involved in the biogenesis of DMVs and the MW are still ill-defined and future analyses are needed to get further insight into their mechanisms of action. HCV RNA synthesis takes place in tight association with the MW. It is mainly driven by the viral polymerase NS5B and regulated by *cis*-acting replication elements with the help of other viral NS proteins and numerous host factors. The mechanistic details of viral RNA replication are still enigmatic due to the close connection between translation, polyprotein processing, morphogenesis of membrane alterations and RNA synthesis. However, the process of RNA replication must be tightly regulated, since it generates a five to tenfold excess of positive-strand RNA molecules compared to negative-strand RNA (Lohmann et al. 1999a; Quinkert et al. 2005; Keum et al. 2012). In addition, HCV RNA levels in vivo (Lanford et al. 2011) and in cell culture (Keum et al. 2012; Quinkert et al. 2005) are quite limited compared to other positive- strand RNA viruses, which might be an important determinant to maintain persistence.

We have learned much about the functions of individual components of the replicase by reverse genetics, crystal structures of viral proteins, and biochemical analyses on purified proteins in vitro. This knowledge will be summarized in the next sections.

7 Linkage Between HCV RNA Synthesis and Membrane Alterations

For positive-strand RNA viruses the process of viral RNA synthesis is tightly linked to the biogenesis of virus-induced membrane alterations. However, surprisingly little is known about how these structures are functionally connected to RNA synthesis itself, because it has yet not been possible to generate replication competent mutants in absence of membrane alterations, whereas the morphology of the membrane alterations is not dependent on RNA replication in the case of HCV (Romero-Brey et al. 2012). Still, as detailed above, most current evidence suggests that RNA synthesis occurs within vesicular structures, most likely DMVs: (i) Most of the viral RNA, in particular almost all of the negative strand RNA, is nuclease-protected but detergent sensitive in cell extracts containing biochemically active viral replication complexes, so called CRCs (Quinkert et al. 2005; Miyanari et al. 2003), suggesting that viral replication intermediates are indeed shielded by membranes. (ii) The numbers and kinetics of appearance of DMVs coincides with RNA synthesis (Romero-Brey et al. 2012). (iii) Affinity purified membrane fractions from HCV replicon cells mainly consist of DMVs and retain the ability to synthesize viral RNA in vitro (Paul et al. 2013; Ferraris et al. 2010). (iv) Nascent RNA has been found associated to vesicular structures in EM studies (Gosert et al. 2003; Sir et al. 2012). However, at the moment it cannot be ruled out that RNA replication may be associated with alternative membrane structures, which could be less prominent in EM studies and, hence, DMVs might represent late stages of viral replication factories (Romero-Brey et al. 2012).

The reasons for the enclosure of RNA synthesis into vesicular structures is not fully clear yet. Viral induced membrane alterations might be a way to concentrate the active viral replicase proteins, RNA and host factors at specific sites. The replication vesicles might thereby also contribute to the template specificity of the replicase, by excluding cellular RNAs. This might represent an evolutionarily conserved mechanism, paralleling the encapsidation of the reverse transcription machinery in retroviruses (Schwartz et al. 2002). The membrane alterations might furthermore facilitate the separation of different steps of the viral replication cycle (translation, replication, assembly) by compartmentalization. Thus, after a certain amount of proteins has been synthesized from a viral genome, a membrane invagination process driven by the NS proteins might stop translation by preventing access of new ribosomes (Quinkert et al. 2005). Another example is the separation of the viral RNA from the capsid protein, which is localized on LDs (Barba et al. 1997; Moradpour et al. 1996), to avoid packaging of genomes prior RNA replication. In addition, viral replication intermediates like dsRNA are potent molecular patterns recognized by cytosolic pattern recognition receptors (RIG-I, MDA5; reviewed in Reikine et al. 2014) and replication vesicles might therefore have a role in shielding them from detection by innate immune responses.

8 *Cis*-Acting Replication Elements

A number of *cis*-acting elements essential for RNA replication have been identified within or close to the 5′ and 3′ NTR by secondary structure predictions and characterized in detail by reverse genetic studies and biochemical/structural analyses (Fig. 5). However, their function in RNA replication likely involves regulating the initiation of negative and positive strand synthesis, which is mainly accomplished by the 3′ends of the positive and negative strand genome, respectively (Fig. 5b, c).

8.1 The 3′ End of the Positive Strand

Several important regulatory sequences are localized close to the 3′ end of the HCV genome: the 3′NTR, a *cis*-acting replication element (CRE) within the C-terminal NS5B coding region (designated SL3.2) and a stem-loop structure further upstream, termed 9005 (Fig. 5b). The 3′NTR is comprised of a so-called variable region, a polyU/UC tract of variable length, and a highly conserved 98-nt element designated X-tail or 3′X (Kolykhalov et al. 1996; Tanaka et al. 1995). The variable region encompasses predicted stem-loop structures surrounding the stop codon of the polyprotein. The variable region is not essential for viral replication, since deletion mutants in the replicon model and in vivo are still replication competent, albeit with strongly reduced efficiency (Yanagi et al. 1999; Friebe and Bartenschlager 2002; Yi and Lemon 2003a). The polyU/UC tract is composed of homopolymeric uridine stretches interspersed by individual cytosines and varies between 30 and 90 nucleotides in length between viral isolates (Kolykhalov et al. 1996). A minimal of 26–33 homopolymeric uridine residues, which must not be interrupted by cytosines, is essential and sufficient for efficient RNA replication in cell culture (Friebe and Bartenschlager 2002; You and Rice 2008). The position of this polyU stretch within the polyU/UC region is flexible (You and Rice 2008; Yi and Lemon 2003a). The functional significance of the polyU/UC tract is not clear yet, but it might provide an assembly platform for the viral replicase, since the helicase domain of

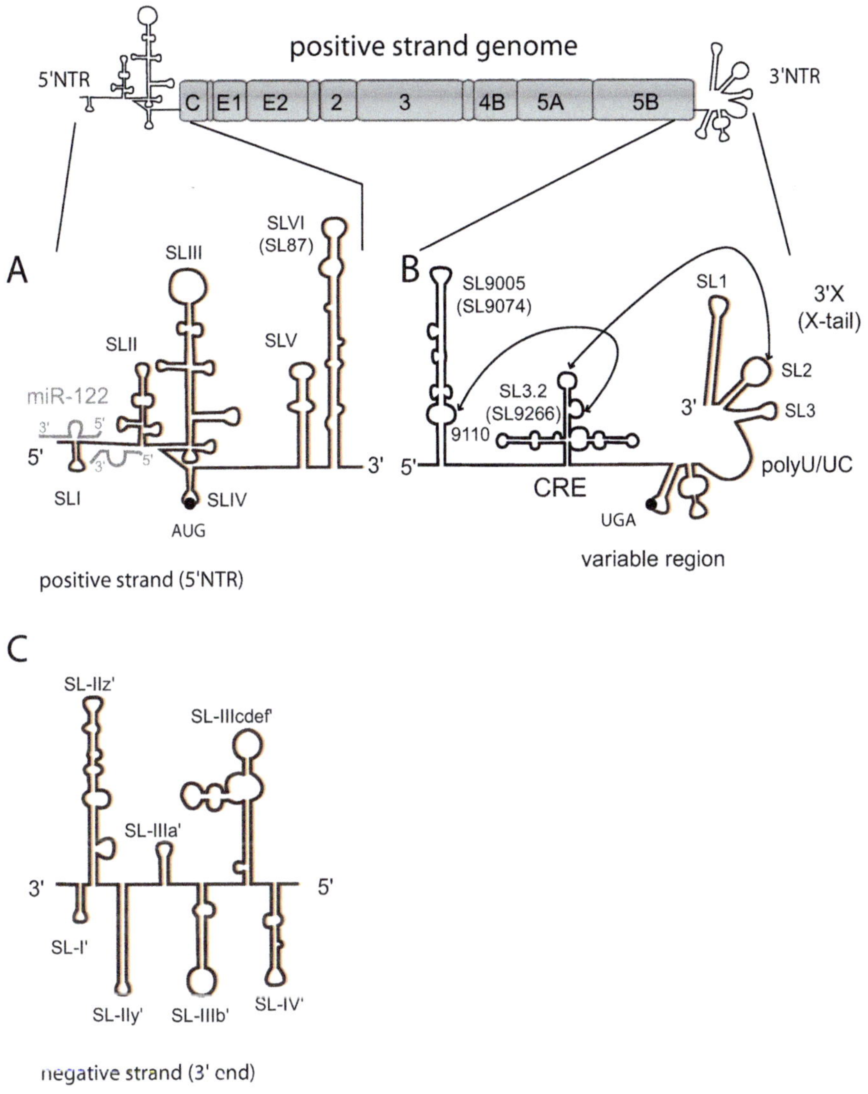

Fig. 5 Schematic representation of *cis*-acting replication elements. (**a**) 5′ end of the viral positive-strand including the IRES (Honda et al. 1996) and stem-loop structures in the core coding region (McMullan et al. 2007). Two copies of miR-122 binding to the 5′NTR are shown in grey (Machlin et al. 2011). (**b**) 3′ end of the viral positive strand (Blight and Rice 1997). Long range interactions of SL3.2 with sequences around 9110 (Tuplin et al. 2012), which is part of SL9005 (Chu et al. 2013), and with the loop region of SL2 (Friebe et al. 2005) are indicated by *arrows*. (**c**) 3′ end of the viral negative strand (Smith et al. 2002; Schuster et al. 2002). Alternative nomenclatures are given in brackets (Modified from Lohmann 2013)

NS3, NS5A and NS5B have been shown to efficiently bind to polyU (Gwack et al. 1996; Huang et al. 2005; Lohmann et al. 1997). The X-tail (Fig. 5b) was identified years after the cloning of the first HCV genome (Kolykhalov et al. 1996; Tanaka et al. 1995) and comprises three experimentally validated stem-loop structures (Blight and Rice 1997), which are all essential for RNA replication (Friebe and Bartenschlager 2002) and barely tolerate any mutations (Yi and Lemon 2003a, b). This region therefore is likely to be the main regulatory element for the initiation of negative strand synthesis; however, the precise mechanisms remain to be determined.

SL3.2 (Fig. 5b) is part of a cruciform-like RNA secondary structure located in the C-terminal coding sequence of NS5B and has been shown to be essential for viral replication by mutational analysis (You et al. 2004). The terminal loop of SL3.2 is engaged in a kissing-loop interaction with SL2 in the 3′NTR, which is mandatory for RNA replication (Friebe et al. 2005). The complementary of the loop-loop interaction is more important than the actual sequence, arguing for a functional role of the pseudoknot structure (Friebe et al. 2005). The bulge region of SL3.2 undergoes another long-range interaction with a *cis*-acting element around nt 9110 of the HCV genome (Tuplin et al. 2012). Mutual interaction of SL3.2 with this region and the X-tail is likely to play a regulatory role, e.g. possibly regulating a shift from translation to replication (Tuplin et al. 2012). A recent study found that region 9110 is part of an extended stem-loop-structure termed SL9074 (or SL9005 according to the reference isolate H77) and also identified additional stem-loop structures in the NS5B coding region that are critical for replication and assembly (Chu et al. 2013).

8.2 The 3′ End of the Negative Strand and 5′NTR of the Positive Strand

The 3′ end of the negative strand is complementary to the 5′NTR of the positive strand, and therefore the sequence of this part of the genome must fulfill two functions. In the positive strand, nt 44-350 form the IRES, which is essential for cap-independent translation of the polyprotein (Honda et al. 1996) (Fig. 5a). The complementary negative-strand sequence contains essential replication signals for positive-strand synthesis (Fig. 5c), with the 125 terminal nucleotides, encompassing SL-I′ and SL-IIz′, being indispensable for this function (Friebe et al. 2001; Friebe and Bartenschlager 2009). Interestingly, the secondary structures in the positive and negative strand, respectively, are very different, despite their complementarity (compare Fig. 5a–c Honda et al. 1996; Schuster et al. 2002; Smith et al. 2002), reflecting their different functions. This is particularly evident in the sequence following SL-I and SL-I′ respectively, which is single stranded in the positive strand, containing two miR-122 binding sites important for replication and translation (Jopling et al. 2005; Machlin et al. 2011), whereas the complementary sequence is embedded in a strong stem-loop-structure (SL-IIz′; Fig. 5c Schuster et al. 2002; Smith et al. 2002). miR-122 binding therefore might protect the 5′NTR sequence from the formation of alternative secondary structures, which could interfere with IRES function, particularly with the folding of SL-II.

Additional conserved stem-loop structures have been predicted within the core coding region (Smith and Simmonds 1997). Indeed, disruption of SL-VI impaired viral RNA replication in cell culture and in vivo (McMullan et al. 2007; Vassilaki et al. 2008). This is surprising, since these sequences are included in replication-competent subgenomic replicons (Lohmann et al. 1999a). This stem-loop thus seems to have a regulatory role in the context of a full-length genome only.

9 The Viral Polymerase NS5B

9.1 Structure of NS5B

The RdRp NS5B is the central catalytic component of the viral replication complex and has been studied extensively in vitro and structurally, since the active enzyme can be expressed heterologously in insect cells and in *E. coli* (Al et al. 1998; Behrens et al. 1996). NS5B encompasses a catalytic domain, followed by a linker sequence and a C-terminal membrane insertion sequence, which tethers the polymerase to membranes (Fig. 6a). The membrane insertion sequence is essential for RNA replication in cell culture (Ivashkina et al. 2002; Moradpour et al. 2004), but not required for RNA synthesis in vitro using heterologously expressed, purified NS5B (Lohmann et al. 1997; Yamashita et al. 1998). The structure of the catalytic domain comprises a so-called right hand shape, common to many single-subunit polymerases, with fingers, thumb and palm subdomains (Fig. 6b, Ago et al. 1999; Lesburg et al. 1999; Bressanelli et al. 1999). Most published structures reveal a closed conformation, encircled on one side by the fingertips and on the other side by a linker and a so-called beta-flap (or β-hairpin). The linker or a variation thereof is common to *de novo* initiating RdRps (Butcher et al. 2001) whereas the beta-flap is restricted to polymerases from the *Flaviviridae*. Such closed NS5B structures imply two functional consequences. (i) The closed structure has only space for a single stranded template and priming nucleotides. It is therefore supposed to represent the initiation state of the polymerase (Simister et al. 2009). However, after RNA synthesis switches from initiation to elongation, the linker has to be removed to allow egress of the primer-template duplex, causing a major structural rearrangement towards an open conformation (Fig. 6c). Such an "open" conformation with a large cavity capable of binding dsRNA was indeed found with a genotype 2a NS5B (Biswal et al. 2005), in a structure of NS5B complexed with a primer-template following deletion of the beta-flap (Mosley et al. 2012), and upon SAXS analysis in solution (Harrus et al. 2010). A recent structural analysis of NS5B further supports this concept (Appleby et al. 2015, comment by Bressanelli 2015).

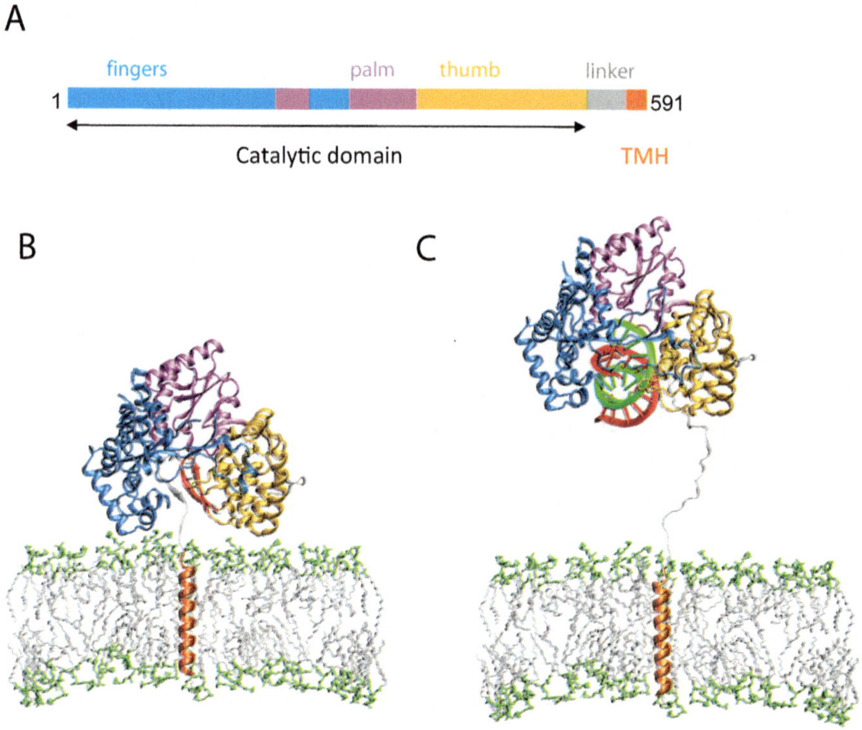

Fig. 6 Schematic subdomain composition of NS5B (**a**) and structure of full length NS5B (PDB entry 1GX6) and its association with the membrane via its C-terminal transmembrane helix (*TMH*) in its closed (panel **b**) and open conformation (panel **c**). The finger, thumb and palm subdomains are given in *blue*, *magenta* and *gold*, respectively. The C-terminal linker (*grey*) connects the core of the enzyme with the membrane insertion sequence (*orange*). (**b**) Structure of NS5B and proposed membrane topology in the so-called closed conformation, which is believed to represent the initiation state of the polymerase. Note that in this model the RNA-binding groove is hidden by the membrane, the so-called beta-flap is shown in red. (**c**) Model of NS5B in a hypothetical elongation mode, which releases the RNA binding groove by moving the linker sequence out of the active center upon switching from initiation to elongation (F. Penin, unpublished data). A double stranded RNA replication intermediate is modeled into the active site according to Mosley et al. (2012). Panels **b** and **c** are a generous gift from F. Penin

Several different stalled NS5B ternary complexes with polymerase, template and incoming nucleotides were generated using elegant strategies to the arrest the enzyme in distinct intermediate states during initiation and elongation. The resulting structures revealed that the beta-flap (here "beta-loop") indeed provides the major priming platform for *de novo* initiation and, together with the C-terminal linker, moves out of the active center, guiding the dsRNA upon opening of the structure. (ii) The enzymatic core of NS5B in its closed conformation appears to be tightly tethered to the membrane, when modeled to the C-terminal linker (Fig. 6b). It is therefore supposed that the switch from initiation to elongation also changes the position of the elongating polymerase relative to the membrane due to distortion of the linker from the catalytic cleft (Fig. 6c, reviewed in Lohmann 2013). However, the functional significance of this altered topology of NS5B remains to be determined and the entire concept requires further experimental validation, since all structures published thus far lack the membrane insertion sequence.

Overall the role of the linker is not fully clarified yet in HCV NS5B. On the one hand, the entire linker can be deleted (so called ΔC47 or ΔC60 constructs), giving rise to strongly enhanced polymerase activity in vitro suggesting a negative regulatory function (Leveque et al. 2003). Such a mechanism might make sense since only a minority of NS5B molecules is supposed to be actively engaged in RNA synthesis (Quinkert et al. 2005). Self-inactivation of NS5B molecules not involved in RNA synthesis might therefore represent an immune evasion mechanism, since NS5B expressed in cells is capable of generating dsRNA from cellular templates (Ranjith-Kumar et al. 2011; Yu et al. 2012), inducing an intracellular immune response.

9.2 The Polymerase in Motion

Purified NS5B can initiate RNA synthesis either by a primer-dependent mechanism or *de novo* (Behrens et al. 1996; Lohmann et al. 1997; Luo et al. 2000; Zhong et al. 2000; Sun et al. 2000). Since the HCV genome has no terminal protein, *de novo* initiation is supposed to be the initiation mode of genome replication in vivo. The first step of *de novo* initiation after binding of the polymerase to a template is the synthesis of a di- or trinucleotide primer, which has to start with a purine base and requires high concentrations of the priming nucleotides (Ferrari et al. 2008; Luo et al. 2000). Primer synthesis requires a "platform" to support the first nucleotide, which has to move out of the active center upon addition of the third base. This platform has recently been shown to be provided by the beta-flap (Appleby et al. 2015), in line with the mechanism suggested for pestiviruses (Lescar and Canard 2009; Choi et al. 2004; D'Abramo et al. 2006). In addition, *de novo* initiation of RNA synthesis is stimulated by high GTP concentrations upon binding to allosteric sites (Lohmann et al. 1999b; Harrus et al. 2010). GTP stabilizes the nucleotides involved in primer synthesis (D'Abramo et al. 2006) and facilitates the step from initiation to elongation (Harrus et al. 2010). The closed conformation seen in almost all crystal structures of NS5B represents most likely the initiation state of the enzyme (Simister et al. 2009), a view supported by a recent study of multiple stalled ternary complexes (Appleby et al. 2015). The switch from initiation to elongation seems to be one of the rate limiting steps at least in vitro and a number of excess primers are synthesized, before the enzyme continues to processive elongation (Harrus et al. 2010). This step furthermore requires a major conformational change towards an open conformation, probable driven by the removal of the linker sequence to allow exit of the dsRNA, thereby likely relocating the entire enzymatic core (Fig. 6c). Position 405 in the thumb seems to be critical for efficient primer synthesis as well as for switching from initiation to elongation in vitro and this residue is also an important determinant for replication efficiency in cell culture (Scrima et al. 2012; Schmitt et al. 2011).

Once RNA synthesis is initiated, NS5B elongates nascent RNA by about 100–400 nts per minute and is capable of processively copying an entire RNA genome in vitro (Oh et al. 1999; Lohmann et al. 1998; Simister et al. 2009) at nucleotide concentrations far lower than required for initiation (Lohmann et al. 1998; Jin et al. 2012). Elongation complexes are quite stable in vitro and can even be stalled and purified (Jin et al. 2012). Nothing is known so far about termination of RNA synthesis, and the enzyme might finally just fall off the template. However, it must be considered that NS5B as well as all other NS proteins involved in RNA synthesis are tightly linked to the membrane and buried in membrane alterations (Figs. 1 and 4). In vivo the template might move along the polymerase rather than the polymerase along the template; therefore, termination and re-initiation might be regulated processes as well.

RNA synthesis by NS5B is error prone and provides the molecular basis of the high genetic variability of HCV isolates, although a nucleotide excision mechanism has recently been reported for NS5B, which might allow limited error correction (Jin et al. 2013). An error rate of ca. 10^{-3} per site was determined in vitro with a strong bias towards G:U/U:G mismatches, which was mirrored by a 75-fold difference in transitions over transversions in vivo (Powdrill et al. 2011).

The natural initiation sites for RNA synthesis in vivo are supposed to be the 3′ ends of the viral positive and negative strand RNAs. In vitro, *de novo* initiation of RNA synthesis has no distinct sequence specificity and can even take place on circular templates (Ranjith-Kumar and Kao 2006) or homopolymers (poly C and polyU, Luo et al. 2000). However, the initiating nucleotide must be a purine (Zhong et al. 2000) and initiation seems most efficient on a single-stranded region of at least 3 nts adjacent to a stem-loop structure (Kao et al. 2000). This is exactly the configuration found at the 3′ end of the negative strand (Fig. 5c), whereas the 3′ end of the positive strand is buried in a strong stem-structure, therefore hardly giving rise to terminal *de novo* initiation in vitro (Fig. 5b Reigadas et al. 2001; Binder et al. 2007). This might suggest that additional viral or host factors are required for initiation of negative-strand synthesis, thereby allowing a tight regulation of RNA synthesis. In principle, primer-synthesis by the polymerase could also take place at a different site (e.g. at an internal CRE as in case of poliovirus Paul et al. 1998, 2000) and the polymerase/primer complex might then be moved to the end of the genome. However, there is as yet no experimental evidence for such a scenario.

9.3 *Contribution of Other Viral Proteins to RNA Synthesis*

The polymerase is capable of *de novo* initiation and can copy an entire genome without the help of other factors in vitro (Lohmann et al. 1998; Oh et al. 1999). Still, it is obvious that additional viral and cellular proteins substantially contribute to

RNA synthesis. However, we do not know much about the molecular details of regulation of RNA synthesis due to the limited availability of appropriate model systems (reviewed in Lohmann 2013). There is still a huge gap between reverse genetic studies, revealing the importance of *cis*-acting elements and viral proteins NS3 to NS5B for RNA replication in cells, and in vitro models that build on purified proteins, and that currently provide only limited mechanistic insight due to a lack of specificity (e.g. of the polymerase for viral templates).

A study using inter-genotypic chimeras revealed that the helicase-domain of NS3 (NS3 helicase), NS5A, NS5B and the NTRs must be derived from the same genotype for efficient RNA replication in cell culture, suggesting that these proteins may form the core of the replicase (Binder et al. 2007). This view is supported by a recently developed in vitro model (Mani et al. 2014) using a circular RNA complexed with human replication protein A as a template, and demonstrating that NS5A might help binding NS5B to the template. The helicase is subsequently recruited by NS5B and facilitates processive elongation (Mani et al. 2014). Stimulation of polymerase activity by the helicase has been found by others (Piccininni et al. 2002), as well as a regulation of helicase activity by the polymerase and the protease (Jennings et al. 2008; Zhang et al. 2005). Regarding NS5A, conflicting results have been published suggesting that low doses of NS5A stimulate NS5B activity, whereas high doses are inhibitory to the polymerase (Shirota et al. 2002; Quezada and Kane 2009).

The distinct role of the NS3 helicase activity has not been clarified yet, but the enzyme (i) might resolve strong stem-loop structures at the 3′ end of the genome to facilitate initiation of RNA synthesis by the polymerase, (ii) unwind double-stranded replication intermediates during RNA synthesis to support NS5B in the elongation phase, and/or (iii) help to strip proteins off the RNA or deliver RNA for packaging into virions by a recently demonstrated ssRNA translocase activity (Gu and Rice 2010). In addition, a recent reverse genetic study suggested that the linker sequence between the protease and helicase domains of NS3 has a regulatory role in replication and assembly (Kohlway et al. 2014).

NS5A is believed to be a central regulator of the viral replication cycle, and it is particularly difficult to separate its contribution to the biogenesis of the MW described earlier in this chapter (Romero-Brey et al. 2012) from essential functions in RNA synthesis and assembly. The RNA binding of NS5A, which is mediated mainly by domain I (Huang et al. 2005) and modulated by domains II and III (Foster et al. 2010) might clearly contribute to RNA synthesis as shown in vitro (Mani et al. 2014). The phosphorylation status of NS5A is also believed to have a major regulatory role in RNA synthesis. However, although a number of phosphorylation sites have recently been identified and analyzed by reverse genetics (Masaki et al. 2014; Ross-Thriepland and Harris 2014) it has not been possible yet to assign distinctly phosphorylated subspecies of NS5A to specific functions. Several replication enhancing mutations identified in replicon cells reduced the relative amount of hyperphosphorylated NS5A (p58 Appel et al. 2005; Blight et al. 2000), which brought up the concept that basally phosphorylated NS5A (p56) is required for RNA replication, whereas p58 favors assembly. Indeed, a recent study showed positive evidence that CKIα mediated phosphorylation events leading to p58 synthesis are required for production of infectious virus (Masaki et al. 2014). In addition, hVAP-A, a cellular protein critical for RNA replication, preferentially binds to p56 and not to p58 (Evans et al. 2004), suggesting that differential phosphorylation of NS5A might regulate the viral life cycle by changing interactions with viral and host factors. Still, a comprehensive concept is missing concerning which distinct NS5A phosphorylation events govern different steps of the replication cycle and by which kinds of mechanisms.

NS4B may have a role in the regulation of replicase activity beyond MW formation. NS4B can bind RNA (Einav et al. 2004, 2008) and also has an NTPase activity. There is also genetic evidence for an interaction of NS4B with NS3 (Paredes and Blight 2008) and NS4B has been shown to inhibit NS5B in vitro (Piccininni et al. 2002).

9.4 Host Factors Involved in RNA Synthesis

Several of the host factors involved in MW formation that have been discussed previously in this chapter may have in addition a more direct role in the regulation of RNA synthesis (e.g. PI4KIIIα, CypA, sphingomyelin).

RNA-binding proteins are of course prime candidates for distinct functions in regulation of RNA synthesis. Twenty-six cellular proteins specifically binding to the IRES in the 5′NTR (Lu et al. 2004) and more than 70 interacting with the 3′NTR (Harris et al. 2006) have been identified in proteomic studies, but their distinct roles remain elusive. Several mechanisms have been brought up mediating the circularization of the viral genome, which might be important to stimulate translation, as in case of cellular messenger RNAs, or to switch between different steps of the viral life cycle, as shown for flaviviruses (reviewed in Villordo and Gamarnik 2009). For HCV, potential circularization sequences have been identified close to the termini of the genome (Romero-Lopez et al. 2014). Genome circulization might further be facilitated by RNA binding proteins, like the NF/NFAR (Isken et al. 2007) or the La protein (Kumar et al. 2013).

The liver specific microRNA miR-122 is a critical host factor in HCV replication in cell culture (Jopling et al. 2005) and in vivo (Lanford et al. 2010) and might substantially contribute to the liver tropism of HCV. miR-122 binds to two sites in the 5′NTR with extensive base pairing outside the seed sequence (Fig. 5a; Machlin et al. 2011), forming an unconventional ternary complex, encompassing the 5′ end of the viral genome (Mortimer and Doudna 2013). The mode of action of miR-122 in the viral replication cycle is not fully clarified yet, but it has been shown to stimulate translation (Henke et al. 2008) and to protect the genome from degradation by the exonucleases Xrn1 (Li et al. 2013) and Xrn2 (Sedano and Sarnow 2014).

10 Dynamics of RNA Synthesis and Open Questions

Mathematic modeling of the decline in circulating virus after the initiation of therapy revealed that about 10^{12} virions are produced per day in infected individuals (Neumann et al. 1998), suggesting a highly dynamic process for RNA replication in the infected liver. Recent studies using two photon microscopy or highly sensitive *in situ* hybridization revealed that 7–20 % (Liang et al. 2009) or 1–54 % (Wieland et al. 2014) of hepatocytes are infected and that the number of infected cells indeed correlates with viral load (Wieland et al. 2014; Mensa et al. 2013).

Detailed intracellular replication kinetics are available from subgenomic replicons (Binder et al. 2007) or after virus infection (Keum et al. 2012), as well as from a quantitative analysis of replication dynamics in persistent replicon cells (Quinkert et al. 2005). These data have been recently used to model replication dynamics *in silico* (Dahari et al. 2007; Binder et al. 2013). Altogether these studies suggest that each of the incoming positive strand RNA molecules is first translated to give rise to ca. 1,000 protein copies (Quinkert et al. 2005), which will then likely induce the formation of the MW to allow initiation of RNA synthesis. However, only a subfraction of less than 5 % of these protein copies seem to be engaged in the formation of viral replication sites and even less in enzymatic replicase activity (Miyanari et al. 2003; Quinkert et al. 2005). It remains elusive which mechanisms render a few replicase copies active and the majority inactive. However, recent *trans*-complementation analyses reveal that some deleterious mutations in NS5A can complement each other, suggesting the existence of several nonstructural protein complexes serving different functions (Fridell et al. 2011).

First viral negative-strand RNAs are detectable 4 h or 6 h after transfection or infection, respectively. At this time point, positive to negative strand ratios are ~1:1 (Binder et al. 2007; Keum et al. 2012), probably reflecting double-stranded replication intermediates (Targett-Adams et al. 2008). The initial log-phase of 4–6 h likely represents the time required for polyprotein translation, generation of the membranous replication compartment and RNA synthesis (100–400 nts/min in vitro) (Lohmann et al. 1998; Simister et al. 2009). After this time point, negative- and positive-strand RNA levels increase exponentially and asymmetrically, reaching a plateau at 24–48 h, with a +/− strand ratio of ~10:1. Roughly 1,000–5,000 positive and 100–500 negative strand RNA molecules per cell were reported for transient and steady state cell cultures and this number might represent a limit of Huh-7-based cell cultures (Quinkert et al. 2005; Keum et al. 2012; Blight et al. 2002), reflecting limiting host factors involved in RNA synthesis as suggested by recent mathematic modelling (Binder et al. 2013). Less efficient genotype 1 replicons exhibit much slower replication kinetics with no clear exponential phase and reach steady state replication levels at later time points (Binder et al. 2007; Krieger et al. 2001). The half-lives of viral NS proteins and viral positive-strand RNA in replicon cells have been shown to be 11–16 h (Pietschmann et al. 2001; Pause et al. 2003). It can therefore be estimated that only about 1,000 positive-strand RNA molecules are synthesized per day per cell by ca. 100 replicase complexes during persistent replication in cell culture (Quinkert et al. 2005).

It is generally not clear, how the progeny positive-strand RNA is released into the cytoplasm. This could involve NS5A and/or the translocase function of the NS3 helicase (Gu and Rice 2010), delivering the RNA through a direct, probably transient connection to the cytoplasm (Romero-Brey et al. 2012) or through a protein pore e.g. nuclear pore like complexes Neufeldt et al. 2013. Due to the low and limiting number of negative-strand RNA copies it seems plausible that negative-strand synthesis can be initiated only once from a positive-strand genome by a *cis*-acting protein complex translated on the same RNA. Initiation of negative-strand RNA would then require a preceding translation of the positive strand RNA, resulting in the formation of a new replication vesicle and each replication site would indeed contain only one negative strand RNA/replication intermediate. Such a model would also explain in part the asymmetry of RNA synthesis, resulting in a strong surplus of positive strands.

Many questions regarding the regulation and dynamics of RNA synthesis are still unresolved and some assumptions are highly speculative. We still have no clear concept how the shift from translation to RNA synthesis is accomplished, how the membrane alterations are functionally linked to RNA synthesis, how negative and positive strand RNA synthesis are initiated to achieve the asymmetric +/− strand ratio and which viral and cellular proteins are involved, just to name a few. Further mechanistic insights will require better defined in vitro models assembled from individual replicase components that allow dissection of the complexities governing HCV RNA synthesis.

11 Conclusions/Perspectives

Twenty five years after its discovery, a new era of HCV research is emerging due to the advent of efficient specific antiviral therapies. This huge success would not have been possible without enormous efforts in basic research, which provided important model systems for drug development as well as an understanding of the viral replication cycle that was essential for definition and evaluation of efficient targets for therapy. However, beyond antiviral drug development, HCV has become an outstandingly important role model for various aspects of virology and virus host interactions in general, including the biogenesis of viral replication sites, interactions with lipids and lipid metabolism, the establishment of persistence, etc. Although our knowledge has increased tremendously over the past years, which is partly reflected by this review, many issues are far from being clarified. This accounts particularly for mechanisms governing viral RNA replication, which remain difficult to address due to a plethora of *cis*-functions regulating translation, induction of membrane alterations, viral RNA synthesis and packaging of newly synthesized genomes into virions. The chances are high that several important questions surrounding the biology of positive-strand RNA viruses as well as of underlying cellular mechanisms can be solved using HCV as a paradigm, now that an overwhelming toolbox has been generated, driven by the importance of HCV as a pathogen. Therefore, we have to keep research efforts on HCV engaged in high gear even in the era of effective antiviral therapy, to reach the next level of mechanistic understanding.

Acknowledgements The authors are grateful to Francois Penin for providing Fig. 6b, c. VL is supported by grants from the Deutsche Forschungsgemeinschaft (DFG) (LO1556/1–2 and LO1556/4–1; FOR1202, TP 3 and TRR77, TP A1)

References

Ago H, Adachi T, Yoshida A, Yamamoto M, Habuka N, Yatsunami K, Miyano M (1999) Crystal structure of the RNA-dependent RNA polymerase of hepatitis C virus. Structure 7:1417–1426

Aizaki H, Lee KJ, Sung VM, Ishiko H, Lai MM (2004) Characterization of the hepatitis C virus RNA replication complex associated with lipid rafts. Virology 324:450–461

Aizaki H, Morikawa K, Fukasawa M, Hara H, Inoue Y, Tani H, Saito K, Nishijima M, Hanada K, Matsuura Y, Lai MM, Miyamura T, Wakita T, Suzuki T (2008) Critical role of virion-associated cholesterol and sphingolipid in hepatitis C virus infection. J Virol 82:5715–5724

Al RH, Xie YP, Wang YH, Hagedorn CH (1998) Expression of recombinant hepatitis C virus non-structural protein 5B in Escherichia coli. Virus Res 53:141–149

Aligo J, Jia S, Manna D, Konan KV (2009) Formation and function of hepatitis C virus replication complexes require residues in the carboxy-terminal domain of NS4B protein. Virology 393:68–83

Alvisi G, Madan V, Bartenschlager R (2011) Hepatitis C virus and host cell lipids: an intimate connection. RNA Biol 8:258–269

Amako Y, Sarkeshik A, Hotta H, Yates J III, Siddiqui A (2009) Role of oxysterol binding protein in hepatitis C virus infection. J Virol 83:9237–9246

Amako Y, Syed GH, Siddiqui A (2011) Protein kinase D negatively regulates hepatitis C virus secretion through phosphorylation of oxysterol-binding protein and ceramide transfer protein. J Biol Chem 286:11265–11274

Amarilio R, Ramachandran S, Sabanay H, Lev S (2005) Differential regulation of endoplasmic reticulum structure through VAP-Nir protein interaction. J Biol Chem 280:5934–5944

Anchisi L, Dessi S, Pani A, Mandas A (2012) Cholesterol homeostasis: a key to prevent or slow down neurodegeneration. Front Physiol 3:486

Appel N, Pietschmann T, Bartenschlager R (2005) Mutational analysis of hepatitis C virus nonstructural protein 5A: potential role of differential phosphorylation in RNA replication and identification of a genetically flexible domain. J Virol 79:3187–3194

Appleby TC, Perry JK, Murakami E, Barauskas O, Feng J, Cho A, Fox D 3rd, Wetmore DR, McGrath ME, Ray AS, Sofia MJ, Swaminathan S, Edwards TE (2015) Structural basis for RNA replication by the hepatitis C virus polymerase. Science 347:771–775

Balla T (2006) Phosphoinositide-derived messengers in endocrine signaling. J Endocrinol 188:135–153

Balla A, Tuymetova G, Tsiomenko A, Varnai P, Balla T (2005) A plasma membrane pool of phosphatidylinositol 4-phosphate is generated by phosphatidylinositol 4-kinase type-III alpha: studies with the PH domains of the oxysterol binding protein and FAPP1. Mol Biol Cell 16:1282–1295

Barba G, Harper F, Harada T, Kohara M, Goulinet S, Matsuura Y, Eder G, Schaff Z, Chapman MJ, Miyamura T, Brechot C (1997) Hepatitis C virus core protein shows a cytoplasmic localization and associates to cellular lipid storage droplets. Proc Natl Acad Sci U S A 94:1200–1205

Bartenschlager R, Cosset FL, Lohmann V (2010) Hepatitis C virus replication cycle. J Hepatol 53:583–585

Bartenschlager R, Penin F, Lohmann V, Andre P (2011) Assembly of infectious hepatitis C virus particles. Trends Microbiol 19:95–103

Bartenschlager R, Lohmann V, Penin F (2013) The molecular and structural basis of advanced antiviral therapy for hepatitis C virus infection. Nat Rev Microbiol 11:482–496

Behrens SE, Tomei L, De Francesco R (1996) Identification and properties of the RNA-dependent RNA polymerase of hepatitis C virus. EMBO J 15:12–22

Berger KL, Cooper JD, Heaton NS, Yoon R, Oakland TE, Jordan TX, Mateu G, Grakoui A, Randall G (2009) Roles for endocytic trafficking and phosphatidylinositol 4-kinase III alpha in hepatitis C virus replication. Proc Natl Acad Sci U S A 106:7577–7582

Berger KL, Kelly SM, Jordan TX, Tartell MA, Randall G (2011) Hepatitis C virus stimulates the phosphatidylinositol 4-kinase III alpha-dependent phosphatidylinositol 4-phosphate production that is essential for its replication. J Virol 85:8870–8883

Berger C, Romero-Brey I, Radujkovic D, Terreux R, Zayas M, Paul D, Harak C, Hoppe S, Gao M, Penin F, Lohmann V, Bartenschlager R (2014) Daclatasvir-like inhibitors of NS5A block early biogenesis of hepatitis C virus-induced membranous replication factories, independent of RNA replication. Gastroenterology 147:1094–1105

Bianco A, Reghellin V, Donnici L, Fenu S, Alvarez R, Baruffa C, Peri F, Pagani M, Abrignani S, Neddermann P, De Francesco R (2012) Metabolism of phosphatidylinositol 4-kinase IIIalpha-dependent PI4P is subverted by HCV and is targeted by a 4-anilino quinazoline with antiviral activity. PLoS Pathog 8:e1002576

Binder M, Quinkert D, Bochkarova O, Klein R, Kezmic N, Bartenschlager R, Lohmann V (2007) Identification of determinants involved in initiation of hepatitis C virus RNA synthesis by using intergenotypic replicase chimeras. J Virol 81:5270–5283

Binder M, Sulaimanov N, Clausznitzer D, Schulze M, Huber CM, Lenz SM, Schloder JP, Trippler M, Bartenschlager R, Lohmann V, Kaderali L (2013) Replication vesicles are load- and choke-points in the hepatitis C virus lifecycle. PLoS Pathog 9:e1003561

Bishe B, Syed GH, Field SJ, Siddiqui A (2012) Role of Phosphatidylinositol 4-Phosphate (PI4P) and its binding protein GOLPH3 in hepatitis C virus secretion. J Biol Chem 287:27637–27647

Biswal BK, Cherney MM, Wang M, Chan L, Yannopoulos CG, Bilimoria D, Nicolas O, Bedard J, James MN (2005) Crystal structures of the RNA-dependent RNA polymerase genotype 2a of hepatitis C virus reveal two conformations and suggest mechanisms of inhibition by non-nucleoside inhibitors. J Biol Chem 280:18202–18210

Blight KJ, Rice CM (1997) Secondary structure determination of the conserved 98-base sequence at the 3′ terminus of hepatitis C virus genome RNA. J Virol 71:7345–7352

Blight KJ, Kolykhalov AA, Rice CM (2000) Efficient initiation of HCV RNA replication in cell culture. Science 290:1972–1974

Blight KJ, McKeating JA, Rice CM (2002) Highly permissive cell lines for subgenomic and genomic hepatitis C virus RNA replication. J Virol 76:13001–13014

Bonifacino JS, Glick BS (2004) The mechanisms of vesicle budding and fusion. Cell 116:153–166

Borawski J, Troke P, Puyang X, Gibaja V, Zhao S, Mickanin C, Leighton-Davies J, Wilson CJ, Myer V, Cornellataracido I, Baryza J, Tallarico J, Joberty G, Bantscheff M, Schirle M, Bouwmeester T, Mathy JE, Lin K, Compton T, Labow M, Wiedmann B, Gaither LA (2009) Class III phosphatidylinositol 4-kinase alpha and beta are novel host factor regulators of hepatitis C virus replication. J Virol 83:10058–10074

Bressanelli S (2015) Kickstarting a viral RNA polymerase. Science 347:715–716

Bressanelli S, Tomei L, Roussel A, Incitti I, Vitale RL, Mathieu M, De FR, Rey FA (1999) Crystal structure of the RNA-dependent RNA polymerase of hepatitis C virus. Proc Natl Acad Sci U S A 96:13034–13039

Butcher SJ, Grimes JM, Makeyev EV, Bamford DH, Stuart DI (2001) A mechanism for initiating RNA-dependent RNA polymerization. Nature 410:235–240
Chao TC, Su WC, Huang JY, Chen YC, Jeng KS, Wang HD, Lai MM (2012) Proline-serine-threonine phosphatase-interacting protein 2 (PSTPIP2), a host membrane-deforming protein, is critical for membranous web formation in hepatitis C virus replication. J Virol 86:1739–1749
Chatterji U, Bobardt M, Selvarajah S, Yang F, Tang H, Sakamoto N, Vuagniaux G, Parkinson T, Gallay P (2009) The isomerase active site of cyclophilin a is critical for hepatitis C virus replication. J Biol Chem 284:16998–17005
Choi KH, Groarke JM, Young DC, Kuhn RJ, Smith JL, Pevear DC, Rossmann MG (2004) The structure of the RNA-dependent RNA polymerase from bovine viral diarrhea virus establishes the role of GTP in de novo initiation. Proc Natl Acad Sci U S A 101:4425–4430
Chu D, Ren S, Hu S, Wang WG, Subramanian A, Contreras D, Kanagavel V, Chung E, Ko J, Amirtham Jacob Appadorai RS, Sinha S, Jalali Z, Hardy DW, French SW, Arumugaswami V (2013) Systematic analysis of enhancer and critical cis-acting RNA elements in the protein-encoding region of the hepatitis C virus genome. J Virol 87:5678–5696
Chung KM, Lee J, Kim JE, Song OK, Cho S, Lim J, Seedorf M, Hahm B, Jang SK (2000) Nonstructural protein 5A of hepatitis C virus inhibits the function of karyopherin beta3. J Virol 74:5233–5241
D'Abramo CM, Deval J, Cameron CE, Cellai L, Gotte M (2006) Control of template positioning during de novo initiation of RNA synthesis by the bovine viral diarrhea virus NS5B polymerase. J Biol Chem 281:24991–24998
Dahari H, Ribeiro RM, Rice CM, Perelson AS (2007) Mathematical modeling of subgenomic hepatitis C virus replication in Huh-7 cells. J Virol 81:750–760
Dansako H, Hiramoto H, Ikeda M, Wakita T, Kato N (2014) Rab18 is required for viral assembly of hepatitis C virus through trafficking of the core protein to lipid droplets. Virology 462–463:166–174
de Chassey B, Navratil V, Tafforeau L, Hiet MS, Aublin-Gex A, Agaugue S, Meiffren G, Pradezynski F, Faria BF, Chantier T, Le Breton M, Pellet J, Davoust N, Mangeot PE, Chaboud A, Penin F, Jacob Y, Vidalain PO, Vidal M, Andre P, Rabourdin-Combe C, Lotteau V (2008) Hepatitis C virus infection protein network. Mol Syst Biol 4:230
Diamond DL, Syder AJ, Jacobs JM, Sorensen CM, Walters KA, Proll SC, McDermott JE, Gritsenko MA, Zhang Q, Zhao R, Metz TO, Camp DG, Waters KM, Smith RD, Rice CM, Katze MG (2010) Temporal proteome and lipidome profiles reveal hepatitis C virus-associated reprogramming of hepatocellular metabolism and bioenergetics. PLoS Pathog 6:e1000719
Donaldson JG, Jackson CL (2011) ARF family G proteins and their regulators: roles in membrane transport, development and disease. Nat Rev Mol Cell Biol 12:362–375
Dreux M, Gastaminza P, Wieland SF, Chisari FV (2009) The autophagy machinery is required to initiate hepatitis C virus replication. Proc Natl Acad Sci U S A 106:14046–14051
Dufner-Beattie J, O'Guin A, O'Guin S, Briley A, Wang B, Balsarotti J, Roth R, Starkey G, Slomczynska U, Noueiry A, Olivo PD, Rice CM (2014) Identification of AP80978, a novel small-molecule inhibitor of hepatitis C virus replication that targets NS4B. Antimicrob Agents Chemother 58:3399–3410
Dvory-Sobol H, Pang PS, Glenn JS (2010) The future of HCV therapy: NS4B as an antiviral target. Viruses 2:2481–2492
Egger D, Wolk B, Gosert R, Bianchi L, Blum HE, Moradpour D, Bienz K (2002) Expression of hepatitis C virus proteins induces distinct membrane alterations including a candidate viral replication complex. J Virol 76:5974–5984
Einav S, Elazar M, Danieli T, Glenn JS (2004) A nucleotide binding motif in hepatitis C virus (HCV) NS4B mediates HCV RNA replication. J Virol 78:11288–11295
Einav S, Gerber D, Bryson PD, Sklan EH, Elazar M, Maerkl SJ, Glenn JS, Quake SR (2008) Discovery of a hepatitis C target and its pharmacological inhibitors by microfluidic affinity analysis. Nat Biotechnol 26:1019–1027

Esser-Nobis K, Romero-Brey I, Ganten TM, Gouttenoire J, Harak C, Klein R, Schemmer P, Binder M, Schnitzler P, Moradpour D, Bartenschlager R, Polyak SJ, Stremmel W, Penin F, Eisenbach C, Lohmann V (2013) Analysis of hepatitis C virus resistance to silibinin in vitro and in vivo points to a novel mechanism involving nonstructural protein 4B. Hepatology 57:953–963

Evans MJ, Rice CM, Goff SP (2004) Phosphorylation of hepatitis C virus nonstructural protein 5A modulates its protein interactions and viral RNA replication. Proc Natl Acad Sci U S A 101:13038–13043

Farhat R, Goueslain L, Wychowski C, Belouzard S, Feneant L, Jackson CL, Dubuisson J, Rouille Y (2013) Hepatitis C virus replication and Golgi function in brefeldin a-resistant hepatoma-derived cells. PLoS ONE 8:e74491

Ferrari E, He Z, Palermo RE, Huang HC (2008) Hepatitis C virus NS5B polymerase exhibits distinct nucleotide requirements for initiation and elongation. J Biol Chem 283:33893–33901

Ferraris P, Blanchard E, Roingeard P (2010) Ultrastructural and biochemical analyses of hepatitis C virus-associated host cell membranes. J Gen Virol 91:2230–2237

Ferraris P, Beaumont E, Uzbekov R, Brand D, Gaillard J, Blanchard E, Roingeard P (2013) Sequential biogenesis of host cell membrane rearrangements induced by hepatitis C virus infection. Cell Mol Life Sci 70:1297–1306

Foster TL, Belyaeva T, Stonehouse NJ, Pearson AR, Harris M (2010) All three domains of the hepatitis C virus nonstructural NS5A protein contribute to RNA binding. J Virol 84:9267–9277

Fridell RA, Qiu D, Valera L, Wang C, Rose RE, Gao M (2011) Distinct functions of NS5A in hepatitis C virus RNA replication uncovered by studies with the NS5A inhibitor BMS-790052. J Virol 85:7312–7320

Friebe P, Bartenschlager R (2002) Genetic analysis of sequences in the 3′ nontranslated region of hepatitis C virus that are important for RNA replication. J Virol 76:5326–5338

Friebe P, Bartenschlager R (2009) Role of RNA structures in genome terminal sequences of the hepatitis C virus for replication and assembly. J Virol 83:11989–11995

Friebe P, Lohmann V, Krieger N, Bartenschlager R (2001) Sequences in the 5′ nontranslated region of hepatitis C virus required for RNA replication. J Virol 75:12047–12057

Friebe P, Boudet J, Simorre JP, Bartenschlager R (2005) Kissing-loop interaction in the 3′ end of the hepatitis C virus genome essential for RNA replication. J Virol 79:380–392

Gao L, Aizaki H, He JW, Lai MM (2004) Interactions between viral nonstructural proteins and host protein hVAP-33 mediate the formation of hepatitis C virus RNA replication complex on lipid raft. J Virol 78:3480–3488

Gao M, Nettles RE, Belema M, Snyder LB, Nguyen VN, Fridell RA, Serrano-Wu MH, Langley DR, Sun JH, O'Boyle DR, Lemm JA, Wang C, Knipe JO, Chien C, Colonno RJ, Grasela DM, Meanwell NA, Hamann LG (2010) Chemical genetics strategy identifies an HCV NS5A inhibitor with a potent clinical effect. Nature 465:96–100

Gillespie LK, Hoenen A, Morgan G, Mackenzie JM (2010) The endoplasmic reticulum provides the membrane platform for biogenesis of the flavivirus replication complex. J Virol 84:10438–10447

Gosert R, Egger D, Lohmann V, Bartenschlager R, Blum HE, Bienz K, Moradpour D (2003) Identification of the hepatitis C virus RNA replication complex in huh-7 cells harboring subgenomic replicons. J Virol 77:5487–5492

Goueslain L, Alsaleh K, Horellou P, Roingeard P, Descamps V, Duverlie G, Ciczora Y, Wychowski C, Dubuisson J, Rouille Y (2010) Identification of GBF1 as a cellular factor required for hepatitis C virus RNA replication. J Virol 84:773–787

Gouttenoire J, Roingeard P, Penin F, Moradpour D (2010) Amphipathic alpha-helix AH2 is a major determinant for the oligomerization of hepatitis C virus nonstructural protein 4B. J Virol 84:12529–12537

Granato M, Lacconi V, Peddis M, Di RL, Valia S, Rivanera D, Antonelli G, Frati L, Faggioni A, Cirone M (2014) Hepatitis C virus present in the sera of infected patients interferes with the autophagic process of monocytes impairing their in-vitro differentiation into dendritic cells. Biochim Biophys Acta 1843:1348–1355

Grise H, Frausto S, Logan T, Tang H (2012) A conserved tandem cyclophilin-binding site in hepatitis C virus nonstructural protein 5A regulates Alisporivir susceptibility. J Virol 86:4811–4822

Gu M, Rice CM (2010) Three conformational snapshots of the hepatitis C virus NS3 helicase reveal a ratchet translocation mechanism. Proc Natl Acad Sci U S A 107:521–528

Guevin C, Manna D, Belanger C, Konan KV, Mak P, Labonte P (2010) Autophagy protein ATG5 interacts transiently with the hepatitis C virus RNA polymerase (NS5B) early during infection. Virology 405:1–7

Gwack Y, Kim DW, Han JH, Choe J (1996) Characterization of RNA binding activity and RNA helicase activity of the hepatitis C virus NS3 protein. Biochem Biophys Res Commun 225:654–659

Hamamoto I, Nishimura Y, Okamoto T, Aizaki H, Liu M, Mori Y, Abe T, Suzuki T, Lai MM, Miyamura T, Moriishi K, Matsuura Y (2005) Human VAP-B is involved in hepatitis C virus replication through interaction with NS5A and NS5B. J Virol 79:13473–13482

Hammond GR, Fischer MJ, Anderson KE, Holdich J, Koteci A, Balla T, Irvine RF (2012) PI4P and PI(4,5)P2 are essential but independent lipid determinants of membrane identity. Science 337:727–730

Hanoulle X, Badillo A, Wieruszeski JM, Verdegem D, Landrieu I, Bartenschlager R, Penin F, Lippens G (2009) Hepatitis C virus NS5A protein is a substrate for the peptidyl-prolyl cis/trans isomerase activity of cyclophilins A and B. J Biol Chem 284:13589–13601

Harak C, Radujkovic D, Taveneau C, Reiss S, Klein R, Bressanelli S, Lohmann V (2014) Mapping of functional domains of the lipid kinase phosphatidylinositol 4-kinase type III alpha involved in enzymatic activity and hepatitis C virus replication. J Virol 88:9909–9926

Harris D, Zhang Z, Chaubey B, Pandey VN (2006) Identification of cellular factors associated with the 3′-nontranslated region of the hepatitis C virus genome. Mol Cell Proteomics 5:1006–1018

Harrus D, Ahmed-El-Sayed N, Simister PC, Miller S, Triconnet M, Hagedorn CH, Mahias K, Rey FA, Astier-Gin T, Bressanelli S (2010) Further insights into the roles of GTP and the C-terminus of the hepatitis C virus polymerase in the initiation of RNA synthesis. J Biol Chem 285:32906–32918

Heck JA, Meng X, Frick DN (2009) Cyclophilin B stimulates RNA synthesis by the HCV RNA dependent RNA polymerase. Biochem Pharmacol 77:1173–1180

Henke JI, Goergen D, Zheng J, Song Y, Schuttler CG, Fehr C, Junemann C, Niepmann M (2008) microRNA-122 stimulates translation of hepatitis C virus RNA. EMBO J 27:3300–3310

Hirata Y, Ikeda K, Sudoh M, Tokunaga Y, Suzuki A, Weng L, Ohta M, Tobita Y, Okano K, Ozeki K, Kawasaki K, Tsukuda T, Katsume A, Aoki Y, Umehara T, Sekiguchi S, Toyoda T, Shimotohno K, Soga T, Nishijima M, Taguchi R, Kohara M (2012) Self-enhancement of hepatitis C virus replication by promotion of specific sphingolipid biosynthesis. PLoS Pathog 8:e1002860

Honda M, Ping LH, Rijnbrand RA, Amphlett E, Clarke B, Rowlands D, Lemon SM (1996) Structural requirements for initiation of translation by internal ribosome entry within genome-length hepatitis C virus RNA. Virology 222:31–42

Hsu NY, Ilnytska O, Belov G, Santiana M, Chen YH, Takvorian PM, Pau C, van der Schaar H, Kaushik-Basu N, Balla T, Cameron CE, Ehrenfeld E, van Kuppeveld FJ, Altan-Bonnet N (2010) Viral reorganization of the secretory pathway generates distinct organelles for RNA replication. Cell 141:799–811

Hu J, Troglio F, Mukhopadhyay A, Everingham S, Kwok E, Scita G, Craig AW (2009) F-BAR-containing adaptor CIP4 localizes to early endosomes and regulates epidermal growth factor receptor trafficking and downregulation. Cell Signal 21:1686–1697

Huang L, Hwang J, Sharma SD, Hargittai MR, Chen Y, Arnold JJ, Raney KD, Cameron CE (2005) Hepatitis C virus nonstructural protein 5A (NS5A) is an RNA-binding protein. J Biol Chem 280:36417–36428

Ide Y, Zhang L, Chen M, Inchauspe G, Bahl C, Sasaguri Y, Padmanabhan R (1996) Characterization of the nuclear localization signal and subcellular distribution of hepatitis C virus nonstructural protein NS5A. Gene 182:203–211

Isken O, Baroth M, Grassmann CW, Weinlich S, Ostareck DH, Ostareck-Lederer A, Behrens SE (2007) Nuclear factors are involved in hepatitis C virus RNA replication. RNA 13:1675–1692

Isoyama T, Kuge S, Nomoto A (2002) The core protein of hepatitis C virus is imported into the nucleus by transport receptor Kap123p but inhibits Kap121p-dependent nuclear import of yeast AP1-like transcription factor in yeast cells. J Biol Chem 277:39634–39641

Ivashkina N, Wolk B, Lohmann V, Bartenschlager R, Blum HE, Penin F, Moradpour D (2002) The hepatitis C virus RNA-dependent RNA polymerase membrane insertion sequence is a transmembrane segment. J Virol 76:13088–13093

Jennings TA, Chen Y, Sikora D, Harrison MK, Sikora B, Huang L, Jankowsky E, Fairman ME, Cameron CE, Raney KD (2008) RNA unwinding activity of the hepatitis C virus NS3 helicase is modulated by the NS5B polymerase. Biochemistry 47:1126–1135

Jin Z, Leveque V, Ma H, Johnson KA, Klumpp K (2012) Assembly, purification, and pre-steady-state kinetic analysis of active RNA-dependent RNA polymerase elongation complex. J Biol Chem 287:10674–10683

Jin Z, Leveque V, Ma H, Johnson KA, Klumpp K (2013) NTP-mediated nucleotide excision activity of hepatitis C virus RNA-dependent RNA polymerase. Proc Natl Acad Sci U S A 110: E348–E357

Jopling CL, Yi M, Lancaster AM, Lemon SM, Sarnow P (2005) Modulation of hepatitis C virus RNA abundance by a liver-specific MicroRNA. Science 309:1577–1581

Kao CC, Yang X, Kline A, Wang QM, Barket D, Heinz BA (2000) Template requirements for RNA synthesis by a recombinant hepatitis C virus RNA-dependent RNA polymerase. J Virol 74:11121–11128

Kapadia SB, Chisari FV (2005) Hepatitis C virus RNA replication is regulated by host geranylgeranylation and fatty acids. Proc Natl Acad Sci U S A 102:2561–2566

Kaul A, Stauffer S, Berger C, Pertel T, Schmitt J, Kallis S, Lopez MZ, Lohmann V, Luban J, Bartenschlager R (2009) Essential role of cyclophilin A for hepatitis C virus replication and virus production and possible link to polyprotein cleavage kinetics. PLoS Pathog 5:e1000546

Ke PY, Chen SS (2011) Autophagy: a novel guardian of HCV against innate immune response. Autophagy 7:533–535

Keum SJ, Park SM, Park JH, Jung JH, Shin EJ, Jang SK (2012) The specific infectivity of hepatitis C virus changes through its life cycle. Virology 433:462–470

Khan I, Katikaneni DS, Han Q, Sanchez-Felipe L, Hanada K, Ambrose RL, Mackenzie JM, Konan KV (2014) Modulation of hepatitis C virus genome replication by glycosphingolipids and four-phosphate adaptor protein 2. J Virol 88:12276–12295

Kim JE, Song WK, Chung KM, Back SH, Jang SK (1999) Subcellular localization of hepatitis C viral proteins in mammalian cells. Arch Virol 144:329–343

Kohlway A, Pirakitikulr N, Ding SC, Yang F, Luo D, Lindenbach BD, Pyle AM (2014) The linker region of NS3 plays a critical role in the replication and infectivity of hepatitis C virus. J Virol 88:10970–10974

Kolykhalov AA, Feinstone SM, Rice CM (1996) Identification of a highly conserved sequence element at the 3′ terminus of hepatitis C virus genome RNA. J Virol 70:3363–3371

Krieger N, Lohmann V, Bartenschlager R (2001) Enhancement of hepatitis C virus RNA replication by cell culture- adaptive mutations. J Virol 75:4614–4624

Kumar A, Ray U, Das S (2013) Human La protein interaction with GCAC near the initiator AUG enhances hepatitis C virus RNA replication by promoting linkage between 5′ and 3′ untranslated regions. J Virol 87:6713–6726

Lanford RE, Hildebrandt-Eriksen ES, Petri A, Persson R, Lindow M, Munk ME, Kauppinen S, Orum H (2010) Therapeutic silencing of microRNA-122 in primates with chronic hepatitis C virus infection. Science 327:198–201

Lanford RE, Feng Z, Chavez D, Guerra B, Brasky KM, Zhou Y, Yamane D, Perelson AS, Walker CM, Lemon SM (2011) Acute hepatitis A virus infection is associated with a limited type I interferon response and persistence of intrahepatic viral RNA. Proc Natl Acad Sci U S A 108:11223–11228

Lee JW, Liao PC, Young KC, Chang CL, Chen SS, Chang TT, Lai MD, Wang SW (2011) Identification of hnRNPH1, NF45, and C14orf166 as novel host interacting partners of the mature hepatitis C virus core protein. J Proteome Res 10:4522–4534

Lesburg CA, Cable MB, Ferrari E, Hong Z, Mannarino AF, Weber PC (1999) Crystal structure of the RNA-dependent RNA polymerase from hepatitis C virus reveals a fully encircled active site. Nat Struct Biol 6:937–943

Lescar J, Canard B (2009) RNA-dependent RNA polymerases from flaviviruses and picornaviridae. Curr Opin Struct Biol 19:759–767

Leveque VJ, Johnson RB, Parsons S, Ren J, Xie C, Zhang F, Wang QM (2003) Identification of a C-terminal regulatory motif in hepatitis C virus RNA-dependent RNA polymerase: structural and biochemical analysis. J Virol 77:9020–9028

Li Q, Brass AL, Ng A, Hu Z, Xavier RJ, Liang TJ, Elledge SJ (2009) A genome-wide genetic screen for host factors required for hepatitis C virus propagation. Proc Natl Acad Sci U S A 106:16410–16415

Li Y, Masaki T, Yamane D, McGivern DR, Lemon SM (2013) Competing and noncompeting activities of miR-122 and the 5′ exonuclease Xrn1 in regulation of hepatitis C virus replication. Proc Natl Acad Sci U S A 110:1881–1886

Li H, Yang X, Yang G, Hong Z, Zhou L, Yin P, Xiao Y, Chen L, Chung RT, Zhang L (2014) Hepatitis C virus NS5A hijacks ARFGAP1 to maintain a phosphatidylinositol 4-phosphate-enriched microenvironment. J Virol 88:5956–5966

Liang Y, Shilagard T, Xiao SY, Snyder N, Lau D, Cicalese L, Weiss H, Vargas G, Lemon SM (2009) Visualizing hepatitis C virus infections in human liver by two-photon microscopy. Gastroenterology 137:1448–1458

Liefhebber JM, Hague CV, Zhang Q, Wakelam MJ, McLauchlan J (2014) Modulation of triglyceride and cholesterol ester synthesis impairs assembly of infectious hepatitis C virus. J Biol Chem 289:21276–21288

Lindenbach BD, Rice CM (2013) The ins and outs of hepatitis C virus entry and assembly. Nat Rev Microbiol 11:688–700

Liu Z, Robida JM, Chinnaswamy S, Yi G, Robotham JM, Nelson HB, Irsigler A, Kao CC, Tang H (2009a) Mutations in the hepatitis C virus polymerase that increase RNA binding can confer resistance to cyclosporine A. Hepatology 50:25–33

Liu Z, Yang F, Robotham JM, Tang H (2009b) Critical role of cyclophilin A and its prolyl-peptidyl isomerase activity in the structure and function of the hepatitis C virus replication complex. J Virol 83:6554–6565

Lohmann V (2013) Hepatitis C virus RNA replication. Curr Top Microbiol Immunol 369:167–198

Lohmann V, Korner F, Herian U, Bartenschlager R (1997) Biochemical properties of hepatitis C virus NS5B RNA-dependent RNA polymerase and identification of amino acid sequence motifs essential for enzymatic activity. J Virol 71:8416–8428

Lohmann V, Roos A, Korner F, Koch JO, Bartenschlager R (1998) Biochemical and kinetic analyses of NS5B RNA-dependent RNA polymerase of the hepatitis C virus. Virology 249:108–118

Lohmann V, Körner F, Koch JO, Herian U, Theilmann L, Bartenschlager R (1999a) Replication of subgenomic hepatitis C virus RNAs in a hepatoma cell line. Science 285:110–113

Lohmann V, Overton H, Bartenschlager R (1999b) Selective stimulation of hepatitis C virus and pestivirus NS5B RNA polymerase activity by GTP. J Biol Chem 274:10807–10815

Lu H, Li W, Noble WS, Payan D, Anderson DC (2004) Riboproteomics of the hepatitis C virus internal ribosomal entry site. J Proteome Res 3:949–957

Luo G, Hamatake RK, Mathis DM, Racela J, Rigat KL, Lemm J, Colonno RJ (2000) De novo initiation of RNA synthesis by the RNA-dependent RNA polymerase (NS5B) of hepatitis C virus. J Virol 74:851–863

Machlin ES, Sarnow P, Sagan SM (2011) Masking the 5′ terminal nucleotides of the hepatitis C virus genome by an unconventional microRNA-target RNA complex. Proc Natl Acad Sci U S A 108:3193–3198

Mackenzie JM, Jones MK, Young PR (1996) Immunolocalization of the dengue virus nonstructural glycoprotein NS1 suggests a role in viral RNA replication. Virology 220:232–240

Mackenzie JM, Khromykh AA, Westaway EG (2001) Stable expression of noncytopathic Kunjin replicons simulates both ultrastructural and biochemical characteristics observed during replication of Kunjin virus. Virology 279:161–172

Madan V, Paul D, Lohmann V, Bartenschlager R (2014) Inhibition of HCV replication by cyclophilin antagonists is linked to replication fitness and occurs by inhibition of membranous web formation. Gastroenterology 146:1361–1372

Mani N, Yuzhakov A, Yuzhakov O, Coll JT, Black J, Saxena K, Fulghum JR, Lippke JA, Rao BG, Rijnbrand R, Kwong AD (2014) NS5A and human RPA increase the processivity of HCV NS5B polymerase activity in vitro. J Virol 89:165–180. JVI.01677–14. [Epub ahead of print]

Manna D, Aligo J, Xu C, Park WS, Koc H, Heo WD, Konan KV (2010) Endocytic rab proteins are required for hepatitis C virus replication complex formation. Virology 398:21–37

Masaki T, Matsunaga S, Takahashi H, Nakashima K, Kimura Y, Ito M, Matsuda M, Murayama A, Kato T, Hirano H, Endo Y, Lemon SM, Wakita T, Sawasaki T, Suzuki T (2014) Involvement of hepatitis C virus NS5A hyperphosphorylation mediated by casein kinase I-alpha in infectious virus production. J Virol 88:7541–7555

Matto M, Sklan EH, David N, Melamed-Book N, Casanova JE, Glenn JS, Aroeti B (2011) Role for ADP ribosylation factor 1 in the regulation of hepatitis C virus replication. J Virol 85:946–956

McMullan LK, Grakoui A, Evans MJ, Mihalik K, Puig M, Branch AD, Feinstone SM, Rice CM (2007) Evidence for a functional RNA element in the hepatitis C virus core gene. Proc Natl Acad Sci U S A 104:2879–2884

Mensa L, Perez-del-Pulgar S, Crespo G, Koutsoudakis G, Fernandez-Carrillo C, Coto-Llerena M, Miquel R, Allende H, Castells L, Navasa M, Forns X (2013) Imaging of hepatitis C virus infection in liver grafts after liver transplantation. J Hepatol 59:271–278

Mesmin B, Bigay J, von Moser FJ, Lacas-Gervais S, Drin G, Antonny B (2013) A four-step cycle driven by PI(4)P hydrolysis directs sterol/PI(4)P exchange by the ER-Golgi tether OSBP. Cell 155:830–843

Miller S, Kastner S, Krijnse-Locker J, Buhler S, Bartenschlager R (2007) The non-structural protein 4A of dengue virus is an integral membrane protein inducing membrane alterations in a 2 K-regulated manner. J Biol Chem 282:8873–8882

Miorin L, Romero-Brey I, Maiuri P, Hoppe S, Krijnse-Locker J, Bartenschlager R, Marcello A (2013) Three-dimensional architecture of tick-borne encephalitis virus replication sites and trafficking of the replicated RNA. J Virol 87:6469–6481

Miyanari Y, Hijikata M, Yamaji M, Hosaka M, Takahashi H, Shimotohno K (2003) Hepatitis C virus non-structural proteins in the probable membranous compartment function in viral genome replication. J Biol Chem 278:50301–50308

Mizuno-Yamasaki E, Rivera-Molina F, Novick P (2012) GTPase networks in membrane traffic. Annu Rev Biochem 81:637–659

Moradpour D, Englert C, Wakita T, Wands JR (1996) Characterization of cell lines allowing tightly regulated expression of hepatitis C virus core protein. Virology 222:51–63

Moradpour D, Brass V, Bieck E, Friebe P, Gosert R, Blum HE, Bartenschlager R, Penin F, Lohmann V (2004) Membrane association of the RNA-dependent RNA polymerase is essential for hepatitis C virus RNA replication. J Virol 78:13278–13284

Mortimer SA, Doudna JA (2013) Unconventional miR-122 binding stabilizes the HCV genome by forming a trimolecular RNA structure. Nucleic Acids Res 41:4230–4240

Mosley RT, Edwards TE, Murakami E, Lam AM, Grice RL, Du J, Sofia MJ, Furman PA, Otto MJ (2012) Structure of hepatitis C virus polymerase in complex with primer-template RNA. J Virol 86:6503–6511

Nakagawa M, Sakamoto N, Enomoto N, Tanabe Y, Kanazawa N, Koyama T, Kurosaki M, Maekawa S, Yamashiro T, Chen CH, Itsui Y, Kakinuma S, Watanabe M (2004) Specific inhibition of hepatitis C virus replication by cyclosporin A. Biochem Biophys Res Commun 313:42–47

Nakagawa M, Sakamoto N, Tanabe Y, Koyama T, Itsui Y, Takeda Y, Chen CH, Kakinuma S, Oooka S, Maekawa S, Enomoto N, Watanabe M (2005) Suppression of hepatitis C virus replication by cyclosporin a is mediated by blockade of cyclophilins. Gastroenterology 129:1031–1041

Neufeldt CJ, Joyce MA, Levin A, Steenbergen RH, Pang D, Shields J, Tyrrell DL, Wozniak RW (2013) Hepatitis C virus-induced cytoplasmic organelles use the nuclear transport machinery to establish an environment conducive to virus replication. PLoS Pathog 9:e1003744

Neumann AU, Lam NP, Dahari H, Gretch DR, Wiley TE, Layden TJ, Perelson AS (1998) Hepatitis C viral dynamics in vivo and the antiviral efficacy of interferon-alpha therapy. Science 282:103–107

Nevo-Yassaf I, Yaffe Y, Asher M, Ravid O, Eizenberg S, Henis YI, Nahmias Y, Hirschberg K, Sklan EH (2012) Role for TBC1D20 and Rab1 in hepatitis C virus replication via interaction with lipid droplet-bound nonstructural protein 5A. J Virol 86:6491–6502

Oh JW, Ito T, Lai MC (1999) A recombinant hepatitis C virus RNA-dependent RNA polymerase capable of copying the full-length viral RNA. J Virol 73:7694–7702

Olmstead AD, Knecht W, Lazarov I, Dixit SB, Jean F (2012) Human subtilase SKI-1/S1P is a master regulator of the HCV lifecycle and a potential host cell target for developing indirect-acting antiviral agents. PLoS Pathog 8:e1002468

Overby AK, Popov VL, Niedrig M, Weber F (2010) Tick-borne encephalitis virus delays interferon induction and hides its double-stranded RNA in intracellular membrane vesicles. J Virol 84:8470–8483

Ozeki S, Cheng J, Tauchi-Sato K, Hatano N, Taniguchi H, Fujimoto T (2005) Rab18 localizes to lipid droplets and induces their close apposition to the endoplasmic reticulum-derived membrane. J Cell Sci 118:2601–2611

Paredes AM, Blight KJ (2008) A genetic interaction between hepatitis C virus NS4B and NS3 is important for RNA replication. J Virol 82:10671–10683

Park CY, Jun HJ, Wakita T, Cheong JH, Hwang SB (2009) Hepatitis C virus nonstructural 4B protein modulates sterol regulatory element-binding protein signaling via the AKT pathway. J Biol Chem 284:9237–9246

Paul D, Bartenschlager R (2013) Architecture and biogenesis of plus-strand RNA virus replication factories. World J Virol 2:32–48

Paul AV, van Boon J, Filippov D, Wimmer E (1998) Protein-primed RNA synthesis by purified poliovirus RNA polymerase. Nature 393:280–284

Paul AV, Rieder E, Kim DW, van Boom JH, Wimmer E (2000) Identification of an RNA hairpin in poliovirus RNA that serves as the primary template in the in vitro uridylylation of VPg. J Virol 74:10359–10370

Paul D, Romero-Brey I, Gouttenoire J, Stoitsova S, Krijnse-Locker J, Moradpour D, Bartenschlager R (2011) NS4B self-interaction through conserved C-terminal elements is required for the establishment of functional hepatitis C virus replication complexes. J Virol 85:6963–6976

Paul D, Hoppe S, Saher G, Krijnse-Locker J, Bartenschlager R (2013) Morphological and biochemical characterization of the membranous hepatitis C virus replication compartment. J Virol 87:10612–10627

Pause A, Kukolj G, Bailey M, Brault M, Do F, Halmos T, Lagace L, Maurice R, Marquis M, McKercher G, Pellerin C, Pilote L, Thibeault D, Lamarre D (2003) An NS3 serine protease inhibitor abrogates replication of subgenomic hepatitis C virus RNA. J Biol Chem 278 (22):20374–20380

Pfeifer U, Thomssen R, Legler K, Bottcher U, Gerlich W, Weinmann E, Klinge O (1980) Experimental non-A, non-B hepatitis: four types of cytoplasmic alteration in hepatocytes of infected chimpanzees. Virchows Arch B Cell Pathol Incl Mol Pathol 33:233–243

Piccininni S, Varaklioti A, Nardelli M, Dave B, Raney KD, McCarthy JE (2002) Modulation of the hepatitis C virus RNA-dependent RNA polymerase activity by the non-structural (NS) 3 helicase and the NS4B membrane protein. J Biol Chem 277:45670–45679

Pietschmann T, Lohmann V, Rutter G, Kurpanek K, Bartenschlager R (2001) Characterization of cell lines carrying self-replicating hepatitis C virus RNAs. J Virol 75:1252–1264

Pietschmann T, Kaul A, Koutsoudakis G, Shavinskaya A, Kallis S, Steinmann E, Abid K, Negro F, Dreux M, Cosset FL, Bartenschlager R (2006) Construction and characterization of infectious intragenotypic and intergenotypic hepatitis C virus chimeras. Proc Natl Acad Sci U S A 103:7408–7413

Powdrill MH, Tchesnokov EP, Kozak RA, Russell RS, Martin R, Svarovskaia ES, Mo H, Kouyos RD, Gotte M (2011) Contribution of a mutational bias in hepatitis C virus replication to the genetic barrier in the development of drug resistance. Proc Natl Acad Sci U S A 108:20509–20513

Quezada EM, Kane CM (2009) The hepatitis C virus NS5A stimulates NS5B during in vitro RNA synthesis in a template specific manner. Open Biochem J 3:39–48

Quinkert D, Bartenschlager R, Lohmann V (2005) Quantitative analysis of the hepatitis C virus replication complex. J Virol 79:13594–13605

Ranjith-Kumar CT, Kao CC (2006) Recombinant viral RdRps can initiate RNA synthesis from circular templates. RNA 12:303–312

Ranjith-Kumar CT, Wen Y, Baxter N, Bhardwaj K, Cheng KC (2011) A cell-based assay for RNA synthesis by the HCV polymerase reveals new insights on mechanism of polymerase inhibitors and modulation by NS5A. PLoS ONE 6:e22575

Reigadas S, Ventura M, Sarih-Cottin L, Castroviejo M, Litvak S, Astier-Gin T (2001) HCV RNA-dependent RNA polymerase replicates in vitro the 3′ terminal region of the minus-strand viral RNA more efficiently than the 3′ terminal region of the plus RNA. Eur J Biochem 268:5857–5867

Reikine S, Nguyen JB, Modis Y (2014) Pattern recognition and signaling mechanisms of RIG-I and MDA5. Front Immunol 5:342

Reiss S, Rebhan I, Backes P, Romero-Brey I, Erfle H, Matula P, Kaderali L, Poenisch M, Blankenburg H, Hiet MS, Longerich T, Diehl S, Ramirez F, Balla T, Rohr K, Kaul A, Buhler S, Pepperkok R, Lengauer T, Albrecht M, Eils R, Schirmacher P, Lohmann V, Bartenschlager R (2011) Recruitment and activation of a lipid kinase by hepatitis C virus NS5A is essential for integrity of the membranous replication compartment. Cell Host Microbe 9:32–45

Reiss S, Harak C, Romero-Brey I, Radujkovic D, Klein R, Ruggieri A, Rebhan I, Bartenschlager R, Lohmann V (2013) The lipid kinase phosphatidylinositol-4 kinase III alpha regulates the phosphorylation status of hepatitis C virus NS5A. PLoS Pathog 9:e1003359

Romero-Brey I, Bartenschlager R (2014) Membranous replication factories induced by plus-strand RNA viruses. Viruses 6:2826–2857

Romero-Brey I, Merz A, Chiramel A, Lee JY, Chlanda P, Haselman U, Santarella-Mellwig R, Habermann A, Hoppe S, Kallis S, Walther P, Antony C, Krijnse-Locker J, Bartenschlager R (2012) Three-dimensional architecture and biogenesis of membrane structures associated with hepatitis C virus replication. PLoS Pathog 8:e1003056

Romero-Lopez C, Barroso-Deljesus A, Garcia-Sacristan A, Briones C, Berzal-Herranz A (2014) End-to-end crosstalk within the hepatitis C virus genome mediates the conformational switch of the 3′X-tail region. Nucleic Acids Res 42:567–582

Roosendaal J, Westaway EG, Khromykh A, Mackenzie JM (2006) Regulated cleavages at the West Nile virus NS4A-2 K-NS4B junctions play a major role in rearranging cytoplasmic membranes and Golgi trafficking of the NS4A protein. J Virol 80:4623–4632

Rosnoblet C, Fritzinger B, Legrand D, Launay H, Wieruszeski JM, Lippens G, Hanoulle X (2012) Hepatitis C virus NS5B and host cyclophilin A share a common binding site on NS5A. J Biol Chem 287:44249–44260

Ross-Thriepland D, Harris M (2014) Insights into the complexity and functionality of hepatitis C virus NS5A phosphorylation. J Virol 88:1421–1432

Salloum S, Wang H, Ferguson C, Parton RG, Tai AW (2013) Rab18 binds to hepatitis C virus NS5A and promotes interaction between sites of viral replication and lipid droplets. PLoS Pathog 9:e1003513

Schmeiser S, Mast J, Thiel HJ, Konig M (2014) Morphogenesis of pestiviruses: new insights from ultrastructural studies of strain Giraffe-1. J Virol 88:2717–2724

Schmitt M, Scrima N, Radujkovic D, Caillet-Saguy C, Simister PC, Friebe P, Wicht O, Klein R, Bartenschlager R, Lohmann V, Bressanelli S (2011) A comprehensive structure-function comparison of hepatitis C virus strain JFH1 and J6 polymerases reveals a key residue stimulating replication in cell culture across genotypes. J Virol 85:2565–2581

Schuster C, Isel C, Imbert I, Ehresmann C, Marquet R, Kieny MP (2002) Secondary structure of the 3′ terminus of hepatitis C virus minus-strand RNA. J Virol 76:8058–8068

Schwartz M, Chen J, Janda M, Sullivan M, den Boon J, Ahlquist P (2002) A positive-strand RNA virus replication complex parallels form and function of retrovirus capsids. Mol Cell 9:505–514

Scrima N, Caillet-Saguy C, Ventura M, Harrus D, Astier-Gin T, Bressanelli S (2012) Two crucial early steps in RNA synthesis by the hepatitis C virus polymerase involve a dual role of residue 405. J Virol 86:7107–7117

Sedano CD, Sarnow P (2014) Hepatitis C virus subverts liver-specific miR-122 to protect the viral genome from exoribonuclease Xrn2. Cell Host Microbe 16:257–264

Shi ST, Lee KJ, Aizaki H, Hwang SB, Lai MM (2003) Hepatitis C virus RNA replication occurs on a detergent-resistant membrane that cofractionates with caveolin-2. J Virol 77:4160–4168

Shimizu Y, Hishiki T, Ujino S, Sugiyama K, Funami K, Shimotohno K (2011) Lipoprotein component associated with hepatitis C virus is essential for virus infectivity. Curr Opin Virol 1:19–26

Shirota Y, Luo H, Qin W, Kaneko S, Yamashita T, Kobayashi K, Murakami S (2002) Hepatitis C virus (HCV) NS5A binds RNA-dependent RNA polymerase (RdRP) NS5B and modulates RNA-dependent RNA polymerase activity. J Biol Chem 277:11149–11155

Shrivastava S, Raychoudhuri A, Steele R, Ray R, Ray RB (2011) Knockdown of autophagy enhances the innate immune response in hepatitis C virus-infected hepatocytes. Hepatology 53:406–414

Simister P, Schmitt M, Geitmann M, Wicht O, Danielson UH, Klein R, Bressanelli S, Lohmann V (2009) Structural and functional analysis of hepatitis C virus strain JFH1 polymerase. J Virol 83:11926–11939

Simons K, Sampaio JL (2011) Membrane organization and lipid rafts. Cold Spring Harb Perspect Biol 3:a004697

Sir D, Chen WL, Choi J, Wakita T, Yen TS, Ou JH (2008a) Induction of incomplete autophagic response by hepatitis C virus via the unfolded protein response. Hepatology 48:1054–1061

Sir D, Liang C, Chen WL, Jung JU, Ou JH (2008b) Perturbation of autophagic pathway by hepatitis C virus. Autophagy 4:830–831

Sir D, Kuo CF, Tian Y, Liu HM, Huang EJ, Jung JU, Machida K, Ou JH (2012) Replication of hepatitis C virus RNA on autophagosomal membranes. J Biol Chem 287:18036–18043

Sklan EH, Serrano RL, Einav S, Pfeffer SR, Lambright DG, Glenn JS (2007) TBC1D20 is a Rab1 GTPase-activating protein that mediates hepatitis C virus replication. J Biol Chem 282:36354–36361

Smith DB, Simmonds P (1997) Characteristics of nucleotide substitution in the hepatitis C virus genome: constraints on sequence change in coding regions at both ends of the genome. J Mol Evol 45:238–246

Smith RM, Walton CM, Wu CH, Wu GY (2002) Secondary structure and hybridization accessibility of hepatitis C virus 3′-terminal sequences. J Virol 76:9563–9574

Stewart A, Ghosh M, Spencer DM, Leslie CC (2002) Enzymatic properties of human cytosolic phospholipase A(2)gamma. J Biol Chem 277:29526–29536

Stone M, Jia S, Heo WD, Meyer T, Konan KV (2007) Participation of rab5, an early endosome protein, in hepatitis C virus RNA replication machinery. J Virol 81:4551–4563

Su AI, Pezacki JP, Wodicka L, Brideau AD, Supekova L, Thimme R, Wieland S, Bukh J, Purcell RH, Schultz PG, Chisari FV (2002) Genomic analysis of the host response to hepatitis C virus infection. Proc Natl Acad Sci U S A 99:15669–15674

Su WC, Chao TC, Huang YL, Weng SC, Jeng KS, Lai MM (2011) Rab5 and class III phosphoinositide 3-kinase Vps34 are involved in hepatitis C virus NS4B-induced autophagy. J Virol 85:10561–10571

Sun XL, Johnson RB, Hockman MA, Wang QM (2000) De novo RNA synthesis catalyzed by HCV RNA-dependent RNA polymerase. Biochem Biophys Res Commun 268:798–803

Suzuki R, Sakamoto S, Tsutsumi T, Rikimaru A, Tanaka K, Shimoike T, Moriishi K, Iwasaki T, Mizumoto K, Matsuura Y, Miyamura T, Suzuki T (2005) Molecular determinants for subcellular localization of hepatitis C virus core protein. J Virol 79:1271–1281

Tai AW, Salloum S (2011) The role of the phosphatidylinositol 4-kinase PI4KA in hepatitis C virus-induced host membrane rearrangement. PLoS ONE 6:e26300

Tai AW, Benita Y, Peng LF, Kim SS, Sakamoto N, Xavier RJ, Chung RT (2009) A functional genomic screen identifies cellular cofactors of hepatitis C virus replication. Cell Host Microbe 5:298–307

Takano T, Tsukiyama-Kohara K, Hayashi M, Hirata Y, Satoh M, Tokunaga Y, Tateno C, Hayashi Y, Hishima T, Funata N, Sudoh M, Kohara M (2011) Augmentation of DHCR24 expression by hepatitis C virus infection facilitates viral replication in hepatocytes. J Hepatol 55:512–521

Tanaka T, Kato N, Cho MJ, Shimotohno K (1995) A novel sequence found at the 3′ terminus of hepatitis C virus genome. Biochem Biophys Res Commun 215:744–749

Tanida I, Fukasawa M, Ueno T, Kominami E, Wakita T, Hanada K (2009) Knockdown of autophagy-related gene decreases the production of infectious hepatitis C virus particles. Autophagy 5:937–945

Targett-Adams P, Boulant S, McLauchlan J (2008) Visualization of double-stranded RNA in cells supporting hepatitis C virus RNA replication. J Virol 82:2182–2195

Tsujita K, Suetsugu S, Sasaki N, Furutani M, Oikawa T, Takenawa T (2006) Coordination between the actin cytoskeleton and membrane deformation by a novel membrane tubulation domain of PCH proteins is involved in endocytosis. J Cell Biol 172:269–279

Tuplin A, Struthers M, Simmonds P, Evans DJ (2012) A twist in the tail: SHAPE mapping of long-range interactions and structural rearrangements of RNA elements involved in HCV replication. Nucleic Acids Res 40:6908–6921

Underwood KW, Song C, Kriz RW, Chang XJ, Knopf JL, Lin LL (1998) A novel calcium-independent phospholipase A2, cPLA2-gamma, that is prenylated and contains homology to cPLA2. J Biol Chem 273:21926–21932

Vassilaki N, Friebe P, Meuleman P, Kallis S, Kaul A, Paranhos-Baccala G, Leroux-Roels G, Mavromara P, Bartenschlager R (2008) Role of the hepatitis C virus core+1 open reading frame and core cis-acting RNA elements in viral RNA translation and replication. J Virol 82:11503–11515

Verdegem D, Badillo A, Wieruszeski JM, Landrieu I, Leroy A, Bartenschlager R, Penin F, Lippens G, Hanoulle X (2011) Domain 3 of NS5A protein from the hepatitis C virus has intrinsic alpha-helical propensity and is a substrate of cyclophilin A. J Biol Chem 286:20441–20454

Villordo SM, Gamarnik AV (2009) Genome cyclization as strategy for flavivirus RNA replication. Virus Res 139:230–239

Wang P, Heitman J (2005) The cyclophilins. Genome Biol 6:226

Wang H, Perry JW, Lauring AS, Neddermann P, De FR, Tai AW (2014) Oxysterol-binding protein is a phosphatidylinositol 4-kinase effector required for HCV replication membrane integrity and cholesterol trafficking. Gastroenterology 146:1373–1385

Waris G, Felmlee DJ, Negro F, Siddiqui A (2007) Hepatitis C virus induces proteolytic cleavage of sterol regulatory element binding proteins and stimulates their phosphorylation via oxidative stress. J Virol 81:8122–8130

Watashi K, Hijikata M, Hosaka M, Yamaji M, Shimotohno K (2003) Cyclosporin A suppresses replication of hepatitis C virus genome in cultured hepatocytes. Hepatology 38:1282–1288

Watashi K, Ishii N, Hijikata M, Inoue D, Murata T, Miyanari Y, Shimotohno K (2005) Cyclophilin B is a functional regulator of hepatitis C virus RNA polymerase. Mol Cell 19:111–122

Welsch S, Miller S, Romero-Brey I, Merz A, Bleck CK, Walther P, Fuller SD, Antony C, Krijnse-Locker J, Bartenschlager R (2009) Composition and three-dimensional architecture of the dengue virus replication and assembly sites. Cell Host Microbe 5:365–375

Weng L, Hirata Y, Arai M, Kohara M, Wakita T, Watashi K, Shimotohno K, He Y, Zhong J, Toyoda T (2010) Sphingomyelin activates hepatitis C virus RNA polymerase in a genotype-specific manner. J Virol 84:11761–11770

Wente SR, Rout MP (2010) The nuclear pore complex and nuclear transport. Cold Spring Harb Perspect Biol 2:a000562

Westaway EG, Khromykh AA, Kenney MT, Mackenzie JM, Jones MK (1997a) Proteins C and NS4B of the flavivirus Kunjin translocate independently into the nucleus. Virology 234:31–41

Westaway EG, Mackenzie JM, Kenney MT, Jones MK, Khromykh AA (1997b) Ultrastructure of Kunjin virus-infected cells: colocalization of NS1 and NS3 with double-stranded RNA, and of NS2B with NS3, in virus-induced membrane structures. J Virol 71:6650–6661

Wieland S, Makowska Z, Campana B, Calabrese D, Dill MT, Chung J, Chisari FV, Heim MH (2014) Simultaneous detection of hepatitis C virus and interferon stimulated gene expression in infected human liver. Hepatology 59:2121–2130

Wolk B, Sansonno D, Krausslich HG, Dammacco F, Rice CM, Blum HE, Moradpour D (2000) Subcellular localization, stability, and trans-cleavage competence of the hepatitis C virus NS3-NS4A complex expressed in tetracycline-regulated cell lines. J Virol 74:2293–2304

Wyles JP, McMaster CR, Ridgway ND (2002) Vesicle-associated membrane protein-associated protein-A (VAP-A) interacts with the oxysterol-binding protein to modify export from the endoplasmic reticulum. J Biol Chem 277:29908–29918

Xu S, Pei R, Guo M, Han Q, Lai J, Wang Y, Wu C, Zhou Y, Lu M, Chen X (2012) Cytosolic phospholipase A2 gamma is involved in hepatitis C virus replication and assembly. J Virol 86:13025–13037

Yamanaka T, Kodama T, Doi T (2002) Subcellular localization of HCV core protein regulates its ability for p53 activation and p21 suppression. Biochem Biophys Res Commun 294:528–534

Yamane D, McGivern DR, Wauthier E, Yi M, Madden VJ, Welsch C, Antes I, Wen Y, Chugh PE, McGee CE, Widman DG, Misumi I, Bandyopadhyay S, Kim S, Shimakami T, Oikawa T, Whitmire JK, Heise MT, Dittmer DP, Kao CC, Pitson SM, Merrill AH Jr, Reid LM, Lemon SM (2014) Regulation of the hepatitis C virus RNA replicase by endogenous lipid peroxidation. Nat Med 20:927–935

Yamashita T, Kaneko S, Shirota Y, Qin W, Nomura T, Kobayashi K, Murakami S (1998) RNA-dependent RNA polymerase activity of the soluble recombinant hepatitis C virus NS5B protein truncated at the C-terminal region. J Biol Chem 273:15479–15486

Yanagi M, St CM, Emerson SU, Purcell RH, Bukh J (1999) In vivo analysis of the 3′ untranslated region of the hepatitis C virus after in vitro mutagenesis of an infectious cDNA clone. Proc Natl Acad Sci U S A 96:2291–2295

Yang F, Robotham JM, Nelson HB, Irsigler A, Kenworthy R, Tang H (2008) Cyclophilin A is an essential cofactor for hepatitis C virus infection and the principal mediator of cyclosporine resistance in vitro. J Virol 82:5269–5278

Yang F, Robotham JM, Grise H, Frausto S, Madan V, Zayas M, Bartenschlager R, Robinson M, Greenstein AE, Nag A, Logan TM, Bienkiewicz E, Tang H (2010) A major determinant of cyclophilin dependence and cyclosporine susceptibility of hepatitis C virus identified by a genetic approach. PLoS Pathog 6:e1001118

Yi M, Lemon SM (2003a) 3′ nontranslated RNA signals required for replication of hepatitis C virus RNA. J Virol 77:3557–3568

Yi M, Lemon SM (2003b) Structure-function analysis of the 3′ stem-loop of hepatitis C virus genomic RNA and its role in viral RNA replication. RNA 9:331–345

You S, Rice CM (2008) 3′ RNA elements in hepatitis C virus replication: kissing partners and long poly(U). J Virol 82:184–195

You S, Stump DD, Branch AD, Rice CM (2004) A cis-acting replication element in the sequence encoding the NS5B RNA-dependent RNA polymerase is required for hepatitis C virus RNA replication. J Virol 78:1352–1366

Yu GY, He G, Li CY, Tang M, Grivennikov S, Tsai WT, Wu MS, Hsu CW, Tsai Y, Wang LH, Karin M (2012) Hepatic expression of HCV RNA-dependent RNA polymerase triggers innate immune signaling and cytokine production. Mol Cell 48:313–321

Zhang C, Cai Z, Kim YC, Kumar R, Yuan F, Shi PY, Kao C, Luo G (2005) Stimulation of hepatitis C virus (HCV) nonstructural protein 3 (NS3) helicase activity by the NS3 protease domain and by HCV RNA-dependent RNA polymerase. J Virol 79:8687–8697

Zhang L, Hong Z, Lin W, Shao RX, Goto K, Hsu VW, Chung RT (2012) ARF1 and GBF1 generate a PI4P-enriched environment supportive of hepatitis C virus replication. PLoS ONE 7:e32135

Zhong W, Uss AS, Ferrari E, Lau JY, Hong Z (2000) De novo initiation of RNA synthesis by hepatitis C virus nonstructural protein 5B polymerase. J Virol 74:2017–2022

Hepatitis C Virus-Host Interactions

Ana Shulla and Glenn Randall

Abstract Hepatitis C virus (HCV) is an RNA virus with limited genetic content and a specific tropism for the liver. As such, it depends significantly on its host cell to establish a successful infection. Twenty-five years of studying HCV and its relationship with the host hepatocyte have resulted in the identification of hundreds of HCV host cofactors that facilitate and respond to viral infection. Characterization of these cofactors has provided extensive insight into viral replication strategies and gene functions, in addition to expanding our animal and cell culture systems to study HCV replication and pathogenesis. Some of these host cofactors are also targets for therapeutic development. Although it is impossible to critically discuss every putative HCV-host interaction within the allotted space, we attempt to discuss the well-characterized virus-host interactions that function at each stage of the viral life cycle.

Keywords Host factors • genome wide screens • antiviral targets • replication

1 Entry

HCV gains entry into hepatocytes by engaging multiple host factors in a sequential manner (Fig. 1). The virion has been referred to as a "lipoviral particle" that is highly lipidated and contains components of very low density lipids (VDL), including apolipoproteins (Apo) A, B, and E (Andre et al. 2002; Nielsen et al. 2006; Merz et al. 2011). Initially, HCV circulating in the blood attaches to the basolateral surface of hepatocytes via low affinity interactions with low-density lipoprotein receptor (LDLR) (Agnello et al. 1999; Albecka et al. 2012) and glycosaminoglycans (GAGs) (Germi et al. 2002). Heparan sulfate proteoglycans also aid in attaching virions to cells and syndecan-1 is a major attachment factor (Shi et al. 2013). Apo E may play a role in virion interactions with these low affinity interactors (Owen et al. 2009; Jiang et al. 2012). Following this initial attachment,

A. Shulla • G. Randall, Ph.D. (✉)
Department of Microbiology, The University of Chicago, CLSC 707B, 920 East 58th Street, Chicago, IL 60637, USA
e-mail: grandall@bsd.uchicago.edu

T. Miyamura et al. (eds.), *Hepatitis C Virus I*, DOI 10.1007/978-4-431-56098-2_9

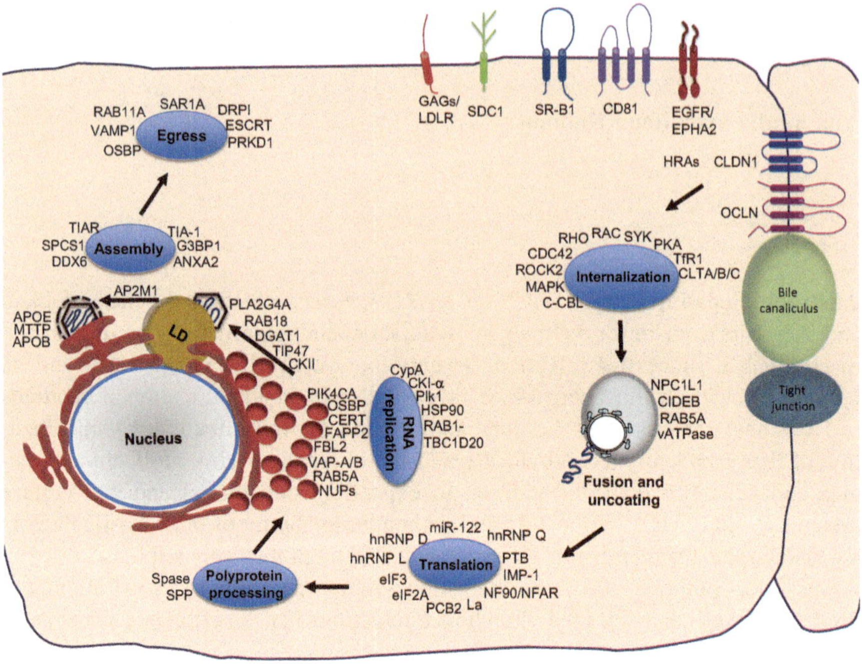

Fig. 1 Map of the HCV life cycle and host cell dependencies. Host cell proteins validated by multiple assays or studies as dependency factors for HCV infection were placed at positions relevant to the particular virus life stage (*blue ovals*). The *arrows* indicate the progression of the viral life cycle from initial attachment, to internalization, uncoating, translation, polyprotein processing, RNA replication, assembly and egress in polarized hepatocytes

the virions bind scavenger receptor class B type I (SR-BI) (Scarselli et al. 2002; Bartosch et al. 2003) and CD81 (Pileri et al. 1998; Cormier et al. 2004; Zhang et al. 2004) cell surface molecules, which are required for infection. SR-BI expression is enriched on hepatocytes, where it functions in binding and mediating cholesterol uptake from serum high-density lipoproteins (HDLs), very-low-density lipoproteins (VLDLs) and oxidized forms of LDLs (Acton et al. 1996). Initially, SR-BI interacts with the HCV-associated surface lipoproteins, and these interactions are thought to alter virion conformation and enable exposure of the buried viral E2 glycoprotein (Catanese et al. 2010; Zeisel et al. 2007). Once the E2 glycoprotein is exposed, SR-BI binds to its hypervariable region 1 (Scarselli et al. 2002; Dao Thi et al. 2012), which structurally reveals the CD81 binding site (Bankwitz et al. 2010), thus allowing for subsequent CD81-E2 interaction.

HCV entry also requires a set of cofactors that localize to tight junctions (TJs), claudin 1 (CLDN1) (Evans et al. 2007) and occludin (OCLN) (Ploss et al. 2009). CLDN1 interacts with CD81 and this is a crucial step for HCV entry, since mutations in CLDN1 that impair CLDN1-CD81 interaction completely abolish virus entry (Harris et al. 2010). Given the different subcellular localizations of the "early" basolateral HCV receptors, CD81 and SR-BI, and the "late" tight junction HCV receptors CLDN and OCLN, it has been proposed that either HCV migrates to the tight junction to enter cells or alternatively, that HCV disrupts cellular polarity to gain access to the tight junction receptors. The CD81-CLDN1 association is regulated by cellular signaling events including protein kinase A (PKA), epidermal growth factor receptor (EGFR) and ephrin receptor A2 (EphA2) (Farquhar et al. 2008; Lupberger et al. 2011). Binding of virus particles to CD81 as well as antibody-mediated crosslinking of CD81 molecules leads to EGFR activation and internalization (Diao et al. 2012). Recently, the GTPase HRas, a component of the EGFR signaling complex, was found to interact with CD81-CLDN1, thus serving as a transducer of EGFR-mediated HCV entry (Zona et al. 2013).

The late steps of HCV entry, including virus internalization, are poorly understood. Several host factors have been proposed to be involved. OCLN acts late in the virus entry process and recent data suggest that OCLN might directly interact with HCV particles and mediate the internalization process (Sourisseau et al. 2013). In addition to playing a role in initial stages of virus entry, SR-BI seems to be involved in late stages as well. Mutations that impair the SR-BI cholesterol uptake function, but not ability to bind the E2 glycoprotein, inhibit HCV entry (Dreux et al. 2009a). Furthermore, using anti-SR-BI monoclonal antibodies that inhibit late steps of HCV entry has revealed that the lipid transfer function of SR-BI, but not its ability to interact with HDL, is crucial for HCV entry (Zahid et al. 2013). PDZK1, a cellular SR-BI interacting protein known to facilitate HCV entry, may be involved in linking SR-BI to the actin cytoskeleton and the endocytic network (Eyre et al. 2010). CD81 also functions late in entry by priming the viral E2 glycoprotein for pH-dependent fusion (Sharma et al. 2011).

Other entry cofactors that are either directly or indirectly involved in HCV entry have recently been identified. Inhibition of the cholesterol uptake receptor Niemann Pick C1 like 1 (NPC1L1) inhibits virus entry (Sainz et al. 2012). Additionally, inhibition of transferrin receptor 1 (TFR1), which is ubiquitously expressed and the main receptor for cellular iron uptake, by silencing, small molecule inhibitors or blocking antibodies leads to impairment of HCV entry (Martin and Uprichard 2013). Recently the liver specific, cell death-inducing DFFA-like effector b (CIDEB) protein was identified as important for HCV entry into hepatocyte-like cells (HLCs) and suggested to act at a late membrane fusion event (Wu et al. 2014).

HCV particles internalize via clathrin-mediated endocytosis (Blanchard et al. 2006; Meertens et al. 2006) and undergo low pH-dependent fusion leading to uncoating from Rab5A positive early endosomes (Tscherne et al. 2006; Hsu et al. 2003; Coller et al. 2009). Host cytoskeletal rearrangements are important for HCV entry. HCV E2 engagement of CD81 leads to activation of Rho GTPase family member proteins Rac, Rho and CDC42 as well as mitogen-activated protein kinase (MAPK) signaling pathways which may contribute to remodeling of the actin cytoskeleton required for lateral movement of virus-receptor complexes and subsequent endocytosis (Brazzoli et al. 2008). Furthermore, E2 binding to CD81 was reported to induce a time-dependent SYK activation and host cytoskeletal ezrin phosphorylation in Huh-7.5 cells (Bukong et al. 2013). Live cell imaging of HCV infection showed that HCV migrates along actin and that actin is required for clathrin-mediated endocytosis (Coller et al. 2009).

2 Translation and Polyprotein Processing

Upon fusion between the viral envelope and the endosomal membrane, the positive-strand RNA genome is released into the cytosol, where it is translated to produce viral proteins. HCV contains an internal ribosome entry site (IRES) element in the 5′-nontranslated region (5′-NTR) of the viral RNA that recruits the cellular translation machinery to initiate translation (Tsukiyama-Kohara et al. 1992). The HCV IRES directly binds the ribosome 40S subunit (Kieft et al. 2001; Lytle et al. 2001) and induces a profound conformational change in 40S as determined by cryo-EM (Spahn et al. 2001). Subsequent binding of eIF3 and the eIF2-GTP-met-tRNA ternary complex (Otto and Puglisi 2004; Ji et al. 2004), followed by recruitment of the 60S subunit yields the complete 80S ribosome positioned at the initiator AUG codon (Fraser et al. 2009). An intermolecular interaction between the HCV IRES and 18S ribosomal RNA consisting of a 3-nucleotide base pairing between the two RNAs is crucial for HCV translation (Hashem et al. 2013). A single loop in domain II of the HCV IRES contacts the 40S-decoding groove and controls the switch from translational initiation to elongation (Filbin et al. 2013).

The HCV 5′NTR also recruits several host RNA-binding proteins that stimulate viral translation such as the La protein (Ali and Siddiqui 1997; Pudi et al. 2004), heterogeneous nuclear RNA interacting protein (hnRNP) Q (Kim et al. 2004), hnRNP D (Paek et al. 2008), hnRNP L (Hahm et al. 1998), polypyrimidine tract-binding protein (PTB) (Gosert et al. 2000), poly(C)-binding protein 2 (PCBP2) (Wang et al. 2011), and insulin-like growth factor 2 mRNA binding protein 1 (IMP-1) (Weinlich et al. 2009).

Many of the cellular proteins that bind the 5′NTR also bind the 3′-NTR segment of the viral RNA, potentially facilitating circularization of the genome and efficient virus replication. These proteins include La (Domitrovich et al. 2005), PTB (Tsuchihara et al. 1997), PCBP2 (Wang et al. 2011) and IMP-1 (Weinlich et al. 2009). Members of the NF90/NFAR protein complex (Isken et al. 2003, 2007) may also regulate genome circularization and the integration of translation and replication. NF90/NFAR proteins bind to both termini of the HCV RNA genome, thus mediating genome circularization. Furthermore, NF90 is recruited to detergent-resistant membranes where HCV replication complexes reside, and associates with NS5A in an RNA-dependent manner (Li et al. 2014b).

Some of these IRES-binding cellular proteins can also interact with each other, thus forming higher order complexes that may help bring together the ends of the viral RNA genome as well as facilitate interaction of the IRES with the ribosomal 40S subunit. As such, hnRNP D interacts with hnRNP L (Kim et al. 2000), hnRNP Q and IMP-1. Furthermore, hnRNP Q interacts with the purified 40S ribosomal subunit, and is thought to facilitate 80S complex formation at the HCV initiation codon (Park et al. 2011).

A key host factor of HCV replication is the liver-specific microRNA 122 (miR-122). miR-122 binds to two closely spaced sites on the 5′ NTR of the HCV genome and facilitates viral replication (Jopling et al. 2005, 2008; Randall et al. 2007) by more than one mode of action. First, miR-122 enhances HCV translational (Henke et al. 2008; Jangra et al. 2010; Roberts et al. 2011). Second, miR-122 binding to the HCV genome has a protective and stabilizing role. miR-122 binds HCV RNA with 3′ overhanging nucleotides that seem to mask the 5′ terminal sequences from nucleolytic degradation or innate immunity cytoplasmic sensors of viral RNA (Machlin et al. 2011). Recent data indicate that miR-122 binding protects the RNA genome from the 5′ to 3′ exoribonuclease Xrn2 (Sedano and Sarnow 2014). Argonaute (Ago) proteins seem to be involved in mediating some of the miR-122 effects on HCV RNA. Knockdown of Ago expression in cells containing miR-122 decreases HCV translation (Roberts et al. 2011; Wilson et al. 2011). Furthermore, miR-122 binds HCV RNA in association with Ago2 protein resulting in slower decay of the viral genome and protection from the cellular exonuclease decay machinery (Shimakami et al. 2012). In support of the stability hypothesis, HCV deleted in part of the miR-122 binding sites can be partially rescued by recombination with stable viral or cellular RNA structures (Li et al. 2011). The ectopic expression of miR-122 in a variety of non-permissive cell types facilitates HCV replication (Chang et al. 2008; Narbus et al. 2011; Long et al. 2011; Kambara et al. 2012; Fukuhara et al. 2012).

The ~3,000 amino acid long polyprotein precursor is co- and post-translationally processed by both host and viral proteases to yield the structural proteins core, E1 and E2 and the nonstructural proteins p7, NS2, NS3, NS4A, NS4B, NS5A and NS5B proteins. Core, E1 and E2 are processed by host cell ER-resident signal peptidase, while cleavage at the NS2/NS3 junction site is mediated by the NS2-NS3 zinc-dependent metalloprotease and the remaining nonstructural proteins are hydrolyzed by the NS3/4A proteinase (Grakoui et al. 1993a, b; Bartenschlager et al. 1994; Penin et al. 2004). Following cleavage by host cellular signal peptidase, immature core protein is additionally cleaved by signal peptide peptidase inside the ER membrane to yield the mature form of the protein (McLauchlan et al. 2002).

3 Immune Recognition/Evasion

To sense invading viruses, host cells rely on membrane bound and cytoplasmic receptors, such as toll-like receptor family (TLRs), the RIG-I-like receptors (RLRs) and the viral DNA sensors. TLR3, present in late endosomal and/or lysosomal membranes has been implicated in sensing HCV double-stranded (ds) RNA, and mutations in TLR3 that disrupt the dsRNA binding capability abolish the antiviral effect of TLR3 (Li et al. 2012; Wang et al. 2009). RLRs, including RIG-I and MDA5, are primary sensors of HCV replication. RIG-I is the best-characterized

HCV sensor (Sumpter et al. 2005; Saito et al. 2008; Eksioglu et al. 2011). It recognizes a free 5′- triphosphate of HCV RNA, in addition to a 34-nucleotide polyuridine core within the poly U/UC region of the 3′-NTR to stimulate the interferon response to infection (Saito et al. 2008; Uzri and Gehrke 2009; Schnell et al. 2012). Recent studies indicate that MDA5 also has antiviral activity against HCV infection (Israelow et al. 2014). Both RIG-I and MDA5 were identified in a large scale screen for antiviral effectors restricting genotype 2a HCV replication in Huh-7.5 cells (Schoggins et al. 2011). Furthermore, polymorphisms in MDA5 correlate with clearance of viral infection, suggesting that MDA5 is involved in the defense against HCV (Hoffmann et al. 2015).

In addition to cleaving viral peptide sequences to process its polyprotein, NS3 has evolved to target cellular substrates, particularly those involved in innate antiviral signaling. As such, three essential components of this response, mitochondrial antiviral-signaling protein (MAVS), TIR-domain-containing adaptor-inducing IFN-b (TRIF), and the Riplet ubiquitin ligase are proteolytically inactivated by NS3 (Li et al. 2005; Meylan et al. 2005; Oshiumi et al. 2013). MAVS is the downstream adaptor of RIG-I and MDA-5 and as such, NS3 cleavage of MAVS prevents RIG-I and MDA5 innate immune signaling. Similarly, Riplet regulates RIG-I activity, while Trif is required for TLR3 signaling. Furthermore, NS3 via its C-terminal helicase domain can directly interact with TANK-binding kinase 1 (TBK1), thus inhibiting the association of TBK1 with IRF-3 and subsequent activation of IRF-3 (Otsuka et al. 2005). Thus, NS3 prevents innate immune signaling at multiple steps.

Other HCV proteins, such as core and NS5A, may modulate cellular antiviral signaling by interfering with the JAK-STAT pathway (Lin et al. 2006; Lan et al. 2007). The core protein interferes with STAT1 nuclear translocation as well as inhibits IFN-alpha-induced transcription of antiviral genes by decreasing binding of ISGF3 to the ISRE (Heim et al. 1999; Hosui et al. 2003; Miller et al. 2004; de Lucas et al. 2005). NS5A blocks the IFN-mediated induction of several genes (Geiss et al. 2003) and this could be due in part to the NS5A ability to bind and inhibit STAT1 phosphorylation (Lan et al. 2007; Kumthip et al. 2012). NS5A also interacts with 2′, 5′ oligoadynylate synthase (2′,5′OAS) and interferes with its antiviral activity (Taguchi et al. 2004). Furthermore, NS5A has been reported to interact with and repress protein kinase R (PKR), a key component of the host innate IFN response, thus avoiding the antiviral effects of IFN (Gale et al. 1997, 1998, 1999).

Despite this antagonism of innate immune sensing pathways, HCV infection of animal models strongly induces ISG mRNA expression (Su et al. 2002; Sheahan et al. 2014). It is possible that HCV is able to persist in the liver by promoting the phosphorylation of the RNA-dependent PKR protein kinase resulting in suppression of ISG protein expression (Garaigorta and Chisari 2009). It has further been suggested that PKR can sense HCV and induce IFN production independently of RIG-I (Arnaud et al. 2010).

4 Replication

4.1 Replication Compartment Formation

HCV infection induces cellular membrane alterations designated as the membranous web (MW), to house viral RNA replication (Gosert et al. 2003). These membranes consist of an accumulation of vesicles derived from the ER, are detergent resistant, and have features of lipid rafts (Shi et al. 2003). Such modified membrane structures, also called replication factories, are a hallmark of positive strand RNA virus infections (den Boon and Ahlquist 2010). They are thought to play an important role in providing a protective environment for RNA replication, shielding viral RNA from cytosolic nucleases and innate immune sensors (Heaton and Randall 2011). Initial observations suggested a primary role for NS4B in creating the necessary membrane rearrangements for replication compartment formation (Egger et al. 2002). However more recently, a role for NS5A has been appreciated in establishing these viral replication compartments in conjunction with NS4B (Romero-Brey et al. 2012).

Multiple host factors post-translationally modify NS5A and in turn, modulate HCV replication and assembly. In infected cells, NS5A exists both as a basally- (p56) and a hyper-phosphorylated (p58) form (Kaneko et al. 1994) and a critical ratio between the p56 and p58 phospho-forms of NS5A has to be maintained for productive HCV RNA replication (Neddermann et al. 2004). Several cellular kinases belonging to the CMCG group of serine-threonine kinases that includes casein kinases, cyclin-dependent kinases (CDKs), MAPKs, as well as MEK1, MEK6, MEK7, AKT and p70S6K, Plk1 have been identified to phosphorylate NS5A on serine and to a lesser extent threonine residues (Reed et al. 1997, 1998; Coito et al. 2004; Chen et al. 2010). It was initially shown that phosphorylation of NS5A by casein kinase I-α (CKI-α) impacts HCV replication (Quintavalle et al. 2006, 2007) while more recent data indicate that CKI-α-mediated hyperphosphorylation of NS5A facilitates proper localization of NS5A around LDs and virus assembly (Masaki et al. 2014). Several studies have implicated casein kinase II (CKII) as a direct NS5A kinase (Kim et al. 1999; Reed et al. 1997), and CKII-dependent phosphorylation of a single serine residue in domain III of NS5A C-terminus controls the switch between virus replication and assembly (Tellinghuisen et al. 2008).

Another post-translational modification of NS5A required for virus replication is the prolyl cis-trans isomerization mediated by cyclophilin A (CypA). CypA is a host protein with peptidyl-prolyl cis-trans isomerase activity that is ubiquitously expressed in all tissues and it is inhibited by cyclosporine A (CsA). It was first discovered that CsA drastically suppresses HCV replication (Watashi et al. 2003) and follow up studies identified CypA as a critical player in HCV replication (Yang et al. 2008a; Kaul et al. 2009; von Hahn et al. 2012). The enzymatic activity of this cellular chaperone is crucial for HCV (Chatterji et al. 2009) and genetic and biochemical studies indicate that the main target is the viral NS5A protein

(Hanoulle et al. 2009; Yang et al. 2010; Verdegem et al. 2011; Foster et al. 2011; Grise et al. 2012). CypA binds to specific proline rich motifs in domains II and III of NS5A, inducing cis-trans isomerization and proper folding of the protein (Coelmont et al. 2010; Verdegem et al. 2011; Foster et al. 2011; Grise et al. 2012). The CypA-NS5A interaction is thought to stimulate efficient binding of NS5A to NS5B and RNA (Liu et al. 2009; Foster et al. 2011), and it is crucial for the ability of NS5A to establish viral replication compartments within cells (Madan et al. 2014).

HCV replication is closely linked to host hepatocyte lipid metabolic pathways. Several siRNA screens have identified the lipid kinase phosphatidylinositol 4-kinase III α (PI4K-IIIα) as an essential host factor in HCV viral RNA replication (Berger et al. 2009; Li et al. 2009; Tai et al. 2009; Vaillancourt et al. 2009; Borawski et al. 2009; Trotard et al. 2009). PI4K-IIIα is an endoplasmic reticulum (ER) resident enzyme that phosphorylates phosphatidylinositol (PI) in the 4 position of the inositol ring to generate PI(4)P. Silencing PIK-IIIα results in an altered MW morphology and clustering/aggregation of HCV replicase components (Berger et al. 2011; Reiss et al. 2011; Tai and Salloum 2011). PI4K-IIIα interacts with NS5A and this interaction is crucial for maintaining the integrity of the membranous replication compartment (Reiss et al. 2011; Berger et al. 2011; Tai and Salloum 2011; Ahn et al. 2004). The interaction site for PI4K-IIIα on NS5A has been mapped to seven amino acids in domain I of NS5A and mutations in that region result in a hyperphosphorylation of NS5A (Reiss et al. 2013). The enzymatic function of the lipid kinase is required for supporting HCV replication (Harak et al. 2014; Reiss et al. 2011; Berger et al. 2011).

HCV infection leads to a hyper-accumulation of PI(4)P in the cytosol, with enrichment at the MW and a depletion at the plasma membrane (Reiss et al. 2011; Berger et al. 2011; Tai and Salloum 2011). PI(4)P mainly serves as a precursor for the synthesis of PI(4,5)P2, however it can also determine intracellular membrane identity and several proteins that bind PI(4)P have been identified. These PI(4)P-binding proteins are frequently lipid transfer proteins, which can alter the local lipid composition of membranes. Two PI(4)P binding proteins have been identified to play a role in HCV replication. Four-phosphate adaptor protein 2 (FAPP2) is recruited to viral replication complexes where it might serve as a supplier of glycosphingolipids (Khan et al. 2014). Another PI(4)P binding protein, oxysterol-binding protein (OSBP), interacts with NS5A and its silencing leads to inhibition of both HCV replication and particle secretion (Wang et al. 2014; Amako et al. 2009, 2011). A recent study indicates that OSBP is recruited to HCV replication compartments in a PI(4)P-dependent manner and together with PI4K-IIIα mediate transport of cholesterol to the MW (Wang et al. 2014). Indeed, highly purified double membrane vesicles (DMVs) are greatly enriched in cholesterol as revealed by immune-gold labeling studies (Paul et al. 2013). PI(4)P might also recruit ceramide transfer protein (CERT) to the replication complex thus mediating ceramide influx to the replication sites.

OSBP and CERT also contain FFAT motifs that allow them to interact with the human vesicle-associated membrane proteins (VAPs) residing in the ER (Amako et al. 2009). VAP-A and its isoform VAP-B are both important host factors that interact with NS5A and NS5B and thought to assist in the assembly of the replicase complex on detergent-resistant membranes (Evans et al. 2004; Gao et al. 2004; Hamamoto et al. 2005; Tu et al. 1999). Notably, association of VAP-A with NS5A is dependent on the phosphorylation state of NS5A (Evans et al. 2004), which also may control the switch between virus replication and assembly (Tellinghuisen et al. 2008). Hypophosphorylated NS5A binds VAP-A and contributes to viral RNA replication, while the hyperphosphorylated form does not bind VAP-A and contributes to nucleocapsid assembly. One model is that hyperphosphorylated NS5A that cannot bind VAP-A is released from replication complexes and shuttled to sites of assembly. The antiviral protein viperin inhibits HCV replication by disrupting the interaction between viral NS5A and VAP-A (Helbig et al. 2011).

Annexin A2 (ANXA2) is recruited to HCV replication sites and associates with HCV proteins in lipid raft enriched microdomains (Backes et al. 2010; Saxena et al. 2012). ANXA2 is known to bind to negatively charged phospholipids, with a preference for phosphatidylinositol (4,5) bisphosphate (Rescher et al. 2004), suggesting that PI4P enriched in sites of viral replication might be important in recruiting ANXA2. The role of ANXA2 in HCV replication is unclear; however, its depletion has a strong effect on virus assembly (Backes et al. 2010).

HCV replication is also sensitive to inhibition of geranylgeranylation, a lipid modification that allows proteins to associate with membranes, suggesting that geranylgeranylated host proteins are involved in HCV replication (Ye et al. 2003). One such geranylgeranylated protein, F-box/LRR-repeat protein 2 (FBL2), was found to interact with NS5A and promote HCV replication (Wang et al. 2005; Yang et al. 2008a). It is thought that geranylgeranylation targets FBL2 to membranes where NS5A resides thus promoting protein-protein interactions.

Other cellular compartments, including endosomes and autophagosomes have been reported to localize in the proximity of replication compartments. Early endosomal proteins, including Rab5, were associated with replication sites and the viral NS4B protein (Stone et al. 2007). The role of Rab5 in replication compartment formation is not clear, but interestingly enteroviruses, which also remodel host cell cytoplasm for their replication, exploit the cellular endocytic machinery for supplying cholesterol to sites of RNA replication (Ilnytska et al. 2013). Other Rab complexes have been implicated in HCV replication, including Rab7 and Rab1-TBC1D20 (Manna et al. 2010; Sklan et al. 2007a, b; Nevo-Yassaf et al. 2012). Nuclear pore complex proteins have also been found to accumulate in the cytoplasm of infected cells and to be recruited at sites of viral replication and assembly where they interact with the viral NS5A and core proteins and may regulate trafficking into and from the membranous web (Neufeldt et al. 2013).

4.2 RNA Synthesis

The nonstructural proteins NS3 to NS5B are the minimal viral protein components required for RNA replication (Lohmann et al. 1999; Blight et al. 2000). The NS5B protein is the HCV RNA-dependent RNA polymerase and constitutes the catalytic core of the HCV replication machinery (Behrens et al. 1996). The NS5B C-terminal tail is anchored on the membrane. Purified NS5B is not very efficient at completing the RNA replication cycle by itself in vitro, suggesting that viral and maybe cellular cofactors are required for HCV RNA synthesis in vivo (Lohmann et al. 1997). Several host factors have been proposed to interact with NS5B and impact viral RNA replication.

NS5B can be phosphorylated by the interacting cellular protein kinase C-related kinase 2 (PRK2), and depletion of PRK2 inhibits HCV replication suggesting that phosphorylation is important for NS5B function (Kim et al. 2004). The cellular chaperonin TRiC/CCT also interacts with NS5B and its depletion suppresses HCV RNA replication, however the precise role of the chaperone is not clear yet (Inoue et al. 2011). In a novel in vitro rolling circle RNA replication system, the human replication protein A (RPA) together with NS5A help recruit NS5B to the template RNA and mediate processive NS5B-dependent HCV RNA replication (Mani et al. 2015).

5 Modulation of Cellular Metabolism

Persistent HCV infection greatly interferes with hepatic lipid metabolism, often leading to an increased fat content in the liver, otherwise known as steatosis (Rubbia-Brandt et al. 2000). Steatosis is thought to be triggered, at least in part, by HCV core, since core transgenic mice develop steatosis (Barba et al. 1997). Additionally, the expression of several cellular genes involved in intracellular lipid biogenesis, degradation and transport is significantly altered during infection (Blackham et al. 2010; Diamond et al. 2010; Woodhouse et al. 2010). The major transcriptional factors regulating lipogenic gene expression are sterol regulatory element binding proteins (SREBPs) (Horton et al. 2002). Inhibiting SREBPs cleavage and activation with 25-hydroxycholesterol leads to a decrease in lipid biosynthesis and a block in HCV replication (Su et al. 2002). HCV infection as well as individual expression of core, NS2 or NS4B induces the expression and post-translational activation of SREBPs (Waris et al. 2007; Oem et al. 2008; Park et al. 2009). One mechanism by which HCV induces transcription of SREBPs and subsequent lipogenic gene expression involves interaction of the viral 3′NTR with cellular DEAD box polypeptide 3, X-linked (DDX3X) to activate IKK-α, which translocates to the nucleus and activates transcription (Li et al. 2013).

One of the genes upregulated during HCV infection encodes fatty acid synthase (FASN), the multidomain enzyme catalyzing the de novo synthesis of fatty acids.

FASN associates with lipid raft microdomains indicative of HCV replication sites and interacts with the viral RNA-dependent RNA-polymerase NS5B as determined by biochemical and microscopy assays (Huang et al. 2013). The increase in FASN abundance and activity during HCV infection might be sufficient to allow virus replication without the need for recruiting the enzyme to sites of HCV replication (Nasheri et al. 2013). Inhibition of FASN results in decreased HCV replication and virus production. FASN was found to increase NS5B activity in vitro suggesting that in vivo it might impact HCV replication by stimulating NS5B activity. In addition to the replication defect, inhibition of FASN interferes with expression of CLDN1 on the cell surface and subsequent virus entry (Yang et al. 2008b).

Chronic HCV infection is also known to elicit ER and oxidative stress in infected cells, which may promote apoptosis and subsequent liver injury (Okuda et al. 2002; Benali-Furet et al. 2005; Joyce et al. 2009). ER stress activates a signaling network known as the unfolded protein response (UPR) which consists of three separate pathways that are named after the regulating protein: the activating transcription factor 6 (ATF6), the inositol-requiring enzyme 1 (IRE1) and the double-stranded RNA-activated protein kinase-like ER kinase (PERK), respectively. Oxidative stress and disruption of mitochondrial function play a role in chronic HCV-related liver injury. Accumulation of lipid peroxides can be easily detected in chronically infected patient samples (Paradis et al. 1997) and both core and NS5A proteins have been implicated in the induction of HCV-related oxidative stress. NS5A protein was found to alter intracellular calcium levels and elevate the levels of reactive oxygen species (ROS) (Gong et al. 2001) and expression of HCV core results in increased ROS and lipid peroxidation products both in transgenic mice (Moriya et al. 2001) as well as in hepatoma cell lines (Okuda et al. 2002). Intriguing recent findings indicate that the lipid peroxides induced during infection inhibit the NS5B replicase suggesting that HCV could be limiting its own replication to promote persistence in the liver (Yamane et al. 2014).

Induction of stress granule (SG) formation is another known host cell response to HCV infection (Ariumi et al. 2011a; Jones et al. 2010). SGs are cytoplasmic structures induced in response to cellular stress, including viral infections, that contain stalled translation complexes as well as RNA binding proteins (Buchan and Parker 2009). The SGs induced during HCV infection accumulate T-cell-restricted intracellular antigen 1 (TIA-1), TIA1-related protein (TIAR) and RasGAP-SH3 domain binding protein 1 (G3BP1) proteins, which are required for efficient HCV RNA and particle production (Garaigorta et al. 2012). HCV recruits the SG-protein G3BP1 as well as the processing body component DDX6 around LDs to facilitate virus assembly (Ariumi et al. 2011a). During chronic infection SGs oscillate between cycles of assembly (translational arrest) and disassembly (active translation), and this dynamics is dramatically increased with the addition of IFNα (Ruggieri et al. 2012).

6 Autophagy

Autophagy is a cellular process that promotes the turnover of damaged organelles and aggregated proteins via lysosomal degradation (Xie and Klionsky 2007). Some viruses and other pathogens can manipulate this host pathway to benefit their own replication cycle (Jordan and Randall 2012). In the case of HCV, initial studies reported that a proviral autophagy is induced in infected cells in culture (Ait-Goughoulte et al. 2008; Dreux et al. 2009b; Sir et al. 2008) as well as in infected patient liver samples (Rautou et al. 2011; Vescovo et al. 2012). Although the studies agree on a proviral role for autophagy in HCV infection, there has not been a consensus on the role of autophagy in HCV replication and whether HCV inhibits the progression of autophagy to autolysosomal degradation. This may reflect differing experimental conditions, or alternatively, distinct autophagies with differing kinetics may play multiple roles in HCV infection. Multiple studies indicate an early role for autophagy in establishing HCV replication (Dreux et al. 2009b; Guevin et al. 2010). Indeed, double membrane vesicles (DMVs) and multi-membrane vesicles (MMVs) induced by HCV accumulate at the MW and morphologically resemble autophagosomes (Ferraris et al. 2010). However, cryo-EM tomography indicates that the DMVs are contiguous with the ER and that autophagosome-like structures associate with the MW with a delayed kinetics relative to HCV replication (Paul et al. 2013). The viral RNA-dependent RNA polymerase NS5B was also found to interact with the autophagy related protein ATG5 early in infection (Guevin et al. 2010), suggesting that the interaction between HCV and autophagy might be a transient one and only required for early stages of the virus life cycle (Dreux et al. 2009b; Guevin et al. 2010). However, other studies suggest that the autophagic machinery might also be required for the production of HCV particles (Tanida et al. 2009).

Autophagy may be involved in repressing the anti-HCV innate immune response (Ke and Chen 2011; Shrivastava et al. 2011; Chandra et al. 2014). Silencing of the autophagy related genes Beclin or ATG7 activates IFN and ISG expression in HCV infected cells (Shrivastava et al. 2011). Autophagy may also regulate lipid metabolism in HCV infected cells, since there is an inverse correlation between activation of autophagy and steatosis in liver biopsies of patients chronically infected with HCV (Vescovo et al. 2012).

While it is true that HCV infection alters cellular metabolic pathways, including mitochondrial functions, there are conflicting results in the literature on whether selective mitochondrial autophagy (mitophagy) is induced or inhibited during HCV infection. Initial reports indicated that HCV infection promotes the translocation of the E3 ubiquitin ligase, Parkin, to mitochondria and subsequent induction of mitophagy (Kim et al. 2013). Furthermore, HCV infection upregulated expression and phosphorylation of dynamin-related protein 1 (Drp1) leading to mitochondrial fission and mitophagy that could contribute to persistent infection (Kim et al. 2014). However, another study suggests that HCV infection might be interfering with mitophagy via the core protein interacting with Parkin and disrupting its translocation to mitochondria, thus suppressing ubiqutination of mitochondria and subsequent mitophagy (Hara et al. 2014).

7 Assembly

HCV assembly initiates with the core (capsid) protein accumulating on the surface of cytosolic lipid droplets (LDs), which are cellular lipid storage organelles located in close proximity to virus replication sites (Barba et al. 1997; Boulant et al. 2006; Miyanari et al. 2007). Besides serving as a platform for core and other nonstructural proteins to coalesce, the precise role of LDs in virus assembly is still being elucidated. Disrupting the pathways that synthesize the major lipid species (triacylglycerides and cholesterol esters) of LDs reduces production of infectious HCV (Liefhebber et al. 2014). Trafficking of core proteins to LDs involves the concerted action of host and viral factors. First, proteolysis at the core C-terminus by the intramembrane protease signal peptide peptidase is required for core maturation and its targeting to LDs (Targett-Adams et al. 2008). The cytosolic phospholipase A2 (PLA2G4A) was recently identified as an important host factor of HCV assembly and its inhibition specifically reduces core protein abundance on LDs as well as downstream envelopment (Menzel et al. 2012). Another host factor that regulates core trafficking is diacylglycerol *O*-acetyltransferase 1 (DGAT1) (Herker et al. 2010). DGAT1 and DGAT2 are cellular enzymes that catalyze the last step in triglyceride biosynthesis and are required for LD biogenesis (Yen et al. 2008). Notably, inhibiting DGAT1 but not DGAT2 activity via siRNA knockdown or by using a small molecule inhibitor disrupts core trafficking and virion production (Herker et al. 2010). In addition to binding and recruiting core to LDs, DGAT1 was recently found to interact with NS5A and facilitate the interaction between core and NS5A on the surface of LDs (Camus et al. 2013).

Several studies suggest that the NS5A protein might provide the link between replication and assembly sites. Rab18, a LD-associated protein, was found to interact with NS5A and promote the juxtaposition of replication and assembly sites (Salloum et al. 2013). TIP47, another LD-associated protein, interacts with NS5A and might also serve to link sites of replication and assembly (Vogt et al. 2013). NS2 is a master regulator of viral assembly bringing together the structural and nonstructural proteins required for particle formation. The cellular signal peptidase complex subunit 1 (SPCS1) interacts with both NS2 and the E2 glycoprotein and controls HCV assembly by mediating membrane association of the NS2-E2 complex (Suzuki et al. 2013).

Following capsid assembly on the surface of LDs, virions bud into the lumen of the ER where the glycoproteins E1/E2 reside, in addition to the VLDL secretion machinery. The clathrin assembly protein complex 2 medium chain (AP2M1) has been implicated in trafficking nucleocapsids to sites of envelopment via an interaction with a conserved YXXϕ (Neveu et al. 2012). HCV assembly and envelopment is linked to the hepatocyte VLDL synthesis pathway (Huang et al. 2007;

Gastaminza et al. 2008), and components of this pathway such as microsomal triglyceride transfer protein (MTTP) (Huang et al. 2007), apolipoprotein B (apoB) (Icard et al. 2009) and especially apolipoprotein E (apoE) (Chang et al. 2007; Benga et al. 2010; Hishiki et al. 2010) have been implicated in HCV assembly and infectivity. HCV particles circulate in patient serum as lipoviroparticles (Andre et al. 2002) and have very low buoyant densities due to their interaction with serum lipoproteins (Hijikata et al. 1993).

ApoE is well established as an HCV-specific infectivity factor (Chang et al. 2007). Notably, enforced expression of apoE in cells lacking this protein makes them capable of producing infectious HCV (Long et al. 2011; Hueging et al. 2014). Affinity purified virus particles contain apoE, in addition to apo A and B, as determined by immune-electron microscopy (Merz et al. 2011; Catanese et al. 2013b). Depletion of apoE reduces titers of both intra- and extra-cellular infectious virus, and recent data indicate that apoE is required for maturation of viral particles in a post-envelopment step of the assembly process (Hueging et al. 2014; Lee et al. 2014). NS5A was found to interact with apoE via its C-terminal alpha-helical domain and this interaction was deemed necessary for HCV assembly (Benga et al. 2010; Cun et al. 2010), thus suggesting that NS5A was responsible for recruiting apoE to sites of assembly. However, a direct NS5A-apoE interaction is hard to envision given their respective membrane topologies, with NS5A being on the ER cytoplasmic side and apoE in the ER lumenal side. Recently, it was shown that apoE interacts with the viral envelope glycoproteins, more specifically with the transmembrane domain of E2, facilitating association with newly enveloped particles (Boyer et al. 2014; Lee et al. 2014). Assembled virus particles traffic through the secretory pathway in association with apoE (Coller et al. 2012) and apoE present on virus particles might also aid the attachment of virions during cell entry by virtue of its binding to LDLR and heparan sulfates (Owen et al. 2009; Jiang et al. 2012).

8 Egress

Following envelopment and apoE association in the ER, HCV is infectious (Gastaminza et al. 2008; Coller et al. 2012). HCV particles then traffic to Golgi likely within COPII secretory vesicles, given the dependence of virion release on SAR1A (Coller et al. 2012). As is true for many enveloped viruses, the secretion of infectious HCV particles depends in part on components of the endosomal-sorting complex required for transport (ESCRT) pathway (Corless et al. 2010; Ariumi et al. 2011b; Tamai et al. 2012). This pathway could be required for envelopment and/or budding of virions into an intermediate secretory compartment. An RNAi screen in conjunction with live cell imaging experiments of HCV core trafficking has shown that nascent viral particles are released from the *trans*-Golgi network in a protein kinase D1-dependent manner and traffic to the recycling endosomes and plasma membrane in association with Vamp1 vesicles, followed by extracellular release (Coller et al. 2012).

The restricted spread of HCV RNA in hepatocytes in vivo suggests that HCV might also be propagating via cell-to-cell transmission (Liang et al. 2009; Kandathil et al. 2013), which allows for efficient virion delivery to neighboring cells as well as avoidance of antibody-mediated neutralization. The host requirements for HCV cell-to-cell transmission are still being elucidated. Recent data suggest that SR-BI plays a critical role in the cell-to-cell spread of HCV (Brimacombe et al. 2011; Catanese et al. 2013a), while CD81 might be dispensable for this route of transmission (Jones et al. 2010; Witteveldt et al. 2009). The tight junction entry factors CLDN and OCLN are also required for HCV cell-cell spread (Brimacombe et al. 2011). The role of ApoE in cell-to-cell transmission is somewhat unclear. While its ectopic expression in non-hepatic cells appears to promote cell-cell spread (Hueging et al. 2014), interference with VLDL secretion in infected Huh7 cell derivatives inhibited cell-free HCV infection, but not cell-cell spread of HCV (Barretto et al. 2014).

Transmission of HCV RNA and possibly virions between hepatocytes can also be mediated by exosomes, which are small membrane vesicles normally used for intercellular communication. Exosomes derived from HCV infected cells contain HCV RNA and can activate innate immune cells (Dreux et al. 2012). Furthermore, exosomes containing positive and negative sense viral RNAs, in addition to Ago2 and miR-122, could also be isolated from sera of chronic HCV infected patients and they were capable of mediating receptor-independent viral transmission to hepatocytes in culture (Bukong et al. 2014). Purified exosomes derived from HCV infected hepatoma cells have also been reported to contain viral RNA, proteins and fully assembled particles capable of infecting naïve cells (Ramakrishnaiah et al. 2013). The importance of exosome-mediated HCV spread is unclear. It may facilitate receptor-independent HCV spread that is resistant to neutralizing antibodies. The relative abundance of exosome-dependent HCV spread versus more canonical HCV release pathways is not well characterized, although it is likely to be quite small given the ability of anti-HCV antibodies to efficiently block HCV entry and the HCV receptor requirements in cell culture and in vivo.

9 Genome Wide Screens

To identify host factors associated with productive virus infection, many groups have successfully utilized siRNA screens (Randall et al. 2007; Ng et al. 2007; Supekova et al. 2008; Berger et al. 2009; Li et al. 2009, 2014a; Tai et al. 2009; Borawski et al. 2009; Trotard et al. 2009; Vaillancourt et al. 2009; Reiss et al. 2011). The functional significance of most of these host factors in HCV infection is unknown and there is frequently minimal overlap between the genes identified in these studies. The host genes that have been validated with some mechanistic analysis have been discussed in this chapter. Other genes with a minimal validation, such as a phenotype in at least two distinct HCV replication assays, are listed in Table 1.

Table 1 Pro-viral host genes identified in siRNA screens and validated by one phenotypic assay

	Host factor	Function	References		Host factor	Function	References
ENTRY	AP2M1	AP2 adaptor	Coller et al. (2009)	REPLICATION	ABCB10	ATPase function	Vaillancourt et al. (2009)
	ATP6VOA1	Endosome acidification	Coller et al. (2009)		ACTN1	Actin remodeling	Randall et al. (2007)
	CBL	Ubiquitin ligase	Coller et al. (2009)		AEBP1	Transcriptional modulator	Vaillancourt et al. (2009)
	CD81	Entry receptor	Randall et al. (2007)		CACNB2	Calcium channel	Vaillancourt et al. (2009) and Borawski et al. (2009)
	CDC42	Actin polymerization	Tai et al. (2009) and Berger et al. (2009)		CEBPD	Transcriptional modulator	Li et al. (2009)
	CHKA	Phosphatidylcholine biosynthesis	Li et al. (2014a)		CHUK	Kinase inhibitor of NFkB	Li et al. (2009)
	CFL1	Actin polymerization	Coller et al. (2009)		CKAP5	Cytoskeletal protein	Tai et al. (2009)
	CLTB	Clathrin coat	Coller et al. (2009)		CMPK1	Nucleic acid biosynthesis	Borawski et al. (2009)
	CLTC	Clathrin coat	Trotard et al. (2009)		COPA	ER/Golgi transport	Tai et al. (2009)
	CLTCL1	Clathrin coat	Coller et al. (2009)		COPB1	ER/Golgi transport	Tai et al. (2009)
	EEA1	Early endosome	Berger et al. (2009)		COPB2	ER/Golgi transport	Borawski et al. (2009) and Tai et al. (2009)
	EPN1	Clathrin/actin	Coller et al. (2009)		CSK	Protein tyrosine kinase	Supekova et al. (2008)
	EPN3	Clathrin/actin	Coller et al. (2009)		CTGF	Tissue growth factor	Li et al. (2009)
	HGS	Receptor sorting	Coller et al. (2009)		CTSF	Lysosomal protease	Li et al. (2009) and Tai et al. (2009)
	HIP1	Clathrin coat	Coller et al. (2009)		CTSL1	Lysosomal protease	Borawski et al. (2009)
	HIP1R	Clathrin coat	Coller et al. (2009)		DDX3X	RNA helicase	Randall et al. (2007) and Li et al. (2009)
	MAP4	Microtubule associated	Li et al. (2014a)		DICER1	miRNA processing	Randall et al. (2007), Li et al. (2009)
	PI4K2A	PtdInsPs biosynthesis	Lupberger et al. (2011)		DHX30	RNA helicase	Vaillancourt et al. (2009)
	RAB5A	Early endosome	Coller et al. (2009)		DNAJC16	hsp40 homolog	Borawski et al. (2009)
	RAB7A	Late endosome	Tai et al. (2009)		EIF2AK2	Translation	Randall et al. (2007)
	RAB7L1	Late endosome	Coller et al. (2009)		ETF1	Translation termination	Li et al. (2009)
	RAB34	Lysosome to Golgi trafficking	Li et al. (2014a)		EWSR1	RNA binding protein	Li et al. (2009)
	RAC1	Ras GTPase	Li et al. (2014a)		FAU	Translation	Li et al. (2009) and Borawski et al. (2009)
	RAF1	Kinase in the MAPK cascade	Randall et al. (2007) and Li et al. (2009)		FBLN5	Cell adhesion	Li et al. (2009)
	ROCK2	Actin polymerization			GAPDH	Metabolism	Randall et al. (2007)
	SMAD6	Transcriptional modulator	Li et al. (2014a)		GPD2	Metabolism	Vaillancourt et al. (2009)
	STAU	ER/Golgi trafficking	Coller et al. (2009)		GRB2	EGFR signal transduction	Randall et al. (2007)
	SYT1	AP2/clathrin recruiting	Coller et al. (2009)		GUCA1A	CA dependent signaling	Borawski et al. (2009)

(continued)

Table 1 (continued)

ASSEMBLY/SECRETION	AP1M1	Clathrin coat associated	Collet et al. (2012)
	ARF3	Golgi trafficking	Collet et al. (2012)
	CLINT1	Golgi sorting	Collet et al. (2012)
	CYTH3	Golgi function	Collet et al. (2012)
	DDOST	Protein glycosylation	Li et al. (2009)
	DGAT1	Lipid droplet associated protein	Li et al. (2009) and Herker et al. (2010)
	ELAVL1	RNA binding protein	Randall et al. (2007)
	GIT1	Actin remodeling	Collet et al. (2012)
	RAB3D	Exocytosis	Collet et al. (2012)
	RAB11A	Exocytosis	Collet et al. (2012)
	RHOA	Actin remodeling	Collet et al. (2012)
	PACSIN3	Vesicle trafficking	Collet et al. (2012)
	PI4KB	Golgi sorting	Collet et al. (2012)
	PRKD1	Golgi sorting	Collet et al. (2012)
	SAR1A	ER/Golgi trafficking	Collet et al. (2012)
	SIK2	Serine/threonine kinase	Borawski et al. (2009)
	STAT3	Transcriptional modulator	Randall et al. (2007) and Li et al. (2009)
	VAMP1	Exocytosis	Collet et al. (2012)
	WAS	Actin remodeling	Collet et al. (2012)
	NOP2	Nucleolus	Li et al. (2009)
	NOP56	Nucleolus	Li et al. (2009)
	NOP58	Nucleolus	Li et al. (2009) and Tai et al. (2009)
	NUAK2	Serine/threonine kinase	Ng et al. (2007), Li et al. (2009), and Tai et al. (2009)
	PARP1	Chromatin remodeling	Vaillancourt et al. (2009)
	PCBP2	Nuclear RNA binding	Randall et al. (2007) and Borawski et al. (2009)
	PECR	Fatty acid biosynthesis	Li et al. (2009)
	PIK3C2G	PtdInsPs biosynthesis	Berger et al. (2009)
	PIK3CA	PtdInsPs biosynthesis	Berger et al. (2009), Li et al. (2009), Tai et al. (2009), Vaillancourt et al. (2009), Borawski et al. (2009), and Trotard et al. (2009)
	PKN3	Serine/threonine kinase	Borawski et al. (2009)
	POU3F2	Transcription factor	Tai et al. (2009)
	PPIC	Prolyl cis/trans isomerase	Borawski et al. (2009)
	PRPF8	Pre-mRNA splicing	Borawski et al. (2009)
	PTBP1	Nuclear RNA binding	Randall et al. (2007) and Xue et al. (2007)
	HADH	Metabolism	Vaillancourt et al. (2009)
	HAS1	Hyaluronan synthesis	Li et al. (2009)
	HCCS	Cytochrome c synthesis	Li et al. (2009)
	HM13	Signal peptidase	Randall et al. (2007)
	HSPA1A	Heat shock protein	Li et al. (2009)
	IPO5	Nuclear import	Randall et al. (2007)
	ITGA7	Integrin alpha 7	Li et al. (2009), Borawski et al. (2009), and Tai et al. (2009)
	JAK1	Tyrosine protein kinase	Supekova et al. (2008)
	KRT31	Keratin	Tai et al. (2009)
	KRT4	Keratin	Li et al. (2009)
	MAP3K14	Serine/threonin kinase	Borawski et al. (2009) and Tai et al. (2009)
	MAPK1	Serine/threonin kinase	Randall et al. (2007) and Li et al. (2009)
	MKK7	Serine/threonin kinase	Ng et al. (2007)
	MRPL15	Mitochondrial ribosome	Li et al. (2009)
	MRPL48	Mitochondrial ribosome	Li et al. (2009)
	MYOF	Membrane fusion	Li et al. (2009)
	NAPA	Vesicle fusion	Tai et al. (2009)
	NEFL	Neurofilaments	Borawski et al. (2009)
	NOL6	Nucleolar protein	Li et al. (2009)
	SLC12A5	**K/Cl transport**	Ng et al. (2007) and Vaillancourt et al. (2009)
	SMURF2	**E3 ubiquitin ligase**	Li et al. (2009)
	SNX4	**Intracellular trafficking**	Tai et al. (2009)
	SPCS2	**Signal peptidase subunit**	Li et al. (2009)
	SPCS3	**Signal peptidase subunit**	Li et al. (2009)
	SRCAP	**Chromatin remodeling**	Randall et al. (2007)
	SSB	**RNA chaperone**	Xue et al. (2007)
	SURF1	**Cytochrome c biogenesis**	Li et al. (2009)
	TBK1	**Serine/threonine kinase**	Tai et al. (2009)
	TBXA2R	**G-protein coupled receptor**	Ng et al. (2007)
	TECR	**Fatty acid biosynthesis**	Li et al. (2009)
	TLK1	**Chromatin assembly**	Reiss et al. (2011)
	TRAF2	**Signal transduction**	Ng et al. (2007)
	TRIM62	**E3 ubiquitin ligase**	Li et al. (2009) and Tai et al. (2009)

(continued)

Table 1 (continued)

	RAB5A	Early endosome	Berger et al. (2009)		UBE2J1	**E2 ubiquitin conjugating**	Li et al. (2014a)
	RABEPK	Endosome to trans Golgi	Tai et al. (2009)		UBB	**Ubiquitin**	Borawski et al. (2009)
	RAN	Nuclear pore translocation	Tai et al. (2009)		VAPA	**Vesicle trafficking**	Xue et al. (2007)
	RELA	NFkB subunit	Randall et al. (2007) and Ng et al. (2007)		VAPB	**Vesicle trafficking**	Randall et al. (2007)
	RNF31	E3 ubiquitin ligase	Li et al. (2009)		VRK1	**Serine/threonine kinase**	Supekova et al. (2008)
	RPS27A	Ribosomal protein	Borawski et al. (2009)		VPS35	**Endosome to trans Golgi**	Randall et al. (2007)
	RPS6KL1	Ribosomal protein	Li et al. (2009) and Borawski et al. (2009)		VWF	**Blood coagulation**	Borawski et al. (2009)
	SDHA	Succinate dehydrogenase	Vaillancourt et al. (2009)		XPNPEP1	**Aminopeptidase**	Li et al. (2009) and Vaillancourt et al. (2009)
	SLC12A4	K/Cl transport	Ng et al. (2007)		ZC3H15	**Erythropoietin**	Tai et al. (2009)

Depicted in *green*—host factors whose function is implicated in HCV entry, *blue*—host factors whose function is implicated in HCV assembly/exit and *red*—host factors whose function is implicated in HCV replication

10 Virus-Host Interactions Expand Our HCV Model Systems

Most studies validating host cell factors important for HCV infection were carried out in Huh-7 cell derivatives, including Huh-7.5 (Blight et al. 2002), Huh-7.5.1 (Zhong et al. 2005) and Huh-7 Lunet cells (Koutsoudakis et al. 2007). These cell lines lack many features of human hepatocytes in vivo, including the ability to polarize and produce normal serum lipoproteins. Efforts in defining the minimal hepatic host factors required for HCV infection has led to an expansion of our cell culture systems. HepG2 hepatocytes are refractory to infection because they lack miR-122 and CD81 expression. HepG2 cells engineered to express miR-122 and CD81 support the entire HCV life cycle infection (Narbus et al. 2011). HepG2-CD81-miR122 cells are of value for at least two reasons. First, they provide a distinct cell line to validate HCV replication studies. Secondly, HepG2 cells polarize to a greater degree than Huh-7 cells in standard two-dimensional cell cultures. This is important in studies of HCV entry and cell-cell spread, wherein the entry factors have distinct subcellular localizations. Indeed, the ectopic expression of a subset of host factors can sensitize non-hepatic cells to support the entire HCV life cycle. Expression of the four entry receptor molecules in combination with miR-122 and apoE is sufficient to reconstitute the entire HCV life cycle in human derived non-hepatic cells (Da Costa et al. 2012), while the same set of host factors in combination with innate immune ablating mutations recapitulates the virus life cycle in non-hepatic mouse cells (Frentzen et al. 2014).

Perhaps the greatest impact of identifying virus-host factor interactions is the ongoing efforts on developing a genetically tractable small animal model for HCV infection. Most species, except humans and chimpanzees, are resistant to HCV infection and this restriction occurs partially at the level of virus entry. Species-specific differences in the entry receptors are mostly responsible for the inefficient virus uptake in nonhuman cells. Introduction of human-specific CD81 and OCLN are sufficient for viral entry into mouse or hamster cells (Ploss et al. 2009). A mouse model system has been developed for HCV by introducing human CD81 and OCLN in mice ablated for part of their immune system (Dorner et al. 2011, 2013).

11 Host Factors as Antiviral Targets

Although traditional antiviral drug development has focused on virally encoded functions, there are potential merits to targeting host factors. These potential advantages include a lower propensity to develop resistance, as a virus would need to evolve to replace a need cofactor with a different strategy, and pan-genotypic or broad-spectrum antiviral activity with host factors required by all HCV genotypes and even unrelated viruses. In the case of HCV drug development, the major drug classes that will cure the vast majority of treated patients target NS3, NS5A, and NS5B (Lange et al. 2014). However, there may be a place for host factor-based therapies for patients that fail to respond to these therapies. Additionally, identifying and developing host factor-based HCV therapeutics will inform future drug development for other viral infections. Some of these therapies in development are listed in Table 2.

Table 2 HCV host factor therapeutic targets

Inhibitor name/company	Host target	Clinical phase	References
ITX-5061	SR-BI	Phase Ib	Syder et al. (2011) and Sulkowski et al. (2014)
Tarceva (Erlotinib/OSI Pharmaceuticals)	EGFR	Phase I	Lupberger et al. (2011)
Sprycel (Dasatinib/Bristol Myers Squibb)	EphA2	In vitro	Lupberger et al. (2011)
Zetia (Ezetimibe)	NPC1L1	Phase I	Sainz et al. (2012)
RG-101 (Regulus)	miR-122	Phase II	www.regulusrx.com
Miravirsen (SPC3649/Santaris Pharma)	miR-122	Phase II	Lanford et al. (2010) and Janssen et al. (2013)
Alisporivir (Debio-025/Novartis)	Cyp A	Phase II	Coelmont et al. (2009)
SCY-635/Scynexis	Cyp A	Phase II	Hopkins et al. (2010)
Ganetespib (STA-9090/SyntaPharma)	Hsp90	Phase I	Goyal et al. (2015)
AL-9	PIK-IIIa	In vitro	Bianco et al. (2012)
OSW-1	OSBP	In vitro	Wang et al. (2014)
LCQ908 (Pradigastat/Novartis)	DGAT-I	In vitro	Yen et al. (2008)

Entry There are several monoclonal antibodies (mAbs) developed to host entry factors that inhibit HCV infection, such as anti-CLDN1 (OM-7D3-B3) (Fofana et al. 2010), anti-SR-BI (NK-8H5-E3) (Zahid et al. 2013), and anti-CD81 (QV-6A8-F2C4) (Fofana et al. 2013; Meuleman et al. 2008). Inhibitors of EGFR (erlotinib), EphA2 (dasatinib) and NPC1L1 (ezetimibe) are already FDA-approved molecules shown to inhibit HCV entry in vitro. There is also a small molecule inhibitor of SR-BI (ITX 5061), which is the most advanced HCV entry inhibitor in clinical trials (Syder et al. 2011; Sulkowski et al. 2014).

Replication RG-101 (Regulus) is a GalNAc-conjugated anti-miR targeting miR-122 for the treatment of HCV and is currently being evaluated in a study being conducted in the Netherlands. Miravirsen (SPC3649/Santaris Pharma) is a locked nucleic-acid modified oligonucleotide complimentary to miR-122, capable of potently antagonizing multiple HCV genotypes in vitro (Li et al. 2011). When administered in chronically infected chimpanzees, this inhibitor resulted in long-lasting suppression of HCV viremia with no evidence of viral resistance and minimal side effects (Lanford et al. 2010). A recent phase II study showed that miravirsen monotherapy could achieve a significant HCV RNA drop in patient serum (Janssen et al. 2013). However, recent data associate depletion of miR-122 with steatohepatitis, fibrosis, and hepatocellular carcinoma, raising concerns about safety in humans (Tsai et al. 2012). Furthermore, as this compound is injected it might not be favored in the all-oral therapy regimen. AL-9, a member of the 4-anilino quinazoline-containing kinase inhibitor family, inhibits HCV replication in vitro by direct inhibition of PI4K-IIIα (Bianco et al. 2012). AL-9 as well as the OSBP inhibitor OSW-1 interfere with the OSBP-dependent delivery of cholesterol to sites of viral replication, resulting in efficient inhibition of HCV infection in vitro (Wang et al. 2014).

Chemical modifications of the initial cyclophilin inhibitor, CsA, resulted in non-immunosuppressive analogs that potently suppress HCV replication in cell culture, such as alisporivir (Debio-025/Novartis) and SCY-635 (Scynexis), which are currently in Phase II clinical trials (Coelmont et al. 2009; Hopkins et al. 2010). An inhibitor of Hsp90 (Ganetespib/STA-9090) is currently in Phase I clinical trials for administration to patients with advanced hepatocellular carcinoma (Goyal et al. 2015).

Assembly DGAT1 inhibitors currently in clinical trials for obesity-related diseases could be potential candidates against HCV (Yen et al. 2008).

12 Conclusions and Outlook

The past 25 years of research studying HCV-host interactions have resulted in the identification of hundreds of cellular factors important for various stages of the virus life cycle. Charactering the significance of these host factors in HCV infection has been challenging, due initially to limiting HCV replication systems, in addition to the difficulty of distinguishing a virus-specific host gene function from its cellular function. Despite these obstacles, a comprehensive view of the roles of many host factors in HCV replication has developed. This has informed our understanding of the viral life cycle, expanded our model systems, and provided viable antiviral therapeutic targets. Despite these impressive advances, much remains to be understood about HCV-host interactions. The mechanistic details of even the best-characterized virus-host interactions are still incomplete. Small animal model development is a work in progress, as current models are immune-compromised. Finally, the list of host cofactors of HCV infection without a known function (Table 1) suggests that we are just scratching the surface of understanding the complex interactions between HCV and it host cell.

References

Acton S, Rigotti A, Landschulz KT, Xu S, Hobbs HH, Krieger M (1996) Identification of scavenger receptor SR-BI as a high density lipoprotein receptor. Science 271:518–520

Agnello V, Abel G, Elfahal M, Knight GB, Zhang QX (1999) Hepatitis C virus and other flaviviridae viruses enter cells via low density lipoprotein receptor. Proc Natl Acad Sci U S A 96:12766–12771

Ahn J, Chung KS, Kim DU, Won M, Kim L, Kim KS, Nam M, Choi SJ, Kim HC, Yoon M, Chae SK, Hoe KL (2004) Systematic identification of hepatocellular proteins interacting with NS5A of the hepatitis C virus. J Biochem Mol Biol 37:741–748

Ait-Goughoulte M, Kanda T, Meyer K, Ryerse JS, Ray RB, Ray R (2008) Hepatitis C virus genotype 1a growth and induction of autophagy. J Virol 82:2241–2249

Albecka A, Belouzard S, Op De Beeck A, Descamps V, Goueslain L, Bertrand-Michel J, Terce F, Duverlie G, Rouille Y, Dubuisson J (2012) Role of low-density lipoprotein receptor in the hepatitis C virus life cycle. Hepatology 55:998–1007

Ali N, Siddiqui A (1997) The La antigen binds 5′ noncoding region of the hepatitis C virus RNA in the context of the initiator AUG codon and stimulates internal ribosome entry site-mediated translation. Proc Natl Acad Sci U S A 94:2249–2254

Amako Y, Sarkeshik A, Hotta H, Yates J 3rd, Siddiqui A (2009) Role of oxysterol binding protein in hepatitis C virus infection. J Virol 83:9237–9246

Amako Y, Syed GH, Siddiqui A (2011) Protein kinase D negatively regulates hepatitis C virus secretion through phosphorylation of oxysterol-binding protein and ceramide transfer protein. J Biol Chem 286:11265–11274

Andre P, Komurian-Pradel F, Deforges S, Perret M, Berland JL, Sodoyer M, Pol S, Brechot C, Paranhos-Baccala G, Lotteau V (2002) Characterization of low- and very-low-density hepatitis C virus RNA-containing particles. J Virol 76:6919–6928

Ariumi Y, Kuroki M, Kushima Y, Osugi K, Hijikata M, Maki M, Ikeda M, Kato N (2011a) Hepatitis C virus hijacks P-body and stress granule components around lipid droplets. J Virol 85:6882–6892

Ariumi Y, Kuroki M, Maki M, Ikeda M, Dansako H, Wakita T, Kato N (2011b) The ESCRT system is required for hepatitis C virus production. PLoS One 6:e14517

Arnaud N, Dabo S, Maillard P, Budkowska A, Kalliampakou KI, Mavromara P, Garcin D, Hugon J, Gatignol A, Akazawa D, Wakita T, Meurs EF (2010) Hepatitis C virus controls interferon production through PKR activation. PLoS One 5:e10575

Backes P, Quinkert D, Reiss S, Binder M, Zayas M, Rescher U, Gerke V, Bartenschlager R, Lohmann V (2010) Role of annexin A2 in the production of infectious hepatitis C virus particles. J Virol 84:5775–5789

Bankwitz D, Steinmann E, Bitzegeio J, Ciesek S, Friesland M, Herrmann E, Zeisel MB, Baumert TF, Keck ZY, Foung SK, Pecheur EI, Pietschmann T (2010) Hepatitis C virus hypervariable region 1 modulates receptor interactions, conceals the CD81 binding site, and protects conserved neutralizing epitopes. J Virol 84:5751–5763

Barba G, Harper F, Harada T, Kohara M, Goulinet S, Matsuura Y, Eder G, Schaff Z, Chapman MJ, Miyamura T, Brechot C (1997) Hepatitis C virus core protein shows a cytoplasmic localization and associates to cellular lipid storage droplets. Proc Natl Acad Sci U S A 94:1200–1205

Barretto N, Sainz B Jr, Hussain S, Uprichard SL (2014) Determining the involvement and therapeutic implications of host cellular factors in hepatitis C virus cell-to-cell spread. J Virol 88:5050–5061

Bartenschlager R, Ahlborn-Laake L, Mous J, Jacobsen H (1994) Kinetic and structural analyses of hepatitis C virus polyprotein processing. J Virol 68:5045–5055

Bartosch B, Vitelli A, Granier C, Goujon C, Dubuisson J, Pascale S, Scarselli E, Cortese R, Nicosia A, Cosset FL (2003) Cell entry of hepatitis C virus requires a set of co-receptors that include the CD81 tetraspanin and the SR-B1 scavenger receptor. J Biol Chem 278:41624–41630

Behrens SE, Tomei L, DE Francesco R (1996) Identification and properties of the RNA-dependent RNA polymerase of hepatitis C virus. EMBO J 15:12–22

Benali-Furet NL, Chami M, Houel L, DE Giorgi F, Vernejoul F, Lagorce D, Buscail L, Bartenschlager R, Ichas F, Rizzuto R, Paterlini-Brechot P (2005) Hepatitis C virus core triggers apoptosis in liver cells by inducing ER stress and ER calcium depletion. Oncogene 24:4921–4933

Benga WJ, Krieger SE, Dimitrova M, Zeisel MB, Parnot M, Lupberger J, Hildt E, Luo G, Mclauchlan J, Baumert TF, Schuster C (2010) Apolipoprotein E interacts with hepatitis C virus nonstructural protein 5A and determines assembly of infectious particles. Hepatology 51:43–53

Berger KL, Cooper JD, Heaton NS, Yoon R, Oakland TE, Jordan TX, Mateu G, Grakoui A, Randall G (2009) Roles for endocytic trafficking and phosphatidylinositol 4-kinase III alpha in hepatitis C virus replication. Proc Natl Acad Sci U S A 106:7577–7582

Berger KL, Kelly SM, Jordan TX, Tartell MA, Randall G (2011) Hepatitis C virus stimulates the phosphatidylinositol 4-kinase III alpha-dependent phosphatidylinositol 4-phosphate production that is essential for its replication. J Virol 85:8870–8883

Bianco A, Reghellin V, Donnici L, Fenu S, Alvarez R, Baruffa C, Peri F, Pagani M, Abrignani S, Neddermann P, DE Francesco R (2012) Metabolism of phosphatidylinositol 4-kinase IIIalpha-dependent PI4P Is subverted by HCV and is targeted by a 4-anilino quinazoline with antiviral activity. PLoS Pathog 8:e1002576

Blackham S, Baillie A, Al-Hababi F, Remlinger K, You S, Hamatake R, Mcgarvey MJ (2010) Gene expression profiling indicates the roles of host oxidative stress, apoptosis, lipid metabolism, and intracellular transport genes in the replication of hepatitis C virus. J Virol 84:5404–5414

Blanchard E, Belouzard S, Goueslain L, Wakita T, Dubuisson J, Wychowski C, Rouille Y (2006) Hepatitis C virus entry depends on clathrin-mediated endocytosis. J Virol 80:6964–6972

Blight KJ, Kolykhalov AA, Rice CM (2000) Efficient initiation of HCV RNA replication in cell culture. Science 290:1972–1974

Blight KJ, Mckeating JA, Rice CM (2002) Highly permissive cell lines for subgenomic and genomic hepatitis C virus RNA replication. J Virol 76:13001–13014

Borawski J, Troke P, Puyang X, Gibaja V, Zhao S, Mickanin C, Leighton-Davies J, Wilson CJ, Myer V, Cornellataracido I, Baryza J, Tallarico J, Joberty G, Bantscheff M, Schirle M, Bouwmeester T, Mathy JE, Lin K, Compton T, Labow M, Wiedmann B, Gaither LA (2009)

Class III phosphatidylinositol 4-kinase alpha and beta are novel host factor regulators of hepatitis C virus replication. J Virol 83:10058–10074
Boulant S, Montserret R, Hope RG, Ratinier M, Targett-Adams P, Lavergne JP, Penin F, Mclauchlan J (2006) Structural determinants that target the hepatitis C virus core protein to lipid droplets. J Biol Chem 281:22236–22247
Boyer A, Dumans A, Beaumont E, Etienne L, Roingeard P, Meunier JC (2014) The association of hepatitis C virus glycoproteins with apolipoproteins E and B early in assembly is conserved in lipoviral particles. J Biol Chem 289:18904–18913
Brazzoli M, Bianchi A, Filippini S, Weiner A, Zhu Q, Pizza M, Crotta S (2008) CD81 is a central regulator of cellular events required for hepatitis C virus infection of human hepatocytes. J Virol 82:8316–8329
Brimacombe CL, Grove J, Meredith LW, Hu K, Syder AJ, Flores MV, Timpe JM, Krieger SE, Baumert TF, Tellinghuisen TL, Wong-Staal F, Balfe P, Mckeating JA (2011) Neutralizing antibody-resistant hepatitis C virus cell-to-cell transmission. J Virol 85:596–605
Buchan JR, Parker R (2009) Eukaryotic stress granules: the ins and outs of translation. Mol Cell 36:932–941
Bukong TN, Kodys K, Szabo G (2013) Human ezrin-moesin-radixin proteins modulate hepatitis C virus infection. Hepatology 58:1569–1579
Bukong TN, Momen-Heravi F, Kodys K, Bala S, Szabo G (2014) Exosomes from hepatitis C infected patients transmit HCV infection and contain replication competent viral RNA in complex with Ago2-miR122-HSP90. PLoS Pathog 10:e1004424
Camus G, Herker E, Modi AA, Haas JT, Ramage HR, Farese RV Jr, Ott M (2013) Diacylglycerol acyltransferase-1 localizes hepatitis C virus NS5A protein to lipid droplets and enhances NS5A interaction with the viral capsid core. J Biol Chem 288:9915–9923
Catanese MT, Ansuini H, Graziani R, Huby T, Moreau M, Ball JK, Paonessa G, Rice CM, Cortese R, Vitelli A, Nicosia A (2010) Role of scavenger receptor class B type I in hepatitis C virus entry: kinetics and molecular determinants. J Virol 84:34–43
Catanese MT, Loureiro J, Jones CT, Dorner M, VON Hahn T, Rice CM (2013a) Different requirements for scavenger receptor class B type I in hepatitis C virus cell-free versus cell-to-cell transmission. J Virol 87:8282–8293
Catanese MT, Uryu K, Kopp M, Edwards TJ, Andrus L, Rice WJ, Silvestry M, Kuhn RJ, Rice CM (2013b) Ultrastructural analysis of hepatitis C virus particles. Proc Natl Acad Sci U S A 110:9505–9510
Chandra PK, Bao L, Song K, Aboulnasr FM, Baker DP, Shores N, Wimley WC, Liu S, Hagedorn CH, Fuchs SY, Wu T, Balart LA, Dash S (2014) HCV infection selectively impairs type I but not type III IFN signaling. Am J Pathol 184:214–229
Chang KS, Jiang J, Cai Z, Luo G (2007) Human apolipoprotein e is required for infectivity and production of hepatitis C virus in cell culture. J Virol 81:13783–13793
Chang J, Guo JT, Jiang D, Guo H, Taylor JM, Block TM (2008) Liver-specific microRNA miR-122 enhances the replication of hepatitis C virus in nonhepatic cells. J Virol 82:8215–8223
Chatterji U, Bobardt M, Selvarajah S, Yang F, Tang H, Sakamoto N, Vuagniaux G, Parkinson T, Gallay P (2009) The isomerase active site of cyclophilin A is critical for hepatitis C virus replication. J Biol Chem 284:16998–17005
Chen YC, Su WC, Huang JY, Chao TC, Jeng KS, Machida K, Lai MM (2010) Polo-like kinase 1 is involved in hepatitis C virus replication by hyperphosphorylating NS5A. J Virol 84:7983–7993
Coelmont L, Kaptein S, Paeshuyse J, Vliegen I, Dumont JM, Vuagniaux G, Neyts J (2009) Debio 025, a cyclophilin binding molecule, is highly efficient in clearing hepatitis C virus (HCV) replicon-containing cells when used alone or in combination with specifically targeted antiviral therapy for HCV (STAT-C) inhibitors. Antimicrob Agents Chemother 53:967–976
Coelmont L, Hanoulle X, Chatterji U, Berger C, Snoeck J, Bobardt M, Lim P, Vliegen I, Paeshuyse J, Vuagniaux G, Vandamme AM, Bartenschlager R, Gallay P, Lippens G, Neyts J

(2010) DEB025 (Alisporivir) inhibits hepatitis C virus replication by preventing a cyclophilin A induced cis-trans isomerisation in domain II of NS5A. PLoS One 5:e13687

Coito C, Diamond DL, Neddermann P, Korth MJ, Katze MG (2004) High-throughput screening of the yeast kinome: identification of human serine/threonine protein kinases that phosphorylate the hepatitis C virus NS5A protein. J Virol 78:3502–3513

Coller KE, Berger KL, Heaton NS, Cooper JD, Yoon R, Randall G (2009) RNA interference and single particle tracking analysis of hepatitis C virus endocytosis. PLoS Pathog 5:e1000702

Coller KE, Heaton NS, Berger KL, Cooper JD, Saunders JL, Randall G (2012) Molecular determinants and dynamics of hepatitis C virus secretion. PLoS Pathog 8:e1002466

Corless L, Crump CM, Griffin SD, Harris M (2010) Vps4 and the ESCRT-III complex are required for the release of infectious hepatitis C virus particles. J Gen Virol 91:362–372

Cormier EG, Tsamis F, Kajumo F, Durso RJ, Gardner JP, Dragic T (2004) CD81 is an entry coreceptor for hepatitis C virus. Proc Natl Acad Sci U S A 101:7270–7274

Cun W, Jiang J, Luo G (2010) The C-terminal alpha-helix domain of apolipoprotein E is required for interaction with nonstructural protein 5A and assembly of hepatitis C virus. J Virol 84:11532–11541

Da Costa D, Turek M, Felmlee DJ, Girardi E, Pfeffer S, Long G, Bartenschlager R, Zeisel MB, Baumert TF (2012) Reconstitution of the entire hepatitis C virus life cycle in nonhepatic cells. J Virol 86(21):11919–11925

Dao Thi VL, Granier C, Zeisel MB, Guerin M, Mancip J, Granio O, Penin F, Lavillette D, Bartenschlager R, Baumert TF, Cosset FL, Dreux M (2012) Characterization of hepatitis C virus particle subpopulations reveals multiple usage of the scavenger receptor BI for entry steps. J Biol Chem 287:31242–31257

DE Lucas S, Bartolome J, Carreno V (2005) Hepatitis C virus core protein down-regulates transcription of interferon-induced antiviral genes. J Infect Dis 191:93–99

DEN Boon JA, Ahlquist P (2010) Organelle-like membrane compartmentalization of positive-strand RNA virus replication factories. Annu Rev Microbiol 64:241–256

Diamond DL, Syder AJ, Jacobs JM, Sorensen CM, Walters KA, Proll SC, McDermott JE, Gritsenko MA, Zhang Q, Zhao R, Metz TO, Camp DG 2nd, Waters KM, Smith RD, Rice CM, Katze MG (2010) Temporal proteome and lipidome profiles reveal hepatitis C virus-associated reprogramming of hepatocellular metabolism and bioenergetics. PLoS Pathog 6: e1000719

Diao J, Pantua H, Ngu H, Komuves L, Diehl L, Schaefer G, Kapadia SB (2012) Hepatitis C virus induces epidermal growth factor receptor activation via CD81 binding for viral internalization and entry. J Virol 86:10935–10949

Domitrovich AM, Diebel KW, Ali N, Sarker S, Siddiqui A (2005) Role of La autoantigen and polypyrimidine tract-binding protein in HCV replication. Virology 335(1):72–86

Dorner M, Horwitz JA, Robbins JB, Barry WT, Feng Q, Mu K, Jones CT, Schoggins JW, Catanese MT, Burton DR, Law M, Rice CM, Ploss A (2011) A genetically humanized mouse model for hepatitis C virus infection. Nature 474:208–211

Dorner M, Horwitz JA, Donovan BM, Labitt RN, Budell WC, Friling T, Vogt A, Catanese MT, Satoh T, Kawai T, Akira S, Law M, Rice CM, Ploss A (2013) Completion of the entire hepatitis C virus life cycle in genetically humanized mice. Nature 501:237–241

Dreux M, Dao Thi VL, Fresquet J, Guerin M, Julia Z, Verney G, Durantel D, Zoulim F, Lavillette D, Cosset FL, Bartosch B (2009a) Receptor complementation and mutagenesis reveal SR-BI as an essential HCV entry factor and functionally imply its intra- and extra-cellular domains. PLoS Pathog 5:e1000310

Dreux M, Gastaminza P, Wieland SF, Chisari FV (2009b) The autophagy machinery is required to initiate hepatitis C virus replication. Proc Natl Acad Sci U S A 106:14046–14051

Dreux M, Garaigorta U, Boyd B, Decembre E, Chung J, Whitten-Bauer C, Wieland S, Chisari FV (2012) Short-range exosomal transfer of viral RNA from infected cells to plasmacytoid dendritic cells triggers innate immunity. Cell Host Microbe 12:558–570

Egger D, Wolk B, Gosert R, Bianchi L, Blum HE, Moradpour D, Bienz K (2002) Expression of hepatitis C virus proteins induces distinct membrane alterations including a candidate viral replication complex. J Virol 76:5974–5984
Eksioglu EA, Zhu H, Bayouth L, Bess J, Liu HY, Nelson DR, Liu C (2011) Characterization of HCV interactions with Toll-like receptors and RIG-I in liver cells. PLoS One 6:e21186
Evans MJ, Rice CM, Goff SP (2004) Phosphorylation of hepatitis C virus nonstructural protein 5A modulates its protein interactions and viral RNA replication. Proc Natl Acad Sci U S A 101:13038–13043
Evans MJ, VON Hahn T, Tscherne DM, Syder AJ, Panis M, Wolk B, Hatziioannou T, Mckeating JA, Bieniasz PD, Rice CM (2007) Claudin-1 is a hepatitis C virus co-receptor required for a late step in entry. Nature 446:801–805
Eyre NS, Drummer HE, Beard MR (2010) The SR-BI partner PDZK1 facilitates hepatitis C virus entry. PLoS Pathog 6:e1001130
Farquhar MJ, Harris HJ, Diskar M, Jones S, Mee CJ, Nielsen SU, Brimacombe CL, Molina S, Toms GL, Maurel P, Howl J, Herberg FW, VAN Ijzendoorn SC, Balfe P, Mckeating JA (2008) Protein kinase A-dependent step(s) in hepatitis C virus entry and infectivity. J Virol 82:8797–8811
Ferraris P, Blanchard E, Roingeard P (2010) Ultrastructural and biochemical analyses of hepatitis C virus-associated host cell membranes. J Gen Virol 91:2230–2237
Filbin ME, Vollmar BS, Shi D, Gonen T, Kieft JS (2013) HCV IRES manipulates the ribosome to promote the switch from translation initiation to elongation. Nat Struct Mol Biol 20:150–158
Fofana I, Krieger SE, Grunert F, Glauben S, Xiao F, Fafi-Kremer S, Soulier E, Royer C, Thumann C, Mee CJ, Mckeating JA, Dragic T, Pessaux P, Stoll-Keller F, Schuster C, Thompson J, Baumert TF (2010) Monoclonal anti-claudin 1 antibodies prevent hepatitis C virus infection of primary human hepatocytes. Gastroenterology 139:953–964, 964 e1–4
Fofana I, Xiao F, Thumann C, Turek M, Zona L, Tawar RG, Grunert F, Thompson J, Zeisel MB, Baumert TF (2013) A novel monoclonal anti-CD81 antibody produced by genetic immunization efficiently inhibits Hepatitis C virus cell-cell transmission. PLoS One 8:e64221
Foster TL, Gallay P, Stonehouse NJ, Harris M (2011) Cyclophilin A interacts with domain II of hepatitis C virus NS5A and stimulates RNA binding in an isomerase-dependent manner. J Virol 85:7460–7464
Fraser CS, Hershey JW, Doudna JA (2009) The pathway of hepatitis C virus mRNA recruitment to the human ribosome. Nat Struct Mol Biol 16:397–404
Frentzen A, Anggakusuma, Gurlevik E, Hueging K, Knocke S, Ginkel C, Brown RJ, Heim M, Dill MT, Kroger A, Kalinke U, Kaderali L, Kuehnel F, Pietschmann T (2014) Cell entry, efficient RNA replication, and production of infectious hepatitis C virus progeny in mouse liver-derived cells. Hepatology 59:78–88
Fukuhara T, Kambara H, Shiokawa M, Ono C, Katoh H, Morita E, Okuzaki D, Maehara Y, Koike K, Matsuura Y (2012) Expression of microRNA miR-122 facilitates an efficient replication in nonhepatic cells upon infection with hepatitis C virus. J Virol 86:7918–7933
Gale MJ Jr, Korth MJ, Tang NM, Tan SL, Hopkins DA, Dever TE, Polyak SJ, Gretch DR, Katze MG (1997) Evidence that hepatitis C virus resistance to interferon is mediated through repression of the PKR protein kinase by the nonstructural 5A protein. Virology 230:217–227
Gale M Jr, Blakely CM, Kwieciszewski B, Tan SL, Dossett M, Tang NM, Korth MJ, Polyak SJ, Gretch DR, Katze MG (1998) Control of PKR protein kinase by hepatitis C virus nonstructural 5A protein: molecular mechanisms of kinase regulation. Mol Cell Biol 18:5208–5218
Gale M Jr, Kwieciszewski B, Dossett M, Nakao H, Katze MG (1999) Antiapoptotic and oncogenic potentials of hepatitis C virus are linked to interferon resistance by viral repression of the PKR protein kinase. J Virol 73:6506–6516
Gao L, Aizaki H, He JW, Lai MM (2004) Interactions between viral nonstructural proteins and host protein hVAP-33 mediate the formation of hepatitis C virus RNA replication complex on lipid raft. J Virol 78:3480–3488

Garaigorta U, Chisari FV (2009) Hepatitis C virus blocks interferon effector function by inducing protein kinase R phosphorylation. Cell Host Microbe 6:513–522

Garaigorta U, Heim MH, Boyd B, Wieland S, Chisari FV (2012) Hepatitis C virus (HCV) induces formation of stress granules whose proteins regulate HCV RNA replication and virus assembly and egress. J Virol 86:11043–11056

Gastaminza P, Cheng G, Wieland S, Zhong J, Liao W, Chisari FV (2008) Cellular determinants of hepatitis C virus assembly, maturation, degradation, and secretion. J Virol 82:2120–2129

Geiss GK, Carter VS, He Y, Kwieciszewski BK, Holzman T, Korth MJ, Lazaro CA, Fausto N, Bumgarner RE, Katze MG (2003) Gene expression profiling of the cellular transcriptional network regulated by alpha/beta interferon and its partial attenuation by the hepatitis C virus nonstructural 5A protein. J Virol 77:6367–6375

Germi R, Crance JM, Garin D, Guimet J, Lortat-Jacob H, Ruigrok RW, Zarski JP, Drouet E (2002) Cellular glycosaminoglycans and low density lipoprotein receptor are involved in hepatitis C virus adsorption. J Med Virol 68:206–215

Gong G, Waris G, Tanveer R, Siddiqui A (2001) Human hepatitis C virus NS5A protein alters intracellular calcium levels, induces oxidative stress, and activates STAT-3 and NF-kappa B. Proc Natl Acad Sci U S A 98:9599–9604

Gosert R, Chang KH, Rijnbrand R, Yi M, Sangar DV, Lemon SM (2000) Transient expression of cellular polypyrimidine-tract binding protein stimulates cap-independent translation directed by both picornaviral and flaviviral internal ribosome entry sites In vivo. Mol Cell Biol 20:1583–1595

Gosert R, Egger D, Lohmann V, Bartenschlager R, Blum HE, Bienz K, Moradpour D (2003) Identification of the hepatitis C virus RNA replication complex in Huh-7 cells harboring subgenomic replicons. J Virol 77:5487–5492

Goyal L, Wadlow RC, Blaszkowsky LS, Wolpin BM, Abrams TA, McCleary NJ, Sheehan S et al (2015) A phase I and pharmacokinetic study of ganetespib (STA-9090) in advanced hepatocellular carcinoma. Invest New Drugs 33(1):128–137

Grakoui A, McCourt DW, Wychowski C, Feinstone SM, Rice CM (1993a) A second hepatitis C virus-encoded proteinase. Proc Natl Acad Sci U S A 90:10583–10587

Grakoui A, Wychowski C, Lin C, Feinstone SM, Rice CM (1993b) Expression and identification of hepatitis C virus polyprotein cleavage products. J Virol 67:1385–1395

Grise H, Frausto S, Logan T, Tang H (2012) A conserved tandem cyclophilin-binding site in hepatitis C virus nonstructural protein 5A regulates Alisporivir susceptibility. J Virol 86:4811–4822

Guevin C, Manna D, Belanger C, Konan KV, Mak P, Labonte P (2010) Autophagy protein ATG5 interacts transiently with the hepatitis C virus RNA polymerase (NS5B) early during infection. Virology 405:1–7

Hahm B, Kim YK, Kim JH, Kim TY, Jang SK (1998) Heterogeneous nuclear ribonucleoprotein L interacts with the 3′ border of the internal ribosomal entry site of hepatitis C virus. J Virol 72:8782–8788

Hamamoto I, Nishimura Y, Okamoto T, Aizaki H, Liu M, Mori Y, Abe T, Suzuki T, Lai MM, Miyamura T, Moriishi K, Matsuura Y (2005) Human VAP-B is involved in hepatitis C virus replication through interaction with NS5A and NS5B. J Virol 79:13473–13482

Hanoulle X, Badillo A, Wieruszeski JM, Verdegem D, Landrieu I, Bartenschlager R, Penin F, Lippens G (2009) Hepatitis C virus NS5A protein is a substrate for the peptidyl-prolyl cis/trans isomerase activity of cyclophilins A and B. J Biol Chem 284:13589–13601

Hara Y, Yanatori I, Ikeda M, Kiyokage E, Nishina S, Tomiyama Y, Toida K et al (2014) Hepatitis C virus core protein suppresses mitophagy by interacting with parkin in the context of mitochondrial depolarization. Am J Pathol 184(11):3026–3039

Harak C, Radujkovic D, Taveneau C, Reiss S, Klein R, Bressanelli S, Lohmann V (2014) Mapping of functional domains of the lipid kinase phosphatidylinositol 4-kinase type III alpha involved in enzymatic activity and hepatitis C virus replication. J Virol 88:9909–9926

Harris HJ, Davis C, Mullins JG, Hu K, Goodall M, Farquhar MJ, Mee CJ, Mccaffrey K, Young S, Drummer H, Balfe P, Mckeating JA (2010) Claudin association with CD81 defines hepatitis C virus entry. J Biol Chem 285:21092–21102

Hashem Y, Des Georges A, Dhote V, Langlois R, Liao HY, Grassucci RA, Pestova TV, Hellen CU, Frank J (2013) Hepatitis-C-virus-like internal ribosome entry sites displace eIF3 to gain access to the 40S subunit. Nature 503:539–543

Heaton NS, Randall G (2011) Multifaceted roles for lipids in viral infection. Trends Microbiol 19:368–375

Heim MH, Moradpour D, Blum HE (1999) Expression of hepatitis C virus proteins inhibits signal transduction through the Jak-STAT pathway. J Virol 73:8469–8475

Helbig KJ, Eyre NS, Yip E, Narayana S, Li K, Fiches G, McCartney EM, Jangra RK, Lemon SM, Beard MR (2011) The antiviral protein viperin inhibits hepatitis C virus replication via interaction with nonstructural protein 5A. Hepatology 54:1506–1517

Henke JI, Goergen D, Zheng J, Song Y, Schuttler CG, Fehr C, Junemann C, Niepmann M (2008) microRNA-122 stimulates translation of hepatitis C virus RNA. EMBO J 27:3300–3310

Herker E, Harris C, Hernandez C, Carpentier A, Kaehlcke K, Rosenberg AR, Farese RV Jr, Ott M (2010) Efficient hepatitis C virus particle formation requires diacylglycerol acyltransferase-1. Nat Med 16:1295–1298

Hishiki T, Shimizu Y, Tobita R, Sugiyama K, Ogawa K, Funami K, Ohsaki Y, Fujimoto T, Takaku H, Wakita T, Baumert TF, Miyanari Y, Shimotohno K (2010) Infectivity of hepatitis C virus is influenced by association with apolipoprotein E isoforms. J Virol 84:12048–12057

Hijikata M, Shimizu YK, Kato H, Iwamoto A, Shih JW, Alter HJ, Purcell RH, Yoshikura H (1993) Equilibrium centrifugation studies of hepatitis C virus: evidence for circulating immune complexes. J Virol 67(4):1953–1958

Hoffmann FS, Schmidt A, Dittmann Chevillotte M, Wisskirchen C, Hellmuth J, Willms S, Gilmore RH et al (2015) Polymorphisms in melanoma differentiation-associated gene 5 link protein function to clearance of hepatitis C virus. Hepatology 61(2):460–470

Hopkins S, Scorneaux B, Huang Z, Murray MG, Wring S, Smitley C, Harris R, Erdmann F, Fischer G, Ribeill Y (2010) SCY-635, a novel nonimmunosuppressive analog of cyclosporine that exhibits potent inhibition of hepatitis C virus RNA replication in vitro. Antimicrob Agents Chemother 54:660–672

Horton JD, Goldstein JL, Brown MS (2002) SREBPs: activators of the complete program of cholesterol and fatty acid synthesis in the liver. J Clin Invest 109:1125–1131

Hosui A, Ohkawa K, Ishida H, Sato A, Nakanishi F, Ueda K, Takehara T, Kasahara A, Sasaki Y, Hori M, Hayashi N (2003) Hepatitis C virus core protein differently regulates the JAK-STAT signaling pathway under interleukin-6 and interferon-gamma stimuli. J Biol Chem 278:28562–28571

Hsu M, Zhang J, Flint M, Logvinoff C, Cheng-Mayer C, Rice CM, Mckeating JA (2003) Hepatitis C virus glycoproteins mediate pH-dependent cell entry of pseudotyped retroviral particles. Proc Natl Acad Sci U S A 100:7271–7276

Huang H, Sun F, Owen DM, Li W, Chen Y, Gale M Jr, Ye J (2007) Hepatitis C virus production by human hepatocytes dependent on assembly and secretion of very low-density lipoproteins. Proc Natl Acad Sci U S A 104:5848–5853

Huang JT, Tseng CP, Liao MH, Lu SC, Yeh WZ, Sakamoto N, Chen CM, Cheng JC (2013) Hepatitis C virus replication is modulated by the interaction of nonstructural protein NS5B and fatty acid synthase. J Virol 87:4994–5004

Hueging K, Doepke M, Vieyres G, Bankwitz D, Frentzen A, Doerrbecker J, Gumz F, Haid S, Wolk B, Kaderali L, Pietschmann T (2014) Apolipoprotein E codetermines tissue tropism of hepatitis C virus and is crucial for viral cell-to-cell transmission by contributing to a postenvelopment step of assembly. J Virol 88:1433–1446

Icard V, Diaz O, Scholtes C, Perrin-Cocon L, Ramiere C, Bartenschlager R, Penin F, Lotteau V, Andre P (2009) Secretion of hepatitis C virus envelope glycoproteins depends on assembly of apolipoprotein B positive lipoproteins. PLoS One 4:e4233

Ilnytska O, Santiana M, Hsu NY, Du WL, Chen YH, Viktorova EG, Belov G, Brinker A, Storch J, Moore C, Dixon JL, Altan-Bonnet N (2013) Enteroviruses harness the cellular endocytic machinery to remodel the host cell cholesterol landscape for effective viral replication. Cell Host Microbe 14:281–293

Inoue Y, Aizaki H, Hara H, Matsuda M, Ando T, Shimoji T, Murakami K, Masaki T, Shoji I, Homma S, Matsuura Y, Miyamura T, Wakita T, Suzuki T (2011) Chaperonin TRiC/CCT participates in replication of hepatitis C virus genome via interaction with the viral NS5B protein. Virology 410:38–47

Isken O, Grassmann CW, Sarisky RT, Kann M, Zhang S, Grosse F, Kao PN, Behrens SE (2003) Members of the NF90/NFAR protein group are involved in the life cycle of a positive-strand RNA virus. EMBO J 22:5655–5665

Isken O, Baroth M, Grassmann CW, Weinlich S, Ostareck DH, Ostareck-Lederer A, Behrens SE (2007) Nuclear factors are involved in hepatitis C virus RNA replication. RNA 13:1675–1692

Israelow B, Narbus CM, Sourisseau M, Evans MJ (2014) HepG2 cells mount an effective antiviral interferon-lambda based innate immune response to hepatitis C virus infection. Hepatology 60:1170–1179

Jangra RK, Yi M, Lemon SM (2010) Regulation of hepatitis C virus translation and infectious virus production by the microRNA miR-122. J Virol 84:6615–6625

Janssen HL, Reesink HW, Lawitz EJ, Zeuzem S, Rodriguez-Torres M, Patel K, van der Meer AJ, Patick AK, Chen A, Zhou Y, Persson R, King BD, Kauppinen S, Levin AA, Hodges MR (2013) Treatment of HCV infection by targeting microRNA. N Engl J Med 368:1685–1694

Ji H, Fraser CS, Yu Y, Leary J, Doudna JA (2004) Coordinated assembly of human translation initiation complexes by the hepatitis C virus internal ribosome entry site RNA. Proc Natl Acad Sci U S A 101:16990–16995

Jiang J, Cun W, Wu X, Shi Q, Tang H, Luo G (2012) Hepatitis C virus attachment mediated by apolipoprotein E binding to cell surface heparan sulfate. J Virol 86:7256–7267

Jones CT, Catanese MT, Law LM, Khetani SR, Syder AJ, Ploss A, Oh TS, Schoggins JW, MacDonald MR, Bhatia SN, Rice CM (2010) Real-time imaging of hepatitis C virus infection using a fluorescent cell-based reporter system. Nat Biotechnol 28:167–171

Jopling CL, Yi M, Lancaster AM, Lemon SM, Sarnow P (2005) Modulation of hepatitis C virus RNA abundance by a liver-specific MicroRNA. Science 309:1577–1581

Jopling CL, Schutz S, Sarnow P (2008) Position-dependent function for a tandem microRNA miR-122-binding site located in the hepatitis C virus RNA genome. Cell Host Microbe 4:77–85

Jordan TX, Randall G (2012) Manipulation or capitulation: virus interactions with autophagy. Microbes Infect 14:126–139

Joyce MA, Walters KA, Lamb SE, Yeh MM, Zhu LF, Kneteman N, Doyle JS, Katze MG, Tyrrell DL (2009) HCV induces oxidative and ER stress, and sensitizes infected cells to apoptosis in SCID/Alb-uPA mice. PLoS Pathog 5:e1000291

Kambara H, Fukuhara T, Shiokawa M, Ono C, Ohara Y, Kamitani W, Matsuura Y (2012) Establishment of a novel permissive cell line for the propagation of hepatitis C virus by expression of microRNA miR122. J Virol 86:1382–1393

Kandathil AJ, Graw F, Quinn J, Hwang HS, Torbenson M, Perelson AS, Ray SC, Thomas DL, Ribeiro RM, Balagopal A (2013) Use of laser capture microdissection to map hepatitis C virus-positive hepatocytes in human liver. Gastroenterology 145(1404–13):e1–e10

Kaneko T, Tanji Y, Satoh S, Hijikata M, Asabe S, Kimura K, Shimotohno K (1994) Production of two phosphoproteins from the NS5A region of the hepatitis C viral genome. Biochem Biophys Res Commun 205:320–326

Kaul A, Stauffer S, Berger C, Pertel T, Schmitt J, Kallis S, Zayas M, Lohmann V, Luban J, Bartenschlager R (2009) Essential role of cyclophilin A for hepatitis C virus replication and virus production and possible link to polyprotein cleavage kinetics. PLoS Pathog 5:e1000546

Ke PY, Chen SS (2011) Activation of the unfolded protein response and autophagy after hepatitis C virus infection suppresses innate antiviral immunity in vitro. J Clin Invest 121:37–56

Khan I, Katikaneni DS, Han Q, Sanchez-Felipe L, Hanada K, Ambrose RL, Mackenzie JM, Konan KV (2014) Modulation of hepatitis C virus genome replication by glycosphingolipids and four-phosphate adaptor protein 2. J Virol 88:12276–12295
Kieft JS, Zhou K, Jubin R, Doudna JA (2001) Mechanism of ribosome recruitment by hepatitis C IRES RNA. RNA 7:194–206
Kim J, Lee D, Choe J (1999) Hepatitis C virus NS5A protein is phosphorylated by casein kinase II. Biochem Biophys Res Commun 257:777–781
Kim JH, Hahm B, Kim YK, Choi M, Jang SK (2000) Protein-protein interaction among hnRNPs shuttling between nucleus and cytoplasm. J Mol Biol 298:395–405
Kim SJ, Kim JH, Kim YG, Lim HS, Oh JW (2004) Protein kinase C-related kinase 2 regulates hepatitis C virus RNA polymerase function by phosphorylation. J Biol Chem 279:50031–50041
Kim SJ, Syed GH, Siddiqui A (2013) Hepatitis C virus induces the mitochondrial translocation of Parkin and subsequent mitophagy. PLoS Pathog 9:e1003285
Kim SJ, Syed GH, Khan M, Chiu WW, Sohail MA, Gish RG, Siddiqui A (2014) Hepatitis C virus triggers mitochondrial fission and attenuates apoptosis to promote viral persistence. Proc Natl Acad Sci U S A 111:6413–6418
Koutsoudakis G, Herrmann E, Kallis S, Bartenschlager R, Pietschmann T (2007) The level of CD81 cell surface expression is a key determinant for productive entry of hepatitis C virus into host cells. J Virol 81:588–598
Kumthip K, Chusri P, Jilg N, Zhao L, Fusco DN, Zhao H, Goto K, Cheng D, Schaefer EA, Zhang L, Pantip C, Thongsawat S, O'Brien A, Peng LF, Maneekarn N, Chung RT, Lin W (2012) Hepatitis C virus NS5A disrupts STAT1 phosphorylation and suppresses type I interferon signaling. J Virol 86:8581–8591
Lan KH, Lan KL, Lee WP, Sheu ML, Chen MY, Lee YL, Yen SH, Chang FY, Lee SD (2007) HCV NS5A inhibits interferon-alpha signaling through suppression of STAT1 phosphorylation in hepatocyte-derived cell lines. J Hepatol 46:759–767
Lanford RE, Hildebrandt-Eriksen ES, Petri A, Persson R, Lindow M, Munk ME, Kauppinen S, Orum H (2010) Therapeutic silencing of microRNA-122 in primates with chronic hepatitis C virus infection. Science 327:198–201
Lange CM, Jacobson IM, Rice CM, Zeuzem S (2014) Emerging therapies for the treatment of hepatitis C. EMBO Mol Med 6:4–15
Lee JY, Acosta EG, Stoeck IK, Long G, Hiet MS, Mueller B, Fackler OT, Kallis S, Bartenschlager R (2014) Apolipoprotein e likely contributes to a maturation step of infectious hepatitis C virus particles and interacts with viral envelope glycoproteins. J Virol 88:12422–12437
Li K, Foy E, Ferreon JC, Nakamura M, Ferreon AC, Ikeda M, Ray SC, Gale M Jr, Lemon SM (2005) Immune evasion by hepatitis C virus NS3/4A protease-mediated cleavage of the Toll-like receptor 3 adaptor protein TRIF. Proc Natl Acad Sci U S A 102:2992–2997
Li Q, Brass AL, Ng A, Hu Z, Xavier RJ, Liang TJ, Elledge SJ (2009) A genome-wide genetic screen for host factors required for hepatitis C virus propagation. Proc Natl Acad Sci U S A 106:16410–16415
Li YP, Gottwein JM, Scheel TK, Jensen TB, Bukh J (2011) MicroRNA-122 antagonism against hepatitis C virus genotypes 1-6 and reduced efficacy by host RNA insertion or mutations in the HCV 5′ UTR. Proc Natl Acad Sci U S A 108:4991–4996
Li K, Li NL, Wei D, Pfeffer SR, Fan M, Pfeffer LM (2012) Activation of chemokine and inflammatory cytokine response in hepatitis C virus-infected hepatocytes depends on Toll-like receptor 3 sensing of hepatitis C virus double-stranded RNA intermediates. Hepatology 55:666–675
Li Q, Pene V, Krishnamurthy S, Cha H, Liang TJ (2013) Hepatitis C virus infection activates an innate pathway involving IKK-alpha in lipogenesis and viral assembly. Nat Med 19:722–729
Li Q, Zhang YY, Chiu S, Hu Z, Lan KH, Cha H, Sodroski C, Zhang F, Hsu CS, Thomas E, Liang TJ (2014a) Integrative functional genomics of hepatitis C virus infection identifies host dependencies in complete viral replication cycle. PLoS Pathog 10:e1004163

Li Y, Masaki T, Shimakami T, Lemon SM (2014b) hnRNP L and NF90 interact with hepatitis C virus 5′-terminal untranslated RNA and promote efficient replication. J Virol 88:7199–7209

Liang Y, Shilagard T, Xiao SY, Snyder N, Lau D, Cicalese L, Weiss H, Vargas G, Lemon SM (2009) Visualizing hepatitis C virus infections in human liver by two-photon microscopy. Gastroenterology 137:1448–1458

Liefhebber JM, Hague CV, Zhang Q, Wakelam MJ, Mclauchlan J (2014) Modulation of triglyceride and cholesterol ester synthesis impairs assembly of infectious hepatitis C virus. J Biol Chem 289:21276–21288

Lin W, Kim SS, Yeung E, Kamegaya Y, Blackard JT, Kim KA, Holtzman MJ, Chung RT (2006) Hepatitis C virus core protein blocks interferon signaling by interaction with the STAT1 SH2 domain. J Virol 80:9226–9235

Liu Z, Yang F, Robotham JM, Tang H (2009) Critical role of cyclophilin A and its prolyl-peptidyl isomerase activity in the structure and function of the hepatitis C virus replication complex. J Virol 83:6554–6565

Lohmann V, Korner F, Herian U, Bartenschlager R (1997) Biochemical properties of hepatitis C virus NS5B RNA-dependent RNA polymerase and identification of amino acid sequence motifs essential for enzymatic activity. J Virol 71:8416–8428

Lohmann V, Korner F, Koch J, Herian U, Theilmann L, Bartenschlager R (1999) Replication of subgenomic hepatitis C virus RNAs in a hepatoma cell line. Science 285:110–113

Long G, Hiet MS, Windisch MP, Lee JY, Lohmann V, Bartenschlager R (2011) Mouse hepatic cells support assembly of infectious hepatitis C virus particles. Gastroenterology 141:1057–1066

Lupberger J, Zeisel MB, Xiao F, Thumann C, Fofana I, Zona L, Davis C, Mee CJ, Turek M, Gorke S, Royer C, Fischer B, Zahid MN, Lavillette D, Fresquet J, Cosset FL, Rothenberg SM, Pietschmann T, Patel AH, Pessaux P, Doffoel M, Raffelsberger W, Poch O, Mckeating JA, Brino L, Baumert TF (2011) EGFR and EphA2 are host factors for hepatitis C virus entry and possible targets for antiviral therapy. Nat Med 17:589–595

Lytle JR, Wu L, Robertson HD (2001) The ribosome binding site of hepatitis C virus mRNA. J Virol 75:7629–7636

Machlin ES, Sarnow P, Sagan SM (2011) Masking the 5′ terminal nucleotides of the hepatitis C virus genome by an unconventional microRNA-target RNA complex. Proc Natl Acad Sci U S A 108:3193–3198

Madan V, Paul D, Lohmann V, Bartenschlager R (2014) Inhibition of HCV replication by cyclophilin antagonists is linked to replication fitness and occurs by inhibition of membranous web formation. Gastroenterology 146(1361–72):e1–e9

Mani N, Yuzhakov A, Yuzhakov O, Coll JT, Black J, Saxena K, Fulghum JR et al (2015) Nonstructural protein 5A (NS5A) and human replication protein A increase the processivity of hepatitis C virus NS5B polymerase activity in vitro. J Virol 89(1):165–180

Manna D, Aligo J, Xu C, Park WS, Koc H, Heo WD, Konan KV (2010) Endocytic Rab proteins are required for hepatitis C virus replication complex formation. Virology 398:21–37

Martin DN, Uprichard SL (2013) Identification of transferrin receptor 1 as a hepatitis C virus entry factor. Proc Natl Acad Sci U S A 110:10777–10782

Masaki T, Matsunaga S, Takahashi H, Nakashima K, Kimura Y, Ito M, Matsuda M, Murayama A, Kato T, Hirano H, Endo Y, Lemon SM, Wakita T, Sawasaki T, Suzuki T (2014) Involvement of hepatitis C virus NS5A hyperphosphorylation mediated by casein kinase I-alpha in infectious virus production. J Virol 88:7541–7555

Mclauchlan J, Lemberg MK, Hope G, Martoglio B (2002) Intramembrane proteolysis promotes trafficking of hepatitis C virus core protein to lipid droplets. EMBO J 21:3980–3988

Meertens L, Bertaux C, Dragic T (2006) Hepatitis C virus entry requires a critical postinternalization step and delivery to early endosomes via clathrin-coated vesicles. J Virol 80:11571–11578

Menzel N, Fischl W, Hueging K, Bankwitz D, Frentzen A, Haid S, Gentzsch J, Kaderali L, Bartenschlager R, Pietschmann T (2012) MAP-kinase regulated cytosolic phospholipase A2 activity is essential for production of infectious hepatitis C virus particles. PLoS Pathog 8: e1002829

Merz A, Long G, Hiet MS, Brugger B, Chlanda P, Andre P, Wieland F, Krijnse-Locker J, Bartenschlager R (2011) Biochemical and morphological properties of hepatitis C virus particles and determination of their lipidome. J Biol Chem 286:3018–3032

Meuleman P, Hesselgesser J, Paulson M, Vanwolleghem T, Desombere I, Reiser H, Leroux-Roels G (2008) Anti-CD81 antibodies can prevent a hepatitis C virus infection in vivo. Hepatology 48:1761–1768

Meylan E, Curran J, Hofmann K, Moradpour D, Binder M, Bartenschlager R, Tschopp J (2005) Cardif is an adaptor protein in the RIG-I antiviral pathway and is targeted by hepatitis C virus. Nature 437:1167–1172

Miller K, Mcardle S, Gale MJ, Geller DA Jr, Tenoever B, Hiscott J, Gretch DR, Polyak SJ (2004) Effects of the hepatitis C virus core protein on innate cellular defense pathways. J Interferon Cytokine Res 24:391–402

Miyanari Y, Atsuzawa K, Usuda N, Watashi K, Hishiki T, Zayas M, Bartenschlager R, Wakita T, Hijikata M, Shimotohno K (2007) The lipid droplet is an important organelle for hepatitis C virus production. Nat Cell Biol 9:1089–1097

Moriya K, Nakagawa K, Santa T, Shintani Y, Fujie H, Miyoshi H, Tsutsumi T, Miyazawa T, Ishibashi K, Horie T, Imai K, Todoroki T, Kimura S, Koike K (2001) Oxidative stress in the absence of inflammation in a mouse model for hepatitis C virus-associated hepatocarcinogenesis. Cancer Res 61:4365–4370

Narbus CM, Israelow B, Sourisseau M, Michta ML, Hopcraft SE, Zeiner GM, Evans MJ (2011) HepG2 cells expressing microRNA miR-122 support the entire hepatitis C virus life cycle. J Virol 85:12087–12092

Nasheri N, Joyce M, Rouleau Y, Yang P, Yao S, Tyrrell DL, Pezacki JP (2013) Modulation of fatty acid synthase enzyme activity and expression during hepatitis C virus replication. Chem Biol 20:570–582

Neddermann P, Quintavalle M, DI Pietro C, Clementi A, Cerretani M, Altamura S, Bartholomew L, DE Francesco R (2004) Reduction of hepatitis C virus NS5A hyperphosphorylation by selective inhibition of cellular kinases activates viral RNA replication in cell culture. J Virol 78:13306–13314

Neufeldt CJ, Joyce MA, Levin A, Steenbergen RH, Pang D, Shields J, Tyrrell DL, Wozniak RW (2013) Hepatitis C virus-induced cytoplasmic organelles use the nuclear transport machinery to establish an environment conducive to virus replication. PLoS Pathog 9:e1003744

Neveu G, Barouch-Bentov R, Ziv-Av A, Gerber D, Jacob Y, Einav S (2012) Identification and targeting of an interaction between a tyrosine motif within hepatitis C virus core protein and AP2M1 essential for viral assembly. PLoS Pathog 8:e1002845

Nevo-Yassaf I, Yaffe Y, Asher M, Ravid O, Eizenberg S, Henis YI, Nahmias Y, Hirschberg K, Sklan EH (2012) Role for TBC1D20 and Rab1 in hepatitis C virus replication via interaction with lipid droplet-bound nonstructural protein 5A. J Virol 86:6491–6502

Ng TI, Mo H, Pilot-Matias T, He Y, Koev G, Krishnan P, Mondal R, Pithawalla R, He W, Dekhtyar T, Packer J, Schurdak M, Molla A (2007) Identification of host genes involved in hepatitis C virus replication by small interfering RNA technology. Hepatology 45:1413–1421

Nielsen SU, Bassendine MF, Burt AD, Martin C, Pumeechockchai W, Toms GL (2006) Association between hepatitis C virus and very-low-density lipoprotein (VLDL)/LDL analyzed in iodixanol density gradients. J Virol 80:2418–2428

Oem JK, Jackel-Cram C, Li YP, Zhou Y, Zhong J, Shimano H, Babiuk LA, Liu Q (2008) Activation of sterol regulatory element-binding protein 1c and fatty acid synthase transcription by hepatitis C virus non-structural protein 2. J Gen Virol 89:1225–1230

Okuda M, Li K, Beard MR, Showalter LA, Scholle F, Lemon SM, Weinman SA (2002) Mitochondrial injury, oxidative stress, and antioxidant gene expression are induced by hepatitis C virus core protein. Gastroenterology 122:366–375

Oshiumi H, Miyashita M, Matsumoto M, Seya T (2013) A distinct role of Riplet-mediated K63-Linked polyubiquitination of the RIG-I repressor domain in human antiviral innate immune responses. PLoS Pathog 9:e1003533

Otsuka M, Kato N, Moriyama M, Taniguchi H, Wang Y, Dharel N, Kawabe T, Omata M (2005) Interaction between the HCV NS3 protein and the host TBK1 protein leads to inhibition of cellular antiviral responses. Hepatology 41:1004–1012

Otto GA, Puglisi JD (2004) The pathway of HCV IRES-mediated translation initiation. Cell 119:369–380

Owen DM, Huang H, Ye J, Gale M Jr (2009) Apolipoprotein E on hepatitis C virion facilitates infection through interaction with low-density lipoprotein receptor. Virology 394:99–108

Paek KY, Kim CS, Park SM, Kim JH, Jang SK (2008) RNA-binding protein hnRNP D modulates internal ribosome entry site-dependent translation of hepatitis C virus RNA. J Virol 82:12082–12093

Paradis V, Kollinger M, Fabre M, Holstege A, Poynard T, Bedossa P (1997) In situ detection of lipid peroxidation by-products in chronic liver diseases. Hepatology 26:135–142

Park CY, Jun HJ, Wakita T, Cheong JH, Hwang SB (2009) Hepatitis C virus nonstructural 4B protein modulates sterol regulatory element-binding protein signaling via the AKT pathway. J Biol Chem 284:9237–9246

Park SM, Paek KY, Hong KY, Jang CJ, Cho S, Park JH, Kim JH, Jan E, Jang SK (2011) Translation-competent 48S complex formation on HCV IRES requires the RNA-binding protein NSAP1. Nucleic Acids Res 39:7791–7802

Paul D, Hoppe S, Saher G, Krijnse-Locker J, Bartenschlager R (2013) Morphological and biochemical characterization of the membranous hepatitis C virus replication compartment. J Virol 87:10612–10627

Penin F, Dubuisson J, Rey FA, Moradpour D, Pawlotsky JM (2004) Structural biology of hepatitis C virus. Hepatology 39:5–19

Pileri P, Uematsu Y, Campagnoli S, Galli G, Falugi F, Petracca R, Weiner AJ, Houghton M, Rosa D, Grandi G, Abrignani S (1998) Binding of hepatitis C virus to CD81. Science 282:938–941

Ploss A, Evans MJ, Gaysinskaya VA, Panis M, You H, DE Jong YP, Rice CM (2009) Human occludin is a hepatitis C virus entry factor required for infection of mouse cells. Nature 457:882–886

Pudi R, Srinivasan P, Das S (2004) La protein binding at the GCAC site near the initiator AUG facilitates the ribosomal assembly on the hepatitis C virus RNA to influence internal ribosome entry site-mediated translation. J Biol Chem 279:29879–29888

Quintavalle M, Sambucini S, DI Pietro C, DE Francesco R, Neddermann P (2006) The alpha isoform of protein kinase CKI is responsible for hepatitis C virus NS5A hyperphosphorylation. J Virol 80:11305–11312

Quintavalle M, Sambucini S, Summa V, Orsatti L, Talamo F, DE Francesco R, Neddermann P (2007) Hepatitis C virus NS5A is a direct substrate of casein kinase I-alpha, a cellular kinase identified by inhibitor affinity chromatography using specific NS5A hyperphosphorylation inhibitors. J Biol Chem 282:5536–5544

Ramakrishnaiah V, Thumann C, Fofana I, Habersetzer F, Pan Q, De Ruiter PE, Willemsen R, Demmers JA, Stalin Raj V, Jenster G, Kwekkeboom J, Tilanus HW, Haagmans BL, Baumert TF, Van Der Laan LJ (2013) Exosome-mediated transmission of hepatitis C virus between human hepatoma Huh7.5 cells. Proc Natl Acad Sci U S A 110:13109–13113

Randall G, Panis M, Cooper JD, Tellinghuisen TL, Sukhodolets KE, Pfeffer S, Landthaler M, Landgraf P, Kan S, Lindenbach BD, Chien M, Weir DB, Russo JJ, Ju J, Brownstein MJ, Sheridan R, Sander C, Zavolan M, Tuschl T, Rice CM (2007) Cellular cofactors affecting hepatitis C virus infection and replication. Proc Natl Acad Sci U S A 104:12884–12889

Rautou PE, Cazals-Hatem D, Feldmann G, Mansouri A, Grodet A, Barge S, Martinot-Peignoux M, Duces A, Bieche I, Lebrec D, Bedossa P, Paradis V, Marcellin P, Valla D, Asselah T, Moreau R

(2011) Changes in autophagic response in patients with chronic hepatitis C virus infection. Am J Pathol 178:2708–2715
Reed KE, Xu J, Rice CM (1997) Phosphorylation of the hepatitis C virus NS5A protein in vitro and in vivo: properties of the NS5A-associated kinase. J Virol 71:7187–7197
Reed KE, Gorbalenya AE, Rice CM (1998) The NS5A/NS5 proteins of viruses from three genera of the family flaviviridae are phosphorylated by associated serine/threonine kinases. J Virol 72:6199–6206
Reiss S, Rebhan I, Backes P, Romero-Brey I, Erfle H, Matula P, Kaderali L, Poenisch M, Blankenburg H, Hiet MS, Longerich T, Diehl S, Ramirez F, Balla T, Rohr K, Kaul A, Buhler S, Pepperkok R, Lengauer T, Albrecht M, Eils R, Schirmacher P, Lohmann V, Bartenschlager R (2011) Recruitment and activation of a lipid kinase by hepatitis C virus NS5A is essential for integrity of the membranous replication compartment. Cell Host Microbe 9:32–45
Reiss S, Harak C, Romero-Brey I, Radujkovic D, Klein R, Ruggieri A, Rebhan I, Bartenschlager R, Lohmann V (2013) The lipid kinase phosphatidylinositol-4 kinase III alpha regulates the phosphorylation status of hepatitis C virus NS5A. PLoS Pathog 9:e1003359
Rescher U, Ruhe D, Ludwig C, Zobiack N, Gerke V (2004) Annexin 2 is a phosphatidylinositol (4,5)-bisphosphate binding protein recruited to actin assembly sites at cellular membranes. J Cell Sci 117:3473–3480
Roberts AP, Lewis AP, Jopling CL (2011) miR-122 activates hepatitis C virus translation by a specialized mechanism requiring particular RNA components. Nucleic Acids Res 39:7716–7729
Romero-Brey I, Merz A, Chiramel A, Lee JY, Chlanda P, Haselman U, Santarella-Mellwig R, Habermann A, Hoppe S, Kallis S, Walther P, Antony C, Krijnse-Locker J, Bartenschlager R (2012) Three-dimensional architecture and biogenesis of membrane structures associated with hepatitis C virus replication. PLoS Pathog 8:e1003056
Rubbia-Brandt L, Quadri R, Abid K, Giostra E, Male PJ, Mentha G, Spahr L, Zarski JP, Borisch B, Hadengue A, Negro F (2000) Hepatocyte steatosis is a cytopathic effect of hepatitis C virus genotype 3. J Hepatol 33:106–115
Ruggieri A, Dazert E, Metz P, Hofmann S, Bergeest JP, Mazur J, Bankhead P, Hiet MS, Kallis S, Alvisi G, Samuel CE, Lohmann V, Kaderali L, Rohr K, Frese M, Stoecklin G, Bartenschlager R (2012) Dynamic oscillation of translation and stress granule formation mark the cellular response to virus infection. Cell Host Microbe 12:71–85
Sainz B Jr, Barretto N, Martin DN, Hiraga N, Imamura M, Hussain S, Marsh KA, Yu X, Chayama K, Alrefai WA, Uprichard SL (2012) Identification of the Niemann-Pick C1-like 1 cholesterol absorption receptor as a new hepatitis C virus entry factor. Nat Med 18:281–285
Saito T, Owen DM, Jiang F, Marcotrigiano J, Gale M Jr (2008) Innate immunity induced by composition-dependent RIG-I recognition of hepatitis C virus RNA. Nature 454:523–527
Salloum S, Wang H, Ferguson C, Parton RG, Tai AW (2013) Rab18 binds to hepatitis C virus NS5A and promotes interaction between sites of viral replication and lipid droplets. PLoS Pathog 9:e1003513
Saxena V, Lai CK, Chao TC, Jeng KS, Lai MM (2012) Annexin A2 is involved in the formation of hepatitis C virus replication complex on the lipid raft. J Virol 86:4139–4150
Scarselli E, Ansuini H, Cerino R, Roccasecca RM, Acali S, Filocamo G, Traboni C, Nicosia A, Cortese R, Vitelli A (2002) The human scavenger receptor class B type I is a novel candidate receptor for the hepatitis C virus. EMBO J 21:5017–5025
Schnell G, Loo YM, Marcotrigiano J, Gale M Jr (2012) Uridine composition of the poly-U/UC tract of HCV RNA defines non-self recognition by RIG-I. PLoS Pathog 8:e1002839
Schoggins JW, Wilson SJ, Panis M, Murphy MY, Jones CT, Bieniasz P, Rice CM (2011) A diverse range of gene products are effectors of the type I interferon antiviral response. Nature 472:481–485
Sedano CD, Sarnow P (2014) Hepatitis C virus subverts liver-specific miR-122 to protect the viral genome from exoribonuclease Xrn2. Cell Host Microbe 16:257–264

Sharma NR, Mateu G, Dreux M, Grakoui A, Cosset FL, Melikyan GB (2011) Hepatitis C virus is primed by CD81 protein for low pH-dependent fusion. J Biol Chem 286:30361–30376

Sheahan T, Imanaka N, Marukian S, Dorner M, Liu P, Ploss A, Rice CM (2014) Interferon lambda alleles predict innate antiviral immune responses and hepatitis C virus permissiveness. Cell Host Microbe 15:190–202

Shi ST, Lee KJ, Aizaki H, Hwang SB, Lai MM (2003) Hepatitis C virus RNA replication occurs on a detergent-resistant membrane that cofractionates with caveolin-2. J Virol 77:4160–4168

Shi Q, Jiang J, Luo G (2013) Syndecan-1 serves as the major receptor for attachment of hepatitis C virus to the surfaces of hepatocytes. J Virol 87:6866–6875

Shimakami T, Yamane D, Jangra RK, Kempf BJ, Spaniel C, Barton DJ, Lemon SM (2012) Stabilization of hepatitis C virus RNA by an Ago2-miR-122 complex. Proc Natl Acad Sci U S A 109:941–946

Shrivastava S, Raychoudhuri A, Steele R, Ray R, Ray RB (2011) Knockdown of autophagy enhances the innate immune response in hepatitis C virus-infected hepatocytes. Hepatology 53:406–414

Sir D, Chen WL, Choi J, Wakita T, Yen TS, Ou JH (2008) Induction of incomplete autophagic response by hepatitis C virus via the unfolded protein response. Hepatology 48:1054–1061

Sklan EH, Serrano RL, Einav S, Pfeffer SR, Lambright DG, Glenn JS (2007a) TBC1D20 is a Rab1 GTPase-activating protein that mediates hepatitis C virus replication. J Biol Chem 282:36354–36361

Sklan EH, Staschke K, Oakes TM, Elazar M, Winters M, Aroeti B, Danieli T, Glenn JS (2007b) A Rab-GAP TBC domain protein binds hepatitis C virus NS5A and mediates viral replication. J Virol 81:11096–11105

Sourisseau M, Michta ML, Zony C, Israelow B, Hopcraft SE, Narbus CM, Parra Martin A, Evans MJ (2013) Temporal analysis of hepatitis C virus cell entry with occludin directed blocking antibodies. PLoS Pathog 9:e1003244

Spahn CM, Kieft JS, Grassucci RA, Penczek PA, Zhou K, Doudna JA, Frank J (2001) Hepatitis C virus IRES RNA-induced changes in the conformation of the 40s ribosomal subunit. Science 291:1959–1962

Stone M, Jia S, Heo WD, Meyer T, Konan KV (2007) Participation of rab5, an early endosome protein, in hepatitis C virus RNA replication machinery. J Virol 81:4551–4563

Su AI, Pezacki JP, Wodicka L, Brideau AD, Supekova L, Thimme R, Wieland S, Bukh J, Purcell RH, Schultz PG, Chisari FV (2002) Genomic analysis of the host response to hepatitis C virus infection. Proc Natl Acad Sci U S A 99:15669–15674

Sulkowski MS, Kang M, Matining R, Wyles D, Johnson VA, Morse GD, Amorosa V, Bhattacharya D, Coughlin K, Wong-Staal F, Glesby MJ, Team ACTGAP (2014) Safety and antiviral activity of the HCV entry inhibitor ITX5061 in treatment-naive HCV-infected adults: a randomized, double-blind, phase 1b study. J Infect Dis 209:658–667

Sumpter R Jr, Loo YM, Foy E, Li K, Yoneyama M, Fujita T, Lemon SM, Gale M Jr (2005) Regulating intracellular antiviral defense and permissiveness to hepatitis C virus RNA replication through a cellular RNA helicase, RIG-I. J Virol 79:2689–2699

Supekova L, Supek F, Lee J, Chen S, Gray N, Pezacki JP, Schlapbach A, Schultz PG (2008) Identification of human kinases involved in hepatitis C virus replication by small interference RNA library screening. J Biol Chem 283:29–36

Suzuki R, Matsuda M, Watashi K, Aizaki H, Matsuura Y, Wakita T, Suzuki T (2013) Signal peptidase complex subunit 1 participates in the assembly of hepatitis C virus through an interaction with E2 and NS2. PLoS Pathog 9:e1003589

Syder AJ, Lee H, Zeisel MB, Grove J, Soulier E, MacDonald J, Chow S, Chang J, Baumert TF, Mckeating JA, Mckelvy J, Wong-Staal F (2011) Small molecule scavenger receptor BI antagonists are potent HCV entry inhibitors. J Hepatol 54:48–55

Taguchi T, Nagano-Fujii M, Akutsu M, Kadoya H, Ohgimoto S, Ishido S, Hotta H (2004) Hepatitis C virus NS5A protein interacts with 2′,5′-oligoadenylate synthetase and inhibits antiviral activity of IFN in an IFN sensitivity-determining region-independent manner. J Gen Virol 85:959–969

Tai AW, Salloum S (2011) The role of the phosphatidylinositol 4-kinase PI4KA in hepatitis C virus-induced host membrane rearrangement. PLoS One 6:e26300

Tai AW, Benita Y, Peng LF, Kim SS, Sakamoto N, Xavier RJ, Chung RT (2009) A functional genomic screen identifies cellular cofactors of hepatitis C virus replication. Cell Host Microbe 5:298–307

Tamai K, Shiina M, Tanaka N, Nakano T, Yamamoto A, Kondo Y, Kakazu E, Inoue J, Fukushima K, Sano K, Ueno Y, Shimosegawa T, Sugamura K (2012) Regulation of hepatitis C virus secretion by the Hrs-dependent exosomal pathway. Virology 422:377–385

Tanida I, Fukasawa M, Ueno T, Kominami E, Wakita T, Hanada K (2009) Knockdown of autophagy-related gene decreases the production of infectious hepatitis C virus particles. Autophagy 5:937–945

Targett-Adams P, Hope G, Boulant S, Mclauchlan J (2008) Maturation of hepatitis C virus core protein by signal peptide peptidase is required for virus production. J Biol Chem 283:16850–16859

Tellinghuisen TL, Foss KL, Treadaway J (2008) Regulation of hepatitis C virion production via phosphorylation of the NS5A protein. PLoS Pathog 4:e1000032

Trotard M, Lepere-Douard C, Regeard M, Piquet-Pellorce C, Lavillette D, Cosset FL, Gripon P, LE Seyec J (2009) Kinases required in hepatitis C virus entry and replication highlighted by small interference RNA screening. FASEB J 23:3780–3789

Tsai WC, Hsu SD, Hsu CS, Lai TC, Chen SJ, Shen R, Huang Y, Chen HC, Lee CH, Tsai TF, Hsu MT, Wu JC, Huang HD, Shiao MS, Hsiao M, Tsou AP (2012) MicroRNA-122 plays a critical role in liver homeostasis and hepatocarcinogenesis. J Clin Invest 122:2884–2897

Tscherne DM, Jones CT, Evans MJ, Lindenbach BD, Mckeating JA, Rice CM (2006) Time- and temperature-dependent activation of hepatitis C virus for low-pH-triggered entry. J Virol 80:1734–1741

Tsuchihara K, Tanaka T, Hijikata M, Kuge S, Toyoda H, Nomoto A, Yamamoto N, Shimotohno K (1997) Specific interaction of polypyrimidine tract-binding protein with the extreme 3′-terminal structure of the hepatitis C virus genome, the 3′X. J Virol 71(9):6720–6726

Tsukiyama-Kohara K, Iizuka N, Kohara M, Nomoto A (1992) Internal ribosome entry site within hepatitis C virus RNA. J Virol 66:1476–1483

Tu H, Gao L, Shi ST, Taylor DR, Yang T, Mircheff AK, Wen Y, Gorbalenya AE, Hwang SB, Lai MM (1999) Hepatitis C virus RNA polymerase and NS5A complex with a SNARE-like protein. Virology 263:30–41

Uzri D, Gehrke L (2009) Nucleotide sequences and modifications that determine RIG-I/RNA binding and signaling activities. J Virol 83:4174–4184

Vaillancourt FH, Pilote L, Cartier M, Lippens J, Liuzzi M, Bethell RC, Cordingley MG, Kukolj G (2009) Identification of a lipid kinase as a host factor involved in hepatitis C virus RNA replication. Virology 387:5–10

Verdegem D, Badillo A, Wieruszeski JM, Landrieu I, Leroy A, Bartenschlager R, Penin F, Lippens G, Hanoulle X (2011) Domain 3 of NS5A protein from the hepatitis C virus has intrinsic alpha-helical propensity and is a substrate of cyclophilin A. J Biol Chem 286:20441–20454

Vescovo T, Romagnoli A, Perdomo AB, Corazzari M, Ciccosanti F, Alonzi T, Nardacci R, Ippolito G, Tripodi M, Garcia-Monzon C, Lo Iacono O, Piacentini M, Fimia GM (2012) Autophagy protects cells from HCV-induced defects in lipid metabolism. Gastroenterology 142:644–653 e3

Vogt DA, Camus G, Herker E, Webster BR, Tsou CL, Greene WC, Yen TS, Ott M (2013) Lipid droplet-binding protein TIP47 regulates hepatitis C Virus RNA replication through interaction with the viral NS5A protein. PLoS Pathog 9:e1003302

VON Hahn T, Schiene-Fischer C, Van ND, Pfaender S, Karavul B, Steinmann E, Potthoff A, Strassburg C, Hamdi N, Abdelaziz AI, Sarrazin C, Muller T, Berg T, Trepo E, Wedemeyer H,

Manns MP, Pietschmann T, Ciesek S (2012) Hepatocytes that express variants of cyclophilin A are resistant to HCV infection and replication. Gastroenterology 143(439–47):e1
Wang C, Gale M Jr, Keller BC, Huang H, Brown MS, Goldstein JL, Ye J (2005) Identification of FBL2 as a geranylgeranylated cellular protein required for hepatitis C virus RNA replication. Mol Cell 18:425–434
Wang N, Liang Y, Devaraj S, Wang J, Lemon SM, Li K (2009) Toll-like receptor 3 mediates establishment of an antiviral state against hepatitis C virus in hepatoma cells. J Virol 83:9824–9834
Wang L, Jeng KS, Lai MM (2011) Poly(C)-binding protein 2 interacts with sequences required for viral replication in the hepatitis C virus (HCV) 5′ untranslated region and directs HCV RNA replication through circularizing the viral genome. J Virol 85:7954–7964
Wang H, Perry JW, Lauring AS, Neddermann P, DE Francesco R, Tai AW (2014) Oxysterol-binding protein is a phosphatidylinositol 4-kinase effector required for HCV replication membrane integrity and cholesterol trafficking. Gastroenterology 146(1373–85):e1–e11
Waris G, Felmlee DJ, Negro F, Siddiqui A (2007) Hepatitis C virus induces proteolytic cleavage of sterol regulatory element binding proteins and stimulates their phosphorylation via oxidative stress. J Virol 81:8122–8130
Watashi K, Hijikata M, Hosaka M, Yamaji M, Shimotohno K (2003) Cyclosporin A suppresses replication of hepatitis C virus genome in cultured hepatocytes. Hepatology 38:1282–1288
Weinlich S, Huttelmaier S, Schierhorn A, Behrens SE, Ostareck-Lederer A, Ostareck DH (2009) IGF2BP1 enhances HCV IRES-mediated translation initiation via the 3′UTR. RNA 15:1528–1542
Wilson JA, Zhang C, Huys A, Richardson CD (2011) Human Ago2 is required for efficient microRNA 122 regulation of hepatitis C virus RNA accumulation and translation. J Virol 85:2342–2350
Witteveldt J, Evans MJ, Bitzegeio J, Koutsoudakis G, Owsianka AM, Angus AG, Keck ZY, Foung SK, Pietschmann T, Rice CM, Patel AH (2009) CD81 is dispensable for hepatitis C virus cell-to-cell transmission in hepatoma cells. J Gen Virol 90:48–58
Woodhouse SD, Narayan R, Latham S, Lee S, Antrobus R, Gangadharan B, Luo S, Schroth GP, Klenerman P, Zitzmann N (2010) Transcriptome sequencing, microarray, and proteomic analyses reveal cellular and metabolic impact of hepatitis C virus infection in vitro. Hepatology 52:443–453
Wu X, Lee EM, Hammack C, Robotham JM, Basu M, Lang J, Brinton MA, Tang H (2014) Cell death-inducing DFFA-like effector b is required for hepatitis C virus entry into hepatocytes. J Virol 88:8433–8444
Xie Z, Klionsky DJ (2007) Autophagosome formation: core machinery and adaptations. Nat Cell Biol 9:1102–1109
Yamane D, Mcgivern DR, Wauthier E, Yi M, Madden VJ, Welsch C, Antes I, Wen Y, Chugh PE, Mcgee CE, Widman DG, Misumi I, Bandyopadhyay S, Kim S, Shimakami T, Oikawa T, Whitmire JK, Heise MT, Dittmer DP, Kao CC, Pitson SM, Merrill AH Jr, Reid LM, Lemon SM (2014) Regulation of the hepatitis C virus RNA replicase by endogenous lipid peroxidation. Nat Med 20:927–935
Yang F, Robotham JM, Nelson HB, Irsigler A, Kenworthy R, Tang H (2008a) Cyclophilin A is an essential cofactor for hepatitis C virus infection and the principal mediator of cyclosporine resistance in vitro. J Virol 82:5269–5278
Yang W, Hood BL, Chadwick SL, Liu S, Watkins SC, Luo G, Conrads TP, Wang T (2008b) Fatty acid synthase is up-regulated during hepatitis C virus infection and regulates hepatitis C virus entry and production. Hepatology 48:1396–1403
Yang F, Robotham JM, Grise H, Frausto S, Madan V, Zayas M, Bartenschlager R, Robinson M, Greenstein AE, Nag A, Logan TM, Bienkiewicz E, Tang H (2010) A major determinant of cyclophilin dependence and cyclosporine susceptibility of hepatitis C virus identified by a genetic approach. PLoS Pathog 6:e1001118

Ye J, Wang C, Sumpter R Jr, Brown MS, Goldstein JL, Gale M Jr (2003) Disruption of hepatitis C virus RNA replication through inhibition of host protein geranylgeranylation. Proc Natl Acad Sci U S A 100:15865–15870

Yen CL, Stone SJ, Koliwad S, Harris C, Farese RV Jr (2008) Thematic review series: glycerolipids. DGAT enzymes and triacylglycerol biosynthesis. J Lipid Res 49:2283–2301

Zahid MN, Turek M, Xiao F, Thi VL, Guerin M, Fofana I, Bachellier P, Thompson J, Delang L, Neyts J, Bankwitz D, Pietschmann T, Dreux M, Cosset FL, Grunert F, Baumert TF, Zeisel MB (2013) The postbinding activity of scavenger receptor class B type I mediates initiation of hepatitis C virus infection and viral dissemination. Hepatology 57:492–504

Zeisel MB, Koutsoudakis G, Schnober EK, Haberstroh A, Blum HE, Cosset FL, Wakita T, Jaeck D, Doffoel M, Royer C, Soulier E, Schvoerer E, Schuster C, Stoll-Keller F, Bartenschlager R, Pietschmann T, Barth H, Baumert TF (2007) Scavenger receptor class B type I is a key host factor for hepatitis C virus infection required for an entry step closely linked to CD81. Hepatology 46:1722–1731

Zhang J, Randall G, Higginbottom A, Monk P, Rice CM, Mckeating JA (2004) CD81 is required for hepatitis C virus glycoprotein-mediated viral infection. J Virol 78:1448–1455

Zhong J, Gastaminza P, Cheng G, Kapadia S, Kato T, Burton DR, Wieland SF, Uprichard SL, Wakita T, Chisari FV (2005) Robust hepatitis C virus infection in vitro. Proc Natl Acad Sci U S A 102:9294–9299

Zona L, Lupberger J, Sidahmed-Adrar N, Thumann C, Harris HJ, Barnes A, Florentin J, Tawar RG, Xiao F, Turek M, Durand SC, Duong FH, Heim MH, Cosset FL, Hirsch I, Samuel D, Brino L, Zeisel MB, LE Naour F, Mckeating JA, Baumert TF (2013) HRas signal transduction promotes hepatitis C virus cell entry by triggering assembly of the host tetraspanin receptor complex. Cell Host Microbe 13:302–313

Lipid Peroxidation and Hepatitis C Virus Replication

Daisuke Yamane and Stanley M. Lemon

Abstract Unlike many other positive-strand RNA viruses, replication of hepatitis C virus (HCV) is difficult to achieve in cell culture due to barriers that are not fully understood. An exception to this rule is the JFH1 strain of HCV that atypically replicates robustly without the need for adaptation to cell culture and is therefore widely used in laboratories. Recent studies have revealed that the RNA replicase of most HCV isolates is equipped with a highly sensitive "lipid peroxidation sensor" that shuts off RNA synthesis upon exposure to cellular lipid peroxidation. HCV variants that lack the capacity to be suppressed by lipid peroxidation, such as JFH1 and several highly-adapted viral mutants, are capable of replicating to high levels in cell culture. Thus, adaptive mutations that render the replicase of HCV resistant to lipid peroxidation appear to be a pre-requisite for efficient replication in cell culture. This sensitivity to lipid peroxidation is unique to HCV among other positive-strand RNA viruses that similarly form active RNA replicase complexes derived from host endoplasmic reticulum membranes. This unique regulatory mechanism likely promotes the long-term persistence of HCV by limiting viral exposure to the immune system. This review focuses on the mechanism of lipid peroxidation-mediated regulation of replication in HCV-infected cells, including what is currently known about viral determinants of replication fitness in cell culture and how adaptive mutations in the replicase proteins of HCV alter their interactions with host membranes to determine sensitivity or resistance to lipid peroxidation.

Keywords Oxidative stress • Lipid peroxidation • RNA replicase

Abbreviations

DAA	direct-acting antiviral
DRM	detergent-resistant membrane

D. Yamane (✉) • S.M. Lemon
Departments of Medicine and Microbiology & Immunology, Lineberger Comprehensive Cancer Center, University of North Carolina, Chapel Hill, NC 27599-7030, USA
e-mail: yamane@email.unc.edu

T. Miyamura et al. (eds.), *Hepatitis C Virus I*, DOI 10.1007/978-4-431-56098-2_10

EC_{50}	50 % effective concentration
ER	endoplasmic reticulum
FBS	Fetal bovine serum
H_2O_2	hydrogen peroxide
HCV	hepatitis C virus
LCMV	lymphocytic choriomeningitis virus
LOX	lipoxygenase
NOS	nitric oxide synthase
NOX	nicotinamide adenine dinucleotide phosphate oxidase
NS	nonstructural protein
PI4K	phosphatidylinositol-4-kinase
PUFA	polyunsaturated fatty acid
RdRp	RNA-dependent RNA polymerase
SKI	sphingosine kinase inhibitor
SPHK	sphingosine kinase
UTR	untranslated RNA

1 Introduction: The Pro-oxidative Nature of HCV Infection

Products of lipid peroxidation frequently accumulate in the liver of patients with chronic hepatitis C (Farinati et al. 1995; Fierbinteanu-Braticevici et al. 2009; Higueras et al. 1994; Konishi et al. 2006; Paradis et al. 1997). Similar pathological findings were recapitulated in transgenic mice models expressing viral protein (s) (Korenaga et al. 2005; Lerat et al. 2002; Moriya et al. 2001; Okuda et al. 2002). Due to technical barriers in growing hepatitis C virus (HCV) efficiently in vitro prior to discovery of JFH1 strain in 2005, numerous early studies of this phenomenon relied on overexpression of HCV protein(s) in hepatic cell lines and in transgenic mouse models. Although the relevance on protein overexpression systems used in these studies is uncertain as viral proteins are expressed at very low levels in chronic hepatitis C patients and in most cases are difficult to detect by conventional methods (Liang et al. 2009), these studies reveal that expression of certain viral proteins, including core and NS5A, is linked to the induction of oxidative stress which can culminate in oxidative DNA damage, apoptotic cell death and oxidative degradation of lipids (a.k.a. lipid peroxidation).

Given the pro-oxidative nature of this oncogenic virus, oxidative tissue damage is considered to be a key feature of HCV pathogenesis contributing to the induction of hepatic fibrosis and carcinogenesis (McGivern and Lemon 2009). However, there is little understanding of the evolutionary forces that have shaped this pro-oxidative property. Has HCV evolved mechanisms to induce oxidative stress for its own benefit, or is the oxidative stress it induces simply the unintended consequences of viral products or possibly a host antiviral response? HCV maintains this pro-oxidative property in persistently infected hosts despite the fact that the RNA replicase of most strains of HCV possesses an exquisitely sensitive and unique "sensor" function that shuts off RNA synthesis in response to oxidative membrane damage (Yamane et al. 2014). It is tempting to speculate that this unique property of HCV may provide a substantial advantage to HCV in establishing long-term persistence in hosts.

2 Early Studies Investigating Lipid Peroxidation and HCV Replication

The modeling of HCV replication in cell culture became technically feasible following the development of sub-genomic or genome-length RNA replicons expressing markers allowing the selection of human hepatoma cells supporting their replication, typically with accumulation of cell culture-adaptive mutations (Ikeda et al. 2002; Lohmann et al. 1999). In vitro studies with such replicons indicate that oxidative stress is induced by replicating HCV RNA in the absence of infectious virus production or immune responses. In addition to studies linking oxidative stress to RNA replication, multiple studies suggest that oxidative stress may in turn regulate viral replication (Fig. 1). Choi et al. demonstrated that hydrogen peroxide-induced oxidative stress rapidly inhibits HCV replication in Huh-7 cells and also in isolated replicase membranes in vitro (Choi et al. 2004, 2006). Along similar lines, Yano et al. reported that lipid-soluble vitamins with anti-oxidative properties (vitamins E and K) have the ability to promote genotype 1b HCV-O replication through inhibition of lipid peroxidation (Yano et al. 2007, 2009).

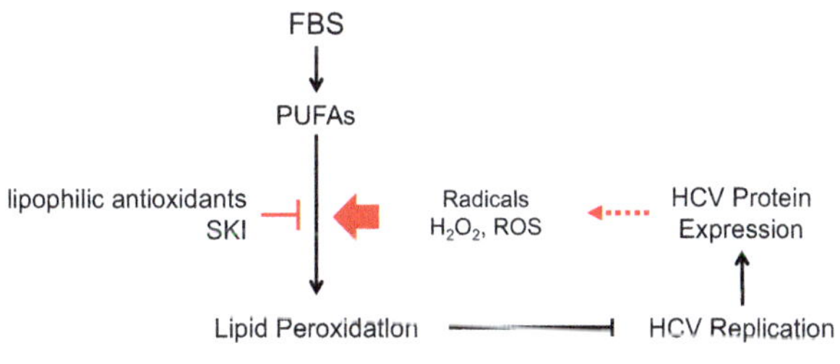

Fig. 1 Pathways leading to lipid peroxidation in HCV-infected cells. As mammalian cells lack desaturase enzymes that are essential for biosynthesis of polyunsaturated fatty acids (PUFAs), essential PUFAs, such as linoleic acid or α-linoleic acid synthesized by lower eukaryotes, are normally provided to cultured cells by fetal bovine serum (FBS). PUFAs are enzymatically or non-enzymatically oxidized by reacting with radicals, such as reactive oxygen species (ROS) and hydrogen peroxide (H_2O_2), to initiate lipid peroxidation, which in turn suppresses HCV replication. In contrast, lipid peroxidation inhibitors, such as lipophilic antioxidants (also typically present in FBS) or sphingosine kinase inhibitor (SKI), increase HCV replication and viral protein expression. HCV proteins induce ROS production followed by lipid peroxidation as a result of mitochondrial injury, an effect that may be magnified by host inflammatory responses, thereby auto-regulating viral replication to low levels and promoting persistence

Polyunsaturated fatty acids (PUFAs) are the dominant substrates for lipid peroxidation as a result of the interaction of their carbon-carbon double bonds with free radicals such as reactive oxygen species (Ayala et al. 2014). This chain reaction can be terminated by lipophilic antioxidants such as α-tocopherol (vitamin E) (Fig. 2). Consistent with this, PUFA-mediated suppression of genotype 1b replicons was completely reversed by supplementing cells with vitamin E (Huang et al. 2007). Importantly, only lipid-soluble, not water-soluble antioxidants have the capacity to protect viral replication from these effects of reactive oxygen species, indicating the importance of limiting oxidative damage within membranes for optimal viral replication.

As described in detail in the chapter by Brey and Lohmann in this volume, the replication factories (replicase complexes) of HCV consist of a complex of viral nonstructural proteins (NS3-NS5B) embedded within a 'web' of membranes formed from the ER. The integrity of this membranous web is fundamentally important for replication of the viral RNA, most likely helping to orchestrate the precise architecture of the replicase complex and the viral proteins that comprise

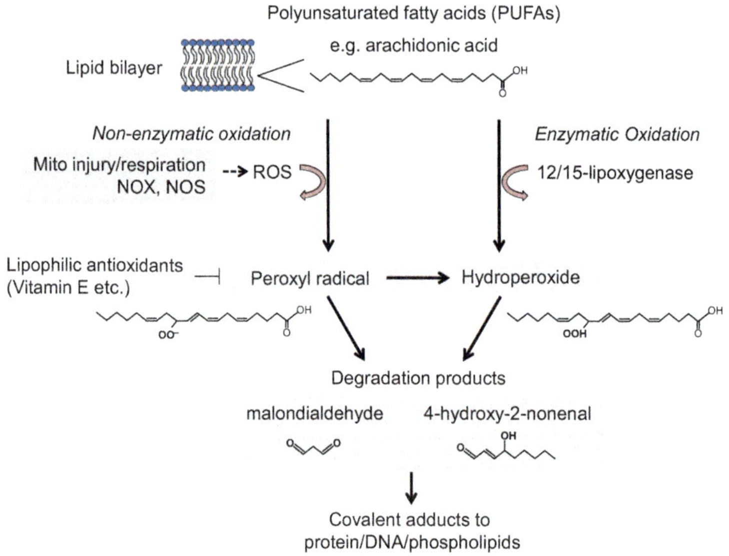

Fig. 2 Peroxidation of polyunsaturated fatty acids (PUFA) esterified to phospholipids in a lipid bilayer. This process is initiated by non-enzymatic reactions with radicals such as reactive oxygen species (ROS) produced as a consequence of mitochondrial (Mito) injury or respiration, by engagement of nicotinamide adenine dinucleotide phosphate oxidase (NOX) or nitric oxide synthase (NOS), or by an enzymatic reaction mediated by 12/15-lipoxygenase (LOX). Peroxyl radicals, a product of non-enzymatic reaction, or enzymatically produced peroxides are subject to degradation to form aldehydes, such as malondialdehyde or 4-hydroxy-2-nonenal, which form adducts with arginine and lysine amino acid side chains in proteins, guanine nucleotides, and phospholipids, thereby altering the function of proteins, nucleic acids and membranes. Lipophilic antioxidants, such as vitamin E or coenzyme Q10, scavenge peroxyl radicals and terminate the peroxidation cascades

it. Thus, it is not surprising that replication might be impaired when these membranes are subject to oxidative damage. Early studies demonstrating this were based on the use of cell culture-adapted genotype 1b replicons. However, it remained unclear whether HCVs of different genotypes as well as wild-type viruses that lack cell culture adaptive mutations are similarly regulated by lipid peroxidation. This question was resolved by a recent study by Yamane et al. that demonstrated that this sensitivity to lipid peroxidation is present in multiple HCV strains (genotypes 1–4), and absent only in a small number of viruses that are extraordinarily well adapted to cell culture, including the unique genotype 2a strain, JFH1, as discussed below (Yamane et al. 2014).

3 Viral Determinants of Robust Replication in Cell Culture

The discovery of the JFH1 strain of HCV, which is capable of robust replication and infectious virus production in human hepatoma Huh-7 cells (Lindenbach et al. 2005; Zhong et al. 2005), was a major breakthrough in the HCV field as it facilitated studies of viral particle secretion pathways and entry receptors, and allowed recapitulation of the entire viral lifecycle in vitro. Several studies have confirmed that JFH1 infection leads to oxidative stress and apoptosis which is associated with mitochondrial injury (Blackham et al. 2010; Deng et al. 2008). Because the replicase proteins (NS3-NS5B) of JFH1 possess a remarkable ability to amplify the JFH1 genome in Huh-7 cells (Kato et al. 2003), chimeric RNAs that express structural proteins derived from other genotypes in the background of the JFH1 genome and replicase have been extensively studied (Chan et al. 2012; Lindenbach et al. 2005; Pietschmann et al. 2006; Yi et al. 2007).

Numerous efforts have been invested in understanding how the JFH1 replicase differs from other HCV replicases that do not typically provide for genome replication in cell culture in the absence of adaptive mutations, and even with such mutations are generally much less efficient in replicating the viral RNA. Murayama et al. analyzed chimeras constructed from the sequences of genotype 2a viruses JFH1 and J6, which by itself replicates poorly in cell culture, and found that inclusion of the NS3 helicase and NS5B-3′ untranslated RNA (UTR) sequences of JFH1 enhances replication fitness of the chimeric RNA (Murayama et al. 2007). Several other studies credit higher RNA-dependent RNA polymerase (RdRp) activity of JFH1 for its robust replication phenotype (Murayama et al. 2010; Ranjith-Kumar et al. 2011; Schmitt et al. 2011). Although a certain amino acid residue in JFH1 NS5B, Ile-405, was shown to be key for this superior RdRp activity (Schmitt et al. 2011), Ile-405 substitutions in different genomic backgrounds increases replication fitness to only a limited extent and fails to fully recapitulate the efficient replication phenotype of JFH1. This reflects the complexity of the viral determinants required for a robust replication phenotype. Following these studies, many questions remained concerning the interaction of JFH1 replicase proteins with host factors and how such interactions influence the efficiency of replication.

4 HCV Replicase and Sphingolipid Metabolism

A unique feature of the JFH1 strain of HCV in terms of its interactions with host lipids is that its replicase activity does not require de novo sphingolipid biosynthesis (Aizaki et al. 2008; Weng et al. 2010), unlike other typical HCV strains (Hirata et al. 2012; Katsume et al. 2013; Sakamoto et al. 2005). Sphingolipids, often coupled with cholesterol, constitute a major component of lipid rafts (or detergent-resistant membranes, DRM) where the replicase complexes responsible for HCV RNA synthesis are presumed to assemble (Aizaki et al. 2004; Shi et al. 2003). Among sphingolipids, sphingomyelin is a key molecule that binds NS5B of genotype 1 HCV strains and regulates its localization on DRM as well as the catalytic activity of a genotype 1b strain (Sakamoto et al. 2005; Weng et al. 2010). Thus, inhibition of sphingolipid biosynthesis depletes the intracellular abundance of sphingomyelin, subsequently releasing NS5B from the DRM to non-DRM fractions and thereby inhibiting catalytic activity of the RdRp. The sphingomyelin-independence of the JFH1 replicase likely explains its unique resistance to inhibition of host sphingolipid biosynthesis.

A detailed comparison of inhibitors targeting different sphingolipid-converting enzymes revealed that the JFH1 and genotype 1a H77S.3 replicases have disparate sphingolipid metabolism requirements for optimal replication (Yamane et al. 2014). H77S.3 is a highly cell culture-adapted, genotype 1a virus with multiple mutations that replicates well in hepatoma cells, producing virus that is infectious in cell culture and chimpanzees (Shimakami et al. 2011; Yi et al. 2014), but that still falls short of JFH1 in its overall replication capacity. Consistent with the requirement for sphingomyelin typical of most HCV strains, inhibitors that lead to depletion of sphingomyelin abundance, such as myriocin, fumonisin B1 and D609, inhibited the genotype 1a H77S.3 replicase, while having very little effect on the JFH1 replicase. Conversely, inhibitors targeting neutral sphingomyelinase, ceramidase, glucosyl ceramide synthase, sphingosine kinase (SPHK), and sphingosine 1-phosphate lyase, all enhanced H77S.3 replication while having no such effect on the JFH1 replicase (Yamane et al. 2014).

This study revealed that a sphingosine kinase (SPHK) inhibitor, SKI, had pronounced differential effect on replication of the two viruses, enhancing H77S.3 replication up to tenfold while inhibiting JFH1 replicase by 50 % (Yamane et al. 2014). Importantly, H77S.3 replicates as efficiently as JFH1 in the presence of 1 μM SKI, indicating that SKI eliminates a cellular restriction that inhibits replication of H77S.3 but not JFH1. Knockdown experiments in which the two SPHK isoforms targeted by SKI (SPHK1 and SPHK2) were depleted by RNA interference revealed that silencing SPHK2 recapitulates the SKI effect on H77S.3 versus JFH1 replication. In contrast, silencing SPHK1 expression had the opposite effect: it reduced H77S.3 but slightly increased JFH1 replication. This is consistent with the observation that 1 μM SKI inhibits only SPHK2 in a cell-free assay (Yamane et al. 2014). Unlike SPHK1, which is predominantly localized to cytosol, the majority of SPHK2 is found present in nuclear and internal membrane fractions

(Hait et al. 2009; Igarashi et al. 2003; Maceyka et al. 2005). SPHK2-catalyzed sphingosine 1-phosphate may localize in close proximity to replicase membranes, but the molecular mechanism by which these different SPHK isoforms function to regulate replicase activity needs further investigation. Importantly, extensive mass spectrometry studies failed to identify a significant difference in any sphingolipid species following treatment of Huh-7 cells with SKI that could be linked to these disparate effects on the H77S.3 and JFH1 replicases (Yamane et al. 2014).

5 SPHK2 Regulates HCV Replication Through Peroxidation of Endogenous PUFAs

Host-targeting compounds, such as cyclosporine A and its derivatives (e.g. alisporivir and SCY-635) (Flisiak et al. 2009; Hopkins et al. 2010; Inoue et al. 2007) and type III phosphatidylinositol-4-kinase (PI4KIIIα) inhibitors (AL-9 and compound-23) (Bianco et al. 2012; Leivers et al. 2014), have been developed and extensively studied for their mechanism of anti-HCV action in vitro and in vivo. In contrast to these compounds that have pan-genotypic antiviral effects, multiple PUFAs, including arachidonic acid, docosahexaenoic acid and linoleic acid, had very little effect against replication driven by the JFH1 replicase, while impairing H77S.3 and genotype 1b HCV-N.2 replicases through induction of lipid peroxidation (Yamane et al. 2014). This observation led to the discovery that the JFH1 replicase has a unique resistance to lipid peroxidation that sets it apart from other typical HCVs. Replication of JFH1 was neither enhanced by lipid peroxidation inhibitors (lipophilic antioxidants) such as vitamin E, nor inhibited by lipid peroxidation inducers such as PUFAs or cumene hydroperoxide. The resistance of JFH1 to lipid peroxidation is not genotype-specific, but rather strain-specific, as JFH-2 (also genotype 2a) shares the lipid peroxidation sensitivity phenotype observed with most HCV strains (Yamane et al. 2014).

Like lipophilic antioxidants, SKI reduces the intracellular abundance of the breakdown products of lipid peroxidation in cell culture such as malondialdehyde and reverses the suppressive effects of PUFAs on genotype 1 viruses, demonstrating a close relationship between the SKI effect and cellular lipid peroxidation. Likewise, silencing SPHK2 expression recapitulates the inhibition of PUFA-induced lipid peroxidation by SKI. In contrast, SPHK1 knockdown increases lipid peroxidation, which likely explains the contrasting effects of SPHK1 and SPHK2 on H77S.3 replication (Yamane et al. 2014). However, it remains unknown whether SPHK2 promotes peroxidation of intracellular PUFAs through modulation of oxidant/antioxidant gene transcription by the nuclear form of SPHK2, or by direct engagement and activity of the kinase within the endoplasmic reticulum (ER) or mitochondrial membranes. Alternatively, SPHK2 may down-regulate HCV replication through alterations in intracellular calcium homeostasis (Maceyka et al. 2005) that have been shown to be required for ROS-mediated suppression of HCV (Choi et al. 2006), or through induction of oxidative stress as a byproduct of aerobic metabolism regulated by mitochondrial SPHK2 (Strub et al. 2011). Although SPHK2 regulation of lipid peroxidation may involve one or more novel mechanism(s), the opposing functions of different SPHK isoforms indicate the importance of localized sphingosine 1-phosphate in redox signaling.

6 Lipid Peroxidation and the HCV Replicase Conformation

How does lipid peroxidation regulate the function of the replicase expressed by typical HCV strains (those other than JFH1) that assemble on membranes? While studying whether lipid peroxidation regulates the stability of H77S.3 RNA by arresting viral RdRp activity using a potent RNA polymerase inhibitor (PSI-6130) and tracing intracellular viral RNA abundance over the ensuing 60 h as described by Shimakami et al. (2012), we unexpectedly found that lipid peroxidation inhibitors (i.e., vitamin E and SKI) completely masked the antiviral effects of the compound on H77S.3 (Yamane et al. 2014). Interestingly, this was observed only with H77S.3, but not with the lipid peroxidation-resistant JFH1 virus that was efficiently inhibited by PSI-6130 regardless of the presence of lipid peroxidation inhibitors. The ability of the peroxidation inhibitors to mask the activity of the polymerase inhibitor against H77S.3 virus was not due only to the increased replication capacity of the H77S.3 replicase under these conditions, but also to a significant, two to threefold increase in the 50 % effective concentration (EC_{50}) of the compound in classic Hill plots. An increased EC_{50} was found with multiple direct-acting antivirals (DAAs) targeting the viral NS3/4A protease (boceprevir and MK-7009) and NS5B polymerase (PSI-6130, MK-0608 and HCV-796), both components of the replicase, whereas this did not occur with host-targeted antivirals directed against cyclophilins, microRNA-122 or PI4KIIIα. Cellular uptake and metabolism of these compounds was not affected by changes in cellular lipid peroxidation status, as the EC_{50} values of the DAAs remained unchanged with the peroxidation-resistant JFH1 replicase.

Theoretically, EC_{50} values calculated from the Hill equation provide a measure of the affinity of a ligand (the DAA) for its receptor (its binding site) (Weiss 1997). Thus, changes in the EC_{50} are likely to reflect alterations in the conformation of the replicase proteins that are bound by the DAAs, due to lipid peroxidation, whereas the conformation of the JFH1 replicase is somehow insensitive to or protected from lipid peroxidation (Fig. 3). Whether these changes in replicase conformation are mediated by altered membrane structure, involving perhaps changes in curvature or fluidity, or by adduct formation involving the viral proteins individually or collectively (see below), is not known.

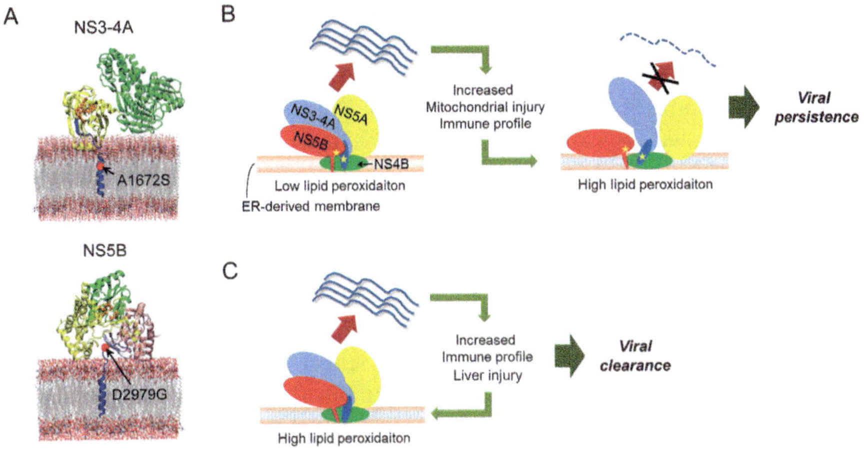

Fig. 3 Models of conformational changes in the replicase complex upon exposure to oxidative membrane damage. (**a**) Structural models showing the putative membrane interactions of NS3-4A (*top*) and NS5B (*bottom*) and the amino acid residues that determine sensitivity to lipid peroxidation. (**b**) A model showing lipid peroxidation-induced changes in the conformation of the replicase of typical HCV, such as H77S and the wildtype HCVs. The replicase activity is turned on in the absence of lipid peroxidation, and down-regulated by oxidative stress due to HCV protein expression or host immune responses driving the accumulation of lipid peroxides in membranes. Yellow stars located in NS4A and NS5B represent amino acid residues that function as "peroxidation sensors". (**c**) Lipid peroxidation-resistant replicases of HCV, such as those expressed by JFH1, H77D, or TNcc viruses that lack a lipid peroxidation sensor. The replicase conformation remains active in the presence of oxidative membrane damage, leading to higher levels of replication that may facilitate host immune recognition and ultimately viral clearance (Figure 3a is reproduced from Yamane et al. 2014)

7 Mutations in HCV Replicase Proteins That Confer Resistance to Lipid Peroxidation

In efforts to identify the specific protein that acts as a lipid peroxidation "sensor" within the viral replicase, sequences encoding individual replicase proteins were exchanged between the lipid peroxidation-sensitive H77S.3 and -resistant JFH1 genomes (Yamane et al. 2014). The results of extensive experiments demonstrated that lipid peroxidation sensitivity is not determined by individual NS4A-4B, NS5A or NS5B proteins alone, but involves multiple NS proteins, probably acting as a multi-protein complex.

A key breakthrough was the discovery that a recently reported, cell culture-adapted and efficiently replicating genotype 1a virus, TNcc, possesses a lipid peroxidation-resistant phenotype similar to that of JFH1. TNcc has eight cell culture adaptive mutations that enable robust replication in Huh-7.5 cells similar to that observed with JFH1 (Li et al. 2012). Strikingly, these eight mutations confer complete resistance to lipid peroxidation when placed in the context of the genotype 1a H77S.3 virus genome in which a key cell culture-adaptive mutation, S2204I, had been reversed to the wild-type sequence. However, despite conferring resistance to lipid peroxidation, the TNcc mutations exerted a strong negative effect on the replication capacity of H77S.3 (Yamane et al. 2014). Subsequent selection of compensatory mutations in the NS4B, NS5A, and NS5B replicase proteins of this virus led to robust replication of H77S.3 carrying the TNcc mutations without loss of resistance to lipid peroxidation. These novel results identified lipid peroxidation resistance as a key factor required for highly efficient HCV replication in cell culture (Yamane et al. 2014).

Additional mutational experiments revealed that TNcc mutations in the NS3 helicase or NS4B do not contribute to the peroxidation resistance phenotype, whereas removal of a mutation in NS4A (A1672S) renders virus susceptible to lipid peroxidation. Additional studies demonstrated that D2979G in NS5B is also required for lipid peroxidation resistance. As neither A1672S nor D2979G alone confers resistance to lipid peroxidation, a combination of these substitutions is necessary for the resistant phenotype. This is in agreement with earlier studies showing that a combination of three mutations F1464L (in NS3 helicase)/A1672S/D2979G (a.k.a. "LSG") provide a robust genetic background for selecting efficiently replicating HCV variants from different genotypes (Li et al. 2012). Although F1464L was found to be dispensable, the lipid peroxidation-resistant phenotype associated with the LSG substitutions likely accounts for the successful selection of cell culture-competent HCV variants reported by this group.

8 Lipid Peroxidation Regulates Conformation of NS3/4A and NS5B

Modeling the A1672S and D2979G mutations within the NS3/4A and NS5B proteins in association with membrane bilayer reveals that both residues are likely to be in close proximity to the membrane (Fig. 3a). Thus, the nature of these residues may directly influence the interaction of the replicase proteins with the membrane under conditions of lipid peroxidation. As the PUFAs that serve as substrates for lipid peroxidation are located within the phospholipid bilayer, A1672S likely alters the mode of NS4A interaction with the membrane. Interestingly, this mutation was recently shown to facilitate dimerization of NS4A (Kohlway et al. 2014). Whether lipid peroxidation alters the oligomerization state of NS4A (and through it, the oligomerization state of NS3) has not been determined. In contrast, D2979G appears to be located within the cytosol in close proximity to the NS5B membrane interaction domain. Given this, it could have an important effect on the conformation of the transmembrane domain of NS5B. These results are consistent with the two to threefold changes in EC_{50} of the direct-acting antiviral agents targeting NS3/4A or NS5A in the context of the lipid peroxidation-sensitive H77S.3, but not resistant H77D variant mentioned above (Yamane et al. 2014).

Exactly how do these mutations in the replicase render it resistant to lipid peroxidation? The breakdown products of lipid peroxidation, such as malondialdehyde and 4-hydroxy-2-nonenal, are capable of passing through membranes because of their hydrophobic nature. They can form adducts with cysteine and lysine residues by alkylating primary amine groups, thereby modulating or inactivating the functions of proteins (Pizzimenti et al. 2013). Thus, it is possible that the substitutions in NS4A and NS5B may alter the conformation of the protein such that it now becomes inaccessible to, or otherwise protected from changes in function due to adduct formation. However, although such protein modifications can be identified by mass spectrometry, studies of adduct formation involving the replicase proteins of HCV have yet to be reported.

An alternative hypothesis would be that lipid peroxidation-sensitive vs. resistant replicase proteins have distinct requirements for specific fatty acids. This is supported by the fact that exogenous fatty acid supplementation increases H77S.3 replication while it has a negligible impact on JFH1 replication (Yamane et al. 2014), consistent with earlier observations in other studies (Kapadia and Chisari 2005; Miyanari et al. 2007). Thus, differential requirements for fatty acids between JFH1 and non-JFH1 replicases might possibly explain their disparate sensitivities to oxidative degradation of cellular fatty acids.

9 Viral Determinants of JFH1 Resistance to Lipid Peroxidation

Although mutations in the transmembrane domain of NS4A (A1672S) and C-terminal linker of NS5B (D2979G) were found to be necessary for lipid peroxidation resistance in the genotype 1a H77D variant, these residues are identical in the peroxidation-sensitive H77S.3 and resistant JFH1 viruses. Thus, alternative amino acid residues in the JFH1 replicase must determine its resistance to lipid peroxidation. Reverse genetics experiments have indicated that the JFH1 NS4A-4B region is not necessary for the resistant phenotype, as JFH1 expressing H77S.3-derived NS4A-4B remains completely resistant to lipid peroxidation (Yamane et al. 2014). Because the NS3 helicase region of JFH1 has been shown to provide for robust replication when placed in the genotype 2a J6 background (Murayama et al. 2007), it is possible that the NS3 helicase sequence of JFH1 influences the mode of membrane interaction of the NS3/4A complex, thereby providing resistance to lipid peroxidation. As the JFH-2 strain, also genotype 2a, is susceptible to lipid peroxidation while being very similar in sequence to JFH1 (87.6 % and 90.6 % at nucleotide and amino acid levels, respectively) (Date et al. 2012), a careful comparison of differences in the membrane interactions of the replicase proteins of these two viruses is likely to allow identification of key amino acid residues that sense oxidative membrane damage.

10 Lipid Peroxidation Sensitivity of Wild-Type HCV

While HCV variants that are adapted to cell culture can be categorized into two distinct phenotypes, lipid peroxidation-sensitive and peroxidation-resistant, based on their response to vitamin E and PUFAs or cumene hydroperoxide in cell culture, a key question is whether lipid peroxidation sensitivity exists in vivo with viruses that contain no cell culture-adaptive mutations. This is a particularly challenging question, as most wild-type HCV strains do not replicate in cell culture with conventional approaches. However, wild-type H77c virus (genotype 1a) can be induced to replicate in cultured cells by suppressing lipid peroxidation, and wild-type HCV-N (genotype 1b) replication is also significantly enhanced when lipid peroxidation is reduced by treatment with vitamin E or SKI (Yamane et al. 2014). These results indicate that these wild-type genotype 1 viruses have a peroxidation-sensitive phenotype, like their cell-culture adapted derivatives, in cell culture (Table 1). The importance of lipid peroxidation in limiting replication of wildtype HCV has been highlighted by the recent discovery that overexpression of SEC14L2, a vitamin E-binding protein that is not expressed in hepatoma cell lines, promotes pan-genotype HCV replication in cell culture through enhancing cellular uptake of vitamin E, thereby protecting replicase membranes from lipid peroxidation (Saeed et al. 2015).

Whether lipid peroxidation acts to restrict replication of HCV within the liver of infected persons is another key question. It has been shown to do so in primary human hepatoblasts (Yamane et al. 2014), but these in vitro studies fall short of the proof one would like to see to confirm the physiologic importance of lipid peroxidation in HCV-infected humans. Studies aimed at answering this question are limited by the lack of availability of robust small animal models supporting the replication of HCV, the presence of substantial amounts of antioxidants in normal mouse feed, and the occurrence of liver disease in inbred strains of laboratory mice when placed on vitamin E-deficient diets (Ibrahim et al. 1997).

While lipid peroxidation-sensitive viruses are capable of being converted to a resistant phenotype by a relatively small number of amino acid substitutions, the majority of HCV strains tested have maintained a peroxidation-sensitive phenotype. This might be interpreted as indicative of a lack of evolutionary pressure having been exerted on the virus by the suppression of replication due to oxidative stress and lipid peroxidation, but it can also be interpreted as suggestive of the virus accruing a survival benefit from maintaining the peroxidation-sensitive phenotype. Importantly, a wide variety of positive-strand viruses classified within the same family as HCV (i.e., so-called "classical" flaviviruses such as dengue virus), or within different families (picornaviruses and alphaviruses) show no evidence of inhibition by lipid peroxidation when cultured under the same conditions as HCV (Yamane et al. 2014). The same is true for lymphocytic choriomeningitis virus (LCMV), an RNA virus that like HCV can persist in infected hosts over long periods. Thus, when one considers the unique "lipid peroxidation-sensitive" as well as "pro-oxidative" features of HCV, it is tempting to speculate that HCV may have evolved these properties to facilitate a negative-feedback mechanism that

Table 1 Sensitivity to lipid peroxidation and infectious titers of different HCVs

Genotype	Virus	Peak infectious titer (FFU ml^{-1})	Enhancement by VE	Susceptibility to cumene hydroperoxide (or PUFA)	Lipid peroxidation sensitivity	References
1a	H77c	<10	Y	Y	Y	Yamane et al. (2014)
	H77S.3	3–5 × 10^3 2 × 10^4 (with VE)	Y	Y	Y	Shimakami et al. (2011) and Yamane et al. (2014)
	H77D	2 × 10^5	N	N	N	Yamane et al. (2014)
	TNcc	6 × 10^4	N	N	N	Li et al. (2012)
1b	HCV-N	<10	Y	Y	Y	Yamane et al. (2014)
	HCV-N.2	1.3 × 10^1 3 × 10^2 (with VE)	Y	Y	Y	Yamane et al. (2014)
	Con1	N/A	Y	Y (PUFA)	Y	Huang et al. (2007) and Kapadia and Chisari (2005)
	HCV-O	N/A	Y	Y (PUFA)	Y	Yano et al. (2007, 2009)
2a	JFH1	4 × 10^5 (Q221L in NS3)	N	N	N	Ma et al. (2008) and Yamane et al. (2014)
	JFH-2	4.3 × 10^4	Y	Y	Y	Date et al. (2012)
3a	S52	N/A	Y	Y	Y	Saeed et al. (2012)
4a	ED43	N/A	Y	Y	Y	Saeed et al. (2012)
JFH1-chimera	HJ3-5	1.5 × 10^6	N	N	N	Yi et al. (2007)

Y Yes, *N* No, *N/A* Not Available, *VE* Vitamin E

limits replication, maintaining it at low levels that may promote its persistence by reducing the silhouette it presents to the immune system.

11 Host Factors That Mediate and Act Downstream of Lipid Peroxidation

Although the underlying mechanism is not understood, SPHK2 is an important mediator of lipid peroxidation (Yamane et al. 2014). It acts to promote oxidation of PUFA, as silencing SPHK2 leads to reduced malondialdehyde synthesis when cells are treated with increasing amounts of PUFA. Multiple studies have indicated that MAPK/ERK signaling may restrict replication of genotype 1b HCV and a serum-derived virus (Ishida et al. 2007; Murata et al. 2005), possibly by mediating oxidative stress-induced antiviral signaling (Yano et al. 2009). However, these results need to be verified by genetic inhibition as these studies were based upon pharmacological inhibition of ERK by PD98059 and in some cases a different inhibitor of ERK, U0126, failed to phenocopy the effect of PD98059 (Yano et al. 2009). Another study has suggested a pro-viral role for MAPK/ERK signaling in the presence of bile acid (Patton et al. 2011). It is noteworthy that MAPK/ERK has been shown to mediate phosphorylation-dependent activation of both SPHK1 and SPHK2 (Hait et al. 2007; Pitson et al. 2003), suggesting the potential for cross-talk between MAPK/ERK and SPHK-regulated redox signaling.

The endogenous substrates for lipid peroxidation (i.e., free- or esterified PUFA) as well as the pathways that mediate oxidation of membrane lipids (enzymatically or non-enzymatically) and are relevant to HCV are yet to be determined. However, replicon cell lines selected with G418 for their capacity to support efficient replication of subgenomic RNA replicons tend to lose vitamin E responsiveness after being passaged for more than 2–3 months (Yano et al. 2007; Yasui et al. 2013). Although it is uncertain whether this is due to the accumulation of viral mutations or selection of cells with minimal lipid peroxidation, the latter may be more likely as we have observed aberrant overexpression of SPHK1 in replicon cells in some cases (Yamane unpublished data). Thus, results from "selected" cells may need to be interpreted with caution as these often represent minority cell populations that have abnormal redox signaling pathways. Overall, this emphasizes the importance of validating results in the context of primary hepatocyte cultures where possible.

12 Conclusive Remarks

HCV is unique among positive strand RNA viruses in that it has an exceptional ability to establish lifelong persistence in hosts without integrating its genome in the host DNA. This may be tightly linked to the unique regulation of its genome amplification by oxidative membrane damage, a feature that sets it apart from other RNA viruses, including hepatitis A virus. This unique regulatory mechanism may largely explain why HCV antigens are so very difficult to detect in the inflamed liver of chronic hepatitis C patients, as well as why wild-type HCV replicate so poorly in cell culture. In contrast, JFH1 virus, which is widely used in research laboratories because of its robust replication capacity, has lost the ability to down-regulate its replicase function by oxidative membrane damage (lipid peroxidation), unlike other, more typical HCV strains. Contrary to its robust replication capacity in vitro, JFH1 or chimeric HCV variants constructed around the JFH1 replicase do not persist in vivo, unlike other HCVs which, in contrast, poorly replicate in cell culture (Zhong et al. 2005; Yi et al. 2014). Collectively, auto-regulation of replicase function by virus-induced oxidative stress may be advantageous for viral persistence.

Despite these advances in our understanding of HCV biology, many questions remain. What happens to the viral replicase upon exposure to lipid peroxidation? Is the replicase subject to covalent adducts of lipid peroxides that inactivate its activity? Are certain lipid species that regulate replicase conformation lost upon oxidative degradation? Are changes in membrane fluidity caused by oxidative membrane damage important for the optimal conformation of the replicase? We believe that comparative analyses of lipids associated with the replicases expressed by lipid peroxidation-sensitive and resistant viruses would provide key insights into these questions, leading to better understanding of a molecular regulatory mechanism that may be important in determining viral persistence.

References

Aizaki H, Lee KJ, Sung VM, Ishiko H, Lai MM (2004) Characterization of the hepatitis C virus RNA replication complex associated with lipid rafts. Virology 324(2):450–461. doi:10.1016/j virol.2004.03.034

Aizaki H, Morikawa K, Fukasawa M, Hara H, Inoue Y, Tani H, Saito K, Nishijima M, Hanada K, Matsuura Y, Lai MM, Miyamura T, Wakita T, Suzuki T (2008) Critical role of virion-associated cholesterol and sphingolipid in hepatitis C virus infection. J Virol 82 (12):5715–5724. doi:10.1128/JVI.02530-07

Ayala A, Munoz MF, Arguelles S (2014) Lipid peroxidation: production, metabolism, and signaling mechanisms of malondialdehyde and 4-hydroxy-2-nonenal. Oxid Med Cell Longev 2014:360438. doi:10.1155/2014/360438

Bianco A, Reghellin V, Donnici L, Fenu S, Alvarez R, Baruffa C, Peri F, Pagani M, Abrignani S, Neddermann P, De Francesco R (2012) Metabolism of phosphatidylinositol 4-kinase IIIalpha-dependent PI4P is subverted by HCV and is targeted by a 4-anilino quinazoline with antiviral activity. PLoS Pathog 8(3):e1002576. doi:10.1371/journal.ppat.1002576

Blackham S, Baillie A, Al-Hababi F, Remlinger K, You S, Hamatake R, McGarvey MJ (2010) Gene expression profiling indicates the roles of host oxidative stress, apoptosis, lipid metabolism, and intracellular transport genes in the replication of hepatitis C virus. J Virol 84 (10):5404–5414. doi:10.1128/JVI.02529-09

Chan K, Cheng G, Beran RK, Yang H, Appleby TC, Pokrovskii MV, Mo H, Zhong W, Delaney WE (2012) An adaptive mutation in NS2 is essential for efficient production of infectious 1b/2a

chimeric hepatitis C virus in cell culture. Virology 422(2):224–234. doi:10.1016/j.virol.2011.10.022
Choi J, Lee KJ, Zheng Y, Yamaga AK, Lai MM, Ou JH (2004) Reactive oxygen species suppress hepatitis C virus RNA replication in human hepatoma cells. Hepatology 39(1):81–89. doi:10.1002/hep.20001
Choi J, Forman HJ, Ou JH, Lai MM, Seronello S, Nandipati A (2006) Redox modulation of the hepatitis C virus replication complex is calcium dependent. Free Radic Biol Med 41(9):1488–1498. doi:10.1016/j.freeradbiomed.2006.08.008
Date T, Kato T, Kato J, Takahashi H, Morikawa K, Akazawa D, Murayama A, Tanaka-Kaneko K, Sata T, Tanaka Y, Mizokami M, Wakita T (2012) Novel cell culture-adapted genotype 2a hepatitis C virus infectious clone. J Virol 86(19):10805–10820. doi:10.1128/JVI.07235-11
Deng L, Adachi T, Kitayama K, Bungyoku Y, Kitazawa S, Ishido S, Shoji I, Hotta H (2008) Hepatitis C virus infection induces apoptosis through a Bax-triggered, mitochondrion-mediated, caspase 3-dependent pathway. J Virol 82(21):10375–10385. doi:10.1128/JVI.00395-08
Farinati F, Cardin R, De Maria N, Della Libera G, Marafin C, Lecis E, Burra P, Floreani A, Cecchetto A, Naccarato R (1995) Iron storage, lipid peroxidation and glutathione turnover in chronic anti-HCV positive hepatitis. J Hepatol 22(4):449–456
Fierbinteanu-Braticevici C, Mohora M, Cretoiu D, Cretoiu S, Petrisor A, Usvat R, Ion DA (2009) Role of oxidative stress in the pathogenesis of chronic hepatitis C (CHC). Rom J Morphol Embryol 50(3):407–412
Flisiak R, Feinman SV, Jablkowski M, Horban A, Kryczka W, Pawlowska M, Heathcote JE, Mazzella G, Vandelli C, Nicolas-Metral V, Grosgurin P, Liz JS, Scalfaro P, Porchet H, Crabbe R (2009) The cyclophilin inhibitor Debio 025 combined with PEG IFNalpha2a significantly reduces viral load in treatment-naive hepatitis C patients. Hepatology 49(5):1460–1468. doi:10.1002/hep.22835
Hait NC, Bellamy A, Milstien S, Kordula T, Spiegel S (2007) Sphingosine kinase type 2 activation by ERK-mediated phosphorylation. J Biol Chem 282(16):12058–12065. doi:10.1074/jbc.M609559200
Hait NC, Allegood J, Maceyka M, Strub GM, Harikumar KB, Singh SK, Luo C, Marmorstein R, Kordula T, Milstien S, Spiegel S (2009) Regulation of histone acetylation in the nucleus by sphingosine-1-phosphate. Science 325(5945):1254–1257. doi:10.1126/science.1176709
Higueras V, Raya A, Rodrigo JM, Serra MA, Roma J, Romero FJ (1994) Interferon decreases serum lipid peroxidation products of hepatitis C patients. Free Radic Biol Med 16(1):131–133
Hirata Y, Ikeda K, Sudoh M, Tokunaga Y, Suzuki A, Weng L, Ohta M, Tobita Y, Okano K, Ozeki K, Kawasaki K, Tsukuda T, Katsume A, Aoki Y, Umehara T, Sekiguchi S, Toyoda T, Shimotohno K, Soga T, Nishijima M, Taguchi R, Kohara M (2012) Self-enhancement of hepatitis C virus replication by promotion of specific sphingolipid biosynthesis. PLoS Pathog 8(8):e1002860. doi:10.1371/journal.ppat.1002860
Hopkins S, Scorneaux B, Huang Z, Murray MG, Wring S, Smitley C, Harris R, Erdmann F, Fischer G, Ribeill Y (2010) SCY-635, a novel nonimmunosuppressive analog of cyclosporine that exhibits potent inhibition of hepatitis C virus RNA replication in vitro. Antimicrob Agents Chemother 54(2):660–672. doi:10.1128/AAC.00660-09
Huang H, Chen Y, Ye J (2007) Inhibition of hepatitis C virus replication by peroxidation of arachidonate and restoration by vitamin E. Proc Natl Acad Sci U S A 104(47):18666–18670. doi:10.1073/pnas.0708423104
Ibrahim W, Lee US, Yeh CC, Szabo J, Bruckner G, Chow CK (1997) Oxidative stress and antioxidant status in mouse liver: effects of dietary lipid, vitamin E and iron. J Nutr 127(7):1401–1406
Igarashi N, Okada T, Hayashi S, Fujita T, Jahangeer S, Nakamura S (2003) Sphingosine kinase 2 is a nuclear protein and inhibits DNA synthesis. J Biol Chem 278(47):46832–46839. doi:10.1074/jbc.M306577200

Ikeda M, Yi M, Li K, Lemon SM (2002) Selectable subgenomic and genome-length dicistronic RNAs derived from an infectious molecular clone of the HCV-N strain of hepatitis C virus replicate efficiently in cultured Huh7 cells. J Virol 76(6):2997–3006

Inoue K, Umehara T, Ruegg UT, Yasui F, Watanabe T, Yasuda H, Dumont JM, Scalfaro P, Yoshiba M, Kohara M (2007) Evaluation of a cyclophilin inhibitor in hepatitis C virus-infected chimeric mice in vivo. Hepatology 45(4):921–928. doi:10.1002/hep.21587

Ishida H, Li K, Yi M, Lemon SM (2007) p21-activated kinase 1 is activated through the mammalian target of rapamycin/p70 S6 kinase pathway and regulates the replication of hepatitis C virus in human hepatoma cells. J Biol Chem 282(16):11836–11848. doi:10.1074/jbc.M610106200

Kapadia SB, Chisari FV (2005) Hepatitis C virus RNA replication is regulated by host geranylgeranylation and fatty acids. Proc Natl Acad Sci U S A 102(7):2561–2566. doi:10.1073/pnas.0409834102

Kato T, Date T, Miyamoto M, Furusaka A, Tokushige K, Mizokami M, Wakita T (2003) Efficient replication of the genotype 2a hepatitis C virus subgenomic replicon. Gastroenterology 125 (6):1808–1817

Katsume A, Tokunaga Y, Hirata Y, Munakata T, Saito M, Hayashi H, Okamoto K, Ohmori Y, Kusanagi I, Fujiwara S, Tsukuda T, Aoki Y, Klumpp K, Tsukiyama-Kohara K, El-Gohary A, Sudoh M, Kohara M (2013) A serine palmitoyltransferase inhibitor blocks hepatitis C virus replication in human hepatocytes. Gastroenterology 145(4):865–873. doi:10.1053/j.gastro.2013.06.012

Kohlway A, Pirakitikulr N, Barrera FN, Potapova O, Engelman DM, Pyle AM, Lindenbach BD (2014) Hepatitis C virus RNA replication and virus particle assembly require specific dimerization of the NS4A protein transmembrane domain. J Virol 88(1):628–642. doi:10.1128/JVI.02052-13

Konishi M, Iwasa M, Yamauchi K, Sugimoto R, Fujita N, Kobayashi Y, Watanabe S, Teraguchi S, Adachi Y, Kaito M (2006) Lactoferrin inhibits lipid peroxidation in patients with chronic hepatitis C. Hepatol Res 36(1):27–32. doi:10.1016/j.hepres.2006.06.005

Korenaga M, Wang T, Li Y, Showalter LA, Chan T, Sun J, Weinman SA (2005) Hepatitis C virus core protein inhibits mitochondrial electron transport and increases reactive oxygen species (ROS) production. J Biol Chem 280(45):37481–37488. doi:10.1074/jbc.M506412200

Leivers AL, Tallant M, Shotwell JB, Dickerson S, Leivers MR, McDonald OB, Gobel J, Creech KL, Strum SL, Mathis A, Rogers S, Moore CB, Botyanszki J (2014) Discovery of selective small molecule type III phosphatidylinositol 4-kinase alpha (PI4KIIIalpha) inhibitors as anti hepatitis C (HCV) agents. J Med Chem 57(5):2091–2106. doi:10.1021/jm400781h

Lerat H, Honda M, Beard MR, Loesch K, Sun J, Yang Y, Okuda M, Gosert R, Xiao SY, Weinman SA, Lemon SM (2002) Steatosis and liver cancer in transgenic mice expressing the structural and nonstructural proteins of hepatitis C virus. Gastroenterology 122(2):352–365

Li YP, Ramirez S, Jensen SB, Purcell RH, Gottwein JM, Bukh J (2012) Highly efficient full-length hepatitis C virus genotype 1 (strain TN) infectious culture system. Proc Natl Acad Sci U S A 109(48):19757–19762. doi:10.1073/pnas.1218260109

Liang Y, Shilagard T, Xiao SY, Snyder N, Lau D, Cicalese L, Weiss H, Vargas G, Lemon SM (2009) Visualizing hepatitis C virus infections in human liver by two-photon microscopy. Gastroenterology 137(4):1448–1458. doi:10.1053/j.gastro.2009.07.050

Lindenbach BD, Evans MJ, Syder AJ, Wolk B, Tellinghuisen TL, Liu CC, Maruyama T, Hynes RO, Burton DR, McKeating JA, Rice CM (2005) Complete replication of hepatitis C virus in cell culture. Science 309(5734):623–626. doi:10.1126/science.1114016

Lohmann V, Korner F, Koch J, Herian U, Theilmann L, Bartenschlager R (1999) Replication of subgenomic hepatitis C virus RNAs in a hepatoma cell line. Science 285(5424):110–113

Ma Y, Yates J, Liang Y, Lemon SM, Yi M (2008) NS3 helicase domains involved in infectious intracellular hepatitis C virus particle assembly. J Virol 82(15):7624–7639. doi:10.1128/JVI.00724-08

Maceyka M, Sankala H, Hait NC, Le Stunff H, Liu H, Toman R, Collier C, Zhang M, Satin LS, Merrill AH Jr, Milstien S, Spiegel S (2005) SphK1 and SphK2, sphingosine kinase isoenzymes with opposing functions in sphingolipid metabolism. J Biol Chem 280(44):37118–37129. doi:10.1074/jbc.M502207200

McGivern DR, Lemon SM (2009) Tumor suppressors, chromosomal instability, and hepatitis C virus-associated liver cancer. Annu Rev Pathol 4:399–415. doi:10.1146/annurev.pathol.4.110807.092202

Miyanari Y, Atsuzawa K, Usuda N, Watashi K, Hishiki T, Zayas M, Bartenschlager R, Wakita T, Hijikata M, Shimotohno K (2007) The lipid droplet is an important organelle for hepatitis C virus production. Nat Cell Biol 9(9):1089–1097. doi:10.1038/ncb1631

Moriya K, Nakagawa K, Santa T, Shintani Y, Fujie H, Miyoshi H, Tsutsumi T, Miyazawa T, Ishibashi K, Horie T, Imai K, Todoroki T, Kimura S, Koike K (2001) Oxidative stress in the absence of inflammation in a mouse model for hepatitis C virus-associated hepatocarcinogenesis. Cancer Res 61(11):4365–4370

Murata T, Hijikata M, Shimotohno K (2005) Enhancement of internal ribosome entry site-mediated translation and replication of hepatitis C virus by PD98059. Virology 340(1):105–115. doi:10.1016/j.virol.2005.06.015

Murayama A, Date T, Morikawa K, Akazawa D, Miyamoto M, Kaga M, Ishii K, Suzuki T, Kato T, Mizokami M, Wakita T (2007) The NS3 helicase and NS5B-to-3′X regions are important for efficient hepatitis C virus strain JFH-1 replication in Huh7 cells. J Virol 81(15):8030–8040. doi:10.1128/JVI.02088-06

Murayama A, Weng L, Date T, Akazawa D, Tian X, Suzuki T, Kato T, Tanaka Y, Mizokami M, Wakita T, Toyoda T (2010) RNA polymerase activity and specific RNA structure are required for efficient HCV replication in cultured cells. PLoS Pathog 6(4):e1000885. doi:10.1371/journal.ppat.1000885

Okuda M, Li K, Beard MR, Showalter LA, Scholle F, Lemon SM, Weinman SA (2002) Mitochondrial injury, oxidative stress, and antioxidant gene expression are induced by hepatitis C virus core protein. Gastroenterology 122(2):366–375

Paradis V, Mathurin P, Kollinger M, Imbert-Bismut F, Charlotte F, Piton A, Opolon P, Holstege A, Poynard T, Bedossa P (1997) In situ detection of lipid peroxidation in chronic hepatitis C: correlation with pathological features. J Clin Pathol 50(5):401–406

Patton JB, George D, Chang KO (2011) Bile acids promote HCV replication through the EGFR/ERK pathway in replicon-harboring cells. Intervirology 54(6):339–348. doi:10.1159/000321452

Pietschmann T, Kaul A, Koutsoudakis G, Shavinskaya A, Kallis S, Steinmann E, Abid K, Negro F, Dreux M, Cosset FL, Bartenschlager R (2006) Construction and characterization of infectious intragenotypic and intergenotypic hepatitis C virus chimeras. Proc Natl Acad Sci U S A 103(19):7408–7413. doi:10.1073/pnas.0504877103

Pitson SM, Moretti PA, Zebol JR, Lynn HE, Xia P, Vadas MA, Wattenberg BW (2003) Activation of sphingosine kinase 1 by ERK1/2-mediated phosphorylation. Embo J 22(20):5491–5500. doi:10.1093/emboj/cdg540

Pizzimenti S, Ciamporcero E, Daga M, Pettazzoni P, Arcaro A, Cetrangolo G, Minelli R, Dianzani C, Lepore A, Gentile F, Barrera G (2013) Interaction of aldehydes derived from lipid peroxidation and membrane proteins. Front Physiol 4:242. doi:10.3389/fphys.2013.00242

Ranjith-Kumar CT, Wen Y, Baxter N, Bhardwaj K, Cheng Kao C (2011) A cell-based assay for RNA synthesis by the HCV polymerase reveals new insights on mechanism of polymerase inhibitors and modulation by NS5A. PLoS One 6(7):e22575. doi:10.1371/journal.pone.0022575

Saeed M, Scheel TK, Gottwein JM, Marukian S, Dustin LB, Bukh J, Rice CM (2012) Efficient replication of genotype 3a and 4a hepatitis C virus replicons in human hepatoma cells. Antimicrob Agents Chemother 56(10):5365–5373. doi:10.1128/AAC.01256-12

Saeed M, Andreo U, Chung HY, Espiritu C, Branch AD, Silva JM, Rice CM (2015) SEC14L2 enables pan-genotype HCV replication in cell culture. Nature 524(7566):471–475. doi:10.1038/nature14899

Sakamoto H, Okamoto K, Aoki M, Kato H, Katsume A, Ohta A, Tsukuda T, Shimma N, Aoki Y, Arisawa M, Kohara M, Sudoh M (2005) Host sphingolipid biosynthesis as a target for hepatitis C virus therapy. Nat Chem Biol 1(6):333–337. doi:10.1038/nchembio742

Schmitt M, Scrima N, Radujkovic D, Caillet-Saguy C, Simister PC, Friebe P, Wicht O, Klein R, Bartenschlager R, Lohmann V, Bressanelli S (2011) A comprehensive structure-function comparison of hepatitis C virus strain JFH1 and J6 polymerases reveals a key residue stimulating replication in cell culture across genotypes. J Virol 85(6):2565–2581. doi:10.1128/JVI.02177-10

Shi ST, Lee KJ, Aizaki H, Hwang SB, Lai MM (2003) Hepatitis C virus RNA replication occurs on a detergent-resistant membrane that cofractionates with caveolin-2. J Virol 77(7):4160–4168

Shimakami T, Welsch C, Yamane D, McGivern DR, Yi M, Zeuzem S, Lemon SM (2011) Protease inhibitor-resistant hepatitis C virus mutants with reduced fitness from impaired production of infectious virus. Gastroenterology 140(2):667–675. doi:10.1053/j.gastro.2010.10.056

Shimakami T, Yamane D, Jangra RK, Kempf BJ, Spaniel C, Barton DJ, Lemon SM (2012) Stabilization of hepatitis C virus RNA by an Ago2-miR-122 complex. Proc Natl Acad Sci U S A 109(3):941–946. doi:10.1073/pnas.1112263109

Strub GM, Paillard M, Liang J, Gomez L, Allegood JC, Hait NC, Maceyka M, Price MM, Chen Q, Simpson DC, Kordula T, Milstien S, Lesnefsky EJ, Spiegel S (2011) Sphingosine-1-phosphate produced by sphingosine kinase 2 in mitochondria interacts with prohibitin 2 to regulate complex IV assembly and respiration. Faseb J 25(2):600–612. doi:10.1096/fj.10-167502

Weiss JN (1997) The Hill equation revisited: uses and misuses. Faseb J 11(11):835–841

Weng L, Hirata Y, Arai M, Kohara M, Wakita T, Watashi K, Shimotohno K, He Y, Zhong J, Toyoda T (2010) Sphingomyelin activates hepatitis C virus RNA polymerase in a genotype-specific manner. J Virol 84(22):11761–11770. doi:10.1128/JVI.00638-10

Yamane D, McGivern DR, Wauthier E, Yi M, Madden VJ, Welsch C, Antes I, Wen Y, Chugh PE, McGee CE, Widman DG, Misumi I, Bandyopadhyay S, Kim S, Shimakami T, Oikawa T, Whitmire JK, Heise MT, Dittmer DP, Kao CC, Pitson SM, Merrill AH Jr, Reid LM, Lemon SM (2014) Regulation of the hepatitis C virus RNA replicase by endogenous lipid peroxidation. Nat Med 20(8):927–935. doi:10.1038/nm.3610

Yano M, Ikeda M, Abe K, Dansako H, Ohkoshi S, Aoyagi Y, Kato N (2007) Comprehensive analysis of the effects of ordinary nutrients on hepatitis C virus RNA replication in cell culture. Antimicrob Agents Chemother 51(6):2016–2027. doi:10.1128/AAC.01426-06

Yano M, Ikeda M, Abe K, Kawai Y, Kuroki M, Mori K, Dansako H, Ariumi Y, Ohkoshi S, Aoyagi Y, Kato N (2009) Oxidative stress induces anti-hepatitis C virus status via the activation of extracellular signal-regulated kinase. Hepatology 50(3):678–688. doi:10.1002/hep.23026

Yasui F, Sudoh M, Arai M, Kohara M (2013) Synthetic lipophilic antioxidant BO-653 suppresses HCV replication. J Med Virol 85(2):241–249. doi:10.1002/jmv.23466

Yi M, Ma Y, Yates J, Lemon SM (2007) Compensatory mutations in E1, p7, NS2, and NS3 enhance yields of cell culture-infectious intergenotypic chimeric hepatitis C virus. J Virol 81(2):629–638. doi:10.1128/JVI.01890-06

Yi M, Hu F, Joyce M, Saxena V, Welsch C, Chavez D, Guerra B, Yamane D, Veselenak R, Pyles R, Walker CM, Tyrrell L, Bourne N, Lanford RE, Lemon SM (2014) Evolution of a cell culture-derived genotype 1a hepatitis C virus (H77S.2) during persistent infection with chronic hepatitis in a chimpanzee. J Virol 88(7):3678–3694. doi:10.1128/JVI.03540-13

Zhong J, Gastaminza P, Cheng G, Kapadia S, Kato T, Burton DR, Wieland SF, Uprichard SL, Wakita T, Chisari FV (2005) Robust hepatitis C virus infection in vitro. Proc Natl Acad Sci U S A 102(26):9294–9299. doi:10.1073/pnas.0503596102

Lipid and Lipoprotein Components Play Important Roles the Egress and Infectivity of Hepatitis C Virions

Takayuki Hishiki, Yuko Shimizu, Saneyuki Ujino, Hironori Nishitsuji, and Kunitada Shimotohno

Abstract Persistent hepatitis C virus (HCV) infection is a leading cause of liver diseases, including chronic hepatitis C, progressive liver fibrosis, cirrhosis, and hepatocellular carcinoma. Chronic hepatitis C is strongly associated with lipid accumulation in the hepatocytes. A significant percentage of chronic HCV patients suffer from diabetes and steatosis. Accumulating evidence indicates a strong relationship between lipid metabolism and HCV proliferation. HCV contains lipoprotein components in its envelope, a unique characteristic that results in the lower buoyant density of HCV virions than those of other enveloped viruses. The incorporation of lipoprotein components is required for viral infection. During the egress of HCV, host factors, such as microsomal triglyceride transfer protein, play some role in the association between lipoprotein and a precursor form of HCV. In this chapter, we summarize how HCV uses the processes of lipid metabolism and discuss the importance of lipoprotein components in the viral lifecycle.

Keywords Lipoprotein • Buoyant density • Apolipoprotein • Lipid droplet • Replication complex • Low density lipoprotein receptor • Scavenger receptor B1

1 Introduction

Even before HCV was isolated as the agent that causes non-A non-B chronic hepatitis (NANBCH), it was known that NANBCH was associated with abnormal lipid metabolism. Notably, the accumulation of neutral lipids in cytosolic lipid droplets within hepatocytes was defined as a pathologic hallmark. HCV is an enveloped positive-stranded RNA virus in the *Flaviviridae* family. Among the

T. Hishiki, Ph.D.
Laboratory of Primate Model, Experimental Research Center for Infectious Diseases, Institute for Virus Research, Kyoto University, Kyoto 606-8507, Japan

Y. Shimizu, Ph.D. • S. Ujino, Ph.D. • H. Nishitsuji, Ph.D. • K. Shimotohno, Ph.D. (✉)
Research Center for Hepatitis and Immunology, National Center for Global Health and Medicine, Ichikawa 272-8516, Chiba, Japan
e-mail: lbshimotohno@hospk.ncgm.go.jp

T. Miyamura et al. (eds.), *Hepatitis C Virus I*, DOI 10.1007/978-4-431-56098-2_11

viruses belonging to this family, propagation of HCV is unique from infection to egress with regards to utilization of the host's processes of lipid metabolism. HCV particles are rich in lipid components that are important for virus infectivity, perhaps including the integration of apolipoprotein E (Apo-E) into HCV virions; this playing a role in the early stages of infection. HCV RNA synthesis occurs in a compartment that is surrounded by cell membrane components. Biochemical analyses have revealed that only a small percentage of the viral nonstructural proteins in infected cells are present in the membranous compartment and contribute to HCV RNA synthesis. The remaining proteins are localized on membranes and exposed directly to the cytoplasm. HCV-infected cells, such as HuH7.5, a subclone derived from the HuH7 hepatocarcinoma cells, also develop lipid droplets. The association between the HCV core and lipid droplets is important for virion production. Virus assembly seems to occur in environments that are rich in core-associating lipid droplets surrounded by membranous components that have been enriched by nonstructural viral proteins.

Here, we outline the role of lipid and lipoprotein components in each step of the HCV lifecycle, including entry, assembly, and egress.

2 The Physical Nature of HCV

The physical and chemical characteristics of the causative agent of post-transfusion non-A, non-B hepatitis (obtained from patient serum or infected chimpanzees) were analyzed using various measures before the discovery of HCV. Microfiltration and ultracentrifugation analyses revealed that the infectious agent has a diameter <80 nm (Bradley et al. 1985). Filtration experiments using the Hutchinson strain (H-strain; i.e., the principal non-A, non-B hepatitis virus) and polycarbonate membranes then revealed that the virus is 30–60 nm in diameter (He et al. 1987). Sensitivity to chloroform suggested that the agent was enveloped (Bradley et al. 1991). This evidence is also well correlated with the nature of Toga- and Flaviviruses, to which non-A, non-B hepatitis agents were classified after the discovery of the HCV genome.

Although initial studies reported that the density of infected chimpanzee-derived HCV was 1.09–1.11 g/mL (Bradley et al. 1991), the buoyant density of HCV was later shown to be discordant between clinical samples (range = 1.08–1.25 g/mL) (Carrick et al. 1992; Miyamoto et al. 1992; Kanto et al. 1994). The H-strain of HCV had a density around 1.06 g/mL but the F-strain formed two peaks at 1.06 and 1.17 g/mL (Hijikata et al. 1993). The difference in densities coincided with association with antihuman immunoglobulin: the high-density HCV fraction was associated with antibody and the low-density virus was not associated with antibody (Thomssen et al. 1993; Choo et al. 1995). However, later studies showed that IgG and IgM are also associated with the serum-derived HCV found in the low-density fractions (Andre et al. 2002). HCV often associates with various blood components, such as plasma lipoproteins, explaining its low density

(Thomssen et al. 1992, 1993). HCV was purified using protein A magnetic beads, which contain Apo-B as the major low-density lipoprotein (LDL) component, and the proposed virus particles were referred to as lipo-viro-particles (LVPs) (Andre et al. 2002).

The ambiguity of HCV structure observed by electron microscopy (EM) may also reflect the nature of the LVPs. EM and immunoabsorption studies of HCV-infected plasma utilizing the anti-HCV envelope reveal spherical virus-like particles that range between 55 and 65 nm in diameter (Kaito et al. 1994). These virus-like particles, which have an electron-dense inner structure, exhibit specific gold labeling when reacted with HCV envelope-derived rabbit polyclonal antibodies. However, these images lack discernible surface features. EM observations of cell culture-derived HCV (HCVcc) also demonstrated similar results (Wakita et al. 2005; Merz et al. 2011). These data suggest that the surface structure of HCV is not sufficiently rigid to be analyzed in detail by EM, which may be due to associations with the lipid structure.

3 HCV Infection Establishes a Specific Cellular Microenvironment That Is More Suited to Viral RNA Replication

In infected cells, positive-stranded RNA viruses form membrane-bound structures that build the structural scaffold for viral RNA replication, which is sometimes called the replication complex. This structure is unique to infected cells. Membrane alterations can be observed in HCV-infected cells using EM (Egger et al. 2002). Distinct patterns of the membranous web of the intracellular membrane are induced by the interaction of HCV proteins with the host cell membranes. The primary membrane alterations are induced in cells expressing HCV NS5B only (Egger et al. 2002). Structural proteins are not essential for establishing a functional web structure because the formation of the membrane-associated replication complex has been observed in cells following the introduction of the autonomously self-replicating HCV subgenome that lacks coding regions for the viral core and envelope proteins (HCV replicons) (Gosert et al. 2003). Nonstructural viral proteins, as well as replicating viral RNA that was identified using metabolic labeling with 5-bromouridine, were also identified within this structure, which indicates that the membranous web is the site for the HCV replication complex (Gosert et al. 2003).

HCV non-structural (NS) proteins are associated with the inner cellular membranes, such as those found in the rough endoplasmic reticulum (ER). However, the proportion of each viral protein in HCV-infected cells that contributes to the active replication complex (which is ultimately responsible for synthesizing the viral genome and mRNA) remains unclear through analysis by EM. To address this question, biochemical analyses of the functional HCV replication complex were conducted on replicon cells (Miyanari et al. 2007). In replicon cells treated with digitonin (which permeabilizes the plasma membrane, although the inner cellular membrane structure remains unchanged), the replication activity of HCV RNA was maintained at the pretreatment level. In the permeabilized cells, most of the HCV proteins were retained inside the plasma membrane, even after thorough washing with certain buffers, indicating that the HCV proteins are retained through their association with insoluble cellular components, possibly membranous components. Importantly, addition of a protease digested most of the HCV proteins in the permeabilized cells to levels that were almost undetectable by western blotting analysis. However, these cells still retained the ability to replicate the subgenomic HCV RNA (Miyanari et al. 2007). The amount of HCV RNA in the permeabilized cells was not drastically reduced by protease treatment. Thus, it is clear that only a small proportion of the HCV proteins plays a role in maintaining HCV RNA synthesis. In fact, very small amounts of the HCV non-structural proteins, NS3, NS4A, NS5A, and NS5B were detectable, following longer exposures in western blotting analysis. However, when the permeabilized cells were treated with detergents such as 1 % NP40, followed by protease treatment, most of the HCV RNA disappeared. This treatment also completely released the HCV proteins from the cells. These observations suggest that small amounts of the external protease-resistant nonstructural HCV proteins and most of the HCV RNA are shielded from external agents that are delivered to the cytoplasmic sites of the permeabilized cells, possibly because of protection by cellular membranous components. The rest of the proteins are positioned on the inner membrane surface without protection by the cellular membrane. Because a very small percentage of the nonstructural HCV proteins are actively responsible for HCV genome replication, the membranous webs observed in EM analyses might reflect the vast majority of the HCV proteins that are not directly involved in active replication complexes. It is possible that the active replication complex is in close proximity to other HCV protein complexes and undergoes dynamic conversion from an "active" to "resting" state. We have included an image of the permeabilized cells that was obtained from our biochemical analysis (Fig. 1).

4 Lipid Droplets Are an Important Organelle for HCV Production

The virus core enhances the formation and accumulation of lipid droplets. This was observed in HCV^{JFH1}-replicating cells. Nearly 50 % of all HCV proteins in HCV^{JFH1}-replicating cells are associated with lipid droplets (Miyanari et al. 2007). Careful confocal microscopic analyses indicate that the NS proteins localize around the lipid droplets as well as in the ER membrane. Fractions that contain lipid droplets, which were partially purified using the floating centrifugation method, were also capable of HCV RNA synthesis. The area that surrounds and encloses the lipid droplets in HCV-replicating cells is rich in membrane-like structures on EM. The accumulation of membranous structures around lipid droplets is a unique characteristic of infected cells and not observed in control, uninfected cells. These data suggest that the active

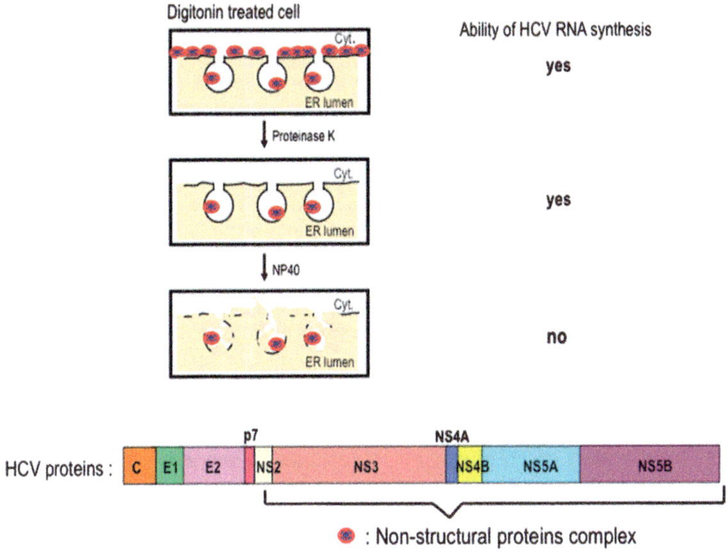

Fig. 1 Subcellular localization of HCV protein complexes. Biochemical analyses showing the distinct localization of HCV nonstructural protein complexes: one is attached to the ER membrane and exposed to the cytoplasmic site and the other resides in the membranous structure. Digitonin +protease treatment destroyed more than 90 % of the total HCV proteins present in these cells. However, the level of HCV RNA synthesis in both digitonin-treated and digitonin+protease-treated cells are almost the same as in untreated cells, indicating that only a small percentage of total HCV proteins contributes to RNA synthesis. However, disruption of the membranous structure entirely abolishes RNA synthesis

HCV replication complex is rich in lipids. In cells bearing a defective HCV genome that does not express the core protein, the NS proteins are mainly localized on the ER and not observed around the lipid droplets: this suggests that the core promotes the localization of NS proteins to lipid droplets. The precise mechanism of the recruitment of NS proteins to the lipid droplets remains unknown. The majority of NS proteins around the lipid droplet are embedded in the membranous structure. The core-coated lipid droplets seem to pull on the membranous structure, via a mechanism that is currently unknown. As described below, NS5A is the likely viral factor involved in interaction between core coated lipid droplets and the membrane because an NS5A-mutant fails to bind to the core and does not associate with lipid droplets. Most core-coated lipid droplets reside near the ER around the nuclear membrane.

Cores with mutations where two proline residues in the central hydrophobic region are converted to alanine (CorePP/AA) are not associated with lipid droplets. When the mutated viral genome that produces CorePP/AA is introduced into cells, the core, as well as the HCV NS proteins, no longer localizes around the lipid droplets. These data indicate that the core, which associates with lipid droplets, recruits other viral proteins to the lipid droplets. Cells expressing this mutant genome do not release virus particles into the culture medium. Virus production may be inhibited by the lack of associations between the viral proteins and lipid droplets, probably a prerequisite for virus production and assembly.

NS5A of HCV-1b is a 466-amino acid protein that can be divided into three domains, domain-I, -II, and -III, from the N-terminus of NS5A. By introducing amino acid substitutions into domain-I using alanine-scanning mutagenesis, NS5A mutants that are not severely affected in terms of genome replication but fail to associate with lipid droplets, were selected. Consequently, NS5A mutants were obtained in which the amino acid residues APK (99–101) and PPT (102–104) were converted to three consecutive alanine residues. The viral RNA genome that encodes each of these NS5A mutants was introduced into cells and virus production was assessed. Only noninfectious virus particles were produced from the cells bearing these replicons. Other mutations in domain-III of NS5A also lacked the interactions between the core and lipid droplets. The HCV genome that carries the mutated NS5A gene cannot produce virus, even though HCV genome replication is unaffected (Masaki et al. 2008). Moreover, cells expressing the mutated genome did not produce virus particles, suggesting NS5A also is involved in the virus assembly process. These observations show the importance of the association between viral proteins and lipid droplets for the production of "infectious" viral particles.

5 Lipid Components Play a Role in HCV Infection

Infectivity studies demonstrate different characteristics in HCV-containing chimpanzee plasma: infectivity is higher in plasma that is rich in low-density HCV (density <1.06 g/mL) than in plasma containing high-density HCV (Bradley et al. 1991; Hijikata et al. 1993). These data suggest that any association with lipoprotein will most likely not only lower the buoyant density of the virus (to between 1.006 and 1.055 g/mL) but also change the infectivity. Infectivity is not always correlated with the fraction profile of HCVcc analyzed for its buoyant density by the amount of the core. A fraction with a low average HCV density demonstrates higher infectivity (Cai et al. 2005). Infection experiments in chimpanzees showed that serum obtained from chimpanzees infected with cultured HCVcc is infectious for cultured cell lines. Moreover, the infectivity of the animal-derived HCV with a low average HCVcc buoyant density is high, which suggests that the physical association with lipid-like factors influences viral infectivity (Lindenbach et al. 2006).

The HCV envelope contains two viral proteins: E1 and E2. Antibodies against epitopes with conserved sequences in the hypervariable E2 region block HCV infection (Kato et al. 1994), indicating the importance of E2 in virus infection. The precise role of E1 in virus infection remains elusive, but the conformational structure of the E1/E2 heterodimer seems necessary for infectivity. The E2 protein binds the tetraspanin, cluster of differentiation (CD) 81 (Pileri et al. 1998). CD81 has also been shown to be involved in the post-binding steps following HCV entry, which result in the internalization of the virion (Bertaux and Dragic 2006; Koutsoudakis et al. 2006; Farquhar et al. 2012). Different infection systems, such as HCV pseudoparticles (HCVpps), recombinant infectious HCVcc, and in vivo infection of human liver grafted into immunodeficient mice (Alb-uPA/SCID), reveal some aspects of the mechanism of HCV infection in hepatocytes (Bartosch et al. 2003; Hsu et al. 2003; Lindenbach et al. 2005; Wakita et al. 2005; Zhong et al. 2005). It is possible that HCV binds to heparan sulfate on the cell surface during the early stages of virus infection. HCV E2 (Barth et al. 2003), or HCV-associated Apo-E, plays a role in this binding (Jiang et al. 2012). Then, HCV associates with cell surface molecules, such as low-density lipoprotein receptor (LDLR), scavenger receptor B1 (SR-BI), and CD81, leading to the hypothesis that these act as viral receptors (Pileri et al. 1998; Monazahian et al. 2000; Petracca et al. 2000; Takikawa et al. 2000; Bartosch et al. 2003). In fact, SR-BI associates with HCV E2 (Scarselli et al. 2002; Albecka et al. 2012). HCV entry is a tightly modulated, multistep process. In addition to those mentioned previously, several other cellular molecules, such as claudin-1 (CLDN1) (Evans et al. 2007) and occludin (OCLN), interact with HCV (Yang et al. 2008; Liu et al. 2009; Ploss et al. 2009). Furthermore, Niemann-Pick C1-like 1 and the epidermal growth factor receptor act as HCV entry factors (Lupberger et al. 2011; Sainz et al. 2012). Very low density lipoprotein receptor (VLDLR) is not expressed in HuH7.5 cells under normoxic culturing condition. However, when the cells are cultured under hypoxic conditions, induction of VLDLR expression can be observed. HCV entry utilizes VLDLR as a receptor without the contribution of CD81 (Ujino et al. 2016), indicating the occurrence of different mechanisms of virus entry. Since human liver specimens express VLDLR, HCV may utilize different entry mechanisms in vivo. The involvement of lipid receptors in virus entry further substantiates the importance of the associations between the lipid components and HCV.

The very-low-density lipoprotein (VLDL) synthesis and secretion pathways in HuH7.5 cells are also correlated with the egress of infectious HCV. Inhibitors of microsomal triglyceride transfer protein significantly attenuate HCV release (Huang et al. 2007; Gastaminza et al. 2008). The components of VLDL, such as Apo-B and Apo-E, have also been implicated in HCV assembly (Chang et al. 2007; Icard et al. 2009). The importance of Apo-E, rather than Apo-B, in HCV secretion has also been reported (Chang et al. 2007). Substantial evidence supports Apo-E as an HCV infectivity factor. Infectious hepatitis C virions that have been purified either from cell culture supernatants (Chang et al. 2007) or infected patients can be specifically immunoprecipitated using anti-Apo-E monoclonal antibodies (Nielsen et al. 2006). Live cell imaging of single HCV particles indicates that mature infectious particles containing Apo-E are transported along secretory pathways (Coller et al. 2012). It has also been suggested that Apo-E may promote HCV infectivity during entry through its interactions with LDLR (Owen et al. 2009) or heparin sulfate (Jiang et al. 2012).

6 Lipoprotein Lipase Disrupts HCV Infectivity

Because the low-density HCVcc fraction infects cultured HuH7.5 cells efficiently, association of low-density component(s), β-lipoprotein according to some reports (Thomssen et al. 1992; Prince et al. 1996), plays a pivotal role in infection. Further support for the role of low-density component(s) in HCV infection is provided by observations of the inhibitory effects of lipoprotein lipase (LPL) treatment on HCV (Thomssen and Bonk 2002). Treating the HCV fraction with *Pseudomonas spp*-derived LPL destroyed the viral structure and reduced the HCV RNA content via virolytic action. Because hepatitis B virus (HBV) is not destroyed by this treatment, the authors of a recent study (Andreo et al. 2007) concluded that the virion structure of HCV is susceptible to LPL treatment. However, other reports (Thomssen and Bonk 2002) contradict these results, suggesting that the inhibitory effects of LPL are due to associations between the virus and cell membrane via bridges with the LPL dimer, rather than via direct virolytic action.

To clarify further the role of LPL in HCV infectivity, we performed a comparative study of LPL, infectivity, and the physical properties of HCV virus particles (Shimizu et al. 2010). In particular, we assessed the effects of LPL treatment on HCV buoyant density (Fig. 2), under the assumption that structural changes in the viral particles were due to LPL treatment-induced lipolytic degradation. First, we observed loss of infectivity by incubating HCV with LPL, indicating that LPL was required to inactivate HCV. After LPL was added to the HCV-containing medium, the medium was used to infect HuH7.5 cells in the presence of an LPL inhibitor,

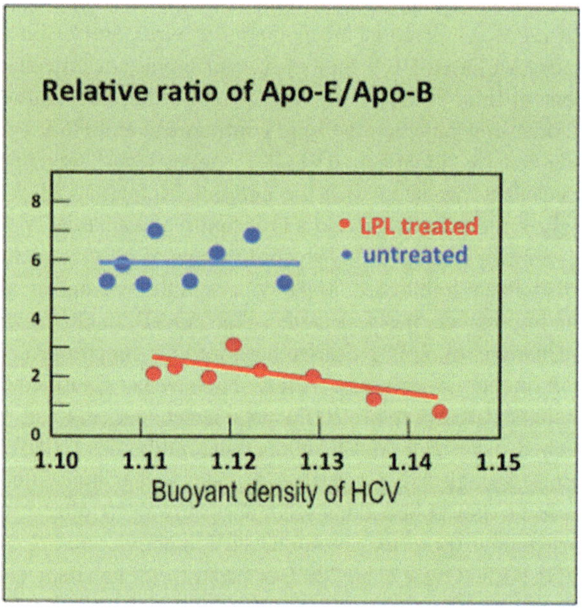

Fig. 2 LPL/HTGL treatment alters the physical and biochemical properties of HCV. HCV released into the culture medium was treated with LPL in the presence of trace amounts of HTGL and the physical properties of HCV were analyzed using density gradient centrifugation. The main average peak shifted from 1.118 to 1.128 g/mL after LPL treatment. Furthermore, the relative proportion of Apo-E to Apo-B was reduced, indicating the deformation of the HCV-associating lipoprotein components by LPL treatment, which may also have diminished infectivity

orlistat. We observed a significant reduction in infectivity, indicating that inhibition directly affects HCV. Since the inhibitory effects of LPL are heat-labile, LPL seems to inhibit infectivity via its lipolytic activities. To confirm that treatment with LPL alters the buoyant density of HCV, we analyzed the physical nature of HCV by ultracentrifugation through an iodixanol gradient. Untreated HCV formed a peak with a density of 1.118 g/mL, as determined by measuring the amount of HCV core protein. Upon treatment with 500 μg/mL LPL, the main peak measured by the amount of core shifted to 1.128 g/mL. A peak shift of HCV RNA also was observed, in parallel with the core shift. Because HCV associates with lipoproteins that have a VLDL- or LDL-like structure, we analyzed the association with Apo-B and Apo-E before and after LPL treatment. Immunoprecipitation using antibodies against Apo-B and Apo-E revealed peak shifts in HCV RNA around 1.13 g/mL; much denser fractions were found following LPL treatment. In fact, the amount of RNA precipitated using anti-Apo-E antibody was less than that before LPL treatment; however, the amount of RNA precipitated with anti-Apo-B antibody before and after LPL treatment did not change substantially. These findings suggest release of Apo-E from virus particles during LPL treatment. The total amount of core protein and HCV RNA in the entire fraction did not change significantly with LPL treatment, indicating that the amount of HCV nucleocapsid was not affected by LPL. Therefore, we suggest that LPL hydrolyzes the lipid components of HCV-associated lipoproteins, which shifts HCV to higher buoyant densities without substantially altering the HCV nucleocapsid. HuH7.5 cells produce hepatic triglyceride lipase (HTGL) and smaller amounts of LPL. In general, during lipolytic processing of VLDL, sequential reactions with LPL and HTGL generate LDL through intermediate-density lipoprotein (IDL) as an intermediate. Assuming that the VLDL-like lipoprotein structure is integrated into HCV, the concerted actions of lipolysis are catalyzed by these enzymes. To identify the lipase involved in HCV inactivation under these conditions, HCV activity was monitored against HTGL activity in the presence of neutralizing antibodies. Then, to evaluate further the role of HTGL on the LPL-induced reduction in HCV infectivity, medium containing HCV was incubated with LPL in the presence of neutralizing antibodies against HTGL, and infectivity was measured. There was a substantial reduction in HCV infectivity under these conditions, indicating that LPL itself potentially could reduce HCV infectivity. Furthermore, HCV infected HuH7.5 cells knocked down for HTGL mRNA by siRNA, produced greater of infectious virus, even though the core protein level in the culture medium remained the same. Ultracentrifugation of HCV produced from HTGL-knockdown cells confirmed the shift of the buoyant density in the lower fraction, in comparison to the virus released from the control cells. HCV's sensitivity to LPL and HTGL treatment indicates the importance of the lipid structure in viral infectivity. Furthermore, these observations suggest that the lipid portion of HCV is somehow different from VLDL.

7 The Role of Apo-E in HCV Infection

The drastic reduction in HCV infectivity following LPL and HTGL treatment coincides with loss of associations between Apo-E and the virion. In fact, the importance of the associations between Apo-E and HCV has been documented in many studies (Gastaminza et al. 2006, 2008; Huang et al. 2007; Jiang and Luo 2009; Owen et al. 2009; Hishiki et al. 2010).

Apo-E and HCV associate before virion egress. Apo-E-knockdown HuH7.5 cells produce noninfectious virus particles and incubating Apo-E with these particles did not restore infectivity, indicating the need for the integration of Apo-E and HCV particles before egress.

Apo-E plays a central role in lipid metabolism and neurobiology. Its three major isoforms—Apo-E2, Apo-E3, and Apo-E4—have different effects on lipid and neuronal homeostasis. These isoforms differ by amino acid substitutions at 1 or 2 sites (residues 130 and 176). Apo-E3 is the most common isoform and there are no reports of any Apo-E3-associated diseases. Apo-E2 is a risk factor for type III hyperlipoproteinemia and Apo-E4 is a risk factor for Alzheimer's disease (Mahley and Rall 2000). Furthermore, Apo-E isoforms affect the extent of damage due to diseases caused by viruses, including herpes simplex virus type 1 (HSV1) and human immunodeficiency virus (HIV) (Wozniak et al. 2002). Previous studies suggest an association between the Apo-E4 allele and severity of HCV-related liver damage (Wozniak et al. 2002).

Lipoprotein receptors, including LDLR and SR-BI, act as entry factors or receptors for HCV infection (Bartosch et al. 2003; Molina et al. 2007; Zeisel et al. 2007; Owen et al. 2009). Because Apo-E2 has a lower affinity for LDLR than Apo-E3 and Apo-E4, we hypothesized that the Apo-E isoform may affect HCV infectivity. To test this hypothesis, we analyzed the production of infectious HCV obtained from cells that express different Apo-E isoforms.

Apo-E isoform-expressing plasmids were transfected into HuH7.5 cells in which endogenous Apo-E3 had previously been knocked down. The ability of HCV to replicate was unaffected in these cells. Although core production was half that observed in parental cells, a similar core level was obtained in each cell line transduced by the Apo-E isoforms. However, the amount of infectious HCV produced in cells bearing Apo-E2 was significantly lower than that produced in Apo-E3- and Apo-E4-producing cells. Assuming LDLR plays a role as an HCV entry factor (Wunschmann et al. 2000; Molina et al. 2007; Owen et al. 2009), this observation is consistent with data showing that the binding affinity of Apo-E for LDLR correlates with differences in HCV infectivity in cells with various Apo-E isoforms.

HCV/Apo-E2 is less infectious for HuH7.5 cells than HCV/Apo-E3. Since the infectivity of HCV/Apo-E2 is similar for the parental HuH7.5 cells and cells which have been silenced for SR-BI, LDLR or LDLR/SR-BI (Hishiki et al. 2010), it is possible that Apo-E2 interacts weakly with LDLR, as well as with SR-BI, although the binding regions for these proteins on the Apo-E molecule do not overlap

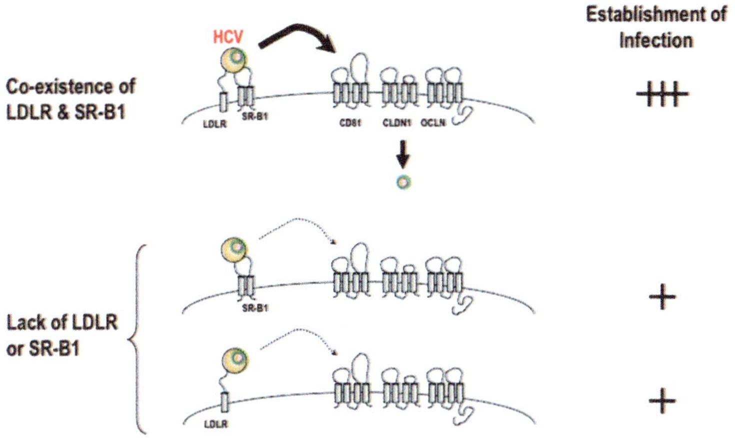

Fig. 3 Requirement of both LDLR and SR-BI expression on target cells for establishment of efficient HCV infection. Suppression of LDLR or SR-BI results in reduced HCV infection

(Li et al. 2002). The importance of expression of both LDLR and SR-BI, in HCV infection was analyzed using HCV/Apo-E3. HCV/Apo-E3 infection studies were carried out in cells in which LDLR, SR-BI, or both had been silenced. As expected from the reported findings of other groups (Owen et al. 2009; Catanese et al. 2010), infectivity was suppressed in LDLR- and SR-BI-silenced HuH7.5 cells. However, to our surprise, infectivity was not reduced further in the doubly silenced cells. This result suggests that both LDLR and SR-BI are required for complete viral entry (Fig. 3). The absence of either of these proteins should reduce infectivity. It is important to determine whether these receptor molecules function independently or cooperatively during HCV entry.

An epidemiological study of Caucasians with persistent chronic hepatitis indicated the notable absence of the Apo-E2/Apo-E2 genotype in HCV antibody-positive individuals (Price et al. 2006), consistent with the observed lower infectivity of Apo-E2-bearing HCVcc. However, our results strongly suggest that HCV produced by Apo-E2-bearing individuals is quickly eliminated due to low infectivity. In the future, it will be important to conduct a virological study on HCV obtained from individuals carrying different Apo-E isoforms.

8 The Role of HCV Proteins in Early-Stage HCV Egress

HCV starts its assembly around lipid droplets that are surrounded by the cell membrane compartments. In this environment, the core needs to associate with lipid droplets in order to produce virus particles. Other HCV proteins mainly localize around the areas surrounding the core-coated lipid droplets (Miyanari et al. 2007). Assembling the virus particle requires the recruitment of NS5A to lipid droplets, where it interacts with core proteins. However, there are no clear immunoelectron microscopic observations showing the direct interactions between the core and NS5A (Miyanari et al. 2007), although biochemical analyses suggest this possibility (Masaki et al. 2008; Hughes et al. 2009; Gawlik et al. 2014).

Domain III of NS5A is important for virus assembly (Appel et al. 2008; Tellinghuisen et al. 2008; Kim et al. 2011). Specifically, some serine residues in domain III need to be phosphorylated in order for assembly to progress (Kim et al. 2011; Masaki et al. 2008). It has been suggested that NS5A associates through domain III with the p7/NS2 complex, another factor required for virus assembly (Jirasko et al. 2010; Ma et al. 2011; Popescu et al. 2011; Scheel et al. 2012), although other researchers argue that this occurs through domain I of NS5A (Gawlik et al. 2014). NS2 plays an important role in assembling other viral proteins, such as E1/E2, p7, and NS3-4A (Phan et al. 2009; Jirasko et al. 2010; Boson et al. 2011; Ma et al. 2011; Popescu et al. 2011). Thus, NS5A may be involved in the formation of the large HCV protein complexes required for the initial stages of virus assembly and may also regulate switches between the replication of the genome and virus assembly via the phosphorylation of domain III. The process of virus budding into the ER lumen is represented in Fig. 4.

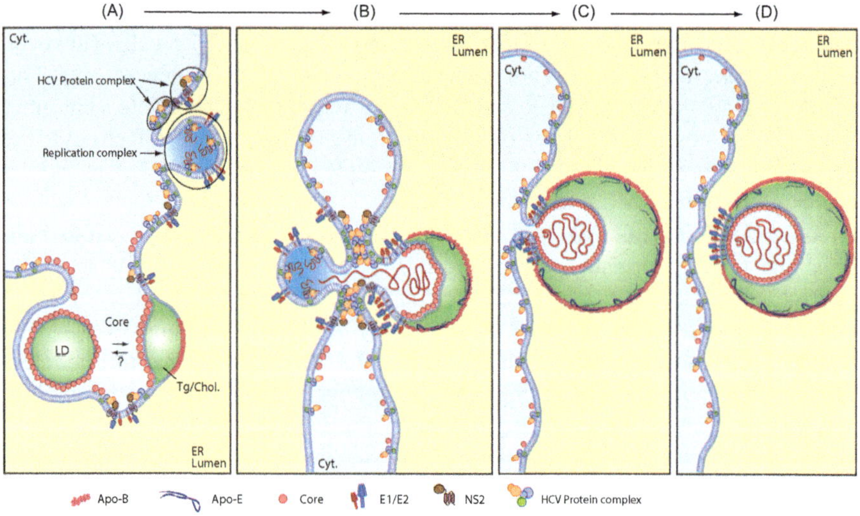

Fig. 4 Model of the HCV assembly process. The HCV core is not always associated with lipid droplets and it may also associate with ER membranes, rich in lipid components in lumen of the bilayer. The HCV replication complex is proximal to the core-coated lipid droplets and the core-associating ER membrane that is rich in lipids may come into contact with the HCV proteins that induce the topology of the structural membrane; therefore, it is easier to encapsidate de novo synthesized HCV RNA when it is associated with the core and lipid-rich capsule. The capsule then starts its maturation process concomitant with the formation of apolipoprotein by the host. Eventually, "pre-virus particles" bud into the luminal area and are possibly released through the ESCRT process via the Golgi apparatus

The fact that the HCV particle density in the intracellular fraction is higher than that in the extracellular fraction suggests that lipidation is not required to initiate particle assembly, but instead follows the formation of virus particles (Gastaminza et al. 2006, 2008). Perhaps, lipidated virus particles use the endosomal sorting complex required for transport (ESCRT) system for virus egress (Corless et al. 2010; Ariumi et al. 2011).

9 Conclusion

Aberrant lipid metabolism is often observed in patients with chronic hepatitis, and HCV infection seems to be related to the development of diabetes and steatohepatitis, which may increase the probability of developing hepatocellular carcinoma. HCV subsumes various lipid metabolism mechanisms during its lifecycle, as described in this chapter. However, the precise mechanisms through which HCV utilizes the components of lipid metabolism have not yet been delineated. Further clarification of such mechanisms may provide new insights into the pathophysiology of HCV infection and help identify therapeutic targets for diseases caused by HCV infection.

References

Albecka A, Belouzard S, Op de Beeck A, Descamps V, Goueslain L, Bertrand-Michel J, Terce F, Duverlie G, Rouille Y, Dubuisson J (2012) Role of low-density lipoprotein receptor in the hepatitis C virus life cycle. Hepatology 55(4):998–1007. doi:10.1002/hep.25501

Andre P, Komurian-Pradel F, Deforges S, Perret M, Berland JL, Sodoyer M, Pol S, Brechot C, Paranhos-Baccala G, Lotteau V (2002) Characterization of low- and very low-density hepatitis C virus RNA-containing particles. J Virol 76(14):6919–6928. doi:10.1128/jvi.76.14.6919-6928.2002

Andreo U, Maillard P, Kalinina O, Walic M, Meurs E, Martinot M, Marcellin P, Budkowska A (2007) Lipoprotein lipase mediates hepatitis C virus (HCV) cell entry and inhibits HCV infection. Cell Microbiol 9(10):2445–2456. doi:10.1111/j.1462-5822.2007.00972.x

Appel N, Zayas M, Miller S, Krijnse-Locker J, Schaller T, Friebe P, Kallis S, Engel U, Bartenschlager R (2008) Essential role of domain III of nonstructural protein 5A for hepatitis C virus infectious particle assembly. PLoS Pathog 4(3):e1000035. doi:10.1371/journal.ppat.1000035

Ariumi Y, Kuroki M, Maki M, Ikeda M, Dansako H, Wakita T, Kato N (2011) The ESCRT system is required for hepatitis C virus production. PLoS One 6(1):e14517. doi:10.1371/journal.pone.0014517

Barth H, Schafer C, Adah MI, Zhang F, Linhardt RJ, Toyoda H, Kinoshita-Toyoda A, Toida T, Van Kuppevelt TH, Depla E, Von Weizsacker F, Blum HE, Baumert TF (2003) Cellular binding of hepatitis C virus envelope glycoprotein E2 requires cell surface heparan sulfate. J Biol Chem 278(42):41003–41012. doi:10.1074/jbc.M302267200

Bartosch B, Vitelli A, Granier C, Goujon C, Dubuisson J, Pascale S, Scarselli E, Cortese R, Nicosia A, Cosset FL (2003) Cell entry of hepatitis C virus requires a set of co-receptors that

include the CD81 tetraspanin and the SR-B1 scavenger receptor. J Biol Chem 278 (43):41624–41630. doi:10.1074/jbc.M305289200
Bertaux C, Dragic T (2006) Different domains of CD81 mediate distinct stages of hepatitis C virus pseudoparticle entry. J Virol 80(10):4940–4948. doi:10.1128/jvi.80.10.4940-4948.2006
Boson B, Granio O, Bartenschlager R, Cosset FL (2011) A concerted action of hepatitis C virus p7 and nonstructural protein 2 regulates core localization at the endoplasmic reticulum and virus assembly. PLoS Pathog 7(7):e1002144. doi:10.1371/journal.ppat.1002144
Bradley DW, McCaustland KA, Cook EH, Schable CA, Ebert JW, Maynard JE (1985) Posttransfusion non-A, non-B hepatitis in chimpanzees. Physicochemical evidence that the tubule-forming agent is a small, enveloped virus. Gastroenterology 88(3):773–779
Bradley D, McCaustland K, Krawczynski K, Spelbring J, Humphrey C, Cook EH (1991) Hepatitis C virus: buoyant density of the factor VIII-derived isolate in sucrose. J Med Virol 34 (3):206–208
Cai Z, Zhang C, Chang KS, Jiang J, Ahn BC, Wakita T, Liang TJ, Luo G (2005) Robust production of infectious hepatitis C virus (HCV) from stably HCV cDNA-transfected human hepatoma cells. J Virol 79(22):13963–13973. doi:10.1128/JVI.79.22.13963-13973.2005
Carrick RJ, Schlauder GG, Peterson DA, Mushahwar IK (1992) Examination of the buoyant density of hepatitis C virus by the polymerase chain reaction. J Virol Methods 39(3):279–289
Catanese MT, Ansuini H, Graziani R, Huby T, Moreau M, Ball JK, Paonessa G, Rice CM, Cortese R, Vitelli A, Nicosia A (2010) Role of scavenger receptor class B type I in hepatitis C virus entry: kinetics and molecular determinants. J Virol 84(1):34–43. doi:10.1128/jvi.02199-08
Chang KS, Jiang J, Cai Z, Luo G (2007) Human apolipoprotein E is required for infectivity and production of hepatitis C virus in cell culture. J Virol 81(24):13783–13793. doi:10.1128/JVI.01091-07
Choo SH, So HS, Cho JM, Ryu WS (1995) Association of hepatitis C virus particles with immunoglobulin: a mechanism for persistent infection. J Gen Virol 76(Pt 9):2337–2341
Coller KE, Heaton NS, Berger KL, Cooper JD, Saunders JL, Randall G (2012) Molecular determinants and dynamics of hepatitis C virus secretion. PLoS Pathog 8(1):e1002466. doi:10.1371/journal.ppat.1002466
Corless L, Crump CM, Griffin SD, Harris M (2010) Vps4 and the ESCRT-III complex are required for the release of infectious hepatitis C virus particles. J Gen Virol 91(Pt 2):362–372. doi:10.1099/vir.0.017285-0
Egger D, Wolk B, Gosert R, Bianchi L, Blum HE, Moradpour D, Bienz K (2002) Expression of hepatitis C virus proteins induces distinct membrane alterations including a candidate viral replication complex. J Virol 76(12):5974–5984
Evans MJ, von Hahn T, Tscherne DM, Syder AJ, Panis M, Wolk B, Hatziioannou T, McKeating JA, Bieniasz PD, Rice CM (2007) Claudin-1 is a hepatitis C virus co-receptor required for a late step in entry. Nature 446(7137):801–805. doi:10.1038/nature05654
Farquhar MJ, Hu K, Harris HJ, Davis C, Brimacombe CL, Fletcher SJ, Baumert TF, Rappoport JZ, Balfe P, McKeating JA (2012) Hepatitis C virus induces CD81 and claudin-1 endocytosis. J Virol 86(8):4305–4316. doi:10.1128/jvi.06996-11
Gastaminza P, Kapadia SB, Chisari FV (2006) Differential biophysical properties of infectious intracellular and secreted hepatitis C virus particles. J Virol 80(22):11074–11081. doi:10.1128/JVI.01150-06
Gastaminza P, Cheng G, Wieland S, Zhong J, Liao W, Chisari FV (2008) Cellular determinants of hepatitis C virus assembly, maturation, degradation, and secretion. J Virol 82(5):2120–2129. doi:10.1128/JVI.02053-07
Gawlik K, Baugh J, Chatterji U, Lim PJ, Bobardt MD, Gallay PA (2014) HCV core residues critical for infectivity are also involved in core-NS5A complex formation. PLoS One 9(2): e88866. doi:10.1371/journal.pone.0088866

Gosert R, Egger D, Lohmann V, Bartenschlager R, Blum HE, Bienz K, Moradpour D (2003) Identification of the hepatitis C virus RNA replication complex in Huh-7 cells harboring subgenomic replicons. J Virol 77(9):5487–5492

He LF, Alling D, Popkin T, Shapiro M, Alter HJ, Purcell RH (1987) Determining the size of non-A, non-B hepatitis virus by filtration. J Infect Dis 156(4):636–640

Hijikata M, Shimizu YK, Kato H, Iwamoto A, Shih JW, Alter HJ, Purcell RH, Yoshikura H (1993) Equilibrium centrifugation studies of hepatitis C virus: evidence for circulating immune complexes. J Virol 67(4):1953–1958

Hishiki T, Shimizu Y, Tobita R, Sugiyama K, Ogawa K, Funami K, Ohsaki Y, Fujimoto T, Takaku H, Wakita T, Baumert TF, Miyanari Y, Shimotohno K (2010) Infectivity of hepatitis C virus is influenced by association with apolipoprotein E isoforms. J Virol 84(22):12048–12057. doi:10.1128/jvi.01063-10

Hsu M, Zhang J, Flint M, Logvinoff C, Cheng-Mayer C, Rice CM, McKeating JA (2003) Hepatitis C virus glycoproteins mediate pH-dependent cell entry of pseudotyped retroviral particles. Proc Natl Acad Sci U S A 100(12):7271–7276. doi:10.1073/pnas.0832180100

Huang H, Sun F, Owen DM, Li W, Chen Y, Gale M Jr, Ye J (2007) Hepatitis C virus production by human hepatocytes dependent on assembly and secretion of very low-density lipoproteins. Proc Natl Acad Sci U S A 104(14):5848–5853. doi:10.1073/pnas.0700760104

Hughes M, Griffin S, Harris M (2009) Domain III of NS5A contributes to both RNA replication and assembly of hepatitis C virus particles. J Gen Virol 90(Pt 6):1329–1334. doi:10.1099/vir.0.009332-0

Icard V, Diaz O, Scholtes C, Perrin-Cocon L, Ramiere C, Bartenschlager R, Penin F, Lotteau V, Andre P (2009) Secretion of hepatitis C virus envelope glycoproteins depends on assembly of apolipoprotein B positive lipoproteins. PLoS One 4(1):e4233. doi:10.1371/journal.pone.0004233

Jiang J, Luo G (2009) Apolipoprotein E but not B is required for the formation of infectious hepatitis C virus particles. J Virol 83(24):12680–12691. doi:10.1128/JVI.01476-09

Jiang J, Cun W, Wu X, Shi Q, Tang H, Luo G (2012) Hepatitis C virus attachment mediated by apolipoprotein E binding to cell surface heparan sulfate. J Virol 86(13):7256–7267. doi:10.1128/JVI.07222-11

Jirasko V, Montserret R, Lee JY, Gouttenoire J, Moradpour D, Penin F, Bartenschlager R (2010) Structural and functional studies of nonstructural protein 2 of the hepatitis C virus reveal its key role as organizer of virion assembly. PLoS Pathog 6(12):e1001233. doi:10.1371/journal.ppat.1001233

Kaito M, Watanabe S, Tsukiyama-Kohara K, Yamaguchi K, Kobayashi Y, Konishi M, Yokoi M, Ishida S, Suzuki S, Kohara M (1994) Hepatitis C virus particle detected by immunoelectron microscopic study. J Gen Virol 75(Pt 7):1755–1760

Kanto T, Hayashi N, Takehara T, Hagiwara H, Mita E, Naito M, Kasahara A, Fusamoto H, Kamada T (1994) Buoyant density of hepatitis C virus recovered from infected hosts: two different features in sucrose equilibrium density-gradient centrifugation related to degree of liver inflammation. Hepatology 19(2):296–302

Kato N, Ootsuyama Y, Sekiya H, Ohkoshi S, Nakazawa T, Hijikata M, Shimotohno K (1994) Genetic drift in hypervariable region 1 of the viral genome in persistent hepatitis C virus infection. J Virol 68(8):4776–4784

Kim S, Welsch C, Yi M, Lemon SM (2011) Regulation of the production of infectious genotype 1a hepatitis C virus by NS5A domain III. J Virol 85(13):6645–6656. doi:10.1128/jvi.02156-10

Koutsoudakis G, Kaul A, Steinmann E, Kallis S, Lohmann V, Pietschmann T, Bartenschlager R (2006) Characterization of the early steps of hepatitis C virus infection by using luciferase reporter viruses. J Virol 80(11):5308–5320. doi:10.1128/JVI.02460-05

Li X, Kan HY, Lavrentiadou S, Krieger M, Zannis V (2002) Reconstituted discoidal ApoE-phospholipid particles are ligands for the scavenger receptor BI. The amino-terminal 1–165 domain of ApoE suffices for receptor binding. J Biol Chem 277(24):21149–21157. doi:10.1074/jbc.M200658200

Lindenbach BD, Evans MJ, Syder AJ, Wolk B, Tellinghuisen TL, Liu CC, Maruyama T, Hynes RO, Burton DR, McKeating JA, Rice CM (2005) Complete replication of hepatitis C virus in cell culture. Science (New York, NY) 309(5734):623–626. doi:10.1126/science.1114016

Lindenbach BD, Meuleman P, Ploss A, Vanwolleghem T, Syder AJ, McKeating JA, Lanford RE, Feinstone SM, Major ME, Leroux-Roels G, Rice CM (2006) Cell culture-grown hepatitis C virus is infectious in vivo and can be recultured in vitro. Proc Natl Acad Sci U S A 103 (10):3805–3809. doi:10.1073/pnas.0511218103

Liu S, Yang W, Shen L, Turner JR, Coyne CB, Wang T (2009) Tight junction proteins claudin-1 and occludin control hepatitis C virus entry and are downregulated during infection to prevent superinfection. J Virol 83(4):2011–2014. doi:10.1128/jvi.01888-08

Lupberger J, Zeisel MB, Xiao F, Thumann C, Fofana I, Zona L, Davis C, Mee CJ, Turek M, Gorke S, Royer C, Fischer B, Zahid MN, Lavillette D, Fresquet J, Cosset FL, Rothenberg SM, Pietschmann T, Patel AH, Pessaux P, Doffoel M, Raffelsberger W, Poch O, McKeating JA, Brino L, Baumert TF (2011) EGFR and EphA2 are host factors for hepatitis C virus entry and possible targets for antiviral therapy. Nat Med 17(5):589–595. doi:10.1038/nm.2341

Ma Y, Anantpadma M, Timpe JM, Shanmugam S, Singh SM, Lemon SM, Yi M (2011) Hepatitis C virus NS2 protein serves as a scaffold for virus assembly by interacting with both structural and nonstructural proteins. J Virol 85(1):86–97. doi:10.1128/jvi.01070-10

Mahley RW, Rall SC Jr (2000) Apolipoprotein E: far more than a lipid transport protein. Annu Rev Genomics Hum Genet 1:507–537. doi:10.1146/annurev.genom.1.1.507

Masaki T, Suzuki R, Murakami K, Aizaki H, Ishii K, Murayama A, Date T, Matsuura Y, Miyamura T, Wakita T, Suzuki T (2008) Interaction of hepatitis C virus nonstructural protein 5A with core protein is critical for the production of infectious virus particles. J Virol 82 (16):7964–7976. doi:10.1128/jvi.00826-08

Merz A, Long G, Hiet MS, Brugger B, Chlanda P, Andre P, Wieland F, Krijnse-Locker J, Bartenschlager R (2011) Biochemical and morphological properties of hepatitis C virus particles and determination of their lipidome. J Biol Chem 286(4):3018–3032. doi:10.1074/jbc.M110.175018

Miyamoto H, Okamoto H, Sato K, Tanaka T, Mishiro S (1992) Extraordinarily low density of hepatitis C virus estimated by sucrose density gradient centrifugation and the polymerase chain reaction. J Gen Virol 73(Pt 3):715–718

Miyanari Y, Atsuzawa K, Usuda N, Watashi K, Hishiki T, Zayas M, Bartenschlager R, Wakita T, Hijikata M, Shimotohno K (2007) The lipid droplet is an important organelle for hepatitis C virus production. Nat Cell Biol 9(9):1089–1097. doi:10.1038/ncb1631

Molina S, Castet V, Fournier-Wirth C, Pichard-Garcia L, Avner R, Harats D, Roitelman J, Barbaras R, Graber P, Ghersa P, Smolarsky M, Funaro A, Malavasi F, Larrey D, Coste J, Fabre JM, Sa-Cunha A, Maurel P (2007) The low-density lipoprotein receptor plays a role in the infection of primary human hepatocytes by hepatitis C virus. J Hepatol 46(3):411–419. doi:10.1016/j.jhep.2006.09.024

Monazahian M, Kippenberger S, Muller A, Seitz H, Bohme I, Grethe S, Thomssen R (2000) Binding of human lipoproteins (low, very low, high density lipoproteins) to recombinant envelope proteins of hepatitis C virus. Med Microbiol Immunol 188(4):177–184

Nielsen SU, Bassendine MF, Burt AD, Martin C, Pumeechockchai W, Toms GL (2006) Association between hepatitis C virus and very-low-density lipoprotein (VLDL)/LDL analyzed in iodixanol density gradients. J Virol 80(5):2418–2428. doi:10.1128/JVI.80.5.2418-2428.2006

Owen DM, Huang H, Ye J, Gale M Jr (2009) Apolipoprotein E on hepatitis C virion facilitates infection through interaction with low-density lipoprotein receptor. Virology 394(1):99–108. doi:10.1016/j.virol.2009.08.037

Petracca R, Falugi F, Galli G, Norais N, Rosa D, Campagnoli S, Burgio V, Di Stasio E, Giardina B, Houghton M, Abrignani S, Grandi G (2000) Structure-function analysis of hepatitis C virus envelope-CD81 binding. J Virol 74(10):4824–4830

Phan T, Beran RK, Peters C, Lorenz IC, Lindenbach BD (2009) Hepatitis C virus NS2 protein contributes to virus particle assembly via opposing epistatic interactions with the E1-E2

glycoprotein and NS3-NS4A enzyme complexes. J Virol 83(17):8379–8395. doi:10.1128/jvi.00891-09
Pileri P, Uematsu Y, Campagnoli S, Galli G, Falugi F, Petracca R, Weiner AJ, Houghton M, Rosa D, Grandi G, Abrignani S (1998) Binding of hepatitis C virus to CD81. Science (New York, NY) 282(5390):938–941
Ploss A, Evans MJ, Gaysinskaya VA, Panis M, You H, de Jong YP, Rice CM (2009) Human occludin is a hepatitis C virus entry factor required for infection of mouse cells. Nature 457 (7231):882–886. doi:10.1038/nature07684
Popescu CI, Callens N, Trinel D, Roingeard P, Moradpour D, Descamps V, Duverlie G, Penin F, Heliot L, Rouille Y, Dubuisson J (2011) NS2 protein of hepatitis C virus interacts with structural and non-structural proteins towards virus assembly. PLoS Pathog 7(2):e1001278. doi:10.1371/journal.ppat.1001278
Price DA, Bassendine MF, Norris SM, Golding C, Toms GL, Schmid ML, Morris CM, Burt AD, Donaldson PT (2006) Apolipoprotein epsilon3 allele is associated with persistent hepatitis C virus infection. Gut 55(5):715–718. doi:10.1136/gut.2005.079905
Prince AM, Huima-Byron T, Parker TS, Levine DM (1996) Visualization of hepatitis C virions and putative defective interfering particles isolated from low-density lipoproteins. J Viral Hepat 3(1):11–17
Sainz B Jr, Barretto N, Martin DN, Hiraga N, Imamura M, Hussain S, Marsh KA, Yu X, Chayama K, Alrefai WA, Uprichard SL (2012) Identification of the Niemann-Pick C1-like 1 cholesterol absorption receptor as a new hepatitis C virus entry factor. Nat Med 18 (2):281–285. doi:10.1038/nm.2581
Scarselli E, Ansuini H, Cerino R, Roccasecca RM, Acali S, Filocamo G, Traboni C, Nicosia A, Cortese R, Vitelli A (2002) The human scavenger receptor class B type I is a novel candidate receptor for the hepatitis C virus. EMBO J 21(19):5017–5025
Scheel TK, Prentoe J, Carlsen TH, Mikkelsen LS, Gottwein JM, Bukh J (2012) Analysis of functional differences between hepatitis C virus NS5A of genotypes 1–7 in infectious cell culture systems. PLoS Pathog 8(5):e1002696. doi:10.1371/journal.ppat.1002696
Shimizu Y, Hishiki T, Sugiyama K, Ogawa K, Funami K, Kato A, Ohsaki Y, Fujimoto T, Takaku H, Shimotohno K (2010) Lipoprotein lipase and hepatic triglyceride lipase reduce the infectivity of hepatitis C virus (HCV) through their catalytic activities on HCV-associated lipoproteins. Virology 407(1):152–159
Takikawa S, Ishii K, Aizaki H, Suzuki T, Asakura H, Matsuura Y, Miyamura T (2000) Cell fusion activity of hepatitis C virus envelope proteins. J Virol 74(11):5066–5074
Tellinghuisen TL, Foss KL, Treadaway J (2008) Regulation of hepatitis C virion production via phosphorylation of the NS5A protein. PLoS Pathog 4(3):e1000032. doi:10.1371/journal.ppat.1000032
Thomssen R, Bonk S (2002) Virolytic action of lipoprotein lipase on hepatitis C virus in human sera. Med Microbiol Immunol 191(1):17–24
Thomssen R, Bonk S, Propfe C, Heermann KH, Kochel HG, Uy A (1992) Association of hepatitis C virus in human sera with beta-lipoprotein. Med Microbiol Immunol 181(5):293–300
Thomssen R, Bonk S, Thiele A (1993) Density heterogeneities of hepatitis C virus in human sera due to the binding of beta-lipoproteins and immunoglobulins. Med Microbiol Immunol 182 (6):329–334
Ujino S, Nishitsuji H, Hishiki T, Sugiyama K, Takaku H, Shimotohno K (2016) Hepatitis C virus utilizes VLDLR as a novel entry pathway. Proc Natl Acad Sci 113(1):188–193
Wakita T, Pietschmann T, Kato T, Date T, Miyamoto M, Zhao Z, Murthy K, Habermann A, Krausslich HG, Mizokami M, Bartenschlager R, Liang TJ (2005) Production of infectious hepatitis C virus in tissue culture from a cloned viral genome. Nat Med 11(7):791–796. doi:10.1038/nm1268
Wozniak MA, Itzhaki RF, Faragher EB, James MW, Ryder SD, Irving WL, Trent HCVSG (2002) Apolipoprotein E-epsilon 4 protects against severe liver disease caused by hepatitis C virus. Hepatology 36(2):456–463. doi:10.1053/jhep.2002.34745

Wunschmann S, Medh JD, Klinzmann D, Schmidt WN, Stapleton JT (2000) Characterization of hepatitis C virus (HCV) and HCV E2 interactions with CD81 and the low-density lipoprotein receptor. J Virol 74(21):10055–10062

Yang W, Qiu C, Biswas N, Jin J, Watkins SC, Montelaro RC, Coyne CB, Wang T (2008) Correlation of the tight junction-like distribution of Claudin-1 to the cellular tropism of hepatitis C virus. J Biol Chem 283(13):8643–8653. doi:10.1074/jbc.M709824200

Zeisel MB, Koutsoudakis G, Schnober EK, Haberstroh A, Blum HE, Cosset FL, Wakita T, Jaeck D, Doffoel M, Royer C, Soulier E, Schvoerer E, Schuster C, Stoll-Keller F, Bartenschlager R, Pietschmann T, Barth H, Baumert TF (2007) Scavenger receptor class B type I is a key host factor for hepatitis C virus infection required for an entry step closely linked to CD81. Hepatology 46(6):1722–1731. doi:10.1002/hep.21994

Zhong J, Gastaminza P, Cheng G, Kapadia S, Kato T, Burton DR, Wieland SF, Uprichard SL, Wakita T, Chisari FV (2005) Robust hepatitis C virus infection in vitro. Proc Natl Acad Sci U S A 102(26):9294–9299. doi:10.1073/pnas.0503596102

Part III
Immune Recognition and Pathogenesis

New Animal Models for Hepatitis C

Markus von Schaewen, Jenna M. Gaska, and Alexander Ploss

Abstract The hepatotropic hepatitis C virus (HCV) belongs to the *Flaviviridae* family and chronically infects 130–150 million people worldwide. The severe consequences the virus has for liver health, especially if left untreated, and the lack of a vaccine continue to make HCV a relevant global health problem. A considerable challenge in studying HCV is the virus' host tropism, which is limited almost exclusively to humans and chimpanzees. The lack of suitable and ethical animal model systems has hindered our abilities to mechanistically decipher interactions of HCV with its mammalian host and to develop vaccines. However, encouraging advances, especially in the refinement of humanized mouse models, have created new opportunities for studying HCV pathogenesis and host antiviral responses in vivo. Additionally, the discovery of hepaciviruses in other organisms and advances in induced pluripotent stem cell technologies have created further avenues for exploration. The ultimate goal is to develop tractable small animal models for HCV, which optimally recapitulate all parts of the viral life cycle and present with clinically relevant manifestations of viral hepatitis. Such new models would undoubtedly shed light on both the biology and clinical consequences of chronic hepatitis C infection.

Keywords Hepatitis C • Hepatitis C virus • Animal model • Host tropism • Vaccines • Immune response

Abbreviations

AFC8	Transgenic construct in which a FK506 binding protein/caspase 8 fusion protein is driven by a mouse albumin promoter
apoE	Apolipoprotein E
Cardif	Caspase activation and recruitment domain adaptor-inducing interferon- β

M. von Schaewen • J.M. Gaska • A. Ploss (✉)
Department of Molecular Biology, Princeton University, 110 Lewis Thomas Laboratory, Washington Road, Princeton, NJ 08544, USA
e-mail: aploss@princeton.edu

T. Miyamura et al. (eds.), *Hepatitis C Virus I*, DOI 10.1007/978-4-431-56098-2_12

CD	Cluster of differentiation
CIDEB	Cell death-inducing DFFA-like effector b
CLND1	Claudin-1
CMV	Cytomegalovirus
DAA	Directly acting antiviral
EGRF	Epidermal growth factor receptor
EphA2	Ephrin receptor A2
ER	Endoplasmatic reticulum
ES	Embryonic stem cell
FAH	Fumarylacetoacetate hydrolase
GAG	Glycosaminoglycan
GBV-B	George Barker virus B
HBV	Hepatitis B virus
HCC	Hepatocellular Carcinoma
HCV	Hepatitis C virus
HEAL	Human ectopic artificial liver
HLA	Human leukocyte antigen
HLC	Hepatocyte-like cell
HSC	Hematopoietic stem cell
HSPG	Heparan sulfate proteoglycan
Huh	Human hepatoma
IFN	Interferon
IL-1	Interleukin 1
iPS	Induced pluripotent stem cell
IPS-1	IFN-β promoter stimulator-1
IRF3	Interferon regulatory factor 3
LDLR	Low-density-lipoprotein receptor
MAVS	Mitochondrial antiviral signal protein
miR-122	MicroRNA-122
MUP	Major urinary protein
NHP	Non-human primate
NPC1L1	Niemann-Pick C1-like 1
NPHV	Non-primate hepacivirus
NS	Non-structural protein
OCLN	occludin
PI4KIIIα	phosphatidylinositol 4 kinase IIIα
PKR	Protein kinase R
SCARB1	Scavenger receptor class B member 1
SCID	Severe combined immunodeficiency
STAT1	Signal transducer and activator of transcribtion 1
TfR1	Transferrin receptor 1
TICAM	Toll/IL-1 receptor domain-containing adaptor molecule
TNFα	Tumor necrosis factor α
TRIF	Toll/IL-1 receptor domain-containing adaptor inducing IFN-β

uPA	Urokinase-type plasminogen activator
VISA	Virus-induced signaling adapter

1 Introduction

Hepatitis C virus (HCV) is an enveloped, positive-sense, single-stranded RNA virus belonging to the genus *Hepacivirus* in the *Flaviviridae* family. While approximately 150 million people are infected with HCV worldwide, this is likely an underestimate as almost twice as many individuals in the United States may carry the virus, many unknowingly (Edlin 2011). HCV causes persistent infection in 70–80 % of those who become exposed to the virus. While the acute disease is usually asymptomatic, chronic carriers left untreated frequently develop fibrosis, cirrhosis and, in some cases, hepatocellular carcinoma. Treatment for HCV has evolved rapidly in recent years and it is now possible to cure the majority of patients with largely well-tolerated therapies that include a combination of pegylated interferon (IFN)-α, ribavirin and direct acting antiviral (DAA) drugs. However, despite their potency, it remains to be seen whether even the newest DAAs will drastically reduce the global burden of disease due to the high associated costs, logistical challenges of mass deployment and risk of drug resistance. A vaccine, which would prevent infection or delay the onset of pathogenesis during a chronic infection, does not exist. Development of effective therapies has been delayed by the lack of both suitable cell culture systems and animal models. While hepaciviruses similar to HCV have been found in a variety of species, including dogs, horses and outbred mice, HCV appears to have a much more limited host range. Robust infection has only been described in humans and experimentally infected chimpanzees, but some studies have provided evidence for transient and intermittent viremia in a more exotic mammal, tree shrews. The narrow host range of HCV is not completely understood but can in part be explained by differences between species in the sequences of essential host factors at the level of entry as well as in innate antiviral responses. This growing understanding of the barriers to interspecies transmission has aided the development of inbred models with inheritable susceptibility to HCV. As will be discussed here, genetic host adaptation has been and continues to be part of a multipronged approach to develop more tractable animal models for studying HCV infection, immunity and pathogenesis (Fig. 1).

2 Hepatitis C Virus Infection in Non-Human Primates

For many years, studies of hepatitis C were limited to experimentally infected chimpanzees or patient volunteers. While chimpanzees have been instrumental in analyzing HCV infection (reviewed in (Bukh 2004), Fig. 2), studies in this species are challenging due to high costs, genetic heterogeneity, small cohort sizes, limited

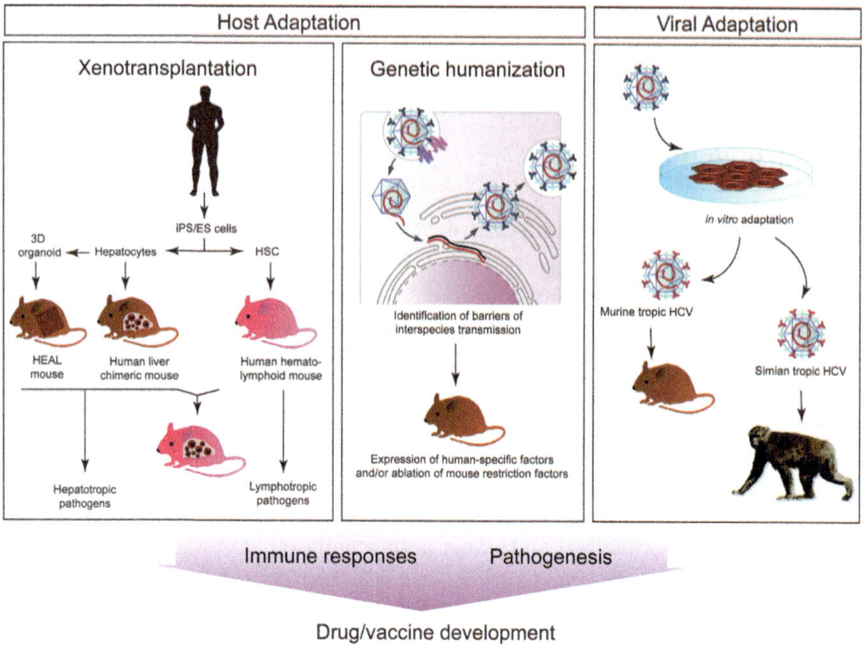

Fig. 1 Host adaptation and viral adaptation approaches to create new animal models for the study of hepatitis C. Host adaptation through transplantation of human hepatocytes to create HEAL mice or human liver chimeric mice and/or hematopoietic stem cells (HSCs) to (co-) engraft components of a human immune system (*left column*). Genetic humanization can be accomplished by identification and expression of human-specific factors or by ablation of restriction factors (*middle column*). Cell culture passaging strategies are used to adapt HCV to murine or simian hosts (*right column*). *iPS* Induced pluripotent stem cells, *ES* embryonic stem cells, *HSC* hematopoietic stem cells, *HEAL* human ectopic artificial liver

access to relevant tissue compartments, the inability to genetically manipulate large apes and growing ethical concerns.

The other non-human primate (NHP) species tested for susceptibility to HCV infection – including cynomolgus monkeys, rhesus monkeys, Japanese monkeys, Green monkeys, Doguera (Abe et al. 1993) and Chacma baboons (Sithebe et al. 2002), cottontop tamarins (Garson et al. 1997) and marmosets – do not seem to support infection. The blocks in HCV transmission in these and potentially other NHP species are not well defined and are likely due to a combination of factors. For example, simian orthologs of essential host factors required for the viral life cycle may be absent or incompatible. Similarly, dominant restriction factors, as observed with HIV, may actively antagonize uptake, replication and/or viral assembly and release. In addition, differences in the kinetics and magnitude of antiviral defenses in many cells may interfere with viral RNA replication. This may be a result of less efficient viral evasion mechanisms that usually enable HCV to establish persistent infection in human cells. For example, in various primate species, differences in the amino acid sequence of the mitochondrial antiviral signal

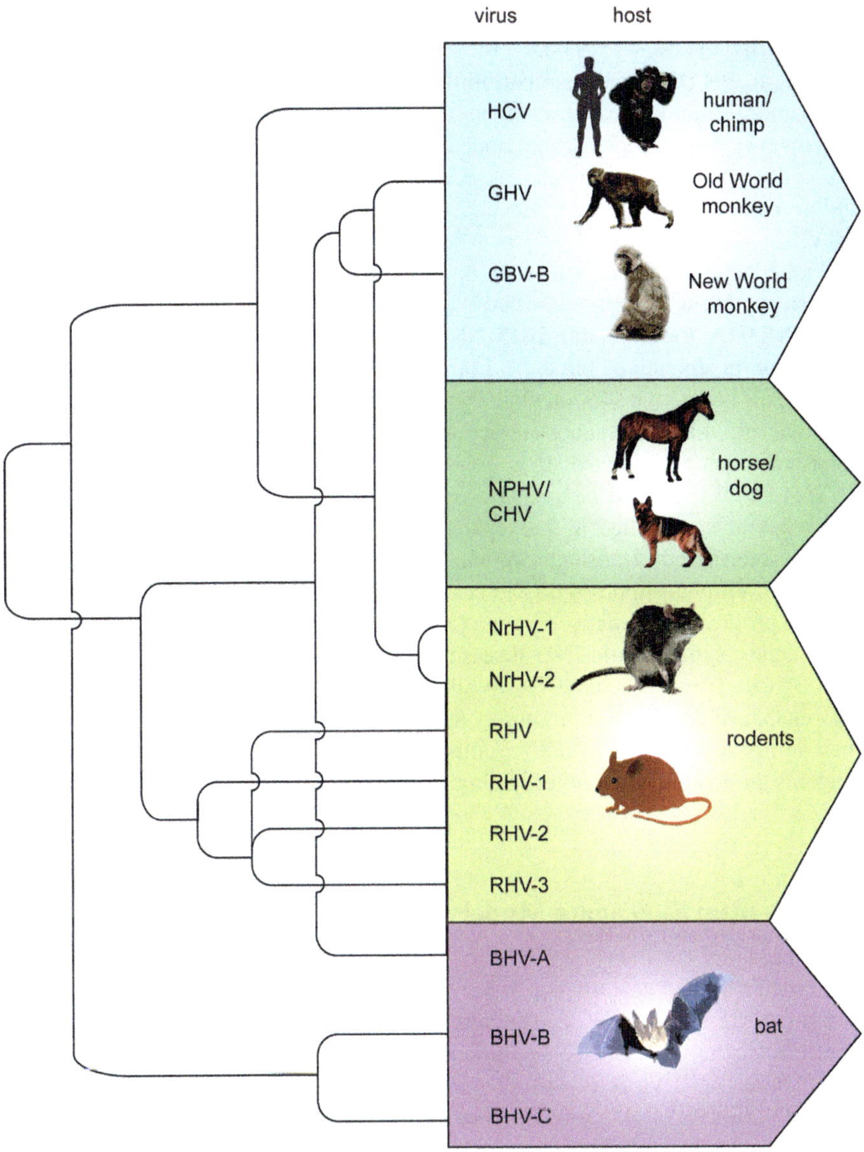

Fig. 2 Phylogenetic relationship of members of the hepacivirus genus and susceptible host species. Phylogenetic tree of the hepaciviruses is adapted from (Pfaender et al. 2014a) and (Firth et al. 2014) and is based on the nucleotide sequence analysis of the NS3 protease domain and the complete NS5B gene. *GHV* Guereza hepacivirus, *GBV-B* George Barker virus B, *NPHV* non-primate hepacivirus, *CHV* canine hepacivirus, *NrHV* Norway rat hepacivirus, *RHV* rodent hepacivirus, *BHV* bat hepacivirus

protein (MAVS, also known as IPS1, VISA or Cardif) prevent its proteolytic cleavage by the HCV NS3/4A protease, leaving host antiviral signaling intact (Patel et al. 2012). These incompatibilities may also result in more effective cellular and humoral immune responses in NHPs, contributing to clearance of HCV.

However, these blocks do not appear absolute as stem cell-derived hepatocyte-like cells from pig-tailed macaques (*Macaca nemestrina*) (Sourisseau et al. 2013) and primary rhesus macaque (*Macaca mulatta*) hepatocytes can support the entire HCV life cycle (Scull et al. 2015). Additionally, HCV RNA replication in rhesus macaque hepatocytes is enhanced upon blunting antiviral immunity, which is consistent with the observation that the rhesus MAVS ortholog is not cleaved by HCV NS3/4A (Scull et al. 2015). Nonetheless, HCV can establish persistent replication in simianized mice, i.e. immunocompromised xenorecipients engrafted with rhesus hepatocytes (Scull et al. 2015). Yet it still remains to be shown whether rhesus or pig-tailed macaques are actually susceptible to HCV in vivo and if viral persistence can be achieved.

Tree shrews (*Tupaia belangeri*) – once designated as small, squirrel-like primates but now classified in the separate order Scandentia – have been shown to support intermittent, transient viremia, becoming more permissive to HCV infection when immunosuppressed (Xie et al. 1998). In follow-up studies, acute infection did progress to persistent viremia (Xu et al. 2007; Amako et al. 2010), resulting in clinically symptomatic liver disease, including steatosis, fibrosis and cirrhosis after 3 years (Amako et al. 2010). While these more recent data are promising and may enable studies of HCV immunity and pathogenesis, there are still limitations to their utility as tree shrews are an outbred, genetically diverse organism and few reagents are available for investigating their immune response to viral infection.

3 Potential Surrogate Models: Non-Primate Hepaciviruses

With such a limited host tropism, other closely related viruses have been considered as a proxy for studying HCV. The best characterized of these viruses, GB viruses, named after the surgeon George Barker (Deinhardt et al. 1967), have been used in NHP studies. GB virus B was able to cause hepatitis in marmosets (*Callithrix jacchus*) (Simons et al. 1995; Lanford et al. 2003) (Fig. 2) as well as other New World monkeys, including tamarins (*Saguinus* spp.) (Karayiannis et al. 1989; Schaluder et al. 1995) and owl monkeys (*Aotus trivirgatus*) (Bukh et al. 2001). GBV-B belongs to the genus *Hepacivirus* in the *Flaviviridae* family and has the same overall genome organization as HCV. However, the polyproteins of HCV and GBV-B share only 28 % amino acid identity and differ even more in their 5′ and 3′ non-coding regions. The discovery of a GBV-B-like virus, called *guereza* hepacivirus (GHV) after the colobus species it was identified in, is the first hepacivirus found in a wild NHP and has led to further questions concerning the evolution of hepaciviruses (Lauck et al. 2013).

Other related viruses of either the *Pegivirus* genus, also a part of the *Flaviviridae* family, or *Hepacivirus* genus have been identified in dogs (Kapoor et al. 2011), horses (Burbelo et al. 2012; Kapoor et al. 2013a), wild mice (Kapoor et al. 2013b), bats(Quan et al. 2013) and rats (Firth et al. 2014). Of the non-primate hepaciviruses (NPHV), those observed in horses are the most genetically similar to HCV (Pfaender et al. 2014b) (Fig. 2). However, it remains to be shown whether these viruses indeed cause hepatitis in experimentally inoculated animals before it can be determined whether they might be surrogates for modeling HCV. Importantly, there is currently no experimental evidence of NPHV transmission between horses and humans (Pfaender et al. 2015).

4 Rodent Models

Mice are widely used in biomedical research, with many existing analytical tools for dissecting their responses to infection. Furthermore, mice of genetically defined backgrounds are available which are amenable to genetic manipulations. In the following sections, we will summarize the previously established rodent models and discuss some of the recent developments that have been explored to model HCV infection and pathogenesis in rodents.

4.1 HCV Transgenic Mice

HCV does not readily infect mice, and thus early attempts to model aspects of HCV pathogenesis in mice were performed by expressing individual or multiple HCV gene products (Table 1 and reviewed in Kremsdorf and Brezillon (2007)). However, depending on the mouse background, the HCV gene product(s) expressed, and the promoter driving expression of these proteins, the histopathological features observed in these mice differed considerably. When HCV core was expressed under the control of a hepatitis B virus (HBV) promoter, animals developed severe liver disease, culminating in hepatocarcinogenesis (Moriya et al. 1997, 1998). In contrast, driving HCV core and/or E1/E2 expression with a major urinary protein (MUP) or CMV promoter produced a less pronounced and more variable disease phenotype (Pasquinelli et al. 1997; Chiyo et al. 2011; Satoh et al. 2010; Naas et al. 2005; Benali-Furet et al. 2005; Chang et al. 2008, 2009; Lerat et al. 2009; Tanaka et al. 2008; Kamegaya et al. 2005; Jeannot et al. 2012). Likewise, NS5A expression was directly cytopathic in some transgenic lines (Wang et al. 2009), but, when under the control of an apoE or MUP promoter, liver pathologies were not observed (Majumder et al. 2003). Similarly, expression of the HCV serine protease NS3/NS4A or NS4B in mouse models has not been shown to induce liver injury (Desai et al. 2011; Frelin et al. 2006; Wang et al. 2006).

Table 1 HCV transgenic mouse models

Transgene C | E1 | E2 | p7 | NS2 | NS3 | NS4A | NS4B | NS5A | NS5B	Phenotype	References
C	Hepatic steatosis, HCC, hepatocyte apoptosis, lipogenesis, cell cycle perturbation and ER stress	Benali-Furet et al. (2005), Chang et al. (2008, 2009), Lerat et al. (2009), Moriya et al. (1997, 1998a), and Tanaka et al. (2008)
C | E2	No evidence for liver disease	Pasquinelli et al. (1997)
C | E1 | E2	Contradicting findings ranging from lacking liver pathology to development of HCC	Kamegaya et al. (2005), and Naas et al. (2005, 2010)
C | E1 | E2 | p7 | NS2	Reduced liver inflammation in transgenic mice compared to controls	Chiyo et al. (2011), Satoh et al. (2010), and Jeannot et al. (2012)
E1 | E2 | p7	Induction of liver tumors by aflatoxin B1	Jeannot et al. (2012)
E1 | E2 | NS2	Liver injury due to induction of CTL responses	Takaku et al. (2003)
NS3 | NS4A	Resistance to TNF-α-induced liver disease, differential IFN-induced autophagy	Ahlen et al. (2009), Desai et al. (2011), and Frelin et al. (2006)
NS4B	No evidence for liver disease	Wang et al. (2006)
NS5A	Contradicting findings regarding the occurrence of liver pathology, inhibition of IFNγ induction	Kanda et al. (2009), Kriegs et al. (2009), Majumder et al. (2003), and Wang et al. (2009a)
C | E1 | E2 | p7 | NS2 | NS3 | NS4A | NS4B | NS5A | NS5B	Impaired clearance of HCV trangsgene-positive hepatocytes, hepatic steatosis and lymphocyte infiltratates, lymphomagenesis, interruption of type1 IFN production, ER stress and hepatocyte apoptosis	Alonzi et al. (2004), Disson et al. (2004), Ernst et al. (2007), Furutani et al. (2006), Tsukiyama-Kohara et al. (2011), Tumurbaatar et al. (2007), and Wegert et al. (2009)

In addition to modeling aspects of HCV-induced liver disease, expression of HCV proteins has been utilized for studying HCV-specific adaptive immune responses in the liver (Disson et al. 2004; Alonzi et al. 2004; Tsukiyama-Kohara et al. 2011; Wegert et al. 2009; Tumurbaatar et al. 2007; Ernst et al. 2007; Furutani et al. 2006; Takaku et al. 2003; Naas et al. 2010; Kriegs et al. 2009; Kanda et al. 2009). Pre-natal expression of HCV proteins in mice causes the murine immune system to become tolerized to viral gene products, but this tolerance can be disrupted via DNA vaccination, which primes CD8 T cells to target hepatocytes expressing HCV NS3/4A (Ahlen et al. 2009). This model lends itself to testing T cell based vaccine candidates.

Undoubtedly, HCV transgenic mice have helped analyze HCV immune responses and viral pathogenesis, but a number of factors still diminish their utility. Transgene copy numbers, and consequently levels and distribution of HCV protein expression, can vary considerably due to random integration in the mouse genome. When driven by strong viral or cellular promoters, expression of HCV gene products can surpass the levels of viral proteins that would be reached by actual infection. Furthermore, interpreting data acquired in transgenic mice is further complicated since HCV proteins are being expressed outside of the inflammatory context of acute and chronic viral infection.

4.2 Xenotransplantation Models

To overcome some of the challenges of HCV transgenic mice, xenotransplantation models have been established in which the murine host is rendered susceptible to HCV infection by xenoengraftment of permissive human cells in the mouse liver (Fig. 1). As described below, a variety of approaches have been taken to accomplish this goal.

4.2.1 Engraftment of Human Hepatoma Cells In Vivo

The simplest mouse models of xenoengraftment are made by intrahepatic injection of human hepatoma cells. To quantify RNA replication and responses to antiviral treatment such as interferon-α (IFN-α) in vivo, Huh7 cells containing an HCV replicon expressing luciferase have been injected into severe combined immunodeficient (SCID)/beige mice and analyzed by whole-body bioluminescence imaging (Zhu et al. 2006). This system is simple with minimal intra- and interexperimental variation but is not a bona fide infection model.

To actually enable the study of anti-HCV immune responses, immunocompetent fetal rats have been tolerized in utero to Huh7 cells and, after birth, transplanted with a larger number of the same cells (Wu et al. 2005). Remarkably, Huh7 cells, which are usually not readily susceptible to HCV infection in vitro, supported viremia of a patient-derived genotype 1a isolate. However, the low levels of observed viremia, complex nature of these experiments, and potential inability of rat T cells to recognize HCV antigens due to the presence of human leukocyte antigen (HLA) on the transplanted Huh7 cells make this model less than ideal.

4.2.2 Ectopic Liver Implantation Models

While it is more desirable to engraft hepatocytes instead of hepatoma cells in the human parenchyma, this is not readily accomplished with primary human hepatocytes. However, pieces of human liver and even artificial human liver organoids have been successfully implanted in ectopic sites. In the so-called "Trimera mouse" (Ilan et al. 2002), small pieces of human liver were maintained under the kidney capsule or the ear of SCID mice. When the liver tissue was taken from HCV positive donors or naïve tissue was infected with HCV prior to transplantation, viremia was maintained for several weeks. This model has been subsequently used to assess the efficacy of neutralizing antibodies (Eren et al. 2006), but the technically and logistically challenging experimental set-up, fairly rapid graft failure and low levels of HCV viremia have hampered the utility of the model.

To overcome the need for primary liver tissue, bioengineering approaches have been undertaken to reconstruct increasingly more complex tissue organoids suitable for transplantation. Human ectopic artificial livers (HEALs) have been created where cryopreserved primary human hepatocytes are supported by polymeric scaffolds, which aid maintenance of the microenvironment and thus stablize these cells. While simpler, polyethylene glycol (PEG)-based polymers were used initially (Chen et al. 2011), newer models allow for even greater control of the scaffold architecture, improving vascularization and, consequently, hepatocyte survival (Miller et al. 2012; Stevens et al. 2013). In mice engrafted intraperitoneally with HEALs (Fig. 1), humanized liver functions could be monitored for several weeks but susceptibility to hepatotropic pathogens, including HCV, has yet to be shown.

4.2.3 Human Liver Chimeric Mice

The most commonly used and best characterized humanized xenotransplantation models for HCV are human liver chimeric mice (Fig. 1). Suitable xenorecipient strains are immunodeficient to avoid graft rejection and also have endogenous liver injury to both promote hepatocyte proliferation and give the donor hepatocytes a growth advantage over the mouse hepatocytes. Donor cells, including hepatoma cell lines, primary hepatocytes and, more recently, stem-cell derived hepatocytes, are injected intrasplenically. Traveling via the portal venous system, the donor cells pass through the liver sinusoidal endothelial cells and form clusters that expand upon induction of liver injury. This can be done via partial hepatectomy or treatment with hepatotoxic chemicals, like retrorsine and carbon tetrachloride. Genetic approaches have also been utilized as they allow for more control over the severity of the liver injury and can limit hepatotoxicity to specifically mouse hepatocytes.

Robust engraftment of human hepatocytes has been shown in a number of immunodeficient liver injury models, including Alb-uPA (Meuleman et al. 2005; Mercer et al. 2001), $FAH^{-/-}$ (Bissig et al. 2010; de Jong et al. 2014), AFC8

(Washburn et al. 2011), MUP-uPA (Tesfaye et al. 2013) and HSV-TK (Kosaka et al. 2013) mice. The resultant human liver chimeric mice are susceptible to several human-tropic pathogens, including HBV, HCV and parasites that cause malaria in humans (reviewed in (Meuleman and Leroux-Roels 2008). Additionally, these mice can be used for monitoring human-like metabolic and toxicological responses in testing antimicrobial compounds.

With the exception of AFC8 mice, all the above-mentioned xenorecipient strains have demonstrated robust human hepatic chimerism when using adult hepatocytes. Due to genetic differences between hepatocyte donors, host responses can differ. To minimize inter-experimental variations and create a renewable resource for human donor cells, stem cell-derived hepatocytes have been explored as a possible solution. Hepatocyte-like cells (HLCs) can now be routinely generated from embryonic stem (ES) or induced pluripotent stem (iPS) cells (Touboul et al. 2010; Si-Tayeb et al. 2010). These HLCs express hepatocyte-specific markers, support hepatocyte-specific metabolic functions and can be infected with HCV (Schwartz et al. 2012; Wu et al. 2012; Roelandt et al. 2012). Recent studies suggest that HLCs can also be engrafted reasonably well in vivo and support persistent HCV infection (Carpentier et al. 2014). However, engraftment efficiency seems to depend strongly on the xenorecipient strain, as immunodeficient MUP-uPA mice, but not other liver injury models, support robust in vivo expansion.

4.2.4 HCV Immunity and Pathogenesis in Humanized Xenotransplantation Models

Human liver chimeric mice are currently the only experimental models besides chimpanzees that are readily susceptible to HCV. These mice have been used to study innate host responses to HCV and for testing the efficacy of novel therapeutic regimens. To expand the use of these mice in analyzing human immune responses to HCV, protocols are being refined so that mice are co-engrafted with human hepatocytes and components of a human immune system (Fig. 1). Initial attempts co-injected a mixture of human fetal hepatoblasts, non-parenchymal cells and hematopoietic stem cells (HSCs) into AFC8 mice, yielding reasonable immune cell engraftment but low human hepatic chimerism (Washburn et al. 2011). Nonetheless, dually engrafted mice did become chronically infected following inoculation with HCV patient isolates and exhibited an HCV-specific T cell response, which appeared to be responsible for observed signs of early liver fibrosis. While these data are encouraging, protocols need to be refined further to improve dual chimerism and minimize inter- and intravariability of experiments. More recent reports have demonstrated that extensive double humanization of both the liver and immune system can be achieved with mature hepatocytes and HSCs (Gutti et al. 2014; Wilson et al. 2014). Long-term dual reconstitution, without any evidence of hepatocyte rejection by the human immune system, was sustained even when the human cells were mismatched in their major histocompatibility complex (MHC, (Gutti et al. 2014). The latter observation is consistent with the limited HLA matching in human liver transplantations, presumably due to the tolerogenic microenvironment of the liver. However, the limited function of the transplanted human immune system at least partially contributes to the lack of allogeneic graft rejection in dually engrafted humanized mice. To improve both the cellular complexity and functionality of the engrafted human immune system, several modifications are being tested. These include, but are not limited to: the expression of human orthologs of non-redundant cytokines with limited biological cross-reactivity to foster development of human immune cell lineages which currently do not develop efficiently in conventional humanized mice; expression of human MHC in the absence of mouse MHC to ensure faithful presentation of self- and virally derived peptides to human T cells and to reduce graft-versus-host-disease; co-transplantation of HSC donor-matched human thymic cortical epithelium to facilitate proper T cell selection; the improvement of lymphoid architecture organization, especially in the spleen and lymph-nodes, to allow for adequate T and B cell priming; genetic replacement of non-compatible immune cell receptors and chemokines expressed on non-hematopoietically derived cells to improve e.g. immune cell trafficking; the introduction of a human microbiome to account for effects of species-specific commensals on the immune system (reviewed in Shultz et al. 2012).

5 Genetically Humanized Mouse Models for HCV Infection

An inbred mouse model with inheritable susceptibility to HCV would overcome the technical difficulties of the xenotransplantation model (Fig. 1). The challenge is to systematically identify and overcome any restrictions to HCV infection in murine cells. HCV's narrow host range is not completely understood, and the viral life cycle is blocked or insufficiently supported in murine cells at multiple steps. Productive HCV uptake into human hepatocytes relies on a large number of human host molecules (reviewed in Ding et al. 2014). These include glycosaminoglycans (GAGs) present on heparan sulfate proteoglycans (HSPGs), low-density-lipoprotein receptor (LDLR) (Agnello et al. 1999), CD81 (Pileri et al. 1998), scavenger receptor class B member 1 (SCARB1) (Scarselli et al. 2002), the tight junction proteins claudin-1 (CLDN1) (Evans et al. 2007) and occludin (OCLN) (Liu et al. 2009; Ploss et al. 2009), the receptor tyrosine kinases epidermal growth factor receptor (EGFR) and ephrin receptor A2 (EphA2) (Lupberger et al. 2011), the cholesterol transporter Niemann-Pick C1-like 1 (NPC1L1) (Sainz et al. 2012), transferrin receptor 1 (TfR1) (Martin and Uprichard 2013) and the cell death-inducing DFFA-like effector b (CIDEB) (Wu et al. 2014). The block of HCV entry in rodent cells can be explained by differences in critical residues in the second extracellular loops of CD81 (Flint et al. 2006; Higginbottom et al. 2000) and OCLN (Michta et al. 2010). Consequently, expression of human CD81 and OCLN, along with human or mouse SCARB1 and CLDN1, facilitates HCV uptake in mouse cells in vitro (Ploss et al. 2009). Other human entry factors appear to contribute minimally to HCV species tropism at the level of entry, but their individual roles still need to be experimentally tested.

Establishing HCV glycoprotein-mediated uptake into mouse livers adenovirally transduced with CD81 and OCLN opened the door for genetically overcoming the barrier to HCV entry in mice. Indeed, expression of human CD81 and OCLN appears sufficient for HCV entry into hepatocytes of fully immunocompetent inbred mice. This genetically humanized mouse model allows dissection of the HCV entry process in the 3D context of the liver in vivo and has been applied to test pre-clinically the efficacy of neutralizing antibodies and vaccine candidates (Giang et al. 2012; Dorner et al. 2011; de Jong et al. 2014). Transgenic mice have also been developed in which human CD81, SCARB1, CLDN1 and OCLN expression are driven by liver-specific promoters (Hikosaka et al. 2011). However, initial reports have suggested that these lines are resistant to HCV infection in vivo (Hikosaka et al. 2011). This observation is likely due to the lower level of entry factor expression in the transgenic mice and the need for a very sensitive reporter system to quantify viral entry (Dorner et al. 2011).

As an alternative to the genetic host adaptations described above, previous studies have shown that the block of HCV at the level of entry can also be overcome through viral adaptation (Fig. 1). Using an in vitro selection approach, mutations within HCV E1 and E2 that increased the affinity of the viral envelope for mouse CD81 were identified. These mutations appeared to more broadly affect the conformation of the viral envelope, as the resulting mouse CD81-adapted strain is also less dependent on human OCLN and can enter cell lines expressing only mouse CD81, SCARB1, CLDN1 and OCLN (Bitzegeio et al. 2010). It has yet to be demonstrated if this mouse-adapted HCV strain can enter mouse primary hepatocytes in vitro or in vivo.

Establishing HCV entry in vivo has some utility, but what is ultimately needed is a model that supports all steps of the viral life cycle. More than a decade ago, it was shown that HCV RNA is translated, but not readily replicated, following entry into murine cells (Dorner et al. 2011; McCaffrey et al. 2002). Subsequent studies in cell culture demonstrated that dominant negative inhibitors are not present and that the murine orthologs of host factors critical for HCV replication cooperate sufficiently with the viral replication machinery, as HCV replicons, i.e. selectable HCV RNA genomes, can replicate in murine cell lines (Zhu et al. 2003; Uprichard et al. 2006; Frentzen et al. 2011). Nevertheless, the efficiency of post-entry steps of the viral life cycle could conceivably be improved with human host factors important for HCV replication, assembly and/or egress. However, previous gain- and loss-of-function studies converged on only a few critical host factors, namely miR-122, cyclophilin A, phosphatidylinositol 4 kinase IIIα (PI4KIIIα), and apolipoprotein E (reviewed in (Bartenschlager et al. 2010)) – all of whose sequences are largely conserved between mice and humans. Thus, additional proviral factors that enhance HCV replication and/or assembly in mouse cells have yet to be identified.

Numerous studies have shown that innate antiviral responses play a critical role in limiting HCV infection in human cells, including hepatoma cell lines and human primary hepatocytes (Andrus et al. 2011; Marukian et al. 2011). Likewise, HCV replication is drastically enhanced in cell lines with strong impairments in type I and III interferon signaling, such as mouse cells lacking MAVS (Frentzen et al. 2014), protein kinase R (PKR; (Chang et al. 2006)), interferon regulatory factor 3 (IRF3; (Lin et al. 2010)) or STAT1 (Vogt et al. 2013). Known mechanisms by which HCV evades antiviral defenses, such as the cleavage of MAVS (Meylan et al. 2005) or Toll/IL-1 receptor domain-containing adaptor inducing IFN-β (TRIF or TICAM; (Li et al. 2005)), appear to function in mouse cells (Vogt et al. 2013). However, differences in the kinetics and/or magnitude of virally induced innate defenses may restrict HCV RNA replication more efficiently in mouse compared to human cells. Consistent with these in vitro data, mice expressing human HCV entry factors crossed to genetic backgrounds impaired in antiviral innate defenses support low level HCV RNA replication (Dorner et al. 2013). In these genetically humanized mice, infectious HCV is detectable in circulation, confirming previous studies that demonstrated late stages of the HCV life cycle are supported in mouse cells if sufficient ApoE is present (Long et al. 2011).

Recapitulating the entire HCV lifecycle in inbred immunocompetent mice is an important next step in developing a mouse model suitable for mechanistic studies of HCV immunity and pathogenesis. The previously published model required immune-suppression to establish low-level viremia. However, more recent work suggests that this may be strain-dependent. In fact, mice expressing human CD81, SCARB1, CLDN1 and OCLN on the fully immunocompetent ICR mouse background not only supported persistent infection with various HCV isolates very efficiently but also developed clinically apparent liver disease (Chen et al. 2014). While these data are somewhat difficult to reconcile with most previously published literature, it is conceivable that a fortuitous allele combination in the genetically variable outbred ICR stock favors susceptibility to HCV.

6 Summary and Outlook

The advent of highly potent DAAs holds promise to effectively treat the great majority of patients. However, current treatments are very expensive and mandate strict adherence to dosing to prevent the outgrowth of resistant viral variants. To provide simpler and more cost-effective interventions and to optimally prevent infections, a HCV vaccine may ultimately be needed. Testing and prioritization of immunotherapies and vaccines is delayed by the lack of (a) readily accessible animal model(s). More tractable in vivo platforms could also be deployed to answer questions of basic virology, HCV pathogenesis, and correlates of protective immunity (Fig. 3). A variety of partially complementary approaches are currently being pursued to develop better small animal models for HCV infection. While most other NHP species besides chimpanzees were thought to be resistant to HCV infection,

	Human	Chimpanzee	Tree Shrew	Humanized mouse xeno-transplantation	Humanized mouse genetic humanization	Mouse viral adaption	Small NHP (e.g. Macaca spp.) viral adaptation

HCV entry	yes	yes	yes	yes	yes	unknown	unknown
HCV replication	yes	yes	yes	yes	only in immuno-compromized strains	unknown	unknown
HCV assembly	yes	yes	yes	yes	yes	unknown	unknown
HCV pathogenesis	fibrosis cirrhosis HCC	milder than in humans	evidence for hepatitis, fibrosis,cirrhosis	evidence for fibrosis	unknown	no	HCV/HIV?
Immune system	human	chimpanzee	tupaia	human	mouse	mouse	NHP
Costs	high	high	medium	medium	low	low	med-high
Throughput	low	low	low	medium	high	high	low
Genetic manipulation	no	no	no	limited	yes	yes	limited
Drug/Vaccine development	yes	yes	unknown	yes (inhibitors)	yes	unknown	yes

Fig. 3 Comparison of different animal models for hepatitis C

recent studies show that the HCV life cycle can be established in HLCs derived from pig-tailed macaques and primary hepatocytes from rhesus macaques. This suggests that certain NHP species may indeed be permissive to HCV infection. In addition, the discovery of hepaciviruses genetically closely related to HCV in outbred mice, rats, dogs and horses may provide further avenues for studying HCV. The barriers of HCV's narrow host tropism are now better understood and have spurred a combination of viral adaptations and/or genetic host humanizations to establish inbred rodent models with inheritable susceptibility to HCV infection. Xenotransplantation approaches are being continuously refined, and it has become possible to reproducibly generate human liver chimeric mice at fairly high throughput. These mice can then be used to analyze all aspects of the viral life cycle with genetically diverse HCV isolates. Improvements in protocols yielding HLCs from directed differentiation of ES and iPS cells hold promise to develop renewable hepatocyte sources of genetically defined backgrounds. Furthermore, these advances may enable the generation of humanized mouse avatars engrafted with patient-specific hepatocytes to model clinically relevant disease phenotypes. In proof-of-concept studies, human liver and components of a human immune system were robustly engrafted in a single xenorecipient, paving the way for modeling HCV-associated hepatitis, including relevant co-infections with HBV and/or HIV. Undoubtedly, as new techniques and protocols are perfected, it will remain important to continue evaluating the ability of any new HCV model to faithfully recapitulate aspects of HCV pathogenesis and its consequences in humans.

Acknowledgements The authors thank Qiang Ding and Florian Douam for edits and critical discussion of the manuscript. Work in the laboratory is in part supported by grants from the National Institutes of Health (2 R01 AI079031-05A1, 1 R01 AI107301-01, 1 R56 AI106005-01), a Research Scholar Award from the American Cancer Society (RSG-15-048-01-MPC) and the Grand Challenge Program of Princeton University. M.v.S. is a recipient of a fellowship from the German Research Foundation (Deutsche Forschungsgemeinschaft). JMG is supported by co-funding from NIAID on iNRSA 5T32GM007388. We apologize to all colleagues whose work could not be cited due to space constraints.

References

Abe K, Kurata T, Teramoto Y, Shiga J, Shikata T (1993) Lack of susceptibility of various primates and woodchucks to hepatitis C virus. J Med Primatol 22(7–8):433–434

Agnello V, Abel G, Elfahal M, Knight GB, Zhang QX (1999) Hepatitis C virus and other flaviviridae viruses enter cells via low density lipoprotein receptor. Proc Natl Acad Sci U S A 96(22):12766–12771

Ahlen G, Derk E, Weiland M, Jiao J, Rahbin N, Aleman S, Peterson DL, Pokrovskaja K, Grander D, Frelin L, Sallberg M (2009) Cleavage of the IPS-1/Cardif/MAVS/VISA does not inhibit T cell-mediated elimination of hepatitis C virus non-structural 3/4A-expressing hepatocytes. Gut 58(4):560–569. doi:gut.2007.147264 [pii] 10.1136/gut.2007.147264

Alonzi T, Agrati C, Costabile B, Cicchini C, Amicone L, Cavallari C, Rocca CD, Folgori A, Fipaldini C, Poccia F, Monica NL, Tripodi M (2004) Steatosis and intrahepatic lymphocyte recruitment in hepatitis C virus transgenic mice. J Gen Virol 85(Pt 6):1509–1520

Amako Y, Tsukiyama-Kohara K, Katsume A, Hirata Y, Sekiguchi S, Tobita Y, Hayashi Y, Hishima T, Funata N, Yonekawa H, Kohara M (2010) Pathogenesis of hepatitis C virus infection in Tupaia belangeri. J Virol 84(1):303–311. doi:JVI.01448-09 [pii] 10.1128/JVI.01448-09

Andrus L, Marukian S, Jones CT, Catanese MT, Sheahan TP, Schoggins JW, Barry WT, Dustin LB, Trehan K, Ploss A, Bhatia SN, Rice CM (2011) Expression of paramyxovirus V proteins promotes replication and spread of hepatitis C virus in cultures of primary human fetal liver cells. Hepatology 54(6):1901–1912. doi:10.1002/hep.24557

Bartenschlager R, Cosset FL, Lohmann V (2010) Hepatitis C virus replication cycle. J Hepatol 53 (3):583–585. doi:10.1016/j.jhep.2010.04.015

Benali-Furet NL, Chami M, Houel L, De Giorgi F, Vernejoul F, Lagorce D, Buscail L, Bartenschlager R, Ichas F, Rizzuto R, Paterlini-Brechot P (2005) Hepatitis C virus core triggers apoptosis in liver cells by inducing ER stress and ER calcium depletion. Oncogene 24 (31):4921–4933

Bissig KD, Wieland SF, Tran P, Isogawa M, Le TT, Chisari FV, Verma IM (2010) Human liver chimeric mice provide a model for hepatitis B and C virus infection and treatment. J Clin Invest 120(3):924–930. doi:40094 [pii] 10.1172/JCI40094

Bitzegeio J, Bankwitz D, Hueging K, Haid S, Brohm C, Zeisel MB, Herrmann E, Iken M, Ott M, Baumert TF, Pietschmann T (2010) Adaptation of hepatitis C virus to mouse CD81 permits infection of mouse cells in the absence of human entry factors. PLoS Pathog 6:e1000978. doi:10.1371/journal.ppat.1000978

Bukh J (2004) A critical role for the chimpanzee model in the study of hepatitis C. Hepatology 39 (6):1469–1475. doi:10.1002/hep.20268

Bukh J, Apgar CL, Govindarajan S, Purcell RH (2001) Host range studies of GB virus-B hepatitis agent, the closest relative of hepatitis C virus, in New World monkeys and chimpanzees. J Med Virol 65(4):694–697. doi:10.1002/jmv.2092 [pii]

Burbelo PD, Dubovi EJ, Simmonds P, Medina JL, Henriquez JA, Mishra N, Wagner J, Tokarz R, Cullen JM, Iadarola MJ, Rice CM, Lipkin WI, Kapoor A (2012) Serology enabled discovery of genetically diverse hepaciviruses in a new host. J Virol. doi:JVI.00250-12 [pii] 10.1128/JVI.00250-12

Carpentier A, Tesfaye A, Chu V, Nimgaonkar I, Zhang F, Lee SB, Thorgeirsson SS, Feinstone SM, Liang TJ (2014) Engrafted human stem cell-derived hepatocytes establish an infectious HCV murine model. J Clin Invest 124(11):4953–4964. doi:10.1172/JCI75456

Chang KS, Cai Z, Zhang C, Sen GC, Williams BR, Luo G (2006) Replication of hepatitis C virus (HCV) RNA in mouse embryonic fibroblasts: protein kinase R (PKR)-dependent and PKR-independent mechanisms for controlling HCV RNA replication and mediating interferon activities. J Virol 80(15):7364–7374. doi:80/15/7364 [pii] 10.1128/JVI.00586-06

Chang ML, Chen JC, Chang MY, Yeh CT, Lin WP, Liang CK, Huang SF, Dang KN, Chiu CT, Lin DY (2008) Acute expression of hepatitis C core protein in adult mouse liver: mitochondrial stress and apoptosis. Scand J Gastroenterol 43(6):747–755. doi:789872799 [pii] 10.1080/00365520701875987

Chang ML, Chen TH, Chang MY, Yeh CT (2009) Cell cycle perturbation in the hepatocytes of HCV core transgenic mice following common bile duct ligation is associated with enhanced p21 expression. J Med Virol 81(3):467–472. doi:10.1002/jmv.21403

Chen AA, Thomas DK, Ong LL, Schwartz RE, Golub TR, Bhatia SN (2011) Humanized mice with ectopic artificial liver tissues. Proc Natl Acad Sci U S A 108(29):11842–11847. doi:10.1073/pnas.1101791108

Chen J, Zhao Y, Zhang C, Chen H, Feng J, Chi X, Pan Y, Du J, Guo M, Cao H, Chen H, Wang Z, Pei R, Wang Q, Pan L, Niu J, Chen X, Tang H (2014) Persistent hepatitis C virus infections and hepatopathological manifestations in immune-competent humanized mice. Cell Res 24(9):1050–1066. doi:10.1038/cr.2014.116

Chiyo T, Sekiguchi S, Hayashi M, Tobita Y, Kanegae Y, Saito I, Kohara M (2011) Conditional gene expression in hepatitis C virus transgenic mice without induction of severe liver injury using a non-inflammatory Cre-expressing adenovirus. Virus Res 160(1–2):89–97. doi:10.1016/j.virusres.2011.05.019

de Jong YP, Dorner M, Mommersteeg MC, Xiao JW, Balazs AB, Robbins JB, Winer BY, Gerges S, Vega K, Labitt RN, Donovan BM, Giang E, Krishnan A, Chiriboga L, Charlton MR, Burton DR, Baltimore D, Law M, Rice CM, Ploss A (2014) Broadly neutralizing antibodies abrogate established hepatitis C virus infection. Sci Transl Med 6(254):254ra129. doi:10.1126/scitranslmed.3009512

Deinhardt F, Holmes AW, Capps RB, Popper H (1967) Studies on the transmission of human viral hepatitis to marmoset monkeys. I. Transmission of disease, serial passages, and description of liver lesions. J Exp Med 125(4):673–688

Desai MM, Gong B, Chan T, Davey RA, Soong L, Kolokoltsov AA, Sun J (2011) Differential, type I interferon-mediated autophagic trafficking of hepatitis C virus proteins in mouse liver. Gastroenterology 141(2):674–685, 685 e671–676. doi:S0016-5085(11)00615-9 [pii] 10.1053/j.gastro.2011.04.060

Ding Q, von Schaewen M, Ploss A (2014) The impact of hepatitis C virus entry on viral tropism. Cell Host Microbe 16(5):562–568. doi:10.1016/j.chom.2014.10.009

Disson O, Haouzi D, Desagher S, Loesch K, Hahne M, Kremer EJ, Jacquet C, Lemon SM, Hibner U, Lerat H (2004) Impaired clearance of virus-infected hepatocytes in transgenic mice expressing the hepatitis C virus polyprotein. Gastroenterology 126(3):859–872. doi:S0016508503020183 [pii]

Dorner M, Horwitz JA, Robbins JB, Barry WT, Feng Q, Mu K, Jones CT, Schoggins JW, Catanese MT, Burton DR, Law M, Rice CM, Ploss A (2011) A genetically humanized mouse model for hepatitis C virus infection. Nature 474(7350):208–211. doi:nature10168 [pii] 10.1038/nature10168

Dorner M, Horwitz JA, Donovan BM, Labitt RN, Budell WC, Friling T, Vogt A, Catanese MT, Satoh T, Kawai T, Akira S, Law M, Rice CM, Ploss A (2013) Completion of the entire hepatitis

C virus life cycle in genetically humanized mice. Nature 501(7466):237–241. doi:10.1038/nature12427
Edlin BR (2011) Perspective: test and treat this silent killer. Nature 474(7350):S18–S19. doi:474S18a [pii] 10.1038/474S18a
Eren R, Landstein D, Terkieltaub D, Nussbaum O, Zauberman A, Ben-Porath J, Gopher J, Buchnick R, Kovjazin R, Rosenthal-Galili Z, Aviel S, Ilan E, Shoshany Y, Neville L, Waisman T, Ben-Moshe O, Kischitsky A, Foung SK, Keck ZY, Pappo O, Eid A, Jurim O, Zamir G, Galun E, Dagan S (2006) Preclinical evaluation of two neutralizing human monoclonal antibodies against hepatitis C virus (HCV): a potential treatment to prevent HCV reinfection in liver transplant patients. J Virol 80(6):2654–2664. doi:80/6/2654 [pii] 10.1128/JVI.80.6.2654-2664.2006
Ernst E, Schonig K, Bugert JJ, Blaker H, Pfaff E, Stremmel W, Encke J (2007) Generation of inducible hepatitis C virus transgenic mouse lines. J Med Virol 79(8):1103–1112. doi:10.1002/jmv.20911
Evans MJ, von Hahn T, Tscherne DM, Syder AJ, Panis M, Wolk B, Hatziioannou T, McKeating JA, Bieniasz PD, Rice CM (2007) Claudin-1 is a hepatitis C virus co-receptor required for a late step in entry. Nature 446(7137):801–805. doi:nature05654 [pii] 10.1038/nature05654
Firth C, Bhat M, Firth MA, Williams SH, Frye MJ, Simmonds P, Conte JM, Ng J, Garcia J, Bhuva NP, Lee B, Che X, Quan PL, Lipkin WI (2014) Detection of zoonotic pathogens and characterization of novel viruses carried by commensal Rattus norvegicus in New York City. mBio 5(5):e01933-01914. doi:10.1128/mBio.01933-14
Flint M, von Hahn T, Zhang J, Farquhar M, Jones CT, Balfe P, Rice CM, McKeating JA (2006) Diverse CD81 proteins support hepatitis C virus infection. J Virol 80(22):11331–11342
Frelin L, Brenndorfer ED, Ahlen G, Weiland M, Hultgren C, Alheim M, Glaumann H, Rozell B, Milich DR, Bode JG, Sallberg M (2006) The hepatitis C virus and immune evasion: non-structural 3/4A transgenic mice are resistant to lethal tumour necrosis factor alpha mediated liver disease. Gut 55(10):1475–1483. doi:gut.2005.085050 [pii] 10.1136/gut.2005.085050
Frentzen A, Huging K, Bitzegeio J, Friesland M, Haid S, Gentzsch J, Hoffmann M, Lindemann D, Zimmer G, Zielecki F, Weber F, Steinmann E, Pietschmann T (2011) Completion of hepatitis C virus replication cycle in heterokaryons excludes dominant restrictions in human non-liver and mouse liver cell lines. PLoS Pathog 7(4):e1002029. doi:10.1371/journal.ppat.1002029
Frentzen A, Anggakusuma, Gurlevik E, Hueging K, Knocke S, Ginkel C, Brown RJ, Heim M, Dill MT, Kroger A, Kalinke U, Kaderali L, Kuehnel F, Pietschmann T (2014) Cell entry, efficient RNA replication, and production of infectious hepatitis C virus progeny in mouse liver-derived cells. Hepatology 59(1):78–88. doi:10.1002/hep.26626
Furutani T, Hino K, Okuda M, Gondo T, Nishina S, Kitase A, Korenaga M, Xiao SY, Weinman SA, Lemon SM, Sakaida I, Okita K (2006) Hepatic iron overload induces hepatocellular carcinoma in transgenic mice expressing the hepatitis C virus polyprotein. Gastroenterology 130(7):2087–2098. doi:10.1053/j.gastro.2006.02.060
Garson JA, Whitby K, Watkins P, Morgan AJ (1997) Lack of susceptibility of the cottontop tamarin to hepatitis C infection. J Med Virol 52(3):286–288. doi:10.1002/(SICI)1096-9071(199707)52:3<286::AID-JMV9>3.0.CO;2-Z [pii]
Giang E, Dorner M, Prentoe JC, Dreux M, Evans MJ, Bukh J, Rice CM, Ploss A, Burton DR, Law M (2012) Human broadly neutralizing antibodies to the envelope glycoprotein complex of hepatitis C virus. Proc Natl Acad Sci USA 109(16):6205–6210. doi:1114927109 [pii] 10.1073/pnas.1114927109
Gutti TL, Knibbe JS, Makarov E, Zhang J, Yannam GR, Gorantla S, Sun Y, Mercer DF, Suemizu H, Wisecarver JL, Osna NA, Bronich TK, Poluektova LY (2014) Human hepatocytes and hematolymphoid dual reconstitution in treosulfan-conditioned uPA-NOG mice. Am J Pathol 184(1):101–109. doi:10.1016/j.ajpath.2013.09.008

Higginbottom A, Quinn ER, Kuo CC, Flint M, Wilson LH, Bianchi E, Nicosia A, Monk PN, McKeating JA, Levy S (2000) Identification of amino acid residues in CD81 critical for interaction with hepatitis C virus envelope glycoprotein E2. J Virol 74(8):3642–3649

Hikosaka K, Noritake H, Kimura W, Sultana N, Sharkar MT, Tagawa Y, Uezato T, Kobayashi Y, Wakita T, Miura N (2011) Expression of human factors CD81, claudin-1, scavenger receptor, and occludin in mouse hepatocytes does not confer susceptibility to HCV entry. Biomed Res 32(2):143–150. doi:JST.JSTAGE/biomedres/32.143 [pii]

Ilan E, Arazi J, Nussbaum O, Zauberman A, Eren R, Lubin I, Neville L, Ben-Moshe O, Kischitzky A, Litchi A, Margalit I, Gopher J, Mounir S, Cai W, Daudi N, Eid A, Jurim O, Czerniak A, Galun E, Dagan S (2002) The hepatitis C virus (HCV)-Trimera mouse: a model for evaluation of agents against HCV. J Infect Dis 185(2):153–161. doi:JID010683 [pii] 10.1086/338266

Jeannot E, Boorman GA, Kosyk O, Bradford BU, Shymoniak S, Tumurbaatar B, Weinman SA, Melnyk SB, Tryndyak V, Pogribny IP, Rusyn I (2012) Increased incidence of aflatoxin B1-induced liver tumors in hepatitis virus C transgenic mice. Int J Cancer J Int Cancer 130 (6):1347–1356. doi:10.1002/ijc.26140

Kamegaya Y, Hiasa Y, Zukerberg L, Fowler N, Blackard JT, Lin W, Choe WH, Schmidt EV, Chung RT (2005) Hepatitis C virus acts as a tumor accelerator by blocking apoptosis in a mouse model of hepatocarcinogenesis. Hepatology 41(3):660–667. doi:10.1002/hep.20621

Kanda T, Steele R, Ray R, Ray RB (2009) Inhibition of intrahepatic gamma interferon production by hepatitis C virus nonstructural protein 5A in transgenic mice. J Virol 83(17):8463–8469. doi:JVI.00751-09 [pii] 10.1128/JVI.00751-09

Kapoor A, Simmonds P, Gerold G, Qaisar N, Jain K, Henriquez JA, Firth C, Hirschberg DL, Rice CM, Shields S, Lipkin WI (2011) Characterization of a canine homolog of hepatitis C virus. Proc Natl Acad Sci U S A 108(28):11608–11613. doi:1101794108 [pii] 10.1073/pnas.1101794108

Kapoor A, Simmonds P, Cullen JM, Scheel TK, Medina JL, Giannitti F, Nishiuchi E, Brock KV, Burbelo PD, Rice CM, Lipkin WI (2013a) Identification of a pegivirus (GB virus-like virus) that infects horses. J Virol 87(12):7185–7190. doi:10.1128/JVI.00324-13

Kapoor A, Simmonds P, Scheel TK, Hjelle B, Cullen JM, Burbelo PD, Chauhan LV, Duraisamy R, Sanchez Leon M, Jain K, Vandegrift KJ, Calisher CH, Rice CM, Lipkin WI (2013b) Identification of rodent homologs of hepatitis C virus and pegiviruses. mBio 4(2):e00216-00213. doi:10.1128/mBio.00216-13

Karayiannis P, Petrovic LM, Fry M, Moore D, Enticott M, McGarvey MJ, Scheuer PJ, Thomas HC (1989) Studies of GB hepatitis agent in tamarins. Hepatology 9(2):186–192

Kosaka K, Hiraga N, Imamura M, Yoshimi S, Murakami E, Nakahara T, Honda Y, Ono A, Kawaoka T, Tsuge M, Abe H, Hayes CN, Miki D, Aikata H, Ochi H, Ishida Y, Tateno C, Yoshizato K, Sasaki T, Chayama K (2013) A novel TK-NOG based humanized mouse model for the study of HBV and HCV infections. Biochem Biophys Res Commun 441(1):230–235. doi:10.1016/j.bbrc.2013.10.040

Kremsdorf D, Brezillon N (2007) New animal models for hepatitis C viral infection and pathogenesis studies. World J Gastroenterol: WJG 13(17):2427–2435

Kriegs M, Burckstummer T, Himmelsbach K, Bruns M, Frelin L, Ahlen G, Sallberg M, Hildt E (2009) The hepatitis C virus non-structural NS5A protein impairs both the innate and adaptive hepatic immune response in vivo. J Biol Chem 284(41):28343–28351. doi:10.1074/jbc.M109.038877

Lanford RE, Chavez D, Notvall L, Brasky KM (2003) Comparison of tamarins and marmosets as hosts for GBV-B infections and the effect of immunosuppression on duration of viremia. Virology 311(1):72–80

Lauck M, Sibley SD, Lara J, Purdy MA, Khudyakov Y, Hyeroba D, Tumukunde A, Weny G, Switzer WM, Chapman CA, Hughes AL, Friedrich TC, O'Connor DH, Goldberg TL (2013) A novel hepacivirus with an unusually long and intrinsically disordered NS5A protein in a wild Old World primate. J Virol 87(16):8971–8981. doi:10.1128/JVI.00888-13

Lerat H, Kammoun HL, Hainault I, Merour E, Higgs MR, Callens C, Lemon SM, Foufelle F, Pawlotsky JM (2009) Hepatitis C virus proteins induce lipogenesis and defective triglyceride secretion in transgenic mice. J Biol Chem 284(48):33466–33474. doi:M109.019810 [pii] 10.1074/jbc.M109.019810

Li K, Foy E, Ferreon JC, Nakamura M, Ferreon AC, Ikeda M, Ray SC, Gale M Jr, Lemon SM (2005) Immune evasion by hepatitis C virus NS3/4A protease-mediated cleavage of the Toll-like receptor 3 adaptor protein TRIF. Proc Natl Acad Sci U S A 102(8):2992–2997

Lin LT, Noyce RS, Pham TN, Wilson JA, Sisson GR, Michalak TI, Mossman KL, Richardson CD (2010) Replication of subgenomic hepatitis C virus replicons in mouse fibroblasts is facilitated by deletion of interferon regulatory factor 3 and expression of liver-specific microRNA 122. J Virol 84(18):9170–9180. doi:JVI.00559-10 [pii] 10.1128/JVI.00559-10

Liu S, Yang W, Shen L, Turner JR, Coyne CB, Wang T (2009) Tight junction proteins claudin-1 and occludin control hepatitis C virus entry and are downregulated during infection to prevent superinfection. J Virol 83(4):2011–2014. doi:JVI.01888-08 [pii] 10.1128/JVI.01888-08

Long G, Hiet MS, Windisch MP, Lee JY, Lohmann V, Bartenschlager R (2011) Mouse hepatic cells support assembly of infectious hepatitis C virus particles. Gastroenterology 141 (3):1057–1066. doi:10.1053/j.gastro.2011.06.010

Lupberger J, Zeisel MB, Xiao F, Thumann C, Fofana I, Zona L, Davis C, Mee CJ, Turek M, Gorke S, Royer C, Fischer B, Zahid MN, Lavillette D, Fresquet J, Cosset FL, Rothenberg SM, Pietschmann T, Patel AH, Pessaux P, Doffoel M, Raffelsberger W, Poch O, McKeating JA, Brino L, Baumert TF (2011) EGFR and EphA2 are host factors for hepatitis C virus entry and possible targets for antiviral therapy. Nat Med 17(5):589–595. doi:nm.2341 [pii] 10.1038/nm.2341

Majumder M, Steele R, Ghosh AK, Zhou XY, Thornburg L, Ray R, Phillips NJ, Ray RB (2003) Expression of hepatitis C virus non-structural 5A protein in the liver of transgenic mice. FEBS Lett 555(3):528–532. doi:S0014579303013371 [pii]

Martin DN, Uprichard SL (2013) Identification of transferrin receptor 1 as a hepatitis C virus entry factor. Proc Natl Acad Sci U S A 110(26):10777–10782. doi:10.1073/pnas.1301764110

Marukian S, Andrus L, Sheahan TP, Jones CT, Charles ED, Ploss A, Rice CM, Dustin LB (2011) Hepatitis C virus induces interferon-lambda and interferon-stimulated genes in primary liver cultures. Hepatology 54(6):1913–1923. doi:10.1002/hep.24580

McCaffrey AP, Ohashi K, Meuse L, Shen S, Lancaster AM, Lukavsky PJ, Sarnow P, Kay MA (2002) Determinants of hepatitis C translational initiation in vitro, in cultured cells and mice. Mol Ther 5(6):676–684

Mercer DF, Schiller DE, Elliott JF, Douglas DN, Hao C, Rinfret A, Addison WR, Fischer KP, Churchill TA, Lakey JR, Tyrrell DL, Kneteman NM (2001) Hepatitis C virus replication in mice with chimeric human livers. Nat Med 7(8):927–933

Meuleman P, Leroux-Roels G (2008) The human liver-uPA-SCID mouse: a model for the evaluation of antiviral compounds against HBV and HCV. Antivir Res 80(3):231–238. doi:10.1016/j.antiviral.2008.07.006

Meuleman P, Libbrecht L, De Vos R, de Hemptinne B, Gevaert K, Vandekerckhove J, Roskams T, Leroux-Roels G (2005) Morphological and biochemical characterization of a human liver in a uPA-SCID mouse chimera. Hepatology 41(4):847–856. doi:10.1002/hep.20657

Meylan E, Curran J, Hofmann K, Moradpour D, Binder M, Bartenschlager R, Tschopp J (2005) Cardif is an adaptor protein in the RIG-I antiviral pathway and is targeted by hepatitis C virus. Nature 437(7062):1167–1172

Michta ML, Hopcraft SE, Narbus CM, Kratovac Z, Israelow B, Sourisseau M, Evans MJ (2010) Species-specific regions of occludin required by hepatitis C virus for cell entry. J Virol 84 (22):11696–11708. doi:JVI.01555-10 [pii] 10.1128/JVI.01555-10

Miller JS, Stevens KR, Yang MT, Baker BM, Nguyen DH, Cohen DM, Toro E, Chen AA, Galie PA, Yu X, Chaturvedi R, Bhatia SN, Chen CS (2012) Rapid casting of patterned vascular networks for perfusable engineered three-dimensional tissues. Nat Mater 11(9):768–774. doi:10.1038/nmat3357

Moriya K, Yotsuyanagi H, Shintani Y, Fujie H, Ishibashi K, Matsuura Y, Miyamura T, Koike K (1997) Hepatitis C virus core protein induces hepatic steatosis in transgenic mice. J Gen Virol 78(Pt 7):1527–1531

Moriya K, Fujie H, Shintani Y, Yotsuyanagi H, Tsutsumi T, Ishibashi K, Matsuura Y, Kimura S, Miyamura T, Koike K (1998) The core protein of hepatitis C virus induces hepatocellular carcinoma in transgenic mice. Nat Med 4(9):1065–1067. doi:10.1038/2053

Naas T, Ghorbani M, Alvarez-Maya I, Lapner M, Kothary R, De Repentigny Y, Gomes S, Babiuk L, Giulivi A, Soare C, Azizi A, Diaz-Mitoma F (2005) Characterization of liver histopathology in a transgenic mouse model expressing genotype 1a hepatitis C virus core and envelope proteins 1 and 2. J Gen Virol 86(Pt 8):2185–2196. doi:86/8/2185 [pii] 10.1099/vir.0.80969-0

Naas T, Ghorbani M, Soare C, Scherling N, Muller R, Ghorbani P, Diaz-Mitoma F (2010) Adoptive transfer of splenocytes to study cell-mediated immune responses in hepatitis C infection using HCV transgenic mice. Comp Hepatol 9:7. doi:1476-5926-9-7 [pii] 10.1186/1476-5926-9-7

Pasquinelli C, Shoenberger JM, Chung J, Chang KM, Guidotti LG, Selby M, Berger K, Lesniewski R, Houghton M, Chisari FV (1997) Hepatitis C virus core and E2 protein expression in transgenic mice. Hepatology 25(3):719–727

Patel MR, Loo YM, Horner SM, Gale M, Jr., Malik HS (2012) Convergent evolution of escape from hepaciviral antagonism in primates. PLoS Biol 10(3):e1001282. doi:10.1371/journal.pbio.1001282 PBIOLOGY-D-11-03907 [pii]

Pfaender S, Brown RJP, Pietschmann T, Steinmann E (2014a) Natural reservoirs for homologs of hepatitis C virus. Emerg Microbes Infec 3 doi:Artn E21. doi:10.1038/Emi.2014.19

Pfaender S, Cavalleri JM, Walter S, Doerrbecker J, Campana B, Brown RJ, Burbelo PD, Postel A, Hahn K, Kusuma A, Riebesehl N, Baumgartner W, Becher P, Heim MH, Pietschmann T, Feige K, Steinmann E (2014b) Clinical course of infection and viral tissue tropism of hepatitis C virus-like non-primate hepaciviruses. Hepatology. doi:10.1002/hep.27440

Pfaender S, Walter S, Todt D, Behrendt P, Doerrbecker J, Wolk B, Engelmann M, Gravemann U, Seltsam A, Steinmann J, Burbelo PD, Klawonn F, Feige K, Pietschmann T, Cavalleri JM, Steinmann E (2015) Assessment of cross-species transmission of hepatitis C virus-related non-primate hepacivirus in a population of humans at high risk of exposure. J Gen Virol 96 (9):2636–2642. doi:10.1099/vir.0.000208

Pileri P, Uematsu Y, Campagnoli S, Galli G, Falugi F, Petracca R, Weiner AJ, Houghton M, Rosa D, Grandi G, Abrignani S (1998) Binding of hepatitis C virus to CD81. Science 282 (5390):938–941

Ploss A, Evans MJ, Gaysinskaya VA, Panis M, You H, de Jong YP, Rice CM (2009) Human occludin is a hepatitis C virus entry factor required for infection of mouse cells. Nature 457 (7231):882–886. doi:nature07684 [pii] 10.1038/nature07684

Quan PL, Firth C, Conte JM, Williams SH, Zambrana-Torrelio CM, Anthony SJ, Ellison JA, Gilbert AT, Kuzmin IV, Niezgoda M, Osinubi MO, Recuenco S, Markotter W, Breiman RF, Kalemba L, Malekani J, Lindblade KA, Rostal MK, Ojeda-Flores R, Suzan G, Davis LB, Blau DM, Ogunkoya AB, Alvarez Castillo DA, Moran D, Ngam S, Akaibe D, Agwanda B, Briese T, Epstein JH, Daszak P, Rupprecht CE, Holmes EC, Lipkin WI (2013) Bats are a major natural reservoir for hepaciviruses and pegiviruses. Proc Natl Acad Sci U S A 110(20):8194–8199. doi:10.1073/pnas.1303037110

Roelandt P, Obeid S, Paeshuyse J, Vanhove J, Van Lommel A, Nahmias Y, Nevens F, Neyts J, Verfaillie CM (2012) Human pluripotent stem cell-derived hepatocytes support complete replication of hepatitis C virus. J Hepatol 57(2):246–251. doi:10.1016/j.jhep.2012.03.030

Sainz B Jr, Barretto N, Martin DN, Hiraga N, Imamura M, Hussain S, Marsh KA, Yu X, Chayama K, Alrefai WA, Uprichard SL (2012) Identification of the Niemann-Pick C1-like 1 cholesterol absorption receptor as a new hepatitis C virus entry factor. Nat Med 18 (2):281–285. doi:10.1038/nm.2581

Satoh K, Takahashi H, Matsuda C, Tanaka T, Miyasaka M, Zeniya M, Kohara M (2010) Natural killer cells target HCV core proteins during the innate immune response in HCV transgenic mice. J Med Virol 82(9):1545–1553. doi:10.1002/jmv.21859

Scarselli E, Ansuini H, Cerino R, Roccasecca RM, Acali S, Filocamo G, Traboni C, Nicosia A, Cortese R, Vitelli A (2002) The human scavenger receptor class B type I is a novel candidate receptor for the hepatitis C virus. EMBO J 21(19):5017–5025

Schaluder GG, Dawson GJ, Simons JN, Pilot-Matias TJ, Gutierrez RA, Heynen CA, Knigge MF, Kurpiewski GS, Buijk SL, Leary TP et al (1995) Molecular and serologic analysis in the transmission of the GB hepatitis agents. J Med Virol 46(1):81–90

Schwartz RE, Trehan K, Andrus L, Sheahan TP, Ploss A, Duncan SA, Rice CM, Bhatia SN (2012) Modeling hepatitis C virus infection using human induced pluripotent stem cells. Proc Natl Acad Sci USA 109(7):2544–2548. doi:1121400109 [pii] 10.1073/pnas.1121400109

Scull MA, Shi C, de Jong YP, Gerold G, Ries M, von Schaewen M, Donovan BM, Labitt RN, Horwitz JA, Gaska JM, Hrebikova G, Xiao JW, Flatley B, Fung C, Chiriboga L, Walker CM, Evans DT, Rice CM, Ploss A (2015) Hepatitis C virus infects rhesus macaque hepatocytes and simianized mice. Hepatology 62(1):57–67. doi:10.1002/hep.27773

Shultz LD, Brehm MA, Garcia-Martinez JV, Greiner DL (2012) Humanized mice for immune system investigation: progress, promise and challenges. Nat Rev Immunol 12(11):786–798. doi:10.1038/nri3311

Simons JN, Pilot-Matias TJ, Leary TP, Dawson GJ, Desai SM, Schlauder GG, Muerhoff AS, Erker JC, Buijk SL, Chalmers ML et al (1995) Identification of two flavivirus-like genomes in the GB hepatitis agent. Proc Natl Acad Sci U S A 92(8):3401–3405

Si-Tayeb K, Noto FK, Nagaoka M, Li J, Battle MA, Duris C, North PE, Dalton S, Duncan SA (2010) Highly efficient generation of human hepatocyte-like cells from induced pluripotent stem cells. Hepatology 51(1):297–305. doi:10.1002/hep.23354

Sithebe NP, Kew MC, Mphahlele MJ, Paterson AC, Lecatsas G, Kramvis A, de Klerk W (2002) Lack of susceptibility of Chacma baboons (Papio ursinus orientalis) to hepatitis C virus infection. J Med Virol 66(4):468–471. doi:10.1002/jmv.2167 [pii]

Sourisseau M, Goldman O, He W, Gori JL, Kiem HP, Gouon-Evans V, Evans MJ (2013) Hepatic cells derived from induced pluripotent stem cells of pigtail macaques support hepatitis C virus infection. Gastroenterology 145(5):966–969 e967. doi:10.1053/j.gastro.2013.07.026

Stevens KR, Ungrin MD, Schwartz RE, Ng S, Carvalho B, Christine KS, Chaturvedi RR, Li CY, Zandstra PW, Chen CS, Bhatia SN (2013) InVERT molding for scalable control of tissue microarchitecture. Nat Commun 4:1847. doi:10.1038/ncomms2853

Takaku S, Nakagawa Y, Shimizu M, Norose Y, Maruyama I, Wakita T, Takano T, Kohara M, Takahashi H (2003) Induction of hepatic injury by hepatitis C virus-specific CD8+ murine cytotoxic T lymphocytes in transgenic mice expressing the viral structural genes. Biochem Biophys Res Commun 301(2):330–337. doi:S0006291X02030188 [pii]

Tanaka N, Moriya K, Kiyosawa K, Koike K, Aoyama T (2008) Hepatitis C virus core protein induces spontaneous and persistent activation of peroxisome proliferator-activated receptor alpha in transgenic mice: implications for HCV-associated hepatocarcinogenesis. Int J Cancer J Int Cancer 122(1):124–131. doi:10.1002/ijc.23056

Tesfaye A, Stift J, Maric D, Cui Q, Dienes HP, Feinstone SM (2013) Chimeric mouse model for the infection of hepatitis B and C viruses. PLoS ONE 8(10):e77298. doi:10.1371/journal.pone.0077298

Touboul T, Hannan NR, Corbineau S, Martinez A, Martinet C, Branchereau S, Mainot S, Strick-Marchand H, Pedersen R, Di Santo J, Weber A, Vallier L (2010) Generation of functional hepatocytes from human embryonic stem cells under chemically defined conditions that recapitulate liver development. Hepatology 51(5):1754–1765. doi:10.1002/hep.23506

Tsukiyama-Kohara K, Sekiguchi S, Kasama Y, Salem NE, Machida K, Kohara M (2011) Hepatitis C virus-related lymphomagenesis in a mouse model. ISRN Hematol 2011:167501. doi:10.5402/2011/167501

Tumurbaatar B, Sun Y, Chan T, Sun J (2007) Cre-estrogen receptor-mediated hepatitis C virus structural protein expression in mice. J Virol Methods 146(1–2):5–13. doi:10.1016/j.jviromet.2007.05.025

Uprichard SL, Chung J, Chisari FV, Wakita T (2006) Replication of a hepatitis C virus replicon clone in mouse cells. Virol J 3:89. doi:1743-422X-3-89 [pii] 10.1186/1743-422X-3-89

Vogt A, Scull MA, Friling T, Horwitz JA, Donovan BM, Dorner M, Gerold G, Labitt RN, Rice CM, Ploss A (2013) Recapitulation of the hepatitis C virus life-cycle in engineered murine cell lines. Virology 444(1–2):1–11. doi:10.1016/j.virol.2013.05.036

Wang AG, Moon HB, Kim JM, Hwang SB, Yu DY, Lee DS (2006) Expression of hepatitis C virus nonstructural 4B in transgenic mice. Exp Mol Med 38(3):241–246. doi:200606307 [pii]

Wang AG, Lee DS, Moon HB, Kim JM, Cho KH, Choi SH, Ha HL, Han YH, Kim DG, Hwang SB, Yu DY (2009) Non-structural 5A protein of hepatitis C virus induces a range of liver pathology in transgenic mice. J Pathol 219(2):253–262. doi:10.1002/path.2592

Washburn ML, Bility MT, Zhang L, Kovalev GI, Buntzman A, Frelinger JA, Barry W, Ploss A, Rice CM, Su L (2011) A humanized mouse model to study hepatitis C virus infection, immune response, and liver disease. Gastroenterology 140(4):1334–1344. doi:10.1053/j.gastro.2011.01.001

Wegert M, La Monica N, Tripodi M, Adler G, Dikopoulos N (2009) Impaired interferon type I signalling in the liver modulates the hepatic acute phase response in hepatitis C virus transgenic mice. J Hepatol 51(2):271–278. doi:S0168-8278(09)00249-9 [pii] 10.1016/j.jhep.2009.03.014

Wilson EM, Bial J, Tarlow B, Bial G, Jensen B, Greiner DL, Brehm MA, Grompe M (2014) Extensive double humanization of both liver and hematopoiesis in FRGN mice. Stem Cell Res 13(3PA):404–412. doi:10.1016/j.scr.2014.08.006

Wu GY, Konishi M, Walton CM, Olive D, Hayashi K, Wu CH (2005) A novel immunocompetent rat model of HCV infection and hepatitis. Gastroenterology 128(5):1416–1423. doi:S0016508505003951 [pii]

Wu X, Robotham JM, Lee E, Dalton S, Kneteman NM, Gilbert DM, Tang H (2012) Productive hepatitis C virus infection of stem cell-derived hepatocytes reveals a critical transition to viral permissiveness during differentiation. PLoS Pathog 8(4):e1002617. doi:10.1371/journal.ppat.1002617

Wu X, Lee EM, Hammack C, Robotham JM, Basu M, Lang J, Brinton MA, Tang H (2014) Cell death-inducing DFFA-like effector b is required for hepatitis C virus entry into hepatocytes. J Virol 88(15):8433–8444. doi:10.1128/JVI.00081-14

Xie ZC, Riezu-Boj JI, Lasarte JJ, Guillen J, Su JH, Civeira MP, Prieto J (1998) Transmission of hepatitis C virus infection to tree shrews. Virology 244(2):513–520

Xu X, Chen H, Cao X, Ben K (2007) Efficient infection of tree shrew (Tupaia belangeri) with hepatitis C virus grown in cell culture or from patient plasma. J Gen Virol 88(Pt 9):2504–2512. doi:88/9/2504 [pii] 10.1099/vir.0.82878-0

Zhu Q, Guo JT, Seeger C (2003) Replication of hepatitis C virus subgenomes in nonhepatic epithelial and mouse hepatoma cells. J Virol 77(17):9204–9210

Zhu Q, Oei Y, Mendel DB, Garrett EN, Patawaran MB, Hollenbach PW, Aukerman SL, Weiner AJ (2006) Novel robust hepatitis C virus mouse efficacy model. Antimicrob Agents Chemother 50(10):3260–3268. doi:50/10/3260 [pii] 10.1128/AAC.00413-06

Innate Immune Recognition of Hepatitis C Virus

Kui Li

Abstract Infection by HCV elicits a relative strong innate immune response in the liver that is characterized by induction of interferon-stimulated genes (ISGs) and inflammatory cytokines. Mysteriously, this intrinsic host defense response is not followed by effective HCV-specific T cell responses in a majority of infected individuals, which progress to persistent infection. Innate immune recognition of HCV is initiated by hepatocytes and other parenchymal and non-parenchymal liver cells as well as specific immune cell subsets infiltrating the liver, signaling through various intracellular pathways that culminate in induction of interferons, ISGs and cytokines. These provide the first line of defense against HCV and also orchestrate the development of subsequent adaptive immunity to the virus. Data suggest that host genetic polymorphisms and immune evasion by HCV alter the status and magnitude of intrahepatic innate immune responses, and in so doing impact the infection outcome. This chapter summarizes recent advances in innate immune responses to HCV, focusing on the mechanisms by which the host detects HCV infection and initiates antiviral and inflammatory responses and by which HCV circumvents and sometimes exploits aspects of these intrinsic immune responses for maximal survival. Also reviewed are host innate immune factors that regulate spontaneous HCV clearance and response to the interferon-based therapy.

Keywords HCV • Innate immunity • Interferon • Cytokine • Interferon-stimulated gene • Pathogen-associated molecular pattern • Pattern recognition receptor • RIG-I • MDA5 • Toll-like receptor • IL28B • Immune evasion

1 Introduction

As the causative agent of the historic entity, post-transfusion non-A and non-B hepatitis, hepatitis C virus (HCV) affects an estimated 130–170 million people worldwide and is transmitted mainly via needle-sharing drug use in industrialized

K. Li, Ph.D. (✉)
Department of Microbiology, Immunology and Biochemistry, University of Tennessee Health Science Center, Memphis, TN 38163, USA
e-mail: kli1@uthsc.edu

T. Miyamura et al. (eds.), *Hepatitis C Virus I*, DOI 10.1007/978-4-431-56098-2_13

nations and unsafe medical procedures in developing countries (Ghany et al. 2009). The virus infects humans and chimpanzees but not other host species, and replicates predominantly in hepatocytes, the main parenchymal cell type in the liver. Acute HCV infection is typically asymptomatic and often goes undiagnosed. However, the virus is spontaneously cleared in only a small fraction of infected individuals, leaving 60–80 % harboring HCV for life. Persistent, intrahepatic HCV infection is a major risk factor for cirrhosis and hepatocellular carcinoma and as such, is a leading cause of liver transplantation (Lemon 2010; Li and Lemon 2013). Unfortunately, the development of a vaccine for HCV prevention remains a challenge, owing to the lack of antibody-based sterilizing immunity (Liang 2013). Instead, combating the HCV "silent epidemic" has largely relied upon treating those infected. Until 2011, pegylated interferon (IFN)-α plus ribavirin (PegIFN/RBV) had been the standard-of-care for hepatitis C patients. This treatment regimen, however, eliminates HCV in only half of the difficult-to-treat patients infected with the genotype 1 or 4 virus. With the recent approvals of direct-acting antivirals (DAAs) targeting the HCV protease, the NS5A protein and the polymerase for clinical use and other promising DAAs in the development pipeline, curing the majority of patients with future, IFN-sparing DAA combinations is no longer an unachievable goal. Nonetheless, IFN likely will remain a component of HCV therapy for a period of time given the high costs of and/or inaccessibility to the DAAs (reviewed in Lemon et al. 2010; Li and Lemon 2013). Thus, studies of HCV interactions with the host, especially those with the immune system, are still warranted as they will reveal critical molecular aspects that could be exploited to improve hepatitis C treatment modalities and to inform strategies for vaccine development.

Classified within the family Flaviviridae, HCV is a small, enveloped virus that possesses a ~10-kb, single-stranded (ss) RNA genome of positive polarity. Initial infection of hepatocytes with blood-borne HCV starts with the sequential interactions of viral envelope glycoproteins with at least four HCV co-receptors, i.e., CD81, scavenger receptor class BI, claudin 1 and occludin, followed by receptor-mediated endocytosis of the virion (Lindenbach and Rice 2013). In addition, cell-to-cell spread is an important route of HCV transmission, facilitating intrahepatic viral dissemination in the presence of neutralizing antibodies (Brimacombe et al. 2011; Timpe et al. 2008). Thanks to advanced imaging and sensitive in situ hybridization techniques, it has been shown that only ~1–50 % of hepatocytes in chronic hepatitis C patients replicate the virus, often seen as infected clusters (Kandathil et al. 2013; Liang et al. 2009; Pal et al. 2006; Wieland et al. 2014), and that the proportion of infected hepatocytes seem to correlate with serum viral load (Wieland et al. 2014). Following uncoating and its release into cytoplasm, HCV genomic RNA is translated into a single polyprotein that is co- and post-translationally processed by a combination of viral and cellular proteases into individual structural and non-structural (NS) proteins. The NS proteins remodel cytoplasmic membranes into highly compartmentalized viral factories in which they assemble replication complexes with the viral genome, transcribe the antigenome and synthesize nascent genomic RNAs (see the chapter in this volume by Brey and Lohmann). This process generates double-stranded (ds) RNA intermediates (Targett-Adams et al. 2008), which constitute a major type of pathogen-associated molecular pattern (PAMP) that elicits host defense responses (see Sects. 2.2 and 3.2). Newly synthesized HCV genomes are packaged into progeny virions by the viral structural proteins and exit the cells to go on infecting naïve hepatocytes.

Research on HCV-host interactions had been impeded by the lack of efficient tissue culture systems fully permissive for the virus until 2005. The establishment of JFH1 or JFH1-based chimeric virus-derived infection systems in human hepatoma Huh7 cells (Lindenbach et al. 2005; Wakita et al. 2005; Zhong et al. 2005) and more recently, in primary human hepatocytes (PHHs) (Marukian et al. 2011) and stem cell-derived, differentiated hepatocyte-like cells (DHHs) (Schwartz et al. 2012; Wu et al. 2012), while having limitations, allows scientists to address questions regarding the interplay between HCV and cellular processes including the innate immune system. Still, a robust, tractable small animal model for HCV is lacking. Thus, probing host responses to HCV in vivo has been conducted mainly using samples collected from experimentally infected chimpanzees and hepatitis C patients. These studies, nonetheless, have been instrumental in revealing that while acute and chronic HCV infections elicit intrahepatic innate immune responses characterized by expression of IFN-stimulated genes (ISGs) and cytokines (Bigger et al. 2001; Su et al. 2002), antigen specific, CD4+ and CD8+ adaptive T cell responses play a decisive role in resolution of HCV infection (Bowen and Walker 2005). The current view is that HCV infection outcome is determined by dynamic interactions of viral factors (genotype, viral load, etc.) and host factors (genetic background, innate and adaptive immune responses, etc.). In this chapter I summarize recent advances in innate immunity to HCV, focusing on the mechanisms by which the host detects HCV infection and initiates antiviral and inflammatory responses and by which HCV circumvents and sometimes co-opts aspects of the innate immune responses for maximal survival. Also reviewed are host innate immune factors that control viral clearance and response to IFN-based therapy. This chapter aims to stimulate discussion on what remains to be learned in this fast-evolving field to improve hepatitis C therapy and develop a long-awaited vaccine for HCV prophylaxis.

2 Overview of Innate Immune Responses to RNA Viruses

2.1 *Interferons*

As the most salient feature of the vertebrate host's innate immune responses to invading viruses, infected cells and sentinel cells of the innate immune system (such as macrophages and dendritic cells) rapidly produce IFNs and inflammatory cytokines/chemokines in efforts to fend off the intruder. Viral products with conserved molecular structures, most notably viral nucleic acids, present major

PAMPs to various host innate immune sensors termed pattern recognition receptors (PRRs) and are sensed by the latter as non-self materials. Subsequently, intracellular signaling cascades ensue, resulting in activation and nuclear translocation of the latent transcription factors, IFN-regulatory factor-3 or-7 (IRF3 or IRF7) and NF-κB. These collaboratively initiate transcription from promoters of IFNs and inflammatory mediators. Secreted IFNs act in an autocrine/paracrine fashion to upregulate the expression of hundreds of IFN-stimulated genes (ISGs) with direct antiviral activities, curbing viral replication in infected cells and establishing an antiviral state in uninfected neighboring cells. They also regulate the activation of various types of innate immune and immune effector cells, orchestrating the transition between innate and adaptive immunity. Inflammatory cytokines and chemokines, while generally not direct antiviral, play key roles in homing of various innate and adaptive immune cells to the infection site and in regulating their activation (Borden et al. 2007; Li and Lemon 2013).

Based on receptor usage, the IFNs are classified into three types. The type I IFNs are made up of IFN-β and about a dozen IFN-α subtypes. These can be expressed by many cell types in the body and act through a ubiquitously expressed receptor complex comprised of IFNAR1 and IFNAR2. IFN-γ is the sole member of type II IFN, and its production is restricted to NK, NKT and activated T cells. This IFN type signals through a distinct IFNGR1/IFNGR2 receptor complex to exert its effects (Borden et al. 2007). The type III IFNs are the latest additions to the IFN family, first characterized in 2003 (Uze and Monneron 2007). This group includes 4 members, IFN-λ1 (IFNL1, a.k.a., IL29), IFN-λ2 (IFNL2, a.k.a., IL28A), IFN-λ3 (IFNL3, a.k.a., IL28B) and the most recently described member of this family, IFN-λ4 (IFNL4) (Hermant and Michiels 2014; O'Brien et al. 2014). Of note, whereas the first three are produced in people from all ethnic groups, IFN-λ4 is a frame-shift product that arises from a specific genetic variation upstream of the IL28B gene locus and is most frequently present in individuals of African ancestry (Prokunina-Olsson et al. 2013). IFN-λ targets a receptor complex composed of IFNLR1 (a.k.a., IL28R1) and IL10R2, with the former having a restricted tissue distribution, being primarily expressed by epithelial cells. As such, in contrast to type I and type II IFNs that usually confer systemic antiviral protection, type III IFNs' effects are confined to the respiratory and digestive tracts (Hermant and Michiels 2014). It is worth noting here that human hepatocytes express functional receptors for all three IFN types. In line with this, all three types of IFNs inhibit HCV replication in cultured hepatocyte cell lines (Frese et al. 2002; Guo et al. 2001; Marcello et al. 2006; Robek et al. 2005; Zhu et al. 2005).

2.2 *RNA Virus-Activated Signaling Pathways Inducing IFNs and Cytokines*

In uninfected cells, the IRF and NF-κB transcription factors pivotal for IFN and cytokine synthesis predominantly reside in the cytoplasm and lack transcriptional activity. Following virus infection, various types of viral PAMPs (RNA or protein) encounter specific cellular PRRs in different compartments and initiate an alarm that foreign materials have intruded. They then relay signals through different pathways, leading to activation of kinases that phosphorylate IRF3/7 (by TBK1 or IKKε) or the NF-κB-sequestering unit, IκBα (by IKKβ in the classical IKK complex). The end result is the activation and nuclear translocation of the IRF and NF-κB transcription factors, followed by transcription of IFNs and cytokines/chemokines. Diverse PRRs have evolved to sense distinct PAMPs from RNA viruses in different locations of the cell. However, the PRRs actively involved in sensing a given virus infection vary by the host cell type and the virus.

2.2.1 Detection of Viral RNA Ligands in Cytoplasm by the RLRs and PKR

Retinoic-inducible gene I (RIG-I), along with melanoma-differentiation associated protein-5 (MDA5) and laboratory of genetics and physiology 2(LGP2), make up the RIG-I-like receptor (RLR) family of PRRs that detect viral RNAs in the cytoplasm. The RLRs share similar structural motifs, containing N-terminal tandem caspase activation (CARD) domains (except LGP2) that elicit downstream signaling, a central helicase domain, and a C-terminal domain (CTD) that confers a regulatory function (Kato et al. 2011). Upon viral infection, the CTD of RIG-I/MDA5 binds to its preferred RNA ligand, relieving the auto-repressed state of the RLR such that they recruit a specific adaptor protein, mitochondrial antiviral signaling protein (MAVS, a.k.a, Cardif, IPS-1 and VISA) and trigger downstream signaling events leading to IRF3/7 and NF-κB activation. The RLR pathway operates in the vast majority of cell types in the body including various non-immune and immune cells. A notable exception is the plasmacytoid dendritic cell (pDC) in which the TLR7 pathway dominates (see Sect. 2.2.2).

Of note, RIG-I and MDA5 do not serve redundant functions, but rather detect cytoplasmic viral RNAs with distinct characteristics and thus sense viruses from different families. The PAMPs that efficiently engage RIG-I are ssRNAs bearing 5′ triphosphates and rich in polyuridine runs or short (<300 bp), blunt-ended dsRNAs (Kato et al. 2011; Loo and Gale 2011). However, MDA5 preferentially recognizes long dsRNAs such as those generated during replication of picornaviruses as well as high molecular mass poly-I:C, a synthetic dsRNA surrogate (Kato et al. 2006, 2008). Interestingly, both RLRs contribute to detection of certain viruses such as the classical flaviviruses (e.g., dengue virus and West Nile virus) (Fredericksen et al. 2008; Kato et al. 2006; Loo et al. 2008), which yield both types of viral RNA ligands during their replication cycles. In contrast to the well-characterized functions of RIG-I and MDA5, the role of the third RLR member, LGP2, is controversial. While in vitro data showed LGP2 is a negative regulator of RIG-I/MDA5 signaling (Saito et al. 2007), data from LGP2-deficient mice support an essential role for this RLR in facilitating RIG-I/MDA5 sensing of selected viruses, e.g., EMCV (Satoh et al. 2010; Venkataraman et al. 2007).

Another PRR that participates in cytoplasmic detection of RNA viruses is the dsRNA-activated protein kinase R (PKR, a.k.a., EIF2AK2). A serine/threonine kinase that binds dsRNAs and short, imperfect stem-loop RNAs bearing 5′-triphophate moieties, PKR is best known for its ability to phosphorylate the translation initiation factor eIF2α, halting protein synthesis (He 2006; Pindel and Sadler 2011). A recent study suggested that PKR acts as a nonconventional PRR to sense structured regions of HCV RNA independent of its kinase activity, triggering the MAVS-dependent pathway leading to IFN and ISG expression (Arnaud et al. 2011). Whether this pathway operates to recognize and respond to other RNA virus infections remains to be determined.

2.2.2 Detection of Viral RNA Ligands in Endosomes by TLR3, TLR7 and TLR8

Toll-like receptors (TLRs) are type I transmembrane proteins that comprise an N-terminal ectodomain encompassing multiple leucine-rich repeats and mediating PAMP recognition, a central transmembrane domain, and a cytoplasmic, C-terminal Toll-interleukin-1 receptor homology (TIR) domain that recruits adaptor protein(s) and activates downstream signaling (Kawasaki et al. 2011). Of the ten known human TLRs, three detect viral RNA ligands with distinct molecular features in endosomal/lysosomal compartments. TLR3 senses dsRNAs either from the genome of dsRNA viruses or as replication intermediates generated during viral replication, whereas TLR7 and TLR8 recognize ssRNAs rich in GU and AU sequences, respectively. The signaling pathways activated downstream of these TLRs also differ. The TLR3 pathway relays signals through TIR-domain containing adaptor inducing IFN-β (TRIF) to activate both NF-κB and IRF3/7, while TLR7 and TLR8 enlist the myeloid differentiation primary response 88 (MyD88) adaptor, culminating in predominantly IRF7- and NF-κB-dependent gene induction, respectively (Kawai and Akira 2008, 2011). The differences in downstream responses between the TLR7 and TLR8 pathways are believed to stem from disparate cell type-specific expression of the two TLRs as well as the expression status of IRF7. While TLR8 is predominantly expressed in conventional dendritic cells (cDCs) and macrophages, TLR7 is the preferential RNA-sensing TLR expressed in pDCs, a "professional IFN-producing" DC type that has the unique capacity to constitutively express IRF7, a master regulator of type I and type III IFNs (Liu 2005).

2.2.3 Detection of Viral Proteins on Plasma Membrane by TLR2 and TLR4

TLR2 and TLR4 reside on the plasma membrane where their primary role is to monitor the extracellular space for bacterial cell wall/membrane components. Ligation of TLR2 with bacterial lipoproteins or engagement of TLR4 by lipopolysaccharide typically leads to production of NF-κB-dependent inflammatory cytokines through the MyD88-dependent pathway. Upon trafficking to endosomes TLR4 can also signal through a TRIF-dependent pathway, resulting in IRF3/7-dependent IFN induction and sustained NF-κB activation. Accumulating data suggest that during some viral infections TLR2 or TLR4 can recognize certain viral envelope glycoproteins or viral proteins released into extracellular milieu, thereby contributing to inflammatory and/or antiviral responses (Lester and Li 2014).

2.3 Signal Transduction Downstream of the IFN Receptors

Once they have been synthesized and secreted, the IFNs exert their biological effects by engaging cognate receptors on the cell surface and subsequently triggering intracellular signaling cascades termed the Jak-STAT signaling pathway (Borden et al. 2007). This leads to activation of the Janus kinases (Jak1, Jak2 and Tyk2), culminating in phosphorylation and nuclear translocation of the STAT family of transcription factors that activate ISG promoters. STAT1 and STAT2, along with IRF9, form a heterotrimeric transcription complex termed IFN-stimulated gene factor-3 (ISGF3). Although other STAT complexes also contribute, ISGF3 is the main mediator of type I and type III IFNs' transcriptional responses. As such, these two types of IFNs induce similar sets of ISGs, albeit with differences in expression kinetics. In contrast, IFN-γ mainly activates STAT1 homodimers, known as the IFN-γ activation factor (GAF). Thus, the ISG induction profile of IFN-γ is different from those of type I and type III IFNs.

Whereas most of the ISGs are thought to be antiviral or immune-stimulatory, a subset serves as negative regulators of Jak-STAT signaling. Examples include the ubiquitin specific peptidase 18 (USP18, a.k.a, UBP43) and PIAS (protein inhibitor of activated STAT) proteins, among others (Borden et al. 2007; Heim 2013). These regulatory ISGs contribute to fine-tuning of the IFN responses and are believed to help limit the harmful, autoimmune and pathological effects due to excessive or prolonged IFN activation.

Although the prevailing view of IFN's role in innate immunity has settled on its antiviral effects and immune activation, this paradigm is being challenged. Several recent studies show that during viral infections IFNs can have opposing dual roles, i.e., antiviral and immuno-suppressive, which vary from virus to virus or even at different stages of a single virus infection (Odorizzi and Wherry 2013; Teijaro et al. 2013; Wilson et al. 2013). While IFN signaling is critical for suppressing viral replication and spread of acute infections (or the acute phase of infection), prolonged IFN activation also inhibits the immune system in ways that can promote viral persistence (see also discussion in Sect. 6).

3 Innate Immune Detection of HCV Infection

3.1 Overview of the Innate Immune Response in HCV-Infected Liver

Gene expression profiling of liver biopsies from hepatitis C patients and experimentally infected chimpanzees has shown that HCV infection is accompanied with relatively robust, intrahepatic upregulation of a wide array of ISGs (Bigger et al. 2001; Su et al. 2002). In addition, there is often elevated expression of numerous intrahepatic chemokines, such as the CCR5 ligands, CCL5, CCL3, and CCL4, and the CXCR3 ligands, CXCL10, CXCL11, and CXCL9, among others (Helbig et al. 2009; Wald et al. 2007; Zeremski et al. 2007). This is in stark contrast to what has been observed in acute HBV- or HAV-infected chimpanzee livers, in which no or very limited ISGs and cytokines are induced (Lanford et al. 2011; Wieland et al. 2004). The in vivo data demonstrate that HCV is under the surveillance of the innate immune program of the liver and that it elicits host defense responses, but leaves open the question as to the type and specific cellular source of the IFNs that drive the ISG response and the origin of the chemokines/cytokines.

The hepatic innate immune responses to HCV are most likely a net result of complex interactions and cross-talks between HCV, hepatocytes and other liver-resident parenchymal (e.g., hepatic stellate cells (HSCs)) and non-parenchymal (e.g., Kupffer cells) cells as well as infiltrating innate immune cells (such as pDCs, NK cells, and BDCA3+ myeloid DCs (mDC2s)) (Horner and Gale 2013; Park and Rehermann 2014). Hepatocytes, Kupffer cells, pDCs and mDC2s all have the ability to produce IFNs, but their exact contribution to the ISG response in HCV-infected liver is unclear. Following stimulation by virus or poly-I:C, cultured human hepatocytes upregulate the expression of IFN-β and IFN-λs (Li et al. 2005a; Marukian et al. 2011). However, IFN-λs constitute the predominant IFN type induced in PHH cultures following acute HCV infection, with little type I IFN expressed (Helle et al. 2013; Marukian et al. 2011; Park et al. 2012). The liver-resident macrophages, Kupffer cells, express IFN-β in the liver of hepatitis C patients in situ (Lau et al. 2013), presumably after phagocytosis of HCV. pDCs and mDC2 secrete type I and III IFNs, respectively, when placed in co-cultures with HCV-infected hepatoma cells (Takahashi et al. 2010; Zhang et al. 2013). In light of the recent revelations that type III but not type I IFNs are strongly induced in the liver of HCV-infected chimpanzees and hepatitis C patients and that type III IFN levels are correlated with hepatic ISG expression (Park et al. 2012; Thomas et al. 2012), it is plausible that IFN-λs produced by hepatocytes and other cell types play a major part in driving the ISG response in HCV-infected liver. However, it is also possible that small amounts of IFN-β and IFN-α are induced and contribute to ISG expression locally, although their roles are likely minor compared with IFN-λ. As to the cellular sources of intrahepatic cytokines/chemokines, in vitro studies have demonstrated that human hepatocytes can sense HCV dsRNA via TLR3, producing CCL5, CCL3, CCL4 and CXCL10, IL6, etc (Li et al. 2012). In addition, cross-talk between HSCs and infected hepatocytes may stimulate the latter to express some of the cytokines/chemokines mentioned above, as illustrated by co-culture studies using HSC cell lines and HCV-infected hepatoma cells (Nishitsuji et al. 2013). Furthermore, phagocytic update of HCV particles by Kupffer cells, as suggested by in vitro experiments using a differentiated THP1 surrogate model, may trigger inflammasome activation that precipitates proinflammatory IL-1β release (Negash et al. 2013). Whether other liver parenchymal and non-parenchymal cells also contribute to hepatic inflammatory responses in HCV infection is not known.

The innate immune responses activated during acute HCV infection allow the host to cope with the virus in the absence of adaptive immune responses, buying time for the latter to be established. IFNs and ISGs induced in the liver inhibit HCV replication and spread, putting the virus in check. Supporting this notion, studies have shown that disruption of IFN signaling by depleting STAT1 or inhibiting the Jak kinase significantly promotes HCV propagation in PHH cultures (Andrus et al. 2011; Helle et al. 2013). Likewise, knockout of STAT1 in humanized mice expressing HCV co-receptors led to measurable viremia over several weeks after HCV infection (Dorner et al. 2013). Proinflammatory cytokines and chemokines dictate homing of immune cells to the liver and regulate their activation, probably collaboratively with the IFNs. However, very little is known about how the innate immune responses to HCV connect with and orchestrate the development of adaptive T cell responses. Innate immunity to HCV alone is unable to clear the infection in humans, and there is no association between the levels of intrahepatic and peripheral IFNs and infection outcome in acutely infected chimpanzees (Park et al. 2012). There is often ongoing upregulation of ISGs and cytokines in the liver of chronic hepatitis C patients and chronically infected chimpanzees, but these responses fail to contain HCV. Paradoxically, pre-activated hepatic ISG expression is linked to IFN refractoriness and poor response to IFN-based therapy (Asselah et al. 2008; Chen et al. 2005, 2011; Feld et al. 2007; Pfeffer et al. 2014; Sarasin-Filipowicz et al. 2008).

3.2 *Recognition of HCV PAMPs by PRRs*

Current evidence, mostly gleaned from in vitro studies, suggests that various HCV PAMPs are sensed by a variety of PRRs in different cellular compartments and

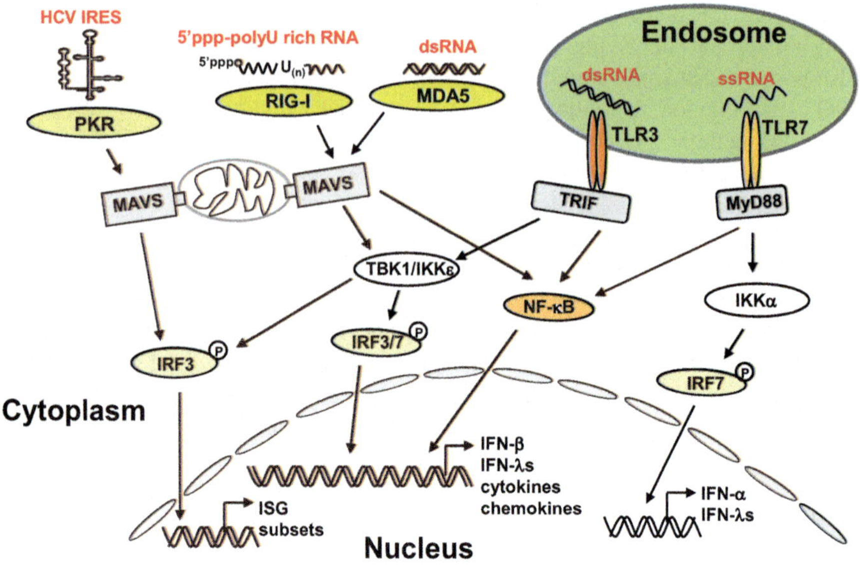

Fig. 1 PRRs and signaling pathways sensing HCV infection and inducing expression of type I and III IFNs, ISGs, and cytokines/chemokines. The poly-U/UC track in the 3′UTR of HCV RNA, along with the 5′-triphosphate moiety, constitute a potent HCV PAMP that engages RIG-I in the cytoplasm and initiates MAVS-dependent signaling cascades, culminating in activation of IRF3/7 and NF-κB and synthesis of type I and III IFNs and inflammatory cytokines/chemokines. This pathway is operative in hepatocytes and Kupffer cells. MDA5 detects viral RNAs, presumably dsRNA intermediates accumulated during HCV RNA replication, in cytoplasm of infected hepatocytes during late phase of infection, and triggers MAVS-dependent signaling. The highly structured HCV IRES is recognized by PKR, which can also activate MAVS-dependent signaling resulting in IRF3 activation and induction of a small subset of ISGs in hepatocytes. dsRNA intermediates produced during HCV RNA replication can be sensed by TLR3 in endolysosomal compartments of hepatocytes and mDC2s and trigger a TRIF-dependent pathway activating IFN-mediated antiviral and inflammatory responses. HCV ssRNA can be sensed by TLR7 in endosomes and activates MyD88-dependent innate immune responses in pDCs and Kupffer cells. The recognition of HCV ssRNA transferred via exosomes from infected hepatocytes to pDCs results primarily in IRF7 activation and subsequent induction of type I and III IFNs, while sensing of phagocytosed HCV ssRNA by TLR7 in Kupffer cells leads to NF-κB-dependent expression of pro-IL1β and, along with changes in intracellular ions, NLRP3 inflammasome activation and IL1β release

elicit innate immune signaling in host hepatocytes as well as in innate immune cells that patrol the liver (Fig. 1).

3.2.1 RIG-I and MDA5

Of the three known RLRs, RIG-I has been identified as a PRR for HCV RNA in the cytoplasm. A potent HCV PAMP for RIG-I has been mapped to the poly-U/UC track RNA in the 3′ untranslated region (UTR), along with the triphosphate moiety present at the 5′ end of the RNA. When the poly-U motif was deleted from in vitro transcribed HCV RNA, the ability to induce ISG expression in mouse liver was diminished (Saito et al. 2008). It was also suggested that a small RNA processed from the NS5B region of HCV RNA by RNase L is capable of activating RIG-I-dependent innate immune signaling to IFN-β (Malathi et al. 2010). It is currently unclear whether in hepatocytes the incoming HCV RNAs released from uncoated virions activate RIG-I (Loo et al. 2006) or those amplified during viral RNA replication are responsible. However, in cultured PHHs infected with HCV, IFN induction was strongly reduced by an inhibitor targeting the NS5B RNA-dependent RNA polymerase (RdRp) and UV-inactivated virions were non-stimulatory (Marukian et al. 2011). Aside from hepatocytes, Kupffer cells may utilize the RIG-I-MAVS pathway, as suggested by a study conducted using the THP1 surrogate model, to sense phagocytosed HCV RNA, resulting in IFN-β expression (Lau et al. 2013). Besides HCV RNA-derived ligands, there is some evidence that overexpression of HCV NS5B alone drives the expression of type I IFNs (Naka et al. 2006) and IL6 in human hepatocytes and mouse liver (Yu et al. 2012). This was shown to be dependent on the RdRp activity of NS5B, which presumably synthesizes small RNA species that activate RIG-I (Yu et al. 2012). Evidence supporting a role for RIG-I in mediating antiviral response to HCV comes from cell culture experiments. The Huh7.5 cell line, which is highly permissive for HCV RNA replication, harbors a lethal T55I mutation in the CARD domain of RIG-I when compared with its parental Huh7 cells. Although other cellular factors may contribute to its highly permissive phenotype, expression of wild-type RIG-I diminished the ability of Huh7.5 cells to support HCV replication (Sumpter et al. 2005).

While early experiments based on transfection of in vitro transcribed HCV RNAs did not suggest a role for MDA5 in detecting HCV, several recent studies have demonstrated that in HCV-infected cell culture settings, MDA5 indeed actively participates in innate immune response to the virus in the hepatocyte and its role is non-redundant to that of RIG-I (Israelow et al. 2014; Cao et al. 2015; Hiet et al. 2015). Whereas RIG-I mediates host responses to HCV early post infection, MDA5 plays a predominant role during late phase of HCV infection (Cao et al. 2015; Hiet et al. 2015). The exact HCV PAMP (presumably viral dsRNAs produced during HCV RNA replication) that engages MDA5 remains to be addressed, as does the question of why this innate sensing mechanism does not kick in during early phase of HCV infection in spite of ongoing, robust viral RNA replication.

3.2.2 PKR

Previous studies have shown that PKR binds to and is activated by the highly structured internal ribosome entry site (IRES) region of HCV RNA (Shimoike et al. 2009; Toroney et al. 2010). In a recent study, PKR was found to bind HCV RNA early post infection in Huh7 cells and recruit MAVS, prior to RIG-I activation, inducing IRF3-dependent expression of a subset of ISGs (Arnaud et al. 2011). In this case, instead of acting as a kinase, PKR serves as a nontraditional PRR (or an adaptor) that elicits innate immune signaling. How activation of this branch of innate immunity regulates HCV pathogenesis is unknown, but paradoxically, in vitro data suggested that HCV might exploit this mechanism to delay RIG-I-dependent antiviral signaling. ISG15 was found to be induced early post infection as a result of activation of the PKR-MAVS-IRF3 axis. This ISG, however, impedes TRIM25-mediated K63-linked ubiquitination of RIG-I (Arnaud et al. 2011), a prerequisite for RIG-I-MAVS signaling (Gack et al. 2007). Overall, the roles that PKR plays in antiviral immunity against HCV are controversial, with both proviral and antiviral effects having been reported in literature (Li and Lemon 2013).

3.2.3 TLR3

Consistent with the fact that the liver is of epithelial origin, many different cell types in this organ express TLR3, including hepatocytes, HSCs, Kupffer cells, biliary endothelial cells and liver sinusoidal endothelial cells (Li and Lemon 2013). TLR3 is also found in mDC2s (Lauterbach et al. 2010), a rare blood DC type that is enriched in the liver. As with cDCs, TLR3 signaling in hepatocytes and mDC2s shows a dependence on endosomal acidification (Li et al. 2012; Zhang et al. 2013). Interestingly, Huh7 cells, which are typically used to propagate HCV in cell culture, lacks TLR3 expression and is thus defective for extracellular dsRNA-induced antiviral gene expression (Li et al. 2005a). When TLR3 signaling was reconstituted, Huh7 cells became significantly less permissive for HCV replication, concomitantly exhibiting HCV-induced IRF3 activation and ISG expression (Wang et al. 2009). Subsequently, it was found that HCV-induced NF-κB activation and expression of inflammatory cytokines and chemokines were also dependent on TLR3 signaling in these cells (Li et al. 2012). Altogether, these data suggest that TLR3 is a genuine PRR for HCV in hepatocytes and that it mediates antiviral as well as inflammatory responses to the virus.

The HCV PAMP for TLR3 has been defined as dsRNA intermediates generated during HCV RNA replication (Li et al. 2012). However, highly structured HCV RNAs with stem-loop motifs, derived from either the viral positive- or negative-strand, are incapable of activating TLR3. Thus, the structured regions within HCV RNA do not appear to present sufficient dsRNA structure for TLR3 signaling. The minimal length of HCV dsRNAs that induce TLR3 signaling is ~80–100 bp (Li et al. 2012); lengthier dsRNAs possess greater stimulating activity (Dansako et al. 2013; Li et al. 2012). However, unlike RIG-I that preferentially senses poly-U rich sequences in the HCV 3′UTR, HCV dsRNAs that effectively activate TLR3 lack both a specific nucleotide composition or specific position within the genome (Li et al. 2012).

Exactly how dsRNAs produced during HCV RNA replication on cytoplasmic membranes are transferred to and engage TLR3 in endolysosomes is unknown.

Intracellular delivery through autophagy or other means of vesicular transport is a possibility; alternatively, HCV dsRNAs released from infected hepatocytes may be sensed by TLR3 in neighboring cells after their endocytosis is facilitated by the class A scavenger receptor type 1, MSR1 (Dansako et al. 2013). These mechanisms are probably not mutual exclusive.

Another cell type that senses HCV dsRNAs via TLR3 in the liver is mDC2, which is not susceptible to HCV but can take up HCV dsRNAs released from infected hepatocytes. These BDCA3+ mDCs, when placed in co-culture with HCV-infected Huh7.5 cells, secreted copious amount of IFN-λs but little, if any, type I IFNs, a process dependent on endocytic transport and endosomal maturation (Zhang et al. 2013). Taken collectively, TLR3-mediated sensing of HCV dsRNAs and subsequent IFN induction in hepatocytes and mDC2 cells may be part of the driving force behind ISG expression in the HCV-infected liver, and TLR3-dependent induction of inflammatory mediators in hepatocytes may contribute to hepatic inflammation and immune responses.

3.2.4 TLR7

Current data suggest that two distinct cell populations in the liver may utilize the TLR7 pathway to sense the presence of HCV infection and elicit innate immune responses. pDCs infiltrate HCV-infected liver (Lau et al. 2008), and those in close contact with infected hepatocytes can sense HCV ssRNA delivered to them via exosomes, activating the TLR7-MyD88-IRF7 pathway and resulting in production of large amounts of type I IFN (Dreux et al. 2012; Takahashi et al. 2010) and lesser quantities of IFN-λ (Zhang et al. 2013). In the liver sinusoids, Kupffer cells can phagocytose HCV particles with the incoming viral RNAs engaging TLR7 in endolysosome compartments. The outcome of TLR7-MyD88 signaling in these liver resident macrophages, however, is NF-κB-dependent pro-IL-1β expression and priming of the NLRP3 inflammasome. Subsequently, changes in cytoplasmic ions due to HCV uptake and perhaps transient protein expression trigger full-activation of the inflammasome, leading to maturation and secretion of IL-1β, which is thought to promote hepatic inflammation (Negash et al. 2013). It is worth noting that in both pDCs and Kupffer cells, the detection of HCV RNA by TLR7 does not rely on active viral replication, consistent with the fact that HCV is unable to productively infect these cells. These studies thus illustrate that the TLR7 pathway is used by two distinct types of nonpermisissive, innate immune sentinel cells in the liver to detect and respond to HCV.

3.2.5 TLR2

Recombinant HCV core and NS3 proteins were shown to engage TLR2 on human monocytes, leading to activation of NF-κB and MAP kinase and increased production of TNFα and IL-10. However, instead of facilitating anti-HCV immunity, these responses seem to impair DC differentiation and IFN-α secretion and promote inflammation (Dolganiuc et al. 2003, 2004). However, it is not clear that the amount of core and NS3 secreted/released from infected hepatocytes approaches the levels that activate TLR2 signaling and whether the recombinant proteins are structurally similar to those detected in HCV infection settings in vivo (Hoffmann et al. 2009).

4 Antiviral Effectors Against HCV

The characterization of the effector mechanisms of the innate immune responses controlling HCV infection, in particular those downstream of the IFNs, represents a robust area of investigation (Li and Lemon 2013; Metz et al. 2013; Schoggins and Rice 2011). Of the >300 known ISGs, a subset has been reported to possess anti-HCV activity in vitro (Table 1). Among this list are PRRs (RIG-I and TLR3) that sense HCV infection, transcription factors (IRF7 and IRF1) that drive IFN and ISG expression and amplify the antiviral response, and ISGs with direct antiviral actions against HCV (IFITM1, ISG20, viperin, etc.). Of the last group (which is also the largest group), the anti-HCV ISGs have been found to act at different steps of the HCV life cycle, including entry (e.g., IFITM1), RNA translation (e.g., ISG56), RNA replication (e.g., viperin and ISG20), and virus production and release (e.g., tetherin). It should be noted, however, that the mechanism of action for many reported anti-HCV ISGs remains to be discovered. Also, data on some of the ISGs from different studies have been inconsistent or contradictory, likely due to differences in the cell systems studied, assays used to measure activity, or perhaps expression levels of HCV proteins that counteract these mechanisms (Li and Lemon 2013; Metz et al. 2013). Conceivably, it is the combinatorial effects of multiple ISGs rather than that of any single ISG that governs the antiviral efficacies of IFNs against HCV (Metz et al. 2013). Apart from the protein-coding ISGs, there is evidence that IFN induces a small subset of non-coding microRNAs, which have been shown to target HCV genome and inhibit HCV replication in cell culture (Pedersen et al. 2007). The biological significance of this finding is unclear given that RNA interference is not deemed to be a key antiviral mechanism in mammalian hosts.

5 Innate Immune Evasion Strategies of HCV

Although multi-faceted innate immune mechanisms are in place to detect and act to contain HCV infection, the virus prevails and persists in a large proportion of infected people. Current data show that HCV has evolved elaborative strategies to not only evade innate recognition but also actively counteract innate immune defenses. These are likely to contribute to the unusual capability of HCV to establish chronic infection.

Table 1 Interferon-stimulated genes reported to possess anti-HCV activity

Gene symbol	Proposed mechanism of action	Reference(s)
ADAR	Viral RNA editing	Taylor et al. (2005)
BST2 (tetherin)	Inhibits virus production and release	Dafa-Berger et al. (2012) and Pan et al. (2013)
DDIT4	Unknown	Schoggins et al. (2011)
DDX58 (RIG-I)	Viral sensing	Schoggins et al. (2011) and Sumpter et al. (2005)
DDX60	Promotes RIG-I signaling	Schoggins et al. (2011)
EIF2AK2 (PKR)	Targets eIF2α to inhibit protein translation	Chang et al. (2006) and Jiang et al. (2008)
GBP1	Unknown	Itsui et al. (2009, 2006)
IFI44L	Unknown	Schoggins et al. (2011)
IFI6	Unknown	Itsui et al. (2009, 2006)
IFIH1 (MDA5)	Viral sensing	Schoggins et al. (2011)
IFIT1 (ISG56)	Inhibits protein translation via IRES	Wang et al. (2003)
IFIT3	Unknown	Metz et al. (2012)
IFITM1	Inhibits viral entry	Wilkins et al. (2013)
IFITM3	Inhibits protein translation	Metz et al. (2012) and Yao et al. (2011)
IFI27 (ISG12)	Unknown	Itsui et al. (2006)
IRF1	Induces IFNs and ISGs	Itsui et al. (2006), Kanazawa et al. (2004), and Schoggins et al. (2011)
IRF7	Induces IFNs and ISGs	Aly et al. (2007) and Schoggins et al. (2011)
ISG20	Exonuclease activity	Jiang et al. (2008) and Zhou et al. (2011)
MAP3K14 (NIK)	Unknown, possibly NF-κB activity	Schoggins et al. (2011)
MOV10	Unknown	Schoggins et al. (2011)
MS4A4A	Unknown	Schoggins et al. (2011)
MX1 (MxA)	Unknown	Itsui et al. (2006)
NOS2	Unknown	Metz et al. (2012)
NT5C3	Unknown	Schoggins et al. (2011)
OAS1	Activates RNase L	Kwon et al. (2013)
OASL	Unknown	Schoggins et al. (2011)
PLSCR1	Unknown	Metz et al. (2012)
RNASEL	Cleaves viral RNA	Han and Barton (2002) and Metz et al. (2012)
RSAD2 (viperin)	Interacts with VAP-A and HCV NS5A	Helbig et al. (2011, 2005), and Wang et al. (2012)
SSBP3	Unknown	Schoggins et al. (2011)
TLR3	Viral sensing	Wang et al. (2009)
TRIM14	Unknown	Metz et al. (2012)

5.1 *Evasion of Innate Immunity by Stealth*

In the blood of infected people, HCV virions are decorated with lipids and lipoproteins, physically resembling very low-density lipoprotein particles (see the chapter in this volume by Macrotrigiano and Catenese). This disguises the virus as self-material and impedes its recognition by innate immune cells (Syed et al. 2010). In infected hepatocytes, HCV replicates its RNA in association with endoplasmic reticulum-derived membrane spherules (Egger et al. 2002; Gosert et al. 2003). This highly compartmentalized microenvironment is believed to protect HCV replication complexes from attacks by cellular proteases and nucleases and to limit its access to and detection by the cytoplasmic PRR, RIG-I. Evasion of RIG-I sensing is also facilitated by HCV-induced autophagy (Ait-Goughoulte et al. 2008; Dreux et al. 2009; Sir et al. 2008; Tanida et al. 2009), which engulfs and sequesters HCV RNAs in autophagosomes (Ferraris et al. 2010) and also inhibits RIG-I signaling (Jounai et al. 2007; Ke and Chen 2011). By binding to miR-122, the most abundant miRNA in the liver, at two sites close to the 5′ end of the HCV genome, the positive HCV RNA strand is not only stabilized but also less likely to be recognized by RIG-I, because the 5′-triphosphate moiety is masked (Jopling et al. 2005; Machlin et al. 2011; Shimakami et al. 2012). Finally, owing to the error-prone NS5B RdRp and continuous pressure from the immune system, HCV propagates and circulates in the host as a "quasispecies" swarm (Martell et al. 1992). This, benefits the virus by allowing it to accumulate mutations that provide for escape from cytotoxic T cells or neutralizing antibodies as well as OAS-RNase L-mediated viral RNA cleavage (Han and Barton 2002).

5.2 *Counteraction of Innate Immune Signaling and Effector Mechanisms*

5.2.1 Inhibition of IFN Induction

HCV has been shown to disrupt IFN induction though either MAVS- or TRIF-dependent pathways (Breiman et al. 2005; Li et al. 2005b; Otsuka et al. 2005). The major HCV protease, NS3/4A complex, proteolytically cleaves both MAVS (Li et al. 2005c; Meylan et al. 2005) and TRIF (Ferreon et al. 2005; Li et al. 2005b). The cleavage of MAVS by NS3/4A displaces this adaptor protein from mitochondrion-associated membranes where it assembles signaling complexes, blocking IFN induction via the RIG/MDA5 pathway (Foy et al. 2003) and likely the recently identified PKR-MAVS-IRF3 axis as well (Arnaud et al. 2011). Remarkably, MAVS cleavage has been detected in liver biopsies of chronic hepatitis C patients (Bellecave et al. 2010; Loo et al. 2006; Stiffler et al. 2009) and found to be weakly correlated with lower hepatic ISG expression (Bellecave et al. 2010). This HCV countermeasure likely facilitates HCV replication and spread and contributes to hepatitis C pathogenesis, but it is unlikely to be the most important factor in HCV chronicity (Lemon 2010). MAVS is also cleaved and inactivated by proteases from GBV-B and HAV (Chen et al. 2007; Yang et al. 2007), two hepatotropic positive-strand RNA viruses that typically cause self-limited infection. TRIF is cleaved in cell-free reactions by purified NS3/4A (Ferreon et al. 2005; Li et al. 2005b). In HCV-infected Huh7.5 cells, TRIF protein level was substantially reduced, but could be partially restored by a specific inhibitor of the NS3/4A protease. In line with this, extracellular poly-I:C induced ISG upregulation was blunted in cells with established HCV infection (Wang et al. 2009). However, it remains to be determined whether TRIF is degraded within the HCV-infected liver.

The NS3 and NS4B proteins have also been implicated in suppressing IFN induction through other mechanisms. NS3 was shown to form a complex with TBK1, disrupting the interaction of this IRF3 kinase with IRF3 (Otsuka et al. 2005). NS4B was reported to bind STING, disrupting its scaffold function in facilitating TBK1-IRF3 interaction and IRF3 activation (Ding et al. 2013; Machlin et al. 2011). The biological relevance of this evasion mechanism is unknown, since the importance of STING in innate immunity to RNA viruses is uncertain.

The diverse HCV countermeasures that hinder the induction of IFN responses undoubtedly confer a survival advantage upon the virus, enabling more efficient viral replication and dissemination. However, HCV can complete its replication cycle in cells with intact innate immune pathways (Frentzen et al. 2011; Sheahan et al. 2014), and ISG transcripts are frequently observed in HCV-infected hepatocytes in patient liver in situ (Wieland et al. 2014). These observations imply that complete inhibition of innate immune responses is not a prerequisite for HCV to progress through its entire life cycle or establish persistence. It may be sufficient if the inhibition is long enough to allow production of progeny viruses capable of initiating infection in new target cells (Sheahan et al. 2014). Alternatively, blocking the synthesis of antiviral effector proteins at translational level may be more critical (Garaigorta and Chisari 2009) (see Sect. 5.2.2).

5.2.2 Inhibition of IFN Effector Mechanisms

There have also been a number of studies suggesting that HCV can interfere with Jak-STAT signaling downstream of the IFN receptors through various mechanisms. Expression of HCV proteins in cell lines showed that core and NS5A physically interact with STAT1 and decrease STAT1 phosphorylation in response to IFN stimulation (Lan et al. 2007; Lin et al. 2006). Core was also reported to inhibit nuclear translocation of STAT1 (Bode et al. 2003; Melen et al. 2004), or interfere with binding of ISGF3 to target ISG promoters (de Lucas et al. 2005). In addition, core induces the expression of suppressor of cytokine signaling-3 (SOCS3) (Bode et al. 2003; Kim et al. 2009), which inhibits STAT1 phosphorylation by binding to and suppressing the activity of Jaks. Impaired Jak-STAT signaling was also observed in mice transgenic for the HCV polyprotein (Blindenbacher et al. 2003) and in liver biopsies of chronic hepatitis C patients (Duong et al. 2004). This was attributed to decreased STAT1 binding to ISG promoters as results of hypomethylation of STAT1 and increased association with PIAS1, a STAT1 inhibitor (Christen et al. 2007; Duong et al. 2006).

In spite of the large body of data documenting the inhibitory effects of HCV proteins on IFN signaling, autonomously replicating, genome-length HCV replicons are sensitive to IFN-induced antiviral actions and can be eliminated from hepatoma cells harboring them after IFN-α or IFN-γ treatment. In addition, HCV infection in hepatoma cells was not found to compromise ISG promoter activation or ISG mRNA induction following IFN stimulation (Cheng et al. 2006; Garaigorta and Chisari 2009). A recent study also found ISG mRNAs were readily detected in HCV-infected cells in liver biopsies of chronic hepatitis C patients (Wieland et al. 2014). Additional studies in the context of HCV infection in vivo are in need to clarify this controversy.

If HCV indeed does not block Jak-STAT signaling, there comes a question – how can HCV continue to replicate in cells with constitutive expression of ISGs? One possibility is that HCV can antagonize the functions of individual ISG products. Examples supporting this mechanism include: NS5A- and E2-mediated blockade on PKR-mediated eIF-2α phosphorylation (Gale et al. 1997, 1998; Taylor et al. 1999), the NS5A association with OAS1 (Taguchi et al. 2004), and inhibition of GBP1 by NS5B (Itsui et al. 2009). However, HCV has a coding capacity of merely 11 proteins. It would thus be very challenging for the virus to develop target-specific counteraction strategies against hundreds of ISG products exerting diverse and sometimes redundant antiviral actions (Park and Rehermann 2014). A more plausible explanation was put forth in a recent study (Garaigorta and Chisari 2009), in which the authors showed that HCV infection activated PKR and subsequent eIF-2α phosphorylation, resulting in suppression of cap-dependent host protein synthesis but having no effect on translation of HCV proteins that is initiated by the IRES, which is resistant to eIF-2α phosphorylation. This mechanism allows for a brake on expression of ISG proteins from their mRNAs, thereby dampening IFN-mediated antiviral actions. This hypothesis will need to be further validated in advanced imaging studies that examine expression of ISG mRNAs and proteins, and HCV RNAs and proteins at a single cell level in liver biopsies from chronic hepatitis C patients (Heim 2013; Wieland et al. 2014).

6 Innate Immune Factors Regulating Spontaneous HCV Clearance and Response to IFN-Based Therapy

As mentioned in the preceding sections, acute HCV infections are typically associated with strong ISG induction in the liver. Why this robust innate immune response fails to usher in successful T cell immunity leading to viral clearance in a majority of infected individuals remains largely a mystery. Clearly, more research on the interface between the innate and adaptive immune responses during HCV infection is warranted. Interestingly, recent studies comparing the early host responses to HAV and HCV in acutely infected chimpanzees have provided novel insights into the role of innate immunity vs CD4+ T cell responses in the control of hepatic RNA virus infections and revealed interesting clues regarding the mechanisms underlying HCV persistence. In the liver of chimpanzees with acute HAV infection, there was very little ISG induction, despite the presence of 100-fold more viral RNA than that observed in chimpanzees with acute resolving HCV infection and harboring similar levels of viremia (Lanford et al. 2011). The near absence of IFN response, however, did not affect the clearance of HAV weeks later, which coincided with a strong HAV-specific CD4+ T cell response (Zhou et al. 2012). This was opposite to what has been observed with HCV infection – which typically is accompanied by strong ISG responses in both acute and chronic phase but with broad CD4+ T cell responses observed only in those who resolve the infection. Thus, the presence of a strong IFN response, especially when it is prolonged and unregulated, could hamper the induction of a successful CD4+ T cell response that is crucial for HCV clearance (Park and Rehermann 2014). While this hypothesis remains to be tested, it fits the findings of two recent studies in which blockade of chronic type I IFN signaling led to restoration of virus-specific CD4+ T cell response and clearance of persistent LCMV infection (Teijaro et al. 2013; Wilson et al. 2013).

Activation of the endogenous IFN system and high basal ISG expression is also the strongest factor that predicts poor response to PegIFN/RBV therapy in chronic hepatitis C patients (Chen et al. 2010; McGilvray et al. 2012; Pfeffer et al. 2014). In chimpanzees and many patients chronically infected with HCV, intrahepatic expression of ISGs is already at peak levels, resulting in a so-called "IFN refractoriness" state. In this scenario, there is no further STAT1 phosphorylation or ISG upregulation following exogenous IFN administration (Lanford et al. 2007; Sarasin-Filipowicz et al. 2008). The mechanism responsible is poorly understood, but evidence suggests the involvement of ISG15 and USP18, two ISGs that participate in the ISGylation/de-ISGylation modification pathway. High levels of ISG15 and USP18 expression prior to therapy consistently correlate with poor response to PegIFN/RBV therapy (Chen et al. 2011). In addition to its de-ISGylation activity, USP18 is known to dampen IFN antiviral action through the Jak-STAT signaling pathway by disrupting the association of IFNAR2 with Jak1 (Malakhova et al. 2006; Sarasin-Filipowicz et al. 2009).

The revelation in 2009 by several research groups that genetic polymorphisms near the IFNL3 (IL28B) gene locus are associated with the response of patients to PegIFN/RBV therapy as well as spontaneous HCV clearance has improved the predictive value of pre-therapy ISG expression levels. Moreover, it helps explain the long-standing mystery of why HCV-infected African-Americans respond poorly to PegIFN/RBV therapy compared with Asians and individuals of European ancestry (Ge et al. 2009; Suppiah et al. 2009; Tanaka et al. 2009; Thomas et al. 2009). The favorable rs12979860 C/C variant (located at 3 kb upstream of IFNL3, or in intron 1 of the newly identified IFNL4) is in strong linkage disequilibrium with the favorable variant at three other sites, i.e., rs8099917 (located 7.5 kb upstream of IFNL3), rs4803217 (located in the 3′UTR of IFNL3), and ss469415590 (located in exon 1 of IFNL4), and is less frequent in people of African ancestry compared with other ethnic groups. Although the mechanism underpinning this association remains to be elucidated, the fact that the polymorphisms are clustered near the IFNL3/IFNL4 gene locus suggests strongly a role for IFNL3/IFNL4 or its regulated innate immune factor(s) in dictating HCV infection outcome and host antiviral responses.

Interestingly, the favorable IFNL3 genotype was found to associate with high initial HCV viremia during the acute infection phase (Ge et al. 2009; Liu et al. 2012) and low baseline ISG during chronic infection (Honda et al. 2010), which represent quantifiable phenotypes in hepatitis C patients that strongly predict spontaneous viral clearance and response to PegIFN/RBV therapy, respectively. The initial hypothesis had been that the polymorphisms would affect the expression of IFN-λ3; indeed, the rs4803217 polymorphism was suggested to influence the stability of IFN-λ3 mRNA via controlling AU-rich element-mediated decay and the binding of two HCV-induced miRNAs to the 3′UTR of the IFN-λ3 transcript (McFarland et al. 2014). However, results have been inconsistent as to whether expression of IFN-λ3 mRNA or protein correlates with the outcome of HCV infection in patients (Park and Rehermann 2014). Of note, a recent study analyzing transcriptional responses of PHHs to HCV infection at a nearly single-cell level has suggested that the IFNL3 genotype does seem to affect innate antiviral responses, with cells from donors with unfavorable polymorphisms infected at greater frequency and exhibiting less focused and less robust antiviral and cell death responses (Sheahan et al. 2014).

Another plausible explanation has come with the identification of IFN-λ4, which is produced only from a frame-shift variant (ΔG) at the ss469415590 polymorphism site that is most frequently carried by those with African ancestry (Prokunina-Olsson et al. 2013). While providing the same information in Asians and individuals of European ancestry as rs12979860, the ss469415590 polymorphism has higher predictive power in persons with African ancestry. Expression of IFN-λ4 may contribute to the pre-activated hepatic ISG expression in the liver and impaired CD4+ T cell response, ultimately leading to HCV persistence and poor treatment response. However, this hypothesis needs to be examined in additional studies; the cellular source and mechanism by which this novel type III IFN is putatively induced remain to be identified.

7 Concluding Remarks and Perspectives

Since the discovery of HCV 25 years ago, exciting progress has been made in understanding the interactions of HCV with innate immune responses of the host. In particular, much has been learned about the PRRs and innate immune signaling pathways that detect and defend against HCV infection, immune evasion strategies of the virus, and host genetic and innate immune factors that regulate infection fate and treatment response. It is becoming clear that the outcome of HCV infection is an end result of complex interactions among multiple viral and host factors, including HCV genotype, host genetic variations, innate and adaptive immune responses, as well as cross-talk between hepatocytes, and other parenchymal and non-parenchymal cells in the liver (Horner 2014; Sheahan et al. 2014), etc.

However, many important questions remain. What is the exact IFN type that drives the ISG response in HCV-infected liver and what are its cellular source and precise induction mechanism(s)? Why does the robust intrahepatic ISG response in chronic hepatitis C patients fail to inhibit replication of the virus? What is the precise molecular basis underlying the association of IFNL3/IFNL4 polymorphisms with spontaneous HCV clearance and treatment response? Are the innate immune responses to HCV differentially regulated among various genotypes of the virus? Obviously, addressing these questions will require more studies in physiologically relevant in vitro culture systems such as PHHs and stem cell-derived differentiated human hepatocytes from donors with different IFNL3/IFNL4 genotypes and in liver biopsies from genetically distinct patient populations with advanced imaging approaches that measure host responses at or close to single-cell resolution. In addition, as the field enters an era without chimpanzees, a tractable, immmunocompetent mouse model allowing complete HCV replication cycle is urgently needed for studies elucidating the molecular details that govern the transition between the innate and adaptive immune responses. Studies in the animal model will also enable the dissection of interactions of hepatocytes with other liver cell types as well as with immune cells infiltrating the liver during HCV infection. Investigations along these lines collectively will yield the much needed new knowledge that will help improve current hepatitis C treatment regimens and guide efforts toward the development of an effective HCV vaccine.

Acknowledgement Work in the author's laboratory was supported by NIH grants AI069285, DA018054, and AI101526.

References

Ait-Goughoulte M, Kanda T, Meyer K, Ryerse JS, Ray RB, Ray R (2008) Hepatitis C virus genotype 1a growth and induction of autophagy. J Virol 82:2241–2249

Aly HH, Watashi K, Hijikata M, Kaneko H, Takada Y, Egawa H, Uemoto S, Shimotohno K (2007) Serum-derived hepatitis C virus infectivity in interferon regulatory factor-7-suppressed human primary hepatocytes. J Hepatol 46:26–36

Andrus L, Marukian S, Jones CT, Catanese MT, Sheahan TP, Schoggins JW, Barry WT, Dustin LB, Trehan K, Ploss A et al (2011) Expression of paramyxovirus V proteins promotes replication and spread of hepatitis C virus in cultures of primary human fetal liver cells. Hepatology 54:1901–1912

Arnaud N, Dabo S, Akazawa D, Fukasawa M, Shinkai-Ouchi F, Hugon J, Wakita T, Meurs EF (2011) Hepatitis C virus reveals a novel early control in acute immune response. PLoS Pathog 7:e1002289

Asselah T, Bieche I, Narguet S, Sabbagh A, Laurendeau I, Ripault MP, Boyer N, Martinot-Peignoux M, Valla D, Vidaud M et al (2008) Liver gene expression signature to predict response to pegylated interferon plus ribavirin combination therapy in patients with chronic hepatitis C. Gut 57:516–524

Bellecave P, Sarasin-Filipowicz M, Donze O, Kennel A, Gouttenoire J, Meylan E, Terracciano L, Tschopp J, Sarrazin C, Berg T et al (2010) Cleavage of mitochondrial antiviral signaling protein in the liver of patients with chronic hepatitis C correlates with a reduced activation of the endogenous interferon system. Hepatology 51:1127–1136

Bigger CB, Brasky KM, Lanford RE (2001) DNA microarray analysis of chimpanzee liver during acute resolving hepatitis C virus infection. J Virol 75:7059–7066

Blindenbacher A, Duong FH, Hunziker L, Stutvoet ST, Wang X, Terracciano L, Moradpour D, Blum HE, Alonzi T, Tripodi M et al (2003) Expression of hepatitis c virus proteins inhibits interferon alpha signaling in the liver of transgenic mice. Gastroenterology 124:1465–1475

Bode JG, Ludwig S, Ehrhardt C, Albrecht U, Erhardt A, Schaper F, Heinrich PC, Haussinger D (2003) IFN-alpha antagonistic activity of HCV core protein involves induction of suppressor of cytokine signaling-3. FASEB J 17:488–490

Borden EC, Sen GC, Uze G, Silverman RH, Ransohoff RM, Foster GR, Stark GR (2007) Interferons at age 50: past, current and future impact on biomedicine. Nat Rev Drug Discov 6:975–990

Bowen DG, Walker CM (2005) Adaptive immune responses in acute and chronic hepatitis C virus infection. Nature 436:946–952

Breiman A, Grandvaux N, Lin R, Ottone C, Akira S, Yoneyama M, Fujita T, Hiscott J, Meurs EF (2005) Inhibition of RIG-I-dependent signaling to the interferon pathway during hepatitis C virus expression and restoration of signaling by IKKepsilon. J Virol 79:3969–3978

Brimacombe CL, Grove J, Meredith LW, Hu K, Syder AJ, Flores MV, Timpe JM, Krieger SE, Baumert TF, Tellinghuisen TL et al (2011) Neutralizing antibody-resistant hepatitis C virus cell-to-cell transmission. J Virol 85:596–605

Cao X, Ding Q, Lu J, Tao W, Huang B, Zhao Y, Niu J, Liu YJ, Zhong J (2015) MDA5 plays a critical role in interferon response during hepatitis C virus infection. J Hepatol 62:771–778

Chang KS, Cai Z, Zhang C, Sen GC, Williams BR, Luo G (2006) Replication of hepatitis C virus (HCV) RNA in mouse embryonic fibroblasts: protein kinase R (PKR)-dependent and PKR-independent mechanisms for controlling HCV RNA replication and mediating interferon activities. J Virol 80:7364–7374

Chen L, Borozan I, Feld J, Sun J, Tannis LL, Coltescu C, Heathcote J, Edwards AM, McGilvray ID (2005) Hepatic gene expression discriminates responders and nonresponders in treatment of chronic hepatitis C viral infection. Gastroenterology 128:1437–1444

Chen Z, Benureau Y, Rijnbrand R, Yi J, Wang T, Warter L, Lanford RE, Weinman SA, Lemon SM, Martin A et al (2007) GB virus B disrupts RIG-I signaling by NS3/4A-mediated cleavage of the adaptor protein MAVS. J Virol 81:964–976

Chen L, Borozan I, Sun J, Guindi M, Fischer S, Feld J, Anand N, Heathcote J, Edwards AM, McGilvray ID (2010) Cell-type specific gene expression signature in liver underlies response to interferon therapy in chronic hepatitis C infection. Gastroenterology 138:1123–1133 e1121–e1123

Chen L, Li S, McGilvray I (2011) The ISG15/USP18 ubiquitin-like pathway (ISGylation system) in hepatitis C virus infection and resistance to interferon therapy. Int J Biochem Cell Biol 43:1427–1431

Cheng G, Zhong J, Chisari FV (2006) Inhibition of dsRNA-induced signaling in hepatitis C virus-infected cells by NS3 protease-dependent and -independent mechanisms. Proc Natl Acad Sci U S A 103:8499–8504

Christen V, Treves S, Duong FH, Heim MH (2007) Activation of endoplasmic reticulum stress response by hepatitis viruses up-regulates protein phosphatase 2A. Hepatology 46:558–565

Dafa-Berger A, Kuzmina A, Fassler M, Yitzhak-Asraf H, Shemer-Avni Y, Taube R (2012) Modulation of hepatitis C virus release by the interferon-induced protein BST-2/tetherin. Virology 428:98–111

Dansako H, Yamane D, Welsch C, McGivern DR, Hu F, Kato N, Lemon SM (2013) Class A scavenger receptor 1 (MSR1) restricts hepatitis C virus replication by mediating toll-like receptor 3 recognition of viral RNAs produced in neighboring cells. PLoS Pathog 9:e1003345

de Lucas S, Bartolome J, Carreno V (2005) Hepatitis C virus core protein down-regulates transcription of interferon-induced antiviral genes. J Infect Dis 191:93–99

Ding Q, Cao X, Lu J, Huang B, Liu YJ, Kato N, Shu HB, Zhong J (2013) Hepatitis C virus NS4B blocks the interaction of STING and TBK1 to evade host innate immunity. J Hepatol 59:52–58

Dolganiuc A, Kodys K, Kopasz A, Marshall C, Do T, Romics L Jr, Mandrekar P, Zapp M, Szabo G (2003) Hepatitis C virus core and nonstructural protein 3 proteins induce pro- and anti-inflammatory cytokines and inhibit dendritic cell differentiation. J Immunol 170:5615–5624

Dolganiuc A, Oak S, Kodys K, Golenbock DT, Finberg RW, Kurt-Jones E, Szabo G (2004) Hepatitis C core and nonstructural 3 proteins trigger toll-like receptor 2-mediated pathways and inflammatory activation. Gastroenterology 127:1513–1524

Dorner M, Horwitz JA, Donovan BM, Labitt RN, Budell WC, Friling T, Vogt A, Catanese MT, Satoh T, Kawai T et al (2013) Completion of the entire hepatitis C virus life cycle in genetically humanized mice. Nature 501:237–241

Dreux M, Gastaminza P, Wieland SF, Chisari FV (2009) The autophagy machinery is required to initiate hepatitis C virus replication. Proc Natl Acad Sci U S A 106:14046–14051

Dreux M, Garaigorta U, Boyd B, Decembre E, Chung J, Whitten-Bauer C, Wieland S, Chisari FV (2012) Short-range exosomal transfer of viral RNA from infected cells to plasmacytoid dendritic cells triggers innate immunity. Cell Host Microbe 12:558–570

Duong FH, Filipowicz M, Tripodi M, La Monica N, Heim MH (2004) Hepatitis C virus inhibits interferon signaling through up-regulation of protein phosphatase 2A. Gastroenterology 126:263–277

Duong FH, Christen V, Filipowicz M, Heim MH (2006) S-Adenosylmethionine and betaine correct hepatitis C virus induced inhibition of interferon signaling in vitro. Hepatology 43:796–806

Egger D, Wolk B, Gosert R, Bianchi L, Blum HE, Moradpour D, Bienz K (2002) Expression of hepatitis C virus proteins induces distinct membrane alterations including a candidate viral replication complex. J Virol 76:5974–5984

Feld JJ, Nanda S, Huang Y, Chen W, Cam M, Pusek SN, Schweigler LM, Theodore D, Zacks SL, Liang TJ et al (2007) Hepatic gene expression during treatment with peginterferon and ribavirin: identifying molecular pathways for treatment response. Hepatology 46:1548–1563

Ferraris P, Blanchard E, Roingeard P (2010) Ultrastructural and biochemical analyses of hepatitis C virus-associated host cell membranes. J Gen Virol 91:2230–2237

Ferreon JC, Ferreon AC, Li K, Lemon SM (2005) Molecular determinants of TRIF proteolysis mediated by the hepatitis C virus NS3/4A protease. J Biol Chem 280:20483–20492

Foy E, Li K, Wang C, Sumpter R Jr, Ikeda M, Lemon SM, Gale M Jr (2003) Regulation of interferon regulatory factor-3 by the hepatitis C virus serine protease. Science 300:1145–1148

Fredericksen BL, Keller BC, Fornek J, Katze MG, Gale M Jr (2008) Establishment and maintenance of the innate antiviral response to West Nile virus involves both RIG-I and MDA5 signaling through IPS-1. J Virol 82:609–616

Frentzen A, Huging K, Bitzegeio J, Friesland M, Haid S, Gentzsch J, Hoffmann M, Lindemann D, Zimmer G, Zielecki F et al (2011) Completion of hepatitis C virus replication cycle in heterokaryons excludes dominant restrictions in human non-liver and mouse liver cell lines. PLoS Pathog 7:e1002029

Frese M, Schwarzle V, Barth K, Krieger N, Lohmann V, Mihm S, Haller O, Bartenschlager R (2002) Interferon-gamma inhibits replication of subgenomic and genomic hepatitis C virus RNAs. Hepatology 35:694–703

Gack MU, Shin YC, Joo CH, Urano T, Liang C, Sun L, Takeuchi O, Akira S, Chen Z, Inoue S et al (2007) TRIM25 RING-finger E3 ubiquitin ligase is essential for RIG-I-mediated antiviral activity. Nature 446:916–920

Gale MJ Jr, Korth MJ, Tang NM, Tan SL, Hopkins DA, Dever TE, Polyak SJ, Gretch DR, Katze MG (1997) Evidence that hepatitis C virus resistance to interferon is mediated through repression of the PKR protein kinase by the nonstructural 5A protein. Virology 230:217–227

Gale M Jr, Blakely CM, Kwieciszewski B, Tan SL, Dossett M, Tang NM, Korth MJ, Polyak SJ, Gretch DR, Katze MG (1998) Control of PKR protein kinase by hepatitis C virus nonstructural 5A protein: molecular mechanisms of kinase regulation. Mol Cell Biol 18:5208–5218

Garaigorta U, Chisari FV (2009) Hepatitis C virus blocks interferon effector function by inducing protein kinase R phosphorylation. Cell Host Microbe 6:513–522

Ge D, Fellay J, Thompson AJ, Simon JS, Shianna KV, Urban TJ, Heinzen EL, Qiu P, Bertelsen AH, Muir AJ et al (2009) Genetic variation in IL28B predicts hepatitis C treatment-induced viral clearance. Nature 461:399–401

Ghany MG, Strader DB, Thomas DL, Seeff LB (2009) Diagnosis, management, and treatment of hepatitis C: an update. Hepatology 49:1335–1374

Gosert R, Egger D, Lohmann V, Bartenschlager R, Blum HE, Bienz K, Moradpour D (2003) Identification of the hepatitis C virus RNA replication complex in Huh-7 cells harboring subgenomic replicons. J Virol 77:5487–5492

Guo JT, Bichko VV, Seeger C (2001) Effect of alpha interferon on the hepatitis C virus replicon. J Virol 75:8516–8523

Han JQ, Barton DJ (2002) Activation and evasion of the antiviral 2′–5′ oligoadenylate synthetase/ribonuclease L pathway by hepatitis C virus mRNA. RNA 8:512–525

He B (2006) Viruses, endoplasmic reticulum stress, and interferon responses. Cell Death Differ 13:393–403

Heim MH (2013) Innate immunity and HCV. J Hepatol 58:564–574

Helbig KJ, Lau DT, Semendric L, Harley HA, Beard MR (2005) Analysis of ISG expression in chronic hepatitis C identifies viperin as a potential antiviral effector. Hepatology 42:702–710

Helbig KJ, Ruszkiewicz A, Lanford RE, Berzsenyi MD, Harley HA, McColl SR, Beard MR (2009) Differential expression of the CXCR3 ligands in chronic hepatitis C virus (HCV) infection and their modulation by HCV in vitro. J Virol 83:836–846

Helbig KJ, Eyre NS, Yip E, Narayana S, Li K, Fiches G, McCartney EM, Jangra RK, Lemon SM, Beard MR (2011) The antiviral protein viperin inhibits hepatitis C virus replication via interaction with nonstructural protein 5A. Hepatology 54:1506–1517

Helle F, Brochot E, Fournier C, Descamps V, Izquierdo L, Hoffmann TW, Morel V, Herpe YE, Bengrine A, Belouzard S et al (2013) Permissivity of primary human hepatocytes and different hepatoma cell lines to cell culture adapted hepatitis C virus. PLoS ONE 8:e70809

Hermant P, Michiels T (2014) Interferon-lambda in the context of viral infections: production, response and therapeutic implications. J Innate Immun 6:563–574

Hiet MS, Bauhofer O, Zayas M, Roth H, Tanaka Y, Schirmacher P, Willemsen J, Grünvogel O, Bender S, Binder M et al (2015) Control of temporal activation of hepatitis C virus-induced interferon response by domain 2 of nonstructural protein 5A. J Hepatol 63:829–837

Hoffmann M, Zeisel MB, Jilg N, Paranhos-Baccala G, Stoll-Keller F, Wakita T, Hafkemeyer P, Blum HE, Barth H, Henneke P et al (2009) Toll-like receptor 2 senses hepatitis C virus core protein but not infectious viral particles. J Innate Immun 1:446–454

Honda M, Sakai A, Yamashita T, Nakamoto Y, Mizukoshi E, Sakai Y, Nakamura M, Shirasaki T, Horimoto K, Tanaka Y et al (2010) Hepatic ISG expression is associated with genetic variation in interleukin 28B and the outcome of IFN therapy for chronic hepatitis C. Gastroenterology 139:499–509

Horner SM (2014) Activation and evasion of antiviral innate immunity by hepatitis C virus. J Mol Biol 426:1198–1209

Horner SM, Gale M Jr (2013) Regulation of hepatic innate immunity by hepatitis C virus. Nat Med 19:879–888

Israelow B, Narbus CM, Sourisseau M, Evans MJ (2014) HepG2 cells mount an effective antiviral interferon-lambda based innate immune response to hepatitis C virus infection. Hepatology 60:1170–1179

Itsui Y, Sakamoto N, Kurosaki M, Kanazawa N, Tanabe Y, Koyama T, Takeda Y, Nakagawa M, Kakinuma S, Sekine Y et al (2006) Expressional screening of interferon-stimulated genes for antiviral activity against hepatitis C virus replication. J Viral Hepat 13:690–700

Itsui Y, Sakamoto N, Kakinuma S, Nakagawa M, Sekine-Osajima Y, Tasaka-Fujita M, Nishimura-Sakurai Y, Suda G, Karakama Y, Mishima K et al (2009) Antiviral effects of the interferon-induced protein guanylate binding protein 1 and its interaction with the hepatitis C virus NS5B protein. Hepatology 50:1727–1737

Jiang D, Guo H, Xu C, Chang J, Gu B, Wang L, Block TM, Guo JT (2008) Identification of three interferon-inducible cellular enzymes that inhibit the replication of hepatitis C virus. J Virol 82:1665–1678

Jopling CL, Yi M, Lancaster AM, Lemon SM, Sarnow P (2005) Modulation of hepatitis C virus RNA abundance by a liver-specific MicroRNA. Science 309:1577–1581

Jounai N, Takeshita F, Kobiyama K, Sawano A, Miyawaki A, Xin KQ, Ishii KJ, Kawai T, Akira S, Suzuki K et al (2007) The Atg5 Atg12 conjugate associates with innate antiviral immune responses. Proc Natl Acad Sci U S A 104:14050–14055

Kanazawa N, Kurosaki M, Sakamoto N, Enomoto N, Itsui Y, Yamashiro T, Tanabe Y, Maekawa S, Nakagawa M, Chen CH et al (2004) Regulation of hepatitis C virus replication by interferon regulatory factor 1. J Virol 78:9713–9720

Kandathil AJ, Graw F, Quinn J, Hwang HS, Torbenson M, Perelson AS, Ray SC, Thomas DL, Ribeiro RM, Balagopal A (2013) Use of laser capture microdissection to map hepatitis C virus-positive hepatocytes in human liver. Gastroenterology 145:1413–1404 e1401–e1410

Kato H, Takeuchi O, Sato S, Yoneyama M, Yamamoto M, Matsui K, Uematsu S, Jung A, Kawai T, Ishii KJ et al (2006) Differential roles of MDA5 and RIG-I helicases in the recognition of RNA viruses. Nature 441:101–105

Kato H, Takeuchi O, Mikamo-Satoh E, Hirai R, Kawai T, Matsushita K, Hiiragi A, Dermody TS, Fujita T, Akira S (2008) Length-dependent recognition of double-stranded ribonucleic acids by retinoic acid-inducible gene-I and melanoma differentiation-associated gene 5. J Exp Med 205:1601–1610

Kato H, Takahasi K, Fujita T (2011) RIG-I-like receptors: cytoplasmic sensors for non-self RNA. Immunol Rev 243:91–98

Kawai T, Akira S (2008) Toll-like receptor and RIG-I-like receptor signaling. Ann N Y Acad Sci 1143:1–20

Kawai T, Akira S (2011) Toll-like receptors and their crosstalk with other innate receptors in infection and immunity. Immunity 34:637–650

Kawasaki T, Kawai T, Akira S (2011) Recognition of nucleic acids by pattern-recognition receptors and its relevance in autoimmunity. Immunol Rev 243:61–73

Ke PY, Chen SS (2011) Activation of the unfolded protein response and autophagy after hepatitis C virus infection suppresses innate antiviral immunity in vitro. J Clin Invest 121:37–56

Kim KA, Lin W, Tai AW, Shao RX, Weinberg E, De Sa Borges CB, Bhan AK, Zheng H, Kamegaya Y, Chung RT (2009) Hepatic SOCS3 expression is strongly associated with non-response to therapy and race in HCV and HCV/HIV infection. J Hepatol 50:705–711

Kwon YC, Kang JI, Hwang SB, Ahn BY (2013) The ribonuclease L-dependent antiviral roles of human 2′,5′-oligoadenylate synthetase family members against hepatitis C virus. FEBS Lett 587:156–164

Lan KH, Lan KL, Lee WP, Sheu ML, Chen MY, Lee YL, Yen SH, Chang FY, Lee SD (2007) HCV NS5A inhibits interferon-alpha signaling through suppression of STAT1 phosphorylation in hepatocyte-derived cell lines. J Hepatol 46:759–767

Lanford RE, Guerra B, Bigger CB, Lee H, Chavez D, Brasky KM (2007) Lack of response to exogenous interferon-alpha in the liver of chimpanzees chronically infected with hepatitis C virus. Hepatology 46:999–1008

Lanford RE, Feng Z, Chavez D, Guerra B, Brasky KM, Zhou Y, Yamane D, Perelson AS, Walker CM, Lemon SM (2011) Acute hepatitis A virus infection is associated with a limited type I interferon response and persistence of intrahepatic viral RNA. Proc Natl Acad Sci U S A 108:11223–11228

Lau DT, Fish PM, Sinha M, Owen DM, Lemon SM, Gale M Jr (2008) Interferon regulatory factor-3 activation, hepatic interferon-stimulated gene expression, and immune cell infiltration in hepatitis C virus patients. Hepatology 47:799–809

Lau DT, Negash A, Chen J, Crochet N, Sinha M, Zhang Y, Guedj J, Holder S, Saito T, Lemon SM et al (2013) Innate immune tolerance and the role of kupffer cells in differential responses to interferon therapy among patients with HCV genotype 1 infection. Gastroenterology 144:402–413 e412

Lauterbach H, Bathke B, Gilles S, Traidl-Hoffmann C, Luber CA, Fejer G, Freudenberg MA, Davey GM, Vremec D, Kallies A et al (2010) Mouse CD8alpha+ DCs and human BDCA3+ DCs are major producers of IFN-lambda in response to poly IC. J Exp Med 207:2703–2717

Lemon SM (2010) Induction and evasion of innate antiviral responses by hepatitis C virus. J Biol Chem 285:22741–22747

Lemon SM, McKeating JA, Pietschmann T, Frick DN, Glenn JS, Tellinghuisen TL, Symons J, Furman PA (2010) Development of novel therapies for hepatitis C. Antiviral Res 86:79–92

Lester SN, Li K (2014) Toll-like receptors in antiviral innate immunity. J Mol Biol 426:1246–1264

Li K, Lemon SM (2013) Innate immune responses in hepatitis C virus infection. Semin Immunopathol 35:53–72

Li K, Chen Z, Kato N, Gale M Jr, Lemon SM (2005a) Distinct poly(I-C) and virus-activated signaling pathways leading to interferon-beta production in hepatocytes. J Biol Chem 280:16739–16747

Li K, Foy E, Ferreon JC, Nakamura M, Ferreon AC, Ikeda M, Ray SC, Gale M Jr, Lemon SM (2005b) Immune evasion by hepatitis C virus NS3/4A protease-mediated cleavage of the toll-like receptor 3 adaptor protein TRIF. Proc Natl Acad Sci U S A 102:2992–2997

Li XD, Sun L, Seth RB, Pineda G, Chen ZJ (2005c) Hepatitis C virus protease NS3/4A cleaves mitochondrial antiviral signaling protein off the mitochondria to evade innate immunity. Proc Natl Acad Sci U S A 102:17717–17722

Li K, Li NL, Wei D, Pfeffer SR, Fan M, Pfeffer LM (2012) Activation of chemokine and inflammatory cytokine response in hepatitis C virus-infected hepatocytes depends on toll-like receptor 3 sensing of hepatitis C virus double-stranded RNA intermediates. Hepatology 55:666–675

Liang TJ (2013) Current progress in development of hepatitis C virus vaccines. Nat Med 19:869–878

Liang Y, Shilagard T, Xiao SY, Snyder N, Lau D, Cicalese L, Weiss H, Vargas G, Lemon SM (2009) Visualizing hepatitis C virus infections in human liver by two-photon microscopy. Gastroenterology 137:1448–1458

Lin W, Kim SS, Yeung E, Kamegaya Y, Blackard JT, Kim KA, Holtzman MJ, Chung RT (2006) Hepatitis C virus core protein blocks interferon signaling by interaction with the STAT1 SH2 domain. J Virol 80:9226–9235

Lindenbach BD, Rice CM (2013) The ins and outs of hepatitis C virus entry and assembly. Nat Rev Microbiol 11:688–700

Lindenbach BD, Evans MJ, Syder AJ, Wolk B, Tellinghuisen TL, Liu CC, Maruyama T, Hynes RO, Burton DR, McKeating JA et al (2005) Complete replication of hepatitis C virus in cell culture. Science 309:623–626

Liu YJ (2005) IPC: professional type 1 interferon-producing cells and plasmacytoid dendritic cell precursors. Annu Rev Immunol 23:275–306

Liu L, Fisher BE, Thomas DL, Cox AL, Ray SC (2012) Spontaneous clearance of primary acute hepatitis C virus infection correlated with high initial viral RNA level and rapid HVR1 evolution. Hepatology 55:1684–1691

Loo YM, Gale M Jr (2011) Immune signaling by RIG-I-like receptors. Immunity 34:680–692

Loo YM, Owen DM, Li K, Erickson AK, Johnson CL, Fish PM, Carney DS, Wang T, Ishida H, Yoneyama M et al (2006) Viral and therapeutic control of IFN-beta promoter stimulator 1 during hepatitis C virus infection. Proc Natl Acad Sci U S A 103:6001–6006

Loo YM, Fornek J, Crochet N, Bajwa G, Perwitasari O, Martinez-Sobrido L, Akira S, Gill MA, Garcia-Sastre A, Katze MG et al (2008) Distinct RIG-I and MDA5 signaling by RNA viruses in innate immunity. J Virol 82:335–345

Machlin ES, Sarnow P, Sagan SM (2011) Masking the 5′ terminal nucleotides of the hepatitis C virus genome by an unconventional microRNA-target RNA complex. Proc Natl Acad Sci U S A 108:3193–3198

Malakhova OA, Kim KI, Luo JK, Zou W, Kumar KG, Fuchs SY, Shuai K, Zhang DE (2006) UBP43 is a novel regulator of interferon signaling independent of its ISG15 isopeptidase activity. EMBO J 25:2358–2367

Malathi K, Saito T, Crochet N, Barton DJ, Gale M Jr, Silverman RH (2010) RNase L releases a small RNA from HCV RNA that refolds into a potent PAMP. RNA 16:2108–2119

Marcello T, Grakoui A, Barba-Spaeth G, Machlin ES, Kotenko SV, MacDonald MR, Rice CM (2006) Interferons alpha and lambda inhibit hepatitis C virus replication with distinct signal transduction and gene regulation kinetics. Gastroenterology 131:1887–1898

Martell M, Esteban JI, Quer J, Genesca J, Weiner A, Esteban R, Guardia J, Gomez J (1992) Hepatitis C virus (HCV) circulates as a population of different but closely related genomes: quasispecies nature of HCV genome distribution. J Virol 66:3225–3229

Marukian S, Andrus L, Sheahan TP, Jones CT, Charles ED, Ploss A, Rice CM, Dustin LB (2011) Hepatitis C virus induces interferon-lambda and interferon-stimulated genes in primary liver cultures. Hepatology 54:1913–1923

McFarland AP, Horner SM, Jarret A, Joslyn RC, Bindewald E, Shapiro BA, Delker DA, Hagedorn CH, Carrington M, Gale M Jr et al (2014) The favorable IFNL3 genotype escapes mRNA decay mediated by AU-rich elements and hepatitis C virus-induced microRNAs. Nat Immunol 15:72–79

McGilvray I, Feld JJ, Chen L, Pattullo V, Guindi M, Fischer S, Borozan I, Xie G, Selzner N, Heathcote EJ et al (2012) Hepatic cell-type specific gene expression better predicts HCV treatment outcome than IL28B genotype. Gastroenterology 142:1122–1131 e1121

Melen K, Fagerlund R, Nyqvist M, Keskinen P, Julkunen I (2004) Expression of hepatitis C virus core protein inhibits interferon-induced nuclear import of STATs. J Med Virol 73:536–547

Metz P, Dazert E, Ruggieri A, Mazur J, Kaderali L, Kaul A, Zeuge U, Windisch MP, Trippler M, Lohmann V et al (2012) Identification of type I and type II interferon-induced effectors controlling hepatitis C virus replication. Hepatology 56:2082–2093

Metz P, Reuter A, Bender S, Bartenschlager R (2013) Interferon-stimulated genes and their role in controlling hepatitis C virus. J Hepatol 59:1331–1341

Meylan E, Curran J, Hofmann K, Moradpour D, Binder M, Bartenschlager R, Tschopp J (2005) Cardif is an adaptor protein in the RIG-I antiviral pathway and is targeted by hepatitis C virus. Nature 437:1167–1172

Naka K, Dansako H, Kobayashi N, Ikeda M, Kato N (2006) Hepatitis C virus NS5B delays cell cycle progression by inducing interferon-beta via toll-like receptor 3 signaling pathway without replicating viral genomes. Virology 346:348–362

Negash AA, Ramos HJ, Crochet N, Lau DT, Doehle B, Papic N, Delker DA, Jo J, Bertoletti A, Hagedorn CH et al (2013) IL-1beta production through the NLRP3 inflammasome by hepatic macrophages links hepatitis C virus infection with liver inflammation and disease. PLoS Pathog 9:e1003330

Nishitsuji H, Funami K, Shimizu Y, Ujino S, Sugiyama K, Seya T, Takaku H, Shimotohno K (2013) Hepatitis C virus infection induces inflammatory cytokines and chemokines mediated by the cross talk between hepatocytes and stellate cells. J Virol 87:8169–8178

O'Brien TR, Prokunina-Olsson L, Donnelly RP (2014) IFN-lambda4: the paradoxical new member of the interferon lambda family. J Interferon Cytokine Res 34:829–838

Odorizzi PM, Wherry EJ (2013) Immunology. An interferon paradox. Science 340:155–156

Otsuka M, Kato N, Moriyama M, Taniguchi H, Wang Y, Dharel N, Kawabe T, Omata M (2005) Interaction between the HCV NS3 protein and the host TBK1 protein leads to inhibition of cellular antiviral responses. Hepatology 41:1004–1012

Pal S, Shuhart MC, Thomassen L, Emerson SS, Su T, Feuerborn N, Kae J, Gretch DR (2006) Intrahepatic hepatitis C virus replication correlates with chronic hepatitis C disease severity in vivo. J Virol 80:2280–2290

Pan XB, Qu XW, Jiang D, Zhao XL, Han JC, Wei L (2013) BST2/Tetherin inhibits hepatitis C virus production in human hepatoma cells. Antiviral Res 98:54–60

Park SH, Rehermann B (2014) Immune responses to HCV and other hepatitis viruses. Immunity 40:13–24

Park H, Serti E, Eke O, Muchmore B, Prokunina-Olsson L, Capone S, Folgori A, Rehermann B (2012) IL-29 is the dominant type III interferon produced by hepatocytes during acute hepatitis C virus infection. Hepatology 56:2060–2070

Pedersen IM, Cheng G, Wieland S, Volinia S, Croce CM, Chisari FV, David M (2007) Interferon modulation of cellular microRNAs as an antiviral mechanism. Nature 449:919–922

Pfeffer LM, Li K, Fleckenstein JF, Marion TN, Diament J, Yang CH, Pfeffer SR, Fan M, Handorf E, Handorf CR (2014) An interferon response gene signature is associated with the therapeutic response of hepatitis C patients. PLoS ONE 9:e104202

Pindel A, Sadler A (2011) The role of protein kinase R in the interferon response. J Interferon Cytokine Res 31:59–70

Prokunina-Olsson L, Muchmore B, Tang W, Pfeiffer RM, Park H, Dickensheets H, Hergott D, Porter-Gill P, Mumy A, Kohaar I et al (2013) A variant upstream of IFNL3 (IL28B) creating a new interferon gene IFNL4 is associated with impaired clearance of hepatitis C virus. Nat Genet 45:164–171

Robek MD, Boyd BS, Chisari FV (2005) Lambda interferon inhibits hepatitis B and C virus replication. J Virol 79:3851–3854

Saito T, Hirai R, Loo YM, Owen D, Johnson CL, Sinha SC, Akira S, Fujita T, Gale M Jr (2007) Regulation of innate antiviral defenses through a shared repressor domain in RIG-I and LGP2. Proc Natl Acad Sci U S A 104:582–587

Saito T, Owen DM, Jiang F, Marcotrigiano J, Gale M Jr (2008) Innate immunity induced by composition-dependent RIG-I recognition of hepatitis C virus RNA. Nature 454:523–527

Sarasin-Filipowicz M, Oakeley EJ, Duong FH, Christen V, Terracciano L, Filipowicz W, Heim MH (2008) Interferon signaling and treatment outcome in chronic hepatitis C. Proc Natl Acad Sci U S A 105:7034–7039

Sarasin-Filipowicz M, Wang X, Yan M, Duong FH, Poli V, Hilton DJ, Zhang DE, Heim MH (2009) Alpha interferon induces long-lasting refractoriness of JAK-STAT signaling in the mouse liver through induction of USP18/UBP43. Mol Cell Biol 29:4841–4851

Satoh T, Kato H, Kumagai Y, Yoneyama M, Sato S, Matsushita K, Tsujimura T, Fujita T, Akira S, Takeuchi O (2010) LGP2 is a positive regulator of RIG-I- and MDA5-mediated antiviral responses. Proc Natl Acad Sci U S A 107:1512–1517

Schoggins JW, Rice CM (2011) Interferon-stimulated genes and their antiviral effector functions. Curr Opin Virol 1:519–525

Schoggins JW, Wilson SJ, Panis M, Murphy MY, Jones CT, Bieniasz P, Rice CM (2011) A diverse range of gene products are effectors of the type I interferon antiviral response. Nature 472:481–485

Schwartz RE, Trehan K, Andrus L, Sheahan TP, Ploss A, Duncan SA, Rice CM, Bhatia SN (2012) Modeling hepatitis C virus infection using human induced pluripotent stem cells. Proc Natl Acad Sci U S A 109:2544–2548

Sheahan T, Imanaka N, Marukian S, Dorner M, Liu P, Ploss A, Rice CM (2014) Interferon lambda alleles predict innate antiviral immune responses and hepatitis C virus permissiveness. Cell Host Microbe 15:190–202

Shimakami T, Yamane D, Jangra RK, Kempf BJ, Spaniel C, Barton DJ, Lemon SM (2012) Stabilization of hepatitis C RNA by an Ago2-miR-122 complex. Proc Natl Acad Sci U S A 109:941–946

Shimoike T, McKenna SA, Lindhout DA, Puglisi JD (2009) Translational insensitivity to potent activation of PKR by HCV IRES RNA. Antiviral Res 83:228–237

Sir D, Chen WL, Choi J, Wakita T, Yen TS, Ou JH (2008) Induction of incomplete autophagic response by hepatitis C virus via the unfolded protein response. Hepatology 48:1054–1061

Stiffler JD, Nguyen M, Sohn JA, Liu C, Kaplan D, Seeger C (2009) Focal distribution of hepatitis C virus RNA in infected livers. PLoS ONE 4:e6661

Su AI, Pezacki JP, Wodicka L, Brideau AD, Supekova L, Thimme R, Wieland S, Bukh J, Purcell RH, Schultz PG et al (2002) Genomic analysis of the host response to hepatitis C virus infection. Proc Natl Acad Sci U S A 99:15669–15674

Sumpter R Jr, Loo YM, Foy E, Li K, Yoneyama M, Fujita T, Lemon SM, Gale M Jr (2005) Regulating intracellular antiviral defense and permissiveness to hepatitis C virus RNA replication through a cellular RNA helicase, RIG-I. J Virol 79:2689–2699

Suppiah V, Moldovan M, Ahlenstiel G, Berg T, Weltman M, Abate ML, Bassendine M, Spengler U, Dore GJ, Powell E et al (2009) IL28B is associated with response to chronic hepatitis C interferon-alpha and ribavirin therapy. Nat Genet 41:1100–1104

Syed GH, Amako Y, Siddiqui A (2010) Hepatitis C virus hijacks host lipid metabolism. Trends Endocrinol Metab 21:33–40

Taguchi T, Nagano-Fujii M, Akutsu M, Kadoya H, Ohgimoto S, Ishido S, Hotta H (2004) Hepatitis C virus NS5A protein interacts with 2′,5′-oligoadenylate synthetase and inhibits antiviral activity of IFN in an IFN sensitivity-determining region-independent manner. J Gen Virol 85:959–969

Takahashi K, Asabe S, Wieland S, Garaigorta U, Gastaminza P, Isogawa M, Chisari FV (2010) Plasmacytoid dendritic cells sense hepatitis C virus-infected cells, produce interferon, and inhibit infection. Proc Natl Acad Sci U S A 107:7431–7436

Tanaka Y, Nishida N, Sugiyama M, Kurosaki M, Matsuura K, Sakamoto N, Nakagawa M, Korenaga M, Hino K, Hige S et al (2009) Genome-wide association of IL28B with response to pegylated interferon-alpha and ribavirin therapy for chronic hepatitis C. Nat Genet 41:1105–1109

Tanida I, Fukasawa M, Ueno T, Kominami E, Wakita T, Hanada K (2009) Knockdown of autophagy-related gene decreases the production of infectious hepatitis C virus particles. Autophagy 5:937–945

Targett-Adams P, Boulant S, McLauchlan J (2008) Visualization of double-stranded RNA in cells supporting hepatitis C virus RNA replication. J Virol 82:2182–2195

Taylor DR, Shi ST, Romano PR, Barber GN, Lai MM (1999) Inhibition of the interferon-inducible protein kinase PKR by HCV E2 protein. Science 285:107–110

Taylor DR, Puig M, Darnell ME, Mihalik K, Feinstone SM (2005) New antiviral pathway that mediates hepatitis C virus replicon interferon sensitivity through ADAR1. J Virol 79:6291–6298

Teijaro JR, Ng C, Lee AM, Sullivan BM, Sheehan KC, Welch M, Schreiber RD, de la Torre JC, Oldstone MB (2013) Persistent LCMV infection is controlled by blockade of type I interferon signaling. Science 340:207–211

Thomas DL, Thio CL, Martin MP, Qi Y, Ge D, O'Huigin C, Kidd J, Kidd K, Khakoo SI, Alexander G et al (2009) Genetic variation in IL28B and spontaneous clearance of hepatitis C virus. Nature 461:798–801

Thomas E, Gonzalez VD, Li Q, Modi AA, Chen W, Noureddin M, Rotman Y, Liang TJ (2012) HCV infection induces a unique hepatic innate immune response associated with robust production of type III interferons. Gastroenterology 142:978–988

Timpe JM, Stamataki Z, Jennings A, Hu K, Farquhar MJ, Harris HJ, Schwarz A, Desombere I, Roels GL, Balfe P et al (2008) Hepatitis C virus cell-cell transmission in hepatoma cells in the presence of neutralizing antibodies. Hepatology 47:17–24

Toroney R, Nallagatla SR, Boyer JA, Cameron CE, Bevilacqua PC (2010) Regulation of PKR by HCV IRES RNA: importance of domain II and NS5A. J Mol Biol 400:393–412
Uze G, Monneron D (2007) IL-28 and IL-29: newcomers to the interferon family. Biochimie 89:729–734
Venkataraman T, Valdes M, Elsby R, Kakuta S, Caceres G, Saijo S, Iwakura Y, Barber GN (2007) Loss of DExD/H box RNA helicase LGP2 manifests disparate antiviral responses. J Immunol 178:6444–6455
Wakita T, Pietschmann T, Kato T, Date T, Miyamoto M, Zhao Z, Murthy K, Habermann A, Krausslich HG, Mizokami M et al (2005) Production of infectious hepatitis C virus in tissue culture from a cloned viral genome. Nat Med 11:791–796
Wald O, Weiss ID, Galun E, Peled A (2007) Chemokines in hepatitis C virus infection: pathogenesis, prognosis and therapeutics. Cytokine 39:50–62
Wang C, Pflugheber J, Sumpter R Jr, Sodora DL, Hui D, Sen GC, Gale M Jr (2003) Alpha interferon induces distinct translational control programs to suppress hepatitis C virus RNA replication. J Virol 77:3898–3912
Wang N, Liang Y, Devaraj S, Wang J, Lemon SM, Li K (2009) Toll-like receptor 3 mediates establishment of an antiviral state against hepatitis C virus in hepatoma cells. J Virol 83:9824–9834
Wang S, Wu X, Pan T, Song W, Wang Y, Zhang F, Yuan Z (2012) Viperin inhibits hepatitis C virus replication by interfering with binding of NS5A to host protein hVAP-33. J Gen Virol 93:83–92
Wieland S, Thimme R, Purcell RH, Chisari FV (2004) Genomic analysis of the host response to hepatitis B virus infection. Proc Natl Acad Sci U S A 101:6669–6674
Wieland S, Makowska Z, Campana B, Calabrese D, Dill MT, Chung J, Chisari FV, Heim MH (2014) Simultaneous detection of hepatitis C virus and interferon stimulated gene expression in infected human liver. Hepatology 59:2121–2130
Wilkins C, Woodward J, Lau DT, Barnes A, Joyce M, McFarlane N, McKeating JA, Tyrrell DL, Gale M Jr (2013) IFITM1 is a tight junction protein that inhibits hepatitis C virus entry. Hepatology 57:461–469
Wilson EB, Yamada DH, Elsaesser H, Herskovitz J, Deng J, Cheng G, Aronow BJ, Karp CL, Brooks DG (2013) Blockade of chronic type I interferon signaling to control persistent LCMV infection. Science 340:202–207
Wu X, Robotham JM, Lee E, Dalton S, Kneteman NM, Gilbert DM, Tang H (2012) Productive hepatitis C virus infection of stem cell-derived hepatocytes reveals a critical transition to viral permissiveness during differentiation. PLoS Pathog 8:e1002617
Yang Y, Liang Y, Qu L, Chen Z, Yi M, Li K, Lemon SM (2007) Disruption of innate immunity due to mitochondrial targeting of a picornaviral protease precursor. Proc Natl Acad Sci U S A 104:7253–7258
Yao L, Dong H, Zhu H, Nelson D, Liu C, Lambiase L, Li X (2011) Identification of the IFITM3 gene as an inhibitor of hepatitis C viral translation in a stable STAT1 cell line. J Viral Hepat 18: e523–e529
Yu GY, He G, Li CY, Tang M, Grivennikov S, Tsai WT, Wu MS, Hsu CW, Tsai Y, Wang LH et al (2012) Hepatic expression of HCV RNA-dependent RNA polymerase triggers innate immune signaling and cytokine production. Mol Cell 48:313–321
Zeremski M, Petrovic LM, Talal AH (2007) The role of chemokines as inflammatory mediators in chronic hepatitis C virus infection. J Viral Hepat 14:675–687
Zhang S, Kodys K, Li K, Szabo G (2013) Human type 2 myeloid dendritic cells produce interferon-lambda and amplify interferon-alpha in response to hepatitis C virus infection. Gastroenterology 144:414–425 e417
Zhong J, Gastaminza P, Cheng G, Kapadia S, Kato T, Burton DR, Wieland SF, Uprichard SL, Wakita T, Chisari FV (2005) Robust hepatitis C virus infection in vitro. Proc Natl Acad Sci U S A 102:9294–9299

Zhou Z, Wang N, Woodson SE, Dong Q, Wang J, Liang Y, Rijnbrand R, Wei L, Nichols JE, Guo JT et al (2011) Antiviral activities of ISG20 in positive-strand RNA virus infections. Virology 409:175–188

Zhou Y, Callendret B, Xu D, Brasky KM, Feng Z, Hensley LL, Guedj J, Perelson AS, Lemon SM, Lanford RE et al (2012) Dominance of the CD4(+) T helper cell response during acute resolving hepatitis A virus infection. J Exp Med 209:1481–1492

Zhu H, Butera M, Nelson DR, Liu C (2005) Novel type I interferon IL-28A suppresses hepatitis C viral RNA replication. Virol J 2:80

B Cell Responses and Control of HCV Infection

Zhen-Yong Keck, Thomas R. Fuerst, Roy A. Mariuzza, and Steven K.H. Foung

Abstract Humoral immunity is the primary correlate of protection for most preventive vaccines. For hepatitis C virus (HCV), early animal studies provided support for the importance of virus neutralizing antibodies to facilitate clearance of infection. The development of retroviral pseudotype particles expressing HCV E1E2 glycoproteins (HCVpp), infectious cell culture-derived HCV virions (HCVcc), and small animal models of acute HCV infection have made possible the measurement of in vitro and in vivo antibody-mediated virus neutralization. Their applications in multiple studies clearly established the importance of neutralizing antibodies in controlling HCV infection. However, the virus has devised a number of decoys of highly immunogenic regions associated with viral escape or non-neutralizing antibodies that deflect the immune response from less immunogenic conserved regions mediating virus neutralization that are not associated with viral escape or escape with compromised fitness. The focus of this review is on the immunogenic determinants on E2, which are roughly segregated into the hypervariable region 1 (HVR1), and five clusters of overlapping epitopes, designated as antigenic domains A-E. A detailed understanding of conserved neutralizing epitopes within these antigenic domains that are not associated with escape and how

Z.-Y. Keck
Department of Pathology, Stanford University School of Medicine, Stanford, CA 94305, USA

T.R. Fuerst
Institute for Bioscience and Biotechnology Research, University of Maryland, Rockville, MD, USA

Department of Cell Biology and Molecular Genetics, University of Maryland, College Park, MD, USA

R.A. Mariuzza
W. M. Keck Laboratory for Structural Biology, Institute for Bioscience and Biotechnology Research, University of Maryland, Rockville, MD, USA

Department of Cell Biology and Molecular Genetics, University of Maryland, College Park, MD, USA

S.K.H. Foung (✉)
Department of Pathology, Stanford University School of Medicine, Stanford, CA 94305, USA

Stanford Blood Center, 3373 Hillview Avenue, Palo Alto, CA 94304, USA
e-mail: sfoung@stanford.edu

T. Miyamura et al. (eds.), *Hepatitis C Virus I*, DOI 10.1007/978-4-431-56098-2_14

other antigenic regions or decoys serve as diversions of the immune response and/or elicit antibodies that negatively modulate broadly neutralizing antibodies will provide a roadmap in rational design for an HCV vaccine.

Keywords Hepatitis C virus • B cell response • Neutralization antibodies • Human monoclonal antibodies • Viral escape • Epitope mapping • Antigenic domains • Vaccine development

Abbreviations

aa	Amino acids
CI	Combination index
DAA	Direct acting antivirals
HCV	Hepatitis C virus
HCVcc	Cell culture-derived HCV virions
HCVpp	HCV pseudoparticles
HDL	High-density lipoprotein
HMAb	Human monoclonal antibody
HVR1	Hypervariable region 1
IC_{50}	50 % inhibitory concentration of the antibody
LDLR	Low density lipoprotein receptor
MOI	Multiplicity of infection
SR-B1	Scavenger receptor class B type 1

1 Introduction

Infection with hepatitis C virus (HCV) is a leading cause of chronic hepatitis, cirrhosis and hepatocellular carcinoma. The World Health Organization estimates an annual increase in the global burden by three to four million new infections (World Health Organization 2010; Shepard et al. 2005). In the United States, HCV has overtaken human immunodeficiency virus as a cause of death (Ly et al. 2012). Encouragingly for patients, advances in in vitro and in vivo HCV infection systems and increased understanding of HCV virology have led to the development of many promising HCV-specific direct acting antivirals (DAA) (Gane et al. 2013; Jacobson et al. 2013; Lawitz et al. 2013; Liang and Ghany 2013; Scheel and Rice 2013). However, the high costs of DAA will limit their access to the large majority of HCV infected patients living in countries with limited resources and it is uncertain if affordable treatment will be available to the majority of patients in more developed countries. Because clinical signs of HCV infection can go unnoticed for decades, transmission may continue to be problematic in high risk populations or under conditions in which unsafe injection drug use practices are common. For these reasons, there is clearly a need for a preventive HCV vaccine. But its development has been technically challenging because of the high variability of the viral genome as well as limitations in animal models (Billerbeck et al. 2013; Houghton and Abrignani 2005). A critical first step in a "rational vaccine design" approach for HCV is to identify mechanisms of immune protection.

2 The Role of Neutralizing Antibodies in Controlling HCV Infection

Humoral immunity is the primary correlate of protection for most preventive vaccines, as shown for hepatitis B, smallpox and other DNA viruses. But for some viruses, such as HCV, the protective role for neutralizing antibodies is not readily apparent because the drop in peak viral load during acute infection is more temporally correlated with cellular immunity rather than the appearance of neutralizing antibodies (Gerlach et al. 1999; Thimme et al. 2001). Robust CD4+ and CD8+ T cell responses correlate with control of acute infection, but are insufficient for preventing long-term persistence in the majority of infected subjects. Nonetheless, the first T cell-based HCV vaccine reported in an early phase clinical study has some evidence of cross-genotype immunity (Barnes et al. 2012). Cumulative evidence at the same time supports the importance of virus neutralizing antibodies to facilitate clearance of infection. Clinical trials with IgG therapy prior to the isolation of HCV prevented transfusion-associated non-A, non-B hepatitis that was due mostly to HCV (Knodell et al. 1976; Conrad and Lemon 1987; Sanchez-Quijano et al. 1988; Sugg et al. 1985). Other clinical studies showed a reduction of infection transmission among sex partners of HCV-infected patients who received gammaglobulin (Piazza et al. 1997). Early animal studies supported this observation as well. An infectious inoculum obtained during acute infection from a patient who eventually developed chronic HCV hepatitis could be neutralized by in vitro incubation with plasma of the same subject collected at 2 years after the initial infection (Farci et al. 1994). Other animal studies observed a delay in the onset of acute infection when chimpanzees were treated with HCV gammaglobulin prior to challenge with infectious virus (Krawczynski et al. 1996).

The development of in vitro cell culture systems, based on retroviral pseudotype particles expressing HCV E1E2 glycoproteins (HCVpp) and infectious cell culture-derived HCV virions (HCVcc), have made possible the measurement of antibody-mediated virus neutralization and the evaluation on the impact of antibodies in the control of infection. (Bartosch et al. 2003a; Hsu et al. 2003; Cai et al. 2005; Zhong et al. 2005; Wakita et al. 2005; Lindenbach et al. 2005). These advances led to the observation that the protective effect provided by IgG preparations in chimpanzees challenged with an infectious inoculum correlated with antibody titers blocking infection of target cells with HCVpp (Bartosch et al. 2003b). In addition, several studies with HCVpp showed a relationship between the control of virus infection and a neutralizing antibody response in single source outbreaks of acute HCV infections (Lavillette et al. 2005; Pestka et al. 2007). One of the studies involved a large cohort of pregnant women who received anti-RhD gammaglobulin contaminated with a single HCV genotype 1b strain, and the other study involved hemodialysis patients with nosocomial acquired HCV infection. Both employing HCVpp established a relationship between the control of virus infection and the neutralizing antibody response during acute HCV infections. Viral clearance was associated with a rapid induction of neutralizing antibodies in the early phase of infection with some evidence that these antibodies are broadly reactive. In contrast, chronic HCV infection was characterized by absent or low-titer neutralizing antibodies in the early phase of infection and the persistence of infection despite the induction of cross-neutralizing antibodies in the late phase of infection. Taken together, these findings suggest that the timing and quality of the neutralizing antibody response during acute infection are potentially the key determinants of infection outcome.

Few animal models of HCV infection have been described, due to the very narrow spectrum of susceptible host species. Only the chimpanzee (*Pan troglodytes*) (Prince and Brotman 1998) and, possibly, tree shrew (*Tupaia belangeri chinensis*) (Xie et al. 1998) are known to be susceptible to HCV infection. However, the National Institutes of Health in the United States beginning in 2013 (http://www.nih.gov/news/health/jun2013/od-26.htm) will no longer fund research involving chimpanzees. An alternative is the development of the Alb-uPA/SCID mouse model with chimeric human liver fragments that has been shown to be susceptible to HCV infection (Mercer et al. 2001). In this model, normal human hepatocytes are transplanted into SCID mice carrying a plasminogen activator transgene, Alb-uPA "normal", resulting in demonstrable production of human albumin for at least 4–5 weeks. When inoculated with infectious HCV serum by IV or IP routes, de novo HCV infection occurs and is maintained for at least 5 weeks with detectable serum HCV RNA levels. In a homologous in vivo protection study, these human-liver chimeric mice were prophylactically treated with chronic phase polyclonal IgG prepared from serum of an individual with chronic HCV infection. The mice were then challenged with infectious virus obtained from the same individual during acute infection. Infection was prevented in five of eight challenged animals. For the non-protected animals, infection was attenuated (Vanwolleghem et al. 2008). In a study of a human monoclonal antibody (HMAb) that neutralized genetically diverse genotype HCVpp, prophylactic treatment provided in vivo protection against infectious HCV quasispecies challenge in this chimeric mouse model (Law et al. 2008). A second model is an immunocompetent humanized mouse model for HCV that exhibited a robust antibody response to vaccination with a recombinant vaccinia virus expressing HCV C-E1-E2-p7-NS2 proteins. Some of the vaccinated animals were protected against a heterologous infectious HCV challenge and protection was correlated with the serum level of antibodies to E2 (Dorner et al. 2011).

3 The Role of the E2 Glycoprotein in HCV Entry

The HCV envelope glycoproteins, E1 and E2, are the natural targets of the protective antibody response. E2 is the main target of neutralizing antibodies but it is also the most variable protein in HCV. E2 forms a heterodimer with E1 to mediate cell entry and fusion. The focus of this review is on the immunogenic determinants on E2 because this viral structure interacts with HCV co-receptors, scavenger receptor class B type 1 (SR-B1) and CD81. There is currently no evidence of E1 interaction with HCV co-receptors. Reported E1-specific neutralizing antibodies are more limited (Meunier et al. 2008; Pietschmann et al. 2006) and probably reflect that this protein is of lower immunogenicity than E2. Nonetheless, there is evidence that E1-specific responses can be invoked following experimental vaccination and that these responses may be protective (Rollier et al. 2004).

HCV entry into target cells requires initial attachment to heparin sulphate aminoglycans (GAGs, Fig. 1) (Barth et al. 2003; Koutsoudakis et al. 2006). Employing insect cell-derived virus-like particles and recombinant E2 proteins, virus attachment is mediated by the hypervariable region-1 (HVR1) in E2 (Barth et al. 2003). The role of GAGs in this early step of entry is also supported by studies with luciferase-reporter HCVcc (Koutsoudakis et al. 2006). In addition, the low density lipoprotein receptor (LDLR) may be involved in attachment that is mediated by virus-associated apolipoproteins (Agnello et al. 1999; Molina et al. 2007; Owen et al. 2009). After initial attachment, the virus interacts with co-receptors, SR-BI (Scarselli et al. 2002) and CD81 (Pileri et al. 1998). SR-B1 facilitates HCV entry by a number of mechanisms that include direct binding to HVR1 (Scarselli et al. 2002), and interactions with high-density lipoprotein (HDL) and virus-associated lipoproteins (Dao Thi et al. 2012) (reviewed in (Dao Thi et al. 2011)). CD81, a member of the tetraspanin superfamily defined by four transmembrane domains, interacts with a binding site on E2 composed of discontinuous segments encompassing amino acids (aa) 410–446, 526–540 and 611–617, that have been identified by alanine substitution studies (Owsianka et al. 2006; Rothwangl et al. 2008; Drummer et al. 2006; Roccasecca et al. 2003). More recently, another highly conserved region in E2, encompassing aa 502–520, has been proposed as being involved in interactions with both CD81 and SR-B1 (Lavie et al. 2014). At least seven other additional host factors (tight junction proteins claudin-1 and occludin, receptor tyrosine kinases, epidermal growth factor receptor, transferrin receptor 1, protein kinase A and the Niemann-Pick C1-like cholesterol absorption receptor) have been reported to be involved in the entry pathway (Benedicto et al. 2008, 2009; Evans et al. 2007; Liu et al. 2009; Meertens et al. 2008; Ploss et al. 2009; Sainz et al. 2012; Martin and Uprichard 2013; Lupberger et al. 2011; Farquhar et al. 2008). However, there is no evidence of direct E2 or E1 interaction with these other host factors.

Aside from virus entry by receptor-mediated endocytosis, HCV spreads by cell-to-cell transmission (Timpe et al. 2008; Brimacombe et al. 2011; Valli et al. 2006; Baldick et al. 2010; Witteveldt et al. 2009). The process appears to be mediated by

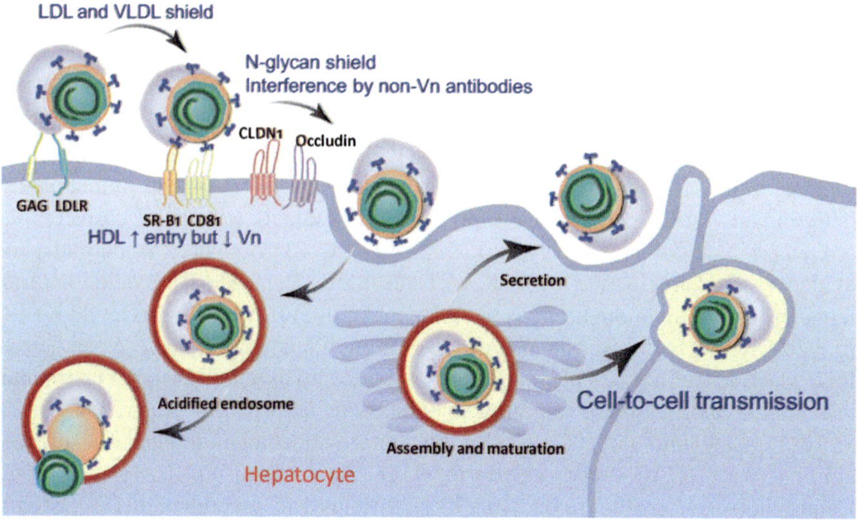

Fig. 1 Hepatitis C virus (HCV) entry in a hepatocyte and factors that impede antibody-mediated virus neutralization. The virus exists in serum in association with apolipoproteins. Initial attachment is by binding to glycosaminoglycans (GAG) and low-density lipoprotein receptor (LDLR). Following binding, the virus interacts with multiple cellular entry factors that include scavenger receptor B1 (SR-B1), CD81, claudin 1 (CLDN1) and occludin. Following clathrin-mediated endocytosis, HCV fusion occurs in the low pH environment in the endosome. After fusion, the viral genome is released into the cytsol. HCV replication, assembly and maturation occur on the membranous web of the endoplasmic reticulum. Maturation and release are the same as for very low-density lipoprotein (VLDL). Cell-to-cell transmission is mediated by HCV receptor molecules, CD81, SR-B1, claudin-1 and occludin. Factors that impede virus neutralization include lipoprotein (LDL and VLDL) shield, specific N-glycans, serum HDL facilitating entry but down modulating virus neutralization (Vn), possible interference by non-neutralizing (non-Vn) antibodies and cell-to-cell transmission

HCV receptor molecules, CD81, SR-B1, claudin-1 and occludin (Timpe et al. 2008; Brimacombe et al. 2011; Witteveldt et al. 2009; Catanese et al. 2013; Fofana et al. 2013). The E2 glycoprotein is likely to be involved. This is supported by the identification of a small molecule inhibitor of HCV entry binding to the transmembrane region of E2 and the isolation of an alpaca nanobody binding to a discontinuous epitope on E2 that block cell-to-cell transmission (Baldick et al. 2010; Tarr et al. 2013).

4 Hypervariable Region 1 on the E2 Glycoprotein

A major determinant of neutralization is the “hyper-variable region 1”, HVR1, which encompasses the first 27 amino acids, aa 384–410, located at the N-terminus of HCV E2 (Farci et al. 1994; Shimizu et al. 1994). This E2 segment is highly immunogenic and antibodies against HVR1 can be detected in the majority of HCV infected individuals (Zibert et al. 1995, 1997; Ray et al. 2010). Animals immunized with synthetic HVR1 peptides elicit high titer serum antibodies to HVR1 (Zucchelli et al. 2001; Shimizu et al. 1996), and these hyperimmune serum antibodies prevent HCV infection in chimpanzees (Farci et al. 1996). But antibodies to HVR1 over time drive replication of viral variants that the existing antibody response does not recognize (Farci et al. 1994; Kato et al. 1993; Weiner et al. 1992). The limited nature of the B cell response to this region is best shown in a study of sequential HCV sequences isolated from one patient over a 26-year period (von Hahn et al. 2007). Patient-specific HCVpp expressing sequential envelope variants were employed to assess virus neutralization by autologous sera. While capable of neutralizing earlier quasispecies, serum antibodies failed to neutralize the contemporaneous dominant HCV E1E2 species that were present in the blood. Escape was associated with mutations within HVR1 leading to decreased binding and neutralization by monoclonal antibodies that were produced to the earliest E2 HVR1 sequence obtained from this patient.

Another study of individuals progressing from the acute phase of infection to chronicity confirmed both the relationship between neutralizing antibodies and virus clearance, and escape from neutralization secondary to HVR1 variants (Dowd et al. 2009). High-titer serum neutralizing antibodies were detected in individuals with spontaneous resolution of infection that peaked at the time of viral clearance. Most individuals progressing to chronic infection had low-titer or the absence of neutralizing antibodies throughout the course of acute infection. When patient-specific HCVpp expressing sequential envelope variants were employed to assess neutralization by contemporaneous autologous sera, neutralization of earlier sequence variants was detected but not to later variants. This pattern is indicative of clearance and evolution of quasispecies variants in response to pressure from neutralizing antibodies. Site directed mutagenesis of the pseudotyped envelope sequence revealed amino acid substitutions within HVR1 that were responsible for the loss of neutralization sensitivity. An interpretation of these findings is that neutralizing antibodies directed at more conserved epitopes outside of HVR1 occur later in the chronic phase of disease progression. Although antibodies to HVR1 are mostly isolate-specific, there are reports of some broadly reactive HVR1 antibodies (Zibert et al. 1995; Zucchelli et al. 2001; Zhang et al. 2003; Puntoricro et al. 1998). The HVR1 region also structurally conceals CD81 binding sites and blocks conserved epitopes mediating virus neutralization on E2. HVR1-deleted viral particles are more sensitive to broadly neutralizing antibodies (Prentoe et al. 2011; Bankwitz et al. 2010). Taken together, these findings suggest that HVR1 serves as an immunologic decoy during acute HCV infection and directs the immune response to HVR1 epitopes associated with rapid viral escape and away from conserved epitopes on E2 mediating broadly neutralizing antibodies.

5 A Highly Conserved Region on the E2 Glycoprotein Encompassing aa 412–423

Immediately downstream of HVR1 is a cluster of overlapping linear epitopes on E2 that are highly conserved across all HCV genotypes and subtypes, encompassing aa 412–423. A number of broadly neutralizing monoclonal antibodies targeting this region have been isolated from experimentally immunized mice (Hsu et al. 2003; Sabo et al. 2011; Flint et al. 1999; Owsianka et al. 2005) and a human monoclonal antibody (HMAb) isolated in a transgenic mouse, designated as HCV1 (Broering et al. 2009). The murine monoclonal antibody, AP33, was the first reported antibody that defined a mostly linear epitope in this region having contact residues within aa 412–423 (Owsianka et al. 2005; Tarr et al. 2006). AP33 broadly neutralized HCVpp bearing E1E2 of the major HCV genotypes 1 through 6 (Tarr et al. 2006). Epitope mapping showed that W420 is a key contact residue shared by these antibodies. W420 also serves as a critical residue for virus binding to CD81 (Owsianka et al. 2006), which explains why this residue is conserved in virtually all HCV isolates, in order to preserve essential viral functions. Furthermore, analysis of over 5500 sequences obtained from the GenBank database showed that the entire AP33 epitope is highly conserved.

Direct crystal structure of E2 peptides in complex with two of these antibodies, AP33 and HCV1, have confirmed their precise contact residues and revealed that this region has a beta-hairpin structure (Kong et al. 2012a, b; Potter et al. 2012). Although aa 412–423 was part of the expressed E2 protein that was successfully determined as mainly globular by crystal studies (Kong et al. 2013), aa 412–420 were disordered in the crystal and could not be visualized. Another murine monoclonal antibody, H77.39, has been described that binds to an epitope within aa 412–423, having contact residues at N415 and N417 (Sabo et al. 2011). This antibody interestingly inhibits both E2 binding to CD81 and SR-B1. Collectively, antibodies to this region hold great promise for immunotherapy and vaccine development.

However, the Asn at 417 is an N-linked glycosylation site that shields this conserved region from being fully exposed to neutralizing antibodies by reducing epitope access (Helle et al. 2007, 2010; Falkowska et al. 2007; Goffard et al. 2005). An adaptive mutation N417S that leads to a glycan shift upstream to N415 blocks virus neutralization by AP33 and HCV1, without a cost in in vitro viral fitness (Pantua et al. 2013; Chung et al. 2013; Dhillon et al. 2010). The N-glycan shift at N417 occurs frequently in passaged HCVcc. The shift can occur in the absence of selection by neutralizing antibodies targeting this region, in the presence of neutralizing antibodies targeting different regions or in the presence of a non-HCV HMAb (Keck et al. 2011, 2012). Surprisingly, cell culture adapted 2a JFH1 HCVcc, containing mostly glycan shifted HCVcc, associated with a N417S mutation, and a minor population of wild-type (wt) HCVcc, displayed an increased sensitivity to neutralization by other HMAbs to aa 412–423, designated as HC33-related antibodies (Keck et al. 2013, 2014). This raised questions whether and how HCV can escape from human antibodies directed against aa 412–423, particularly because the mutation leading to an N-glycan shift from 417 to 415 does not lead to viral escape, but to an increase in sensitivity to these antibodies. We analyzed variants that emerged when 2a HCVcc were passaged under increasing concentrations of one of these HMAbs, HC33.1.

A viral escape selection protocol was designed to maximize the likelihood of escape variants by subjecting wt HCVcc to increasing concentrations of the selection antibody starting at or below the 50 % inhibitory concentration of the antibody (IC_{50}) (μg/ml) value (Keck et al. 2011, 2012). At each antibody concentration, the extracellular virus is passaged repeatedly onto naïve cells until it reaches a titer that infects >80 % of these cells or is greater than 1×10^4 FFU/ml before the virus is subjected to the next higher antibody concentration. This step allows minor variants to be amplified prior to the next round of selective pressure at a higher antibody concentration. As a control virus population, wt HCVcc is subjected to serial passages in increasing concentrations of an isotype-matched control HMAb to provide reference viral variants. This permits specific discrimination between mutations introduced during long-term in vitro propagation of wt HCVcc and those mutations induced under the selective pressure of the tested antibody. This approach identified the same escape mutation in vitro, as observed in vivo (Keck et al. 2008a, 2009, 2011). Thus, this in vitro viral escape selection system mimics the evolution of viral antigenic determinants under immune pressure in humans.

For HC33.1, multiple nonrandom escape pathways were identified (Keck et al. 2014). Two pathways occurred in the context of an N-glycan shift mutation at N417T. At low antibody concentrations, substitutions of two residues outside of the epitope led to variants having improved in vitro viral fitness and reduced sensitivity to HC33.1 binding and neutralization. At moderate concentrations, a S419N mutation occurred within aa 412–423 in escape variants that have greatly reduced sensitivity to HC33.1 but compromised viral fitness. Importantly, the variants generated from these pathways differed in their stability. Variants with mutations outside of the epitope were stable and became dominant as the virions were passaged. The S419N mutation reverted back to N419S when immune pressure was reduced by removing HC33.1. This is the first evidence of a concentration dependent escape mutation. A third mutation at a contact residue, L413I, occurred at higher HC33.1 selection pressure that resulted in variants completely escaping virus neutralization by this antibody. These variants were stable and had in vitro viral fitness similar to wt HCVcc. Although the L413I mutation was elicited in these studies, the Leu in this position is highly conserved in patient sequences (Dhillon et al. 2010). Only 11 of 2108 curated E2 sequences of >100 bases length in the LANL Hepatitis C Virus Database, varied from Leu at this position (Keck et al. 2014; Tamura et al. 2013; Kuiken et al. 2005). All of these variants are L413P, except for a genotype 5a isolate, which is L413F. The elicited L413I mutation is therefore considered not to be found thus far in nature. Overall, the combination of multiple escape pathways enables the virus to persist under a wide range of antibody concentrations. These findings pose a different challenge to vaccine development beyond the identification of highly conserved epitopes. It will be necessary for a vaccine to induce high potency antibodies that prevent the formation of escape variants, which can co-exist with lower potency or levels of neutralizing activities. However, the fact that the aa 412–423 segment on E2 is the target of both AP33-like and HC33.1-like antibodies increases the importance of this region in an effective HCV vaccine. It will be more difficult for the virus to escape simultaneously from both sets of these antibodies.

Finally, epitope mapping of HC33-related antibodies, designated as antigenic domain E antibodies, revealed two subgroups by alanine substitution studies (Fig. 2). One subgroup, e.g., HC33.1, has contact residues similar to murine MAbs to this region, AP33 and H77.39 that are restricted to aa 412–423. The other subgroup, e.g., HC33.4, includes a contact residue within HVR1 located at 408. This raised the possibility of anti-HVR1 antibodies interfering with the functions of antibodies to aa 412–423. As expected, H77.39 inhibited the binding of both subgroups of HC33 HMAbs to 1a H77C E2 proteins by ELISA. Inhibition was also observed against neutralizing HMAbs to two other distinct clusters, designated as antigenic domain B (HC-1 and HC-11) and antigenic domain D (HC84.24 and HC84.26), but not against a neutralizing antibody to a third cluster, designated as antigenic domain C (CBH-7). Antigenic domain B, C and some antigenic domain D epitopes do not have contact residues within aa 412–423. Surprisingly, H77.16, a murine MAb that binds primarily to 1a H77C HVR1 (Sabo et al. 2011) also inhibited all HC33 HMAbs, antigenic domain B and D antibodies, but not antigenic domain C, CBH-7. The H77.16 epitope does not have contact residues within aa 412–423 (Fig. 2).

In summary, the relationship between HVR1 with domains B, D and E is a prime example of an immunogenic decoy. HVR1, as an immunodominant region, diverts the immune response away from conserved domain B, D and E epitopes. While

0–20 % | 21–40 % | 41–60 % | 61–100 % | >100 %

	T404A	P405A	G406A	A407G	K408A	Q409A	N410A	I411A	Q412A	L413A	I414A	N415A	T416A	N417A	G418A	S419A	W42A0	H421A	I422A	N423A	S424A	T425A
AP33	130	106	126	128	99	105	64	92	82	8	69	27	72	87	5	79	0	60	72	80	69	100
H77.39	74	75	81	84	71	84	104	92	87	70	50	1	61	50	1	61	1	92	73	92	79	116
H77.16	69	38	0	61	20	79	0	79	90	98	97	83	86	82	84	82	90	91	83	81	71	93
HC33.1	122	93	120	107	88	98	92	119	110	11	56	137	99	119	1	142	1	100	131	107	93	135
HC33.2	120	82	134	123	120	101	73	133	93	-1	36	92	45	85	0	130	-1	17	109	109	84	147
HC33.3	107	78	117	113	107	102	78	126	89	3	26	34	36	20	3	111	2	46	54	87	84	135
HC33.5	96	80	108	108	98	96	88	126	103	0	32	57	4	47	0	100	0	6	48	85	80	110
HC33.4	89	76	90	96	11	88	76	95	92	7	93	106	104	103	50	127	1	116	103	93	94	111
HC33.8	86	78	90	100	8	84	73	102	94	17	80	134	103	125	1	156	0	136	128	111	99	123
HC33.29	84	86	97	96	11	79	67	90	84	25	78	100	92	101	0	118	0	107	94	87	84	90

Fig. 2 Epitope mapping of antigenic domain E antibodies (HC33 HMAbs), AP33, H77.39 and H77.16. E2 alanine substitution mutant proteins were expressed in 293T cells and cell lysates were analyzed by ELISA. The region of E2 protein that was analyzed: aa 384–425, but only the findings between aa 404–425 are shown. *Red* indicates 0–20 %, *orange* 21–40 %, *brown* 41–60 %, *white* 61–100 % and *green* >100 % binding when the residue was replaced by alanine (or glycine at aa 406), relative to binding to wt E2 protein

virtually all infected subjects continuously produce anti-HVR1, only 2–15 % have antibodies to aa 412–423 (Keck et al. 2013; Tarr et al. 2007, 2012). HVR1 blunts the protective B cell response by a number of mechanisms. HVR1 conceals CD81 binding sites and blocks conserved neutralizing epitopes (Bankwitz et al. 2010), elicits neutralizing antibodies associated with rapid escape throughout the course of infection (von Hahn et al. 2007), and the anti-HVR1 antibodies inhibit the binding by antibodies to domains B, D and E, and interfere with their neutralizing activities. The last mechanism may contribute to the progression of chronic infection with HCV since anti-HVR1 antibodies are continuously induced during the course of infection (Kato et al. 1993; von Hahn et al. 2007).

6 Neutralizing Antibodies to Conformational Epitopes on the E2 Glycoprotein

The majority of antibodies with broad neutralizing activities recognize conformational epitopes on E2 (Law et al. 2008; Keck et al. 2012; Hadlock et al. 2000; Allander et al. 2000; Bugli et al. 2001; Habersetzer et al. 1998; Giang et al. 2012). These HCV HMAbs were isolated from B cells of patients with chronic HCV infection. Cross-competition analyses delineate at least four immunogenic clusters of overlapping conformational epitopes with distinct properties (Keck et al. 2004, 2005, 2012). Non-neutralizing HMAbs fall within one cluster, which is designated as antigenic domain A. It is probable that this cluster of conformational epitopes and other non-neutralizing determinants are highly immunogenic and account for a substantial portion of the antibody response to E2 (Burioni et al. 2001, 2004). Neutralizing HMAbs segregate into three clusters of conformational epitopes, designated as antigenic domain B, C and D. A fifth cluster of linear epitopes mediating neutralization, antigenic domain E, is identified by the HC33-related antibodies.

6.1 A Highly Immunogenic Region on E2, Antigenic Domain B

Of the broadly neutralizing antibodies from different laboratories, many can be included in antigenic domain B. These antibodies display different patterns of neutralizing activity against HCVcc and HCVpp containing glycoproteins of HCV genotypes 1–6, with a subset neutralizing all genotypes (Keck et al. 2008b; Owsianka et al. 2008). Alanine substitution studies identified two conserved E2 residues, G530 and D535, that are required for binding of all antigenic domain B HMAbs. Residues G523 and W529 are also required for some but not all of these antibodies (Keck et al. 2008b; Owsianka et al. 2008). Importantly, W529, G530 and

D535 participate in the interaction of E2 with CD81. Thus, antigenic domain B HMAbs exert potent and broad neutralizing effects on HCV by competing with CD81 for binding to conserved residues on E2 that are necessary for this step in viral entry. This information provides an explanation for why specific N-glycan sites on E2 inhibit antibody-mediated neutralization (Helle et al. 2007, 2010; Falkowska et al. 2007; Goffard et al. 2005). The conserved N-glycan at N532 sits in the middle of critical contact residues for all domain B antibodies and is therefore capable of hindering antibody binding.

These epitopes are highly immunogenic and their antibodies are found in the majority of HCV infected individuals, as shown by increased serum neutralizing titers among different HCV genotype infected subjects against a mutant HCVpp without the glycan at 532 (Falkowska et al. 2007). Consistent with this, studies by multiple groups on broadly neutralizing HMAbs also recognized epitopes containing these residues (Law et al. 2008; Johansson et al. 2007; Perotti et al. 2008). In one study, their antibodies were sorted into three groups, designated as AR1, AR2 and AR3, based on different binding patterns and competition with other defined monoclonal antibodies (Law et al. 2008; Johansson et al. 2007; Perotti et al. 2008). While AR1 and AR2 antibodies had limited potency and breadth of neutralization, AR3 antibodies broadly neutralized genotypes 1–6 HCVpp. In addition, an AR3-specific antibody protected Alb-uPA/SCID mice against the diversity of HCV quasispecies in an infectious inoculum. Epitope mapping identified AR3 epitopes being formed by three discontinuous E2 segments, aa394-424, aa437-447 and aa523-540. Alanine scanning revealed contact residues located at S424, G523, G530, D535, V538 and N540 (Law et al. 2008). Other described broadly neutralizing HMAbs, including A8, 1:7, e137 and e20 also mapped to the E2 segment, aa 523–540, and identified the same conserved contact residues located at W529, G530 or D535 (Johansson et al. 2007; Perotti et al. 2008; Mancini et al. 2009).

Two questions are of concern from a vaccine perspective. First, which of the antigenic domain B epitopes are prone to accumulating mutations under immune pressure, leading to virus escape from neutralization, as observed with the antibody response to HVR-1 and aa 412–423? Second, which domain B epitopes remain relatively invariant so as to accommodate the interactions of E2 with CD81 required for virus viability? Some antigenic domain B antibodies do not neutralize all HCV genotypes, which are indicative of viral escape (Keck et al. 2008b; Owsianka et al. 2008). Other broadly neutralizing antigenic domain B antibodies include CBH-2, HC-11 and HC-1. In an antibody-virus co-culture study that allowed escape variants to be amplified and isolated under increasing concentrations of a neutralizing antibody, three escape patterns were observed with these antibodies. Escape mutants for CBH-2 were isolated containing mutations at D431G or A439E (Keck et al. 2011). The induction of the escape mutant at D431G under the selective pressure of CBH-2 mimics the previously observed naturally occurring variant at this site, D431E, that cannot be neutralized by this antibody (Keck et al. 2008a, 2009). The similarity of the replication rate of infectious CBH-2 escape HCVcc mutants to that of wt HCVcc indicates that escape from CBH-2-like antibodies does not compromise in vitro viral fitness. In escape studies with HMAb HC-11, wt HCVcc was completely eliminated, and no escape mutants were generated at an antibody concentration of 100 μg/ml. Under the condition of progressively increasing the antibody concentration starting at the IC_{50} value of 0.05 μg/ml (0.05, 0.1, 0.5, 1, 5, 10 and 100 μg/ml), escape mutants were identified. An escape mutant with a single substitution at L438F was observed when the antibody concentration reached 10 μg/ml. A variant bearing two mutations, at L438F and N434D, appeared at selection 10 μg/ml. A different variant with double mutations at L438F and T435A appeared when HC-11 was increased to 100 μg/ml. By gradually increasing the antibody concentration, escape mutants were isolated that were not apparent when infectious virus encountered a high antibody concentration. The implication for vaccine design is a requirement for the immunogen, containing an HC-11-like epitope, to be able to elicit a sufficiently high titer antibody response to avoid virus escape. If the antibody response is of lower titers, virus escape can occur. Nonetheless, the escape mutations resulted in a progressive decrease in viral fitness under the selective pressure of HC-11. Huh7.5 cells infected with L438F HC-11 escape mutant at a low multiplicity of infection (MOI = 0.01) yielded eight times less virus than wt HCVcc. The double substitution mutant, L438F and T435A, yielded 256 times less virus. For HC-1, the antibody at a critical concentration completely suppressed viral replication and no escape mutants were isolated. The location of the contact residues for HC-1 appears to be limited to a more invariant segment of E2 encompassing aa 529–535 that interacts with CD81.

Of the three antigenic domain B HMAbs tested, one led to escape mutant viruses without affecting in vitro viral fitness, similar to escape from HVR1; a second led to escape but with compromised in vitro viral fitness; and a third led to complete virus elimination without escape mutants. Sequence analysis of escape mutants revealed a conserved region, aa 529–535, and a region, aa 425–443, on E2 that appears to be associated with escape mutations. Collectively, these findings highlight the substantial challenges inherent in developing HCV vaccines, and show that an effective vaccine will need to induce antibodies to intrinsically conserved epitopes in order to lessen the probability of virus escape.

6.2 Antigenic Domain C

By cross-competition studies, a neutralizing HMAb, designated as CBH-7, does not cross-react with antigenic domain B antibodies and is to a conformational epitope (Hadlock et al. 2000; Keck et al. 2004). It neutralizes genotypes 1 and 2 HCVpp and HCVcc, but not genotypes 3, 4 and 5 HCVpp (Keck et al. 2007, 2008a; Owsianka et al. 2008). Neutralization is mediated by blocking E2 binding to CD81 (Hadlock et al. 2000). Other reported antibodies that are likely to be in this antigenic domain include AR1A, AR1B and AR2A (Law et al. 2008). CBH-7 blocks the binding of these AR1 and AR2 HMAbs to E2 by ≥80 %. CBH-7 also blocks AR5A binding to

E2 by nearly 80 %, an antibody that binds to an epitope with contact residues on both E1 and E2 (Giang et al. 2012). Epitope mapping, however, showed that CBH-7 is definitively related to the AR1 antibodies in that these antibodies shared contact residues within aa 544–549 on E2. It remains possible that CBH-7 shares contact residues with AR2A and AR5A, but additional mapping will need to be performed. Interestingly, CBH-7 has bidirectional competition with a non-neutralizing antigenic domain A HMAb, CBH-4B, but it enhances the binding of a different non-neutralizing domain A HMAb, CBH-4G (Keck et al. 2004). Although CBH-4B blocks CBH-7 binding by nearly 40 % (Keck et al. 2004), neutralization potency of CBH-7 against genotype 1a H77C HCVpp is not inhibited by CBH-4B (Keck et al. 2005).

6.3 Antigenic Domain D

The HVR-1 region and antigenic domain B are clusters of highly immunogenic epitopes that do not elicit the most broadly protective antibodies. More importantly, these antibodies are mainly associated with viral escape. To identify highly conserved epitopes that will be relevant for vaccine design, we used the information gained from epitope mapping of antigenic domain A and B HMAbs to construct soluble E2 mutants that do not bind to their respective antibodies (Keck et al. 2012). Heterologous E2 employed in the screening eliminated antibodies to HVR1. By employing yeast surface display of antibodies, HCV scFvs were isolated that initially bound to a 1a HCV E2 antigenic domain A mutant to minimize the selection of non-neutralizing scFvs and then bound to a second 1a HCV E2 antigenic domain B mutant to minimize the selection of antigenic domain B scFvs. This bias-screening approach led to the selection of a panel of scFvs that showed a broad breadth of reactivity against six different HCV genotypes and subtypes of E1E2 recombinant proteins. They were converted to full IgG_1 molecules and designated as HC84-related HMAbs. Each of these HMAbs is to a conformational epitope and neutralizes HCVcc with genotypes 1–6 envelope proteins with varying profiles (Fig. 3a). Five of these antibodies neutralize representative genotypes 1–6 HCVcc. The patterns of neutralization against HCVcc of different genotypes suggest that these antibodies are directed at highly conserved epitopes. These antibodies show more uniform neutralization against different HCV genotypes and subtypes than antigenic domain B antibodies.

Based on the observation that each HC84 HMAb inhibited E2 binding to CD81 (Fig. 3b), epitope mapping studies by alanine scanning focused on E2 segments encompassing aa 410–446, aa 526–540 and aa 611–617. These regions are known to contain contact residues that form the E2 binding site to CD81. As expected, contact residues were not located within aa 526–540, but within aa 410–446 and aa 611–617 (Fig. 3c). Based on the differences between epitope mapping of previously identified antigenic domain A, B and C HCV HMAbs, the epitopes of the HC84 HMAbs have been designated as antigenic domain D.

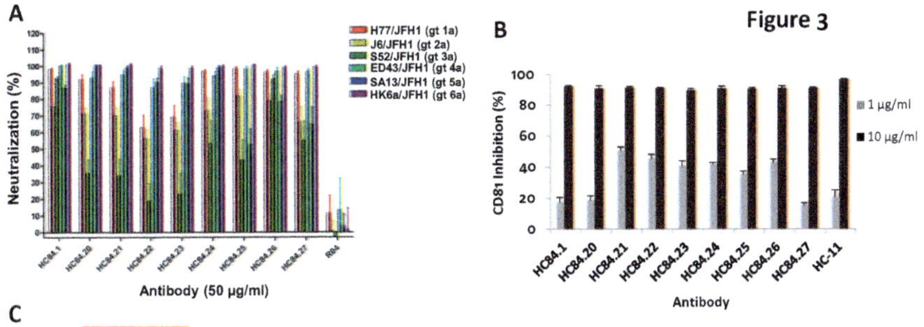

C

0-20% 21-30%

Epitope Map

	W420A	N428A	W437A	L441A	F442A	Y443A	K446A	Y613A	W616A
HC84.1				22	1				
HC84.20				6	26	26		20	0
HC84.21*				28	20	35			
HC84.22	18	19	8	2	0	0			0
HC84.23	20	20	15	2	0	0			0
HC84.24					14	1			
HC84.25				10	1				24
HC84.26				12	0				25
HC84.27				6	0	20	22		11

*epitope mapping was tested at 0.1 ug/ml.

Fig. 3 HC84 HMAb neutralization of JFH1-based genotypes 1–6 HCVcc, inhibition of E2 binding to CD81 and epitope mapping. (**a**) The designation of the viruses is: genotype 1a (H77C/JFH1), 2a (J6/JFH1), 3a (S52/JFH1), 4a (ED43/JFH-1), 5a (SA13/JFH1) and 6a (HK6a/JFH1); all except the 2a virus contained adaptive mutations. R04 is an isotype-matched HMAb negative control. Infectious virus inoculum was incubated with 50 µg/ml each HMAb followed by inoculation onto Huh7.5 cells. Cells were immunostained with a MAb to NS5A antigen and enumerated by FFU. (**b**) Inhibition of E2 binding to CD81-LEL by HC-84 HMAbs. Genotype 1a H77C recombinant E1E2 lysate containing E2 was incubated with each test HMAb, and the antibody-antigen complex was then added onto CD81-LEL-precoated wells. Detection of E2 bound to CD81-LEL was measured with biotinylated CBH-4D (Hadlock et al. 2000). HC-11 was used as a positive control. (**c**) Summary of epitope location of each HC84 HMAb based on alanine substitution studies

The initial expectation was for a minor subset of the antigenic domain D epitopes to be invariant because of functional or structural constraints. Surprisingly, when 2a HCVcc was grown in the presence of HC84 HMAbs, five of five selected antibodies led to no escape variants, under the conditions tested. Of the three central contact residues that form antigenic domain D epitopes (L441, F442 and Y443), L441 and Y443 are absolutely conserved among all known HCV isolates. F442, however, is only 60 % conserved and the remaining substitutions are either Leu or Ile (Fig. 4c). Crystal studies provided an explanation why viral escape was not observed. The epitope forms an alpha-helix (Fig. 4a) (Krey et al. 2013). The key contact residues were confirmed to be aa 441, 442 and 443; 442 and 443 form a hydrophobic protrusion (Fig. 4b) that inserts into a hydrophobic binding pocket of the paratope formed by the heavy chain CDRs. Although 442 is not conserved (Fig. 4c), from a structural perspective and in agreement with no viral escape associated with domain D HMAbs, a F442I or F442L change will not eliminate

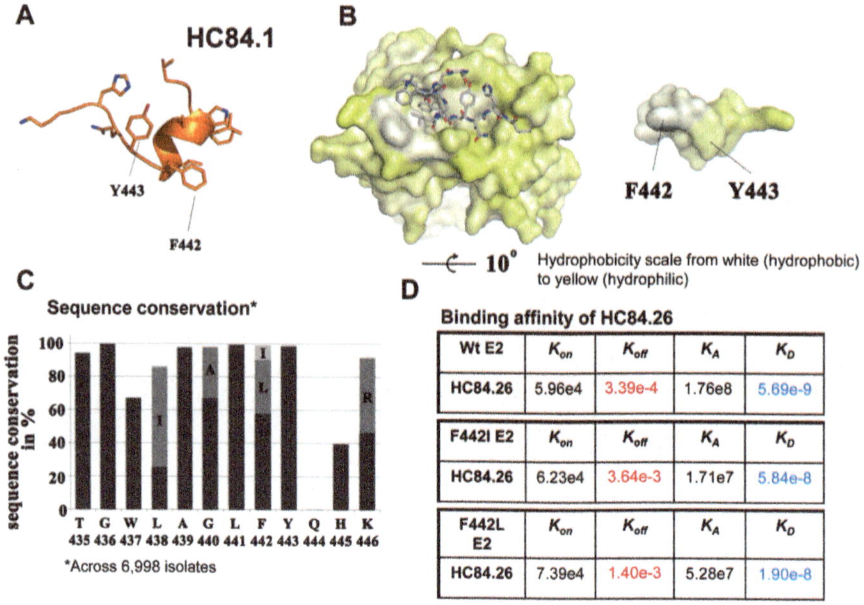

Wt E2	K_{on}	K_{off}	K_A	K_D
HC84.26	5.96e4	3.39e-4	1.76e8	5.69e-9

F442I E2	K_{on}	K_{off}	K_A	K_D
HC84.26	6.23e4	3.64e-3	1.71e7	5.84e-8

F442L E2	K_{on}	K_{off}	K_A	K_D
HC84.26	7.39e4	1.40e-3	5.28e7	1.90e-8

Fig. 4 Crystal structure of antigenic domain D. (**a**) The epitope is an alpha-helix as determined by HC84.1 HMAb. (**b**) Insertion of 442 and 443 forming a hydrophobic protrusion into the hydrophobic paratope binding pocket. (**c**) The degree of residue conservation and (**d**) binding affinity measurements against wt, F442I and F442L E2 with the HC84.26 HMAb

the interaction with domain D HMAbs, but instead slightly decrease the binding energy of the complex. This is in agreement with the faster K_{off} rate measured for HC84.26 for these two mutations leading to a decrease in K_D, which can be overcome by increasing the concentration of antibodies (Fig. 4d). In summary, the HC84 cluster of epitopes, antigenic domain D, is highly conserved among HCV genotypes and subtypes, and mediates broad and potent virus neutralization that is not likely to lead to virus escape. Thus, these epitopes are relevant in vaccine design for this highly diverse virus.

7 Topology of Antigenic Domains of the E2 Glycoprotein

The recently reported crystal structure of the HCV E2 core in complex with a broadly neutralizing human antibody (AR3C) (Kong et al. 2013) enables the mapping of antigenic domains A-E on the surface of this glycoprotein (Fig. 5). These domains form distinct patches with little or no overlap. Domain E (aa 412–423) was modeled by superposing peptide 412–423 from the structure of this peptide in complex with HMAb HCV1 (Kong et al. 2012b) onto aa 421–423 of the E2 structure, in which residues 412–420 are disordered. The known core regions

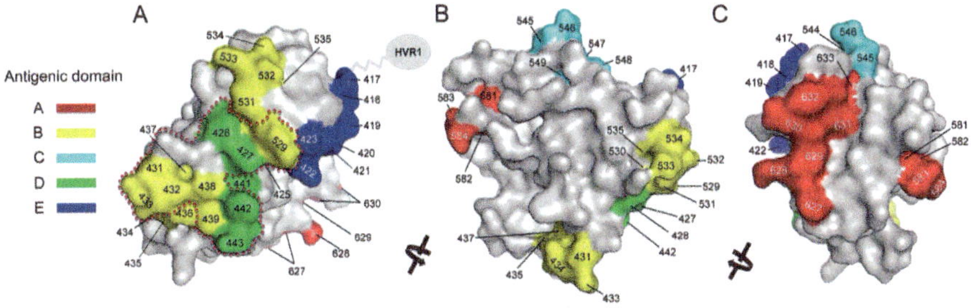

Fig. 5 Location of antigenic domains on the HCV E2 envelope glycoprotein. The E2 structure (PDB accession code 4MWF) (Kong et al. 2013) is shown as a molecular surface colored according to antigenic domains *A–E* (*A*: 581–584 and 627–633; *B*: 431–439 and 529–535; *C*: 544–549; *D*: 420–428, 441–443 and 616; *E*: 412–423). Only those domain residues located on the surface of E2 are numbered. The HVR1 region, which is missing from the structure, is depicted as a gray oval and labeled. Antigenic domain E, comprising residues 412–423, was modeled by superposing peptide 412–423 from the complex of this peptide bound to HMAb HCV1 (4DGY) (Kong et al. 2012b) onto residues 421–423 of the E2 structure, in which residues 412–420 are disordered. (**a**) Face of E2 targeted by neutralizing antibodies to antigenic domains B, D and E. Residues 427, 428 and 442 are implicated in CD81 binding. The footprint of HMAb ARC3 on E2, as defined by the structure of the corresponding complex (4MWF), is marked by *red dots*. (**b**) Face of E2 protected from antibody neutralization by N-linked glycans. (**c**) The non-neutralizing antigenic domain A region

for domains C, D and E are contiguous, as expected for protein epitopes, whereas those for domains A and B are each ‘split’ into two noncontiguous surfaces on E2 (Fig. 5). However, it is important in this regard to distinguish between epitopes defined by mutagenesis studies and epitopes defined by X-ray crystallography. The two do not necessarily coincide, at least not completely. In alanine-scanning mutagenesis, the readout is functional addressing the question, what is the effect of mutations in the antigen on antibody binding. Thus, this method identifies energetically important residues (‘hotspots’) in an antigen-antibody interface. However, hotspot residues invariably represent only a subset of all residues in actual physical contact (defined as atom pair distance <4 Å for carbon-carbon contacts) in crystal structures of antigen-antibody (or other protein-protein) complexes (Sundberg and Mariuzza 2002). Indeed, many (and sometimes most) contacting residues make little or no net energetic contribution to binding, and therefore are not easily detected by alanine-scanning mutagenesis. This is apparent by examining the physical footprint of HMAb ARC3 on the E2 glycoprotein, as defined by the crystal structure of the corresponding complex (Fig. 5a) (Kong et al. 2013). Although ARC3 is mainly within the domain B cluster, it also contacts several residues (N428, F442 and Y443) that are hotspots for antigenic domain D, thereby spanning the seemingly split domain B epitope. Conversely, although no structure is currently available for an antigenic domain D antibody bound to intact E2 (as opposed to a peptide fragment (Krey et al. 2013), it is likely that such an antibody would also contact a number of domain B residues. Therefore, although antigenic domain B and D antibodies may recognize partially overlapping epitopes on E2, they do so in energetically different ways, as manifested by the completely different hotspot residues for antibody binding identified by alanine-scanning mutagenesis. Lastly, some hotspot residues (e.g. W616 of domain D) are not exposed on the surface of E2, as expected for residues that directly contact antibodies, but instead probably influence the conformation of neighboring residues on the E2 surface that are positioned to make direct contacts (e.g., F442).

Strikingly, antigenic domains A–E are not uniformly distributed on the surface of E2 (Fig. 5). Rather, except for domains A and C, they are mainly concentrated on one face of the glycoprotein (Fig. 5a). This face, which is relatively free of N-linked glycans, also includes residues that are critical for the binding of CD81 (420, 441, 443, 529, 530, 535 and 616), and that are part of, or adjacent to, antigenic domain B, D and E epitopes. This readily explains the ability of antibodies to these antigenic domains to block E2 binding to CD81, thereby neutralizing HCV. By contrast, antigenic domain A is distant from the putative CD81 binding site (Fig. 5c), in agreement with the inability of domain A antibodies to neutralize the virus. Opposite the face of E2 targeted by neutralizing antibodies (Fig. 5a) is a face essentially devoid of antigenic determinants (Fig. 5b), but with multiple N-linked glycans. This glycan shield may prevent neutralization by antibodies, while preserving the binding site for CD81. Such a strategy for reducing immunogenicity is analogous to that employed by other viruses with heavily glycosylated envelope proteins, notably influenza and HIV.

8 Negative Modulation of Neutralizing Antibodies

In addition to genetic mutation escape strategy from neutralization, there are other evasion strategies to negatively modulate the neutralizing antibody response (Fig. 1). Glycosylation of the virus envelope, non-neutralizing antibodies or virion-associated lipoproteins may interfere with antibody-mediated neutralization by masking neutralizing epitopes or otherwise limiting access of neutralizing antibodies to their cognate epitopes (Helle et al. 2007; Falkowska et al. 2007; Thomssen et al. 1992). Studies with HCVpp and HCVcc indicate that N-linked glycans at conserved Asn residues of the E2 protein hinder neutralizing activities of HCV-specific polyclonal sera, as well as neutralizing HMAbs (Helle et al. 2007, 2010; Falkowska et al. 2007). The fine mapping of antigenic domain B HMAbs provides an exact explanation why specific N-glycan sites on E2 may modulate antibody-mediated neutralization (Helle et al. 2007; Falkowska et al. 2007). The N-glycan at Asn532 sits in the middle of two critical contact residues for all domain B antibodies and is therefore capable of hindering antibody binding. The substitution of the N-glycan at residue 532 could lead to an improved HCV immunogen.

For non-neutralizing antibodies, it has been proposed that a segment of E2 encompassing aa 434–446 (epitope II) encodes a cluster of epitopes that are associated with non-neutralizing antibodies, and that these antibodies inhibit the neutralizing activities of antibodies directed at an adjacent E2 segment encompassing aa 412–426 (epitope I) (Zhang et al. 2007, 2009). Depletion of the epitope-II antibodies in polyclonal IgG preps led to increase neutralizing titers of these antibody preparations (Zhang et al. 2009). If this is the case, it provides a contributing factor in explaining the existence of viremia in the context of high neutralizing antibodies titers in chronic HCV infection. However, conflicting studies that employed similar approaches of isolating polyclonal antibodies to synthetic peptides encompassing aa 412–423 and aa 434–446 also have been reported (Tarr et al. 2012). Eluted polyclonal antibodies from both peptides neutralized HCV, and when combined showed additive neutralizing activities. In addition the segment aa 434–446 contains a core contact region for antigenic domain D antibodies that are broadly neutralizing and not associated with viral escape (Keck et al. 2012), and when these antibodies are combined with HMAbs to aa 412–423, additive neutralization is observed (Keck et al. 2013). It is therefore unlikely that aa 434–446 induces non-neutralizing antibodies that interfere with neutralizing antibodies to aa 412–423. Nonetheless, the concept of interfering antibodies is a valid concern. It is possible that anti-HVR1 antibodies interfere with other more broadly neutralizing epitopes.

The circulating HCV virions in the blood of patients, associated with host lipoproteins and immunoglobulin in the form of lipo-viro-particles (LVP), are heterogeneous in density and infectivity. The LVP corresponding to low density fractions express high infectivity, which can be blocked by the anti-apolipoprotein B (Apo B) and E antibodies (Andre et al. 2002). HCV envelope glycoproteins have the intrinsic capacity to utilize lipoprotein synthesis and assembly machinery, and HCV virions production is closely connected to lipid metabolism which could serve as an assembly platform (Icard et al. 2009; Chang et al. 2007; Huang et al. 2007). The extent to which the association of HCV with lipoproteins may impede access to the virion surface by neutralizing antibodies is not clear. Infectious HCVcc particles can be readily precipitated with antibody to ApoB, but are nonetheless neutralized with virus-specific HMAbs (Yi et al. 2007). In the presence of high density lipoprotein (HDL) or human serum, the neutralization by anti-E2 neutralizing antibodies of HCVpp and HCVcc could be attenuated (Dreux et al. 2006). Of HCV particles recovered from the liver of an immunodeficient patient, only 27 % of the E2 and 8.5 % of RNA could be immunoprecipitated with AP33 (Nielsen et al. 2006). In contrast, two of our domain B HMAbs precipitated 52–67 % of total E2 and 19–24 % of total RNA (Nielsen et al. 2006). Clearly, not all neutralizing epitopes are masked by lipids, but better quantitation of this potential negative regulator of neutralizing activity will require additional studies of wild-type HCV.

In addition to virus entry by receptor-mediated endocytosis, HCV has been reported to spread from cell-to-cell (Timpe et al. 2008; Brimacombe et al. 2011; Valli et al. 2006; Witteveldt et al. 2008; Baldick et al. 2010). This process appears to be mediated by HCV receptor molecules, CD81, SR-B1, claudin-1 and occludin (Timpe et al. 2008; Brimacombe et al. 2011; Witteveldt et al. 2009). A substantial portion of broadly neutralizing antibodies that block E2 binding to CD81 do not inhibit cell-to-cell spread.

9 Closing Remarks

In order to develop an effective vaccine to HCV, it is necessary to understand the interplay between B cell immunity to this highly diverse virus and optimal immunogen design. The HCV envelope glycoproteins, E1 and E2, are the natural targets of the protective antibody response. However, the virus has devised a number of decoys of highly immunogenic regions, associated with viral escape or non-neutralizing antibodies that deflect the immune response from less immunogenic but more conserved regions mediating virus neutralization that are not associated with viral escape. Moreover, these immunogenic decoys elicit antibodies that may interfere with broadly virus neutralizing antibodies. The regions on the E2 glycoprotein are roughly segregated into five clusters of overlapping epitopes, designated as antigenic domains A-E, and the HVR1. HVR1 is an immunodominant region that is a major decoy. Not only does it elicit isolate-specific antibodies from which the virus can easily escape without compromising viral fitness, the binding of these antibodies prevents the binding of more conserved neutralizing antibodies. This likely contributes to the development of persistent infection since anti-HVR1 antibodies are induced throughout the course of infection. Antigenic domain A is a cluster of overlapping epitopes associated with non-neutralizing antibodies. This region and other regions associated with non-neutralizing antibodies are highly immunogenic and account for a significant portion of the B cell response, and therefore are immunogenic decoys. Antigenic domain B is also highly immunogenic and elicits more broadly neutralizing antibodies. However, a significant portion, if not the majority, of these antibodies are associated with viral escape. Antigenic domains D and E (aa 412–423) are more promising for vaccine development. Antibodies to these regions are broadly neutralizing among the major HCV genotypes and subtypes, but the D and E domains are less immunogenic than HVR1 and antigenic domain B. Taken together, a detailed understanding of conserved virus neutralizing epitopes not associated with escape, and how other antigenic regions or decoys serve as diversions of the immune response and/or elicit antibodies that negatively modulate broadly neutralizing antibodies, will provide a roadmap for rational design for an HCV vaccine. This will guide approaches to down modulate the immune response to immunogenic decoys and to direct the response to conserved E1E2 regions for vaccine development.

Acknowledgements We thank Yili Li and Adam Wang for preparing Fig. 5. This work was supported in part by PHS grants R41-AI108024 and U19-AI123862 (SKHF), and MPower (RAM and TRF).

References

Agnello V, Abel G, Elfahal M, Knight GB, Zhang QX (1999) Hepatitis C virus and other flaviviridae viruses enter cells via low density lipoprotein receptor. Proc Natl Acad Sci U S A 96:12766–12771

Allander T, Drakenberg K, Beyene A, Rosa D, Abrignani S et al (2000) Recombinant human monoclonal antibodies against different conformational epitopes of the E2 envelope glycoprotein of hepatitis C virus that inhibit its interaction with CD81. J Gen Virol 81:2451–2459

Andre P, Komurian-Pradel F, Deforges S, Perret M, Berland JL et al (2002) Characterization of low- and very-low-density hepatitis C virus RNA-containing particles. J Virol 76:6919–6928

Baldick CJ, Wichroski MJ, Pendri A, Walsh AW, Fang J et al (2010) A novel small molecule inhibitor of hepatitis C virus entry. PLoS Pathog 6:e1001086

Bankwitz D, Steinmann E, Bitzegeio J, Ciesek S, Friesland M et al (2010) Hepatitis C virus hypervariable region 1 modulates receptor interactions, conceals the CD81 binding site, and protects conserved neutralizing epitopes. J Virol 84:5751–5763

Barnes E, Folgori A, Capone S, Swadling L, Aston S et al (2012) Novel adenovirus-based vaccines induce broad and sustained T cell responses to HCV in man. Sci Transl Med 4:115ra111

Barth H, Schafer C, Adah MI, Zhang F, Linhardt RJ et al (2003) Cellular binding of hepatitis C virus envelope glycoprotein E2 requires cell surface heparan sulfate. J Biol Chem 278:41003–41012

Bartosch B, Dubuisson J, Cosset FL (2003a) Infectious hepatitis C virus pseudo-particles containing functional E1–E2 envelope protein complexes. J Exp Med 197:633–642

Bartosch B, Bukh J, Meunier JC, Granier C, Engle RE et al (2003b) In vitro assay for neutralizing antibody to hepatitis C virus: evidence for broadly conserved neutralization epitopes. Proc Natl Acad Sci U S A 100:14199–14204

Benedicto I, Molina-Jimenez F, Barreiro O, Maldonado-Rodriguez A, Prieto J et al (2008) Hepatitis C virus envelope components alter localization of hepatocyte tight junction-associated proteins and promote occludin retention in the endoplasmic reticulum. Hepatology 48:1044–1053

Benedicto I, Molina-Jimenez F, Bartosch B, Cosset FL, Lavillette D et al (2009) Tight junction-associated protein occludin is required for a post-binding step in hepatitis C virus entry and infection. J Virol 83:8012–8020

Billerbeck E, de Jong Y, Dorner M, de la Fuente C, Ploss A (2013) Animal models for hepatitis C. Curr Top Microbiol Immunol 369:49–86

Brimacombe CL, Grove J, Meredith LW, Hu K, Syder AJ et al (2011) Neutralizing antibody-resistant hepatitis C virus cell-to-cell transmission. J Virol 85:596–605

Broering TJ, Garrity KA, Boatright NK, Sloan SE, Sandor F et al (2009) Identification and characterization of broadly neutralizing human monoclonal antibodies directed against the E2 envelope glycoprotein of hepatitis C virus. J Virol 83:12473–12482

Bugli F, Mancini N, Kang CY, Di Campli C, Grieco A et al (2001) Mapping B-cell epitopes of hepatitis C virus E2 glycoprotein using human monoclonal antibodies from phage display libraries. J Virol 75:9986–9990

Burioni R, Bugli F, Mancini N, Rosa D, Di Campli C et al (2001) Nonneutralizing human antibody fragments against hepatitis C virus E2 glycoprotein modulate neutralization of binding activity of human recombinant Fabs. Virology 288:29–35

Burioni R, Mancini N, Carletti S, Perotti M, Grieco A et al (2004) Cross-reactive pseudovirus-neutralizing anti-envelope antibodies coexist with antibodies devoid of such activity in persistent hepatitis C virus infection. Virology 327:242–248

Cai Z, Zhang C, Chang KS, Jiang J, Ahn BC et al (2005) Robust production of infectious hepatitis C virus (HCV) from stably HCV cDNA-transfected human hepatoma cells. J Virol 79:13963–13973

Catanese MT, Loureiro J, Jones CT, Dorner M, von Hahn T et al (2013) Different requirements for SR-BI in hepatitis C virus cell-free versus cell-to-cell transmission. J Virol 87:8282

Chang KS, Jiang J, Cai Z, Luo G (2007) Human apolipoprotein e is required for infectivity and production of hepatitis C virus in cell culture. J Virol 81:13783–13793

Chung RT, Gordon FD, Curry MP, Schiano TD, Emre S et al (2013) Human monoclonal antibody MBL-HCV1 delays HCV viral rebound following liver transplantation: a randomized controlled study. Am J Transplant 13:1047–1054

Conrad ME, Lemon SM (1987) Prevention of endemic icteric viral hepatitis by administration of immune serum gamma globulin. J Infect Dis 156:56–63
Dao Thi VL, Dreux M, Cosset FL (2011) Scavenger receptor class B type I and the hypervariable region-1 of hepatitis C virus in cell entry and neutralisation. Expert Rev Mol Med 13:e13
Dao Thi VL, Granier C, Zeisel MB, Guerin M, Mancip J et al (2012) Characterization of hepatitis C virus particle subpopulations reveals multiple usage of the scavenger receptor BI for entry steps. J Biol Chem 287:31242–31257
Dhillon S, Witteveldt J, Gatherer D, Owsianka AM, Zeisel MB et al (2010) Mutations within a conserved region of the hepatitis C virus E2 glycoprotein that influence virus-receptor interactions and sensitivity to neutralizing antibodies. J Virol 84:5494–5507
Dorner M, Horwitz JA, Robbins JB, Barry WT, Feng Q et al (2011) A genetically humanized mouse model for hepatitis C virus infection. Nature 474:208–211
Dowd KA, Netski DM, Wang XH, Cox AL, Ray SC (2009) Selection pressure from neutralizing antibodies drives sequence evolution during acute infection with hepatitis C virus. Gastroenterology 136:2377–2386
Dreux M, Pietschmann T, Granier C, Voisset C, Ricard-Blum S et al (2006) High density lipoprotein inhibits hepatitis C virus-neutralizing antibodies by stimulating cell entry via activation of the scavenger receptor BI. J Biol Chem 281:18285–18295
Drummer HE, Boo I, Maerz AL, Poumbourios P (2006) A conserved Gly436-Trp-Leu-Ala-Gly-Leu-Phe-Tyr motif in hepatitis C virus glycoprotein E2 is a determinant of CD81 binding and viral entry. J Virol 80:7844–7853
Evans MJ, von Hahn T, Tscherne DM, Syder AJ, Panis M et al (2007) Claudin-1 is a hepatitis C virus co-receptor required for a late step in entry. Nature 446:801–805
Falkowska E, Kajumo F, Garcia E, Reinus J, Dragic T (2007) Hepatitis C virus envelope glycoprotein E2 glycans modulate entry, CD81 binding, and neutralization. J Virol 81:8072–8079
Farci P, Alter HJ, Wong DC, Miller RH, Govindarajan S et al (1994) Prevention of hepatitis C virus infection in chimpanzees after antibody-mediated in vitro neutralization. Proc Natl Acad Sci U S A 91:7792–7796
Farci P, Shimoda A, Wong D, Cabezon T, De Gioannis D et al (1996) Prevention of hepatitis C virus infection in chimpanzees by hyperimmune serum against the hypervariable region 1 of the envelope 2 protein. Proc Natl Acad Sci U S A 93:15394–15399
Farquhar MJ, Harris HJ, Diskar M, Jones S, Mee CJ et al (2008) Protein kinase A-dependent step (s) in hepatitis C virus entry and infectivity. J Virol 82:8797–8811
Flint M, Maidens C, Loomis-Price LD, Shotton C, Dubuisson J et al (1999) Characterization of hepatitis C virus E2 glycoprotein interaction with a putative cellular receptor, CD81. J Virol 73:6235–6244
Fofana I, Xiao F, Thumann C, Turek M, Zona L et al (2013) A novel monoclonal anti-CD81 antibody produced by genetic immunization efficiently inhibits Hepatitis C virus cell-cell transmission. PLoS ONE 8:e64221
Gane EJ, Stedman CA, Hyland RH, Ding X, Svarovskaia E et al (2013) Nucleotide polymerase inhibitor sofosbuvir plus ribavirin for hepatitis C. N Engl J Med 368:34–44
Gerlach JT, Diepolder HM, Jung MC, Gruener NH, Schraut WW et al (1999) Recurrence of hepatitis C virus after loss of virus-specific CD4(+) T-cell response in acute hepatitis C. Gastroenterology 117:933–941
Giang E, Dorner M, Prentoe JC, Dreux M, Evans MJ et al (2012) Human broadly neutralizing antibodies to the envelope glycoprotein complex of hepatitis C virus. Proc Natl Acad Sci U S A 109:6205–6210
Goffard A, Callens N, Bartosch B, Wychowski C, Cosset F-L et al (2005) Role of N-linked glycans in the functions of hepatitis C virus envelope glycoproteins. J Virol 79:8400–8409
Habersetzer F, Fournillier A, Dubuisson J, Rosa D, Abrignani S et al (1998) Characterization of human monoclonal antibodies specific to the hepatitis C virus glycoprotein E2 with in vitro binding neutralization properties. Virology 249:32–41

Hadlock KG, Lanford RE, Perkins S, Rowe J, Yang Q et al (2000) Human monoclonal antibodies that inhibit binding of hepatitis C virus E2 protein to CD81 and recognize conserved conformational epitopes. J Virol 74:10407–10416
Helle F, Goffard A, Morel V, Duverlie G, McKeating J et al (2007) The neutralizing activity of anti-hepatitis C virus antibodies is modulated by specific glycans on the E2 envelope protein. J Virol 81:8101–8111
Helle F, Vieyres G, Elkrief L, Popescu CI, Wychowski C et al (2010) Role of N-linked glycans in the functions of hepatitis C virus envelope proteins incorporated into infectious virions. J Virol 84:11905–11915
Houghton M, Abrignani S (2005) Prospects for a vaccine against the hepatitis C virus. Nature 436:961–966
Hsu M, Zhang J, Flint M, Logvinoff C, Cheng-Mayer C et al (2003) Hepatitis C virus glycoproteins mediate pH-dependent cell entry of pseudotyped retroviral particles. Proc Natl Acad Sci U S A 100:7271–7276
Huang H, Sun F, Owen DM, Li W, Chen Y et al (2007) Hepatitis C virus production by human hepatocytes dependent on assembly and secretion of very low-density lipoproteins. Proc Natl Acad Sci U S A 104:5848–5853
Icard V, Diaz O, Scholtes C, Perrin-Cocon L, Ramiere C et al (2009) Secretion of hepatitis C virus envelope glycoproteins depends on assembly of apolipoprotein B positive lipoproteins. PLoS ONE 4:e4233
Jacobson IM, Gordon SC, Kowdley KV, Yoshida EM, Rodriguez-Torres M et al (2013) Sofosbuvir for hepatitis C genotype 2 or 3 in patients without treatment options. N Engl J Med 368:1867–1877
Johansson DX, Voisset C, Tarr AW, Aung M, Ball JK et al (2007) Human combinatorial libraries yield rare antibodies that broadly neutralize hepatitis C virus. Proc Natl Acad Sci U S A 104:16269–16274
Kato N, Sekiya H, Ootsuyama Y, Nakazawa T, Hijikata M et al (1993) Humoral immune response to hypervariable region 1 of the putative envelope glycoprotein (gp70) of hepatitis C virus. J Virol 67:3923–3930
Keck ZY, Op De Beeck A, Hadlock KG, Xia J, Li TK et al (2004) Hepatitis C virus E2 has three immunogenic domains containing conformational epitopes with distinct properties and biological functions. J Virol 78:9224–9232
Keck ZY, Li TK, Xia J, Bartosch B, Cosset FL et al (2005) Analysis of a highly flexible conformational immunogenic domain a in hepatitis C virus E2. J Virol 79:13199–13208
Keck ZY, Xia J, Cai Z, Li TK, Owsianka AM et al (2007) Immunogenic and functional organization of hepatitis C virus (HCV) glycoprotein E2 on infectious HCV virions. J Virol 81:1043–1047
Keck ZY, Olson O, Gal-Tanamy M, Xia J, Patel AH et al (2008a) A point mutation leading to hepatitis C virus escape from neutralization by a monoclonal antibody to a conserved conformational epitope. J Virol 82:6067–6072
Keck ZY, Li TK, Xia J, Gal-Tanamy M, Olson O et al (2008b) Definition of a conserved immunodominant domain on hepatitis C virus E2 glycoprotein by neutralizing human monoclonal antibodies. J Virol 82:6061–6066
Keck ZY, Li SH, Xia J, von Hahn T, Balfe P et al (2009) Mutations in hepatitis C virus E2 located outside the CD81 binding sites lead to escape from broadly neutralizing antibodies but compromise virus infectivity. J Virol 83:6149–6160
Keck ZY, Saha A, Xia J, Wang Y, Lau P et al (2011) Mapping a region of hepatitis C virus E2 that is responsible for escape from neutralizing antibodies and a core CD81-binding region that does not tolerate neutralization escape mutations. J Virol 85:10451–10463
Keck ZY, Xia J, Wang Y, Wang W, Krey T et al (2012) Human monoclonal antibodies to a novel cluster of conformational epitopes on HCV E2 with resistance to neutralization escape in a genotype 2a isolate. PLoS Pathog 8:e1002653

Keck Z, Wang W, Wang Y, Lau P, Carlsen TH et al (2013) Cooperativity in virus neutralization by human monoclonal antibodies to two adjacent regions located at the amino terminus of hepatitis C virus E2 glycoprotein. J Virol 87:37–51

Keck Z, Angus A, Wang W, Lau P, Wang Y, et al (2014) Non-random escape pathways from a broadly neutralizing human monoclonal antibody map to a highly conserved region on the hepatitis C virus E2 glycoprotein encompassing amino acids. PLoS Pathog doi:journal.ppat.1004297

Knodell RG, Conrad ME, Ginsberg AL, Bell CJ (1976) Efficacy of prophylactic gamma-globulin in preventing non-A, non-B post-transfection hepatitis. Lancet 1:557–561

Kong L, Giang E, Nieusma T, Robbins JB, Deller MC et al (2012a) Structure of hepatitis C virus envelope glycoprotein E2 antigenic site 412 to 423 in complex with antibody AP33. J Virol 86:13085–13088

Kong L, Giang E, Robbins JB, Stanfield RL, Burton DR et al (2012b) Structural basis of hepatitis C virus neutralization by broadly neutralizing antibody HCV1. Proc Natl Acad Sci U S A 109:9499–9504

Kong L, Giang E, Nieusma T, Kadam RU, Cogburn KE et al (2013) Hepatitis C virus E2 envelope glycoprotein core structure. Science 342:1090–1094

Koutsoudakis G, Kaul A, Steinmann E, Kallis S, Lohmann V et al (2006) Characterization of the early steps of hepatitis C virus infection by using luciferase reporter viruses. J Virol 80:5308–5320

Krawczynski K, Alter MJ, Tankersley DL, Beach M, Robertson BH et al (1996) Effect of immune globulin on the prevention of experimental hepatitis C virus infection. J Infect Dis 173:822–828

Krey T, Meola A, Keck ZY, Damier-Piolle L, Foung SK et al (2013) Structural basis of HCV neutralization by human monoclonal antibodies resistant to viral neutralization escape. PLoS Pathog 9:e1003364

Kuiken C, Yusim K, Boykin L, Richardson R (2005) The Los Alamos hepatitis C sequence database. Bioinformatics 21:379–384

Lavie M, Sarrazin S, Montserret R, Descamps V, Baumert T et al (2014) Identification of conserved residues in HCV envelope glycoprotein E2 modulating virus dependence on CD81 and SRB1 entry factors. J Virol 88:10584–10597. doi:10.1128/JVI.01402-01414

Lavillette D, Morice Y, Germanidis G, Donot P, Soulier A et al (2005) Human serum facilitates hepatitis C virus infection, and neutralizing responses inversely correlate with viral replication kinetics at the acute phase of hepatitis C virus infection. J Virol 79:6023–6034

Law M, Maruyama T, Lewis J, Giang E, Tarr AW et al (2008) Broadly neutralizing antibodies protect against hepatitis C virus quasispecies challenge. Nat Med 14:25–27

Lawitz E, Mangia A, Wyles D, Rodriguez-Torres M, Hassanein T et al (2013) Sofosbuvir for previously untreated chronic hepatitis C infection. N Engl J Med 368:1878–1887

Liang TJ, Ghany MG (2013) Current and future therapies for hepatitis C virus infection. N Engl J Med 368:1907–1917

Lindenbach BD, Evans MJ, Syder AJ, Wolk B, Tellinghuisen TL et al (2005) Complete replication of hepatitis C virus in cell culture. Science 309:623–626

Liu S, Yang W, Shen L, Turner JR, Coyne CB et al (2009) Tight junction proteins claudin-1 and occludin control hepatitis C virus entry and are downregulated during infection to prevent superinfection. J Virol 83:2011–2014

Lupberger J, Zeisel MB, Xiao F, Thumann C, Fofana I et al (2011) EGFR and EphA2 are host factors for hepatitis C virus entry and possible targets for antiviral therapy. Nat Med 17:589–595

Ly KN, Xing J, Klevens RM, Jiles RB, Ward JW et al (2012) The increasing burden of mortality from viral hepatitis in the United States between 1999 and 2007. Ann Intern Med 156:271–278

Mancini N, Diotti RA, Perotti M, Sautto G, Clementi N et al (2009) Hepatitis C virus (HCV) infection may elicit neutralizing antibodies targeting epitopes conserved in all viral genotypes. PLoS ONE 4:e8254

Martin DN, Uprichard SL (2013) Identification of transferrin receptor 1 as a hepatitis C virus entry factor. Proc Natl Acad Sci U S A 110:10777–10782

Meertens L, Bertaux C, Cukierman L, Cormier E, Lavillette D et al (2008) The tight junction proteins claudin-1, -6, and -9 are entry cofactors for hepatitis C virus. J Virol 82:3555–3560

Mercer DF, Schiller DE, Elliott JF, Douglas DN, Hao C et al (2001) Hepatitis C virus replication in mice with chimeric human livers. Nat Med 7:927–933

Meunier JC, Russell RS, Goossens V, Priem S, Walter H et al (2008) Isolation and characterization of broadly neutralizing human monoclonal antibodies to the E1 glycoprotein of hepatitis C virus. J Virol 82:966–973

Molina S, Castet V, Fournier-Wirth C, Pichard-Garcia L, Avner R et al (2007) The low-density lipoprotein receptor plays a role in the infection of primary human hepatocytes by hepatitis C virus. J Hepatol 46:411–419

Nielsen SU, Bassendine MF, Burt AD, Martin C, Pumeechockchai W et al (2006) Association between hepatitis C virus and very-low-density lipoprotein (VLDL)/LDL analyzed in iodixanol density gradients. J Virol 80:2418–2428

Owen DM, Huang H, Ye J, Gale M Jr (2009) Apolipoprotein E on hepatitis C virion facilitates infection through interaction with low-density lipoprotein receptor. Virology 394:99–108

Owsianka A, Tarr AW, Juttla VS, Lavillette D, Bartosch B et al (2005) Monoclonal antibody AP33 defines a broadly neutralizing epitope on the hepatitis C virus E2 envelope glycoprotein. J Virol 79:11095–11104

Owsianka AM, Timms JM, Tarr AW, Brown RJ, Hickling TP et al (2006) Identification of conserved residues in the E2 envelope glycoprotein of the hepatitis C virus that are critical for CD81 binding. J Virol 80:8695–8704

Owsianka AM, Tarr AW, Keck Z, Li T, Witteveldt J et al (2008) Broadly neutralizing human monoclonal antibodies to Hepatitis C Virus E2 glycoprotein. J Gen Virol 89:653–659

Pantua H, Diao J, Ultsch M, Hazen M, Mathieu M et al (2013) Glycan shifting on hepatitis C virus (HCV) E2 glycoprotein is a mechanism for escape from broadly neutralizing antibodies. J Mol Biol 425:1899–1914

Perotti M, Mancini N, Diotti RA, Tarr AW, Ball JK et al (2008) Identification of a broadly cross-reacting and neutralizing human monoclonal antibody directed against the hepatitis C virus E2 protein. J Virol 82:1047–1052

Pestka JM, Zeisel MB, Blaser E, Schurmann P, Bartosch B et al (2007) Rapid induction of virus-neutralizing antibodies and viral clearance in a single-source outbreak of hepatitis C. Proc Natl Acad Sci U S A 104:6025–6030

Piazza M, Sagliocca L, Tosone G, Guadagnino V, Stazi MA et al (1997) Sexual transmission of the hepatitis C virus and efficacy of prophylaxis with intramuscular immune serum globulin. A randomized controlled trial. Arch Intern Med 157:1537–1544

Pietschmann T, Kaul A, Koutsoudakis G, Shavinskaya A, Kallis S et al (2006) Construction and characterization of infectious intragenotypic and intergenotypic hepatitis C virus chimeras. Proc Natl Acad Sci U S A 103:7408–7413

Pileri P, Uematsu Y, Campagnoli S, Galli G, Falugi F et al (1998) Binding of hepatitis C virus to CD81. Science 282:938–941

Ploss A, Evans MJ, Gaysinskaya VA, Panis M, You H et al (2009) Human occludin is a hepatitis C virus entry factor required for infection of mouse cells. Nature 457:882–886

Potter JA, Owsianka AM, Jeffery N, Matthews DJ, Keck ZY et al (2012) Toward a hepatitis C virus vaccine: the structural basis of hepatitis C virus neutralization by AP33, a broadly neutralizing antibody. J Virol 86:12923–12932

Prentoe J, Jensen TB, Meuleman P, Serre SB, Scheel TK et al (2011) Hypervariable region 1 differentially impacts viability of hepatitis C virus strains of genotypes 1 to 6 and impairs virus neutralization. J Virol 85:2224–2234

Prince AM, Brotman B (1998) Biological and immunological aspects of hepatitis C virus infection in chimpanzees. Curr Stud Hematol Blood Transfus:250–265

Puntoriero G, Meola A, Lahm A, Zucchelli S, Ercole BB et al (1998) Towards a solution for hepatitis C virus hypervariability: mimotopes of the hypervariable region 1 can induce antibodies cross-reacting with a large number of viral variants. Embo J 17:3521–3533

Ray R, Meyer K, Banerjee A, Basu A, Coates S et al (2010) Characterization of antibodies induced by vaccination with hepatitis C virus envelope glycoproteins. J Infect Dis 202:862–866

Roccasecca R, Ansuini H, Vitelli A, Meola A, Scarselli E et al (2003) Binding of the hepatitis C virus E2 glycoprotein to CD81 is strain specific and is modulated by a complex interplay between hypervariable regions 1 and 2. J Virol 77:1856–1867

Rollier C, Depla E, Drexhage JAR, Verschoor EJ, Verstrepen BE et al (2004) Control of heterologous hepatitis C virus infection in chimpanzees is associated with the quality of vaccine-induced peripheral T-helper immune response. J Virol 78:187–196

Rothwangl KB, Manicassamy B, Uprichard SL, Rong L (2008) Dissecting the role of putative CD81 binding regions of E2 in mediating HCV entry: putative CD81 binding region 1 is not involved in CD81 binding. Virol J 5:46

Sabo MC, Luca VC, Prentoe J, Hopcraft SE, Blight KJ et al (2011) Neutralizing monoclonal antibodies against hepatitis C virus E2 protein bind discontinuous epitopes and inhibit infection at a postattachment step. J Virol 85:7005–7019

Sainz B Jr, Barretto N, Martin DN, Hiraga N, Imamura M et al (2012) Identification of the Niemann-Pick C1-like 1 cholesterol absorption receptor as a new hepatitis C virus entry factor. Nat Med 18:281–285

Sanchez-Quijano A, Pineda JA, Lissen E, Leal M, Diaz-Torres MA et al (1988) Prevention of post-transfusion non-A, non-B hepatitis by non-specific immunoglobulin in heart surgery patients. Lancet 1:1245–1249

Scarselli E, Ansuini H, Cerino R, Roccasecca RM, Acali S et al (2002) The human scavenger receptor class B type I is a novel candidate receptor for the hepatitis C virus. Embo J 21:5017–5025

Scheel TK, Rice CM (2013) Understanding the hepatitis C virus life cycle paves the way for highly effective therapies. Nat Med 19:837–849

Shepard CW, Finelli L, Alter MJ (2005) Global epidemiology of hepatitis C virus infection. Lancet Infect Dis 5:558–567

Shimizu YK, Hijikata M, Iwamoto A, Alter HJ, Purcell RH et al (1994) Neutralizing antibodies against hepatitis C virus and the emergence of neutralization escape mutant viruses. J Virol 68:1494–1500

Shimizu YK, Igarashi H, Kiyohara T, Cabezon T, Farci P et al (1996) A hyperimmune serum against a synthetic peptide corresponding to the hypervariable region 1 of hepatitis C virus can prevent viral infection in cell cultures. Virology 223:409–412

Sugg U, Schneider W, Hoffmeister HE, Huth C, Stephan W et al (1985) Hepatitis B immune globulin to prevent non-A, non-B post-transfusion hepatitis. Lancet 1:405–406

Sundberg EJ, Mariuzza RA (2002) Molecular recognition in antibody-antigen complexes. Adv Protein Chem 61:119–160

Tamura K, Stecher G, Peterson D, Filipski A, Kumar S (2013) MEGA6: molecular evolutionary genetics analysis version 6.0. Mol Biol Evol 30:2725–2729

Tarr AW, Owsianka AM, Timms JM, McClure CP, Brown RJ et al (2006) Characterization of the hepatitis C virus E2 epitope defined by the broadly neutralizing monoclonal antibody AP33. Hepatology 43:592–601

Tarr AW, Owsianka AM, Jayaraj D, Brown RJ, Hickling TP et al (2007) Determination of the human antibody response to the epitope defined by the hepatitis C virus-neutralizing monoclonal antibody AP33. J Gen Virol 88:2991–3001

Tarr AW, Urbanowicz RA, Jayaraj D, Brown RJ, McKeating JA et al (2012) Naturally occurring antibodies that recognize linear epitopes in the amino terminus of the hepatitis C virus e2 protein confer noninterfering, additive neutralization. J Virol 86:2739–2749

Tarr AW, Lafaye P, Meredith L, Damier-Piolle L, Urbanowicz RA et al (2013) An alpaca nanobody inhibits hepatitis C virus entry and cell-to-cell transmission. Hepatology 58:932–939

Thimme R, Oldach D, Chang KM, Steiger C, Ray SC et al (2001) Determinants of viral clearance and persistence during acute hepatitis C virus infection. J Exp Med 194:1395–1406

Thomssen R, Bonk S, Propfe C, Heermann KH, Kochel HG et al (1992) Association of hepatitis C virus in human sera with beta-lipoprotein. Med Microbiol Immunol (Berl) 181:293–300

Timpe JM, Stamataki Z, Jennings A, Hu K, Farquhar MJ et al (2008) Hepatitis C virus cell-cell transmission in hepatoma cells in the presence of neutralizing antibodies. Hepatology 47:17–24

Valli MB, Serafino A, Crema A, Bertolini L, Manzin A et al (2006) Transmission in vitro of hepatitis C virus from persistently infected human B-cells to hepatoma cells by cell-to-cell contact. J Med Virol 78:192–201

Vanwolleghem T, Bukh J, Meuleman P, Desombere I, Meunier JC et al (2008) Polyclonal immunoglobulins from a chronic hepatitis C virus patient protect human liver-chimeric mice from infection with a homologous hepatitis C virus strain. Hepatology 47:1846–1855

von Hahn T, Yoon JC, Alter H, Rice CM, Rehermann B et al (2007) Hepatitis C virus continuously escapes from neutralizing antibody and T-cell responses during chronic infection in vivo. Gastroenterology 132:667–678

Wakita T, Pietschmann T, Kato T, Date T, Miyamoto M et al (2005) Production of infectious hepatitis C virus in tissue culture from a cloned viral genome. Nat Med 11:791–796

Weiner AJ, Geysen HM, Christopherson C, Hall JE, Mason TJ et al (1992) Evidence for immune selection of hepatitis C virus (HCV) putative envelope glycoprotein variants: potential role in chronic HCV infections. Proc Natl Acad Sci U S A 89:3468–3472

Witteveldt J, Evans MJ, Bitzegeio J, Koutsoudakis G, Owsianka A et al (2008) CD81 is dispensable for hepatitis C virus cell-to-cell transmission in hepatoma cells. J Gen Virol 90:48–58

Witteveldt J, Evans MJ, Bitzegeio J, Koutsoudakis G, Owsianka AM et al (2009) CD81 is dispensable for hepatitis C virus cell-to-cell transmission in hepatoma cells. J Gen Virol 90:48–58

World Health Organization (2010) Initiative for vaccine research. Hepatitis C. http://www.who.int/vaccine_research/diseases/viral_cancers/en/index2.html. WHO website (online)

Xie ZC, Riezu-Boj JI, Lasarte JJ, Guillen J, Su JH et al (1998) Transmission of hepatitis C virus infection to tree shrews. Virology 244:513–520

Yi M, Ma Y, Yates J, Lemon SM (2007) Compensatory mutations in E1, p7, NS2 and NS3 enhance yields of cell culture-infectious inter-genotypic chimeric hepatitis C virus. J Virol 81:629–638

Zhang XX, Deng Q, Zhang SY, Liu J, Cai Q et al (2003) Broadly cross-reactive mimotope of hypervariable region 1 of hepatitis C virus derived from DNA shuffling and screened by phage display library. J Med Virol 71:511–517

Zhang P, Wu CG, Mihalik K, Virata-Theimer ML, Yu MY et al (2007) Hepatitis C virus epitope-specific neutralizing antibodies in Igs prepared from human plasma. Proc Natl Acad Sci U S A 104:8449–8454

Zhang P, Zhong L, Struble EB, Watanabe H, Kachko A et al (2009) Depletion of interfering antibodies in chronic hepatitis C patients and vaccinated chimpanzees reveals broad cross-genotype neutralizing activity. Proc Natl Acad Sci U S A 106:7537–7541

Zhong J, Gastaminza P, Cheng G, Kapadia S, Kato T et al (2005) Robust hepatitis C virus infection in vitro. Proc Natl Acad Sci U S A 102:9294–9299

Zibert A, Schreier E, Roggendorf M (1995) Antibodies in human sera specific to hypervariable region 1 of hepatitis C virus can block viral attachment. Virology 208:653–661

Zibert A, Dudziak P, Schreier E, Roggendorf M (1997) Characterization of antibody response to hepatitis C virus protein E2 and significance of hypervariable region 1-specific antibodies in viral neutralization. Arch Virol 142:523–534

Zucchelli S, Roccasecca R, Meola A, Ercole BB, Tafi R et al (2001) Mimotopes of the hepatitis C virus hypervariable region 1, but not the natural sequences, induce cross-reactive antibody response by genetic immunization. Hepatology 33:692–703

Index

T. Miyamura et al. (eds.), *Hepatitis C Virus I*, DOI 10.1007/978-4-431-56098-2

Printed by Printforce, the Netherlands